MATERIALS SCIENCE AND TECHNOLOGIES

ANTIMONY

CHARACTERISTICS, COMPOUNDS AND APPLICATIONS

MATERIALS SCIENCE AND TECHNOLOGIES

Additional books in this series can be found on Nova's website under the Series tab.

Additional e-books in this series can be found on Nova's website under the e-book tab.

MATERIALS SCIENCE AND TECHNOLOGIES

ANTIMONY

CHARACTERISTICS, COMPOUNDS AND APPLICATIONS

MANIJEH RAZEGHI
EDITOR

Nova Science Publishers, Inc.
New York

For permission to use material from this book please contact us:
Telephone 631-231-7269; Fax 631-231-8175
Web Site: http://www.novapublishers.com

Library of Congress Cataloging-in-Publication Data

Antimony : characteristics, compounds, and applications / editors, Manijeh Razeghi.
p. cm.
Includes index.
ISBN 978-1-62100-598-8 (hardcover)
1. Antimony. I. Razeghi, M.
TN799.A6A58 2011
669'.75--dc23
2011035665

Published by Nova Science Publishers, Inc. † New York

CONTENTS

Preface		**vii**
Chapter 1	Origin of the Highly Enriched Sb-abundances in Fine-Dust *Gerd Weckwerth*	**1**
Chapter 2	Bridging Antimony Knowledge Gaps by Assessing Its Behavior in a Semi Arid Creek *Jessica L. Ruiz, Jianhong Ren, Thomas C. Vaughan, G. Christopher Shank and Alfred Addo-Mensah*	**19**
Chapter 3	Leishmania Antimony Resistance: Challenges and Opportunities *Denis Sereno*	**43**
Chapter 4	Structure of Liquid Sb and Sb-Based Alloys *A. Yakymovych and S. Mudry*	**63**
Chapter 5	Type-II Antimonide-based Superlattice Photon Detectors and Focal Plane Arrays *Manijeh Razeghi and Binh-Minh Nguyen*	**93**
Chapter 6	Antimony Based Chalcogenide Thin Films for Photovoltaic Applications *B. Krishnan, G. Alan Castillo, T. K. Das Roy and S. Shaji*	**125**
Chapter 7	Antimony Oxide Glasses and their Nanocomposites for Optical, Photonic and Nanophotonic Applications *Tirtha Som and Basudeb Karmakar*	**143**
Chapter 8	High Temperature Interactions of Antimony with Nickel *Olga A. Marina and Larry R. Pederson*	**175**
Chapter 9	Adsorption of Antimony at Metal Electrodes and their Application in Electrocatalysis *Qi-Hui Wu and Shi-Gang Sun*	**201**

Chapter 10 Improved P-type Antimony Telluride Alloys (Sb2Te3) for Thermal Converters **225**
J. P. Carmo, M. F. Silva, J. F. Ribeiro, L. M. Gonçalves and J. H. Correia

Chapter 11 Cobalt Antimonide - Synthesis, Properties and Applications **245**
Elzbieta Godlewska, Kinga Zawadzka and Krzysztof Mars

Chapter 12 Antimony Complexes with o-Quinonato Ligands and Related O,N-, O,N,O-, O,N,O,O-Ligands **271**
Andrey I. Poddel'sky

About the Contributor **307**

Contributor Biographical Sketch **309**

Index **321**

PREFACE

Antimony (Sb) is a semimetal (metalloid) with an atomic number of 51 and an atomic weight of 121.760. Antimony and its minerals have been known and utilized since ancient times by many countries and cultures. The name antimony came from the Greek "anti" and "monos", meaning "not alone" since it is infrequently found native but rather in combined forms with sulfur or heavier metals such as copper, lead, and silver. Another folk etymology of the word is from the French "antimoine", which means "monk killer" as it is toxic and could cause lethal accidents for alchemists, many of whom were monks in the Middle Ages. Antimony's status as an alchemical substance and its high toxicity have obscured understanding of the material, resulting in controversial usage of antimony, especially in medicine.

Today, knowledge about antimony compounds has tremendously increased, and the application range of antimony has become much wider. Due to the diversity of the applications of antimony, researchers may stay focused on their own field while lacking familiarity with antimony-related activities in other fields. Environmentalists dealing with contamination and toxication of antimony-byproducts might not be up to date on significant achievements and applications brought by antimony compounds. Direct users of antimony materials might not be fully aware of the effects of antimony production, usage, and disposal on the environment. Nevertheless, for sustainable development and application of this material, there is a need to raise the awareness of researchers, as well as of the public, on both facets of antimony: the tremendous potential for numerous applications and the severe health and environmental issues that require serious attention.

In an effort to bridge the gap between research fields involving antimony, this edited book aims to cover the diverse aspects of antimony in modern life. From medical, health, and environmental concerns to the latest applications of antimony in science and industry, the book chapters were prepared by and utilized research results from leading scientists in their fields. The participation of researchers from all over the world is proof of both the interest in antimony and its importance, and encourages the idea that the message about antimony should be spread further. Due to space and time limitations, this book was unable to cover many other pioneering activities on antimony. Nevertheless, this will be an initiative linking different antimony-related research directions in future books of this series.

I hope that this book will be a good starter reference for those who are interested in antimony and provide a broad background on activities involving antimony. Crossing the boundary of diverse fields and opening potential multidisciplinary research directions and collaborations are also among the intentions of this work. To that extent, the book will be useful not only for students and scholars, but also practitioners, explorers, industrialists, and environmentalists in diverse fields.

I would like to express our sincere thanks to all the contributing authors for their participation. Without their valuable expertise, dedication, and support, this book would not have been possible. I am grateful to Nova Science publishers for giving me a chance to realize this project. I also would like to acknowledge my assistant, Dr. Binh-Minh Nguyen, for his patience and enthusiasm in accomplishing countless tedious tasks during the preparation.

In: Antimony: Characteristics, Compounds and Applications ISBN 978-1-62100-598-8
Editor: Manijeh Razeghi

Chapter 1

ORIGIN OF THE HIGHLY ENRICHED SB-ABUNDANCES IN FINE-DUST

Gerd Weckwerth
Institut für Geologie und Mineralogie der Universität zu Köln,
Zülpicher Str. 49b, Köln, Germany

ABSTRACT

The mean concentration of antimony (Sb) in the upper continental crust (UCC) is with 0.2 ppm nearly 10 times lower than that of the far better known and stronger observed element arsenic (As). Since the comparable toxic element Sb is in the fine dust (specifically from towns) between 2 and 4 times more abundant than As, it needs a higher environmental attention.

Like no other element Sb belongs in both the coarse grained (2.5-10μm) as well as in the fine grained aerosols (< 2.5μm) to the most highly enriched elements. This proofs presence at least of one widespread anthropogenic Sb-source, which is mechanically produced from a solid and one by condensation from a gas component. After control and disappearance of much higher Sb-rich-components from industry in the 80ties the source of the remained coarse grained component was found soon to come from traffic, according to measured gradients with the distance to frequented streets.

Final evidence was taken from a wide-spread chemical fingerprint found in the coarse grained fraction of urban fine dust (Weckwerth, 2001). The constant ratios of concentration between the elements Sb, Cu and Mo could be verified as the mean ratios of the mainly in Germany 1997 used brake-linings. 10 years later the same ratios could be still measured with a far better statistic figure in a frequently used tunnel at Cologne.

Additionally, a new element could be taken in the chemical fingerprint with tin (Sn), according to a new type of brake lining using Sn-sulfid instead of Sb-sulfide as fine-lubricant. For the first time the measurement in a tunnel 2006 shows an increased enrichment of Sn, according to almost 30% German market share this Sn-rich type of brake lining has reached during 10 years. Meanwhile, similar fingerprints have been measured and used as tracers of traffic emission in many countries, confirming a similar use of Sb-sulfide and a strong dominance of this kind of traffic emissions for the given elements in the coarse grained fraction (2.5-10μm) of the fine dust. In a similar way it could be shown, that also a strong proportion of trains use Sb-rich brake-linings (Weckwerth, 2010).

Keywords: Antimony, arsenic, aerosols, PM10, PM2.5, brake-linings, traffic, sulfides, mining

1. Physical and Chemical Characteristics of Antimony

Antimony is a trace element (symbol Sb, atomic number 51) with two stable isotopes (121,123) and a mean atomic mass of 121.76 amu. It belongs to the 5th main group (V-a) of the periodic system much related with the better known toxic element arsenic, which is placed in the (V-a)-column just above antimony. This indicates an identical occupation of their outer electron shells implying equal oxidation states (+5, +3, -3) and relationship of respective chemical compounds.

With 6.697 g/cm^3 native antimony has an around 1 g/cm^3 higher density than arsenic, while both forming a trigonal crystal structure. Additionally, under specific conditions less stable amorphous structures of both elements can occur. According to a 20 pm higher atomic and covalent radii of 140 pm the atomic density of antimony is more than 50% smaller in comparison with that of arsenic. That means in the opposite a higher place requirement of antimony, causing a more incompatible behavior with the minerals of the terrestrial mantle (→2.).

Above its melting point of 630°C antimony can be easily oxidized producing a white powder of antimony trioxide (Sb_2O_3) and with some more effort it is oxidable to Sb_2O_4 and Sb_2O_5. A slow oxidation of antimony can occur also in the presence of water. Antimony oxides are strongly insoluble in water even more than the already low solubility of native antimony and antimony trisulfide. But they are soluble in specific acids (Merian, 1984) (Antimony, Web 2011).

2. Geochemical Characteristics

Antimony is a silvery lustrous grey metalloid (semi-metal) and can be found at many places in nature as a native mineral. It was already known as one of ten elements in the ancient world. Since that time a wide spectra of utilization of antimony was developed, which had been more and more limited from experiences of toxic and carcinogenic behavior of most of its compounds.

Antimony occurs in nature by far most frequently as stibnite as well called antimonite (Sb_2S_3), a grey mineral, which crystallizes in an orthorhombic space group similar to arsenic trisulfide. Both antimony as well as arsenic belong geochemically to the so called chalcophile elements, what describes in the Goldschmidt classification elements with affinity to form sulphide minerals (Stibnite, Web 2011).

Additionally, antimony is also a siderophile element and likes even more to go into the metal phase, if available. Since metal is on the oxidized surface of our Earth very rare, antimony and arsenic are even stronger enriched in terrestrial sulfides compared e.g. with sulfides in more reduced and metal-rich meteorites. Therefore, abundances of antimony in most natural samples are often correlated with the presence of sulfide (>1%). Since the Sb-compositions in sulfides are varying between 10 ppm and a few percent an estimation of the total proportion of Sb contained in sulfides is at least difficult (Mason, 1985).

If sulfides are not available too, the next geochemically relevant feature of antimony is its high incompatibility with the minerals of the terrestrial mantle and a related enrichment in magmatic melts. Since the siderophile element antimony went during the first 80% of the terrestrial accretion into the core, its initial concentration of ~150 ppb (C1) is reduced to a mean of only 6-8 ppb in the silicate mantle of our present Earth. This was more than compensated during the formation of the upper continental crust (UCC), when the mean concentration of antimony increases up to approximately 200 ppb (Taylor, McLennan, 1985). Together with other incompatible elements antimony is enriched during partial melting of mantle material, from which our actual crust continually was formed. Although, the incompatibility of antimony is according to its bigger ionic radius higher than for arsenic, the enrichment of Sb-concentrations alone by partial melting of mantle silicates does not exceed 1 ppm (Weckwerth, 1989).

A further, most probably the crucial characteristic of antimony on its presence in aerosols, is its volatility according to different temperatures and atmospheric pressures. Under the oxidized metal-free conditions of the terrestrial surface the main carrier of antimony are sulfides and oxides, implying a moderate volatility not far from that under cosmochemical conditions (condensation temperature ~850 K).

The comparison of the relevant trisulfides and trioxides shows that antimony holds under terrestrial conditions with around 930 K/820 K to ~580 K in each case a significant higher melting point than the related arsenic compound. This implies in contrast to the conditions of cosmochemistry a more moderate volatility of antimony, which is important to separation processes. Where both elements have to be separated e.g. in copper mines, the presence of arsenic can even help to lower the melting point of antimony (Clement II, 1989)

New studies on the behavior of antimony shows more and more differences to arsenic so that the earlier assumption of protection agencies in US and Europe that antimony's properties are similar to those of arsenic has to be given up. A relevant point is in specific that antimony oxidizes much more quickly than arsenic when exposed (physorg.com, April 2010).

3. Natural Occurrence and Ecological Risk

Abundances of antimony in the biosphere show strong differences between soil, water and air. Most soils especially from river valleys or coastal areas (Taylor, McLennan, 1985) (Wedepohl, 1995) consist of representative mixed material from sediments. The Sb-concentration is close to the upper continental crust or slightly enriched between 200 -300 ppb. A high natural enrichment of antimony (>1 ppm) in the ground is usually restricted on sulfur-rich areas, often connected to ore deposits, where Sb-concentrations can range between a few ppm and a few percent.

Correlated abundances or at least a similar enrichment can be found for the related element arsenic, which is mostly predominating antimony on a 5-10 times higher level. Therefore, the presence of antimony in environmental studies is up to now often neglected or unconsidered. Although, the toxic effects are known for long on both elements, risks are mainly reported and monitored for arsenic. Nevertheless, there is a high number of localities on the upper continental crust with a specific high enrichment of antimony even up to its native occurrence as a pure mineral found at worldwide more than 270 localities (Antimony, Web 2011).

The locally restricted risk of a high natural enrichment of antimony and arsenic in the ground can be exported and spread out to the marine and the atmospheric environment. In both cases the related mobilization was already part of natural cycles, but it is additionally stimulated by anthropogenic activities and utilization.

The export of antimony to *the marine environment* is limited by a weak water solubility of antimony and its most abundant natural compounds. Nevertheless, small dissolved amounts of antimony in water (<1 μg/g) can be sufficient to produce first symptoms of poisoning in specific, when they are continually present in drinking water of animals or humans. Permanent small damages to the health e.g. to the liver can be added up to life threatening extent or increase cancer.

Since the solubility of antimony is similar or lower than of the related, mostly predominating As-compounds, cases of poisoning from dissolved arsenic are more likely and become more frequently known than for antimony (Merian, 1984). Mysterious diseases and cases of death, which emerged later to be generated from arsenic in drinking water are known e.g. from India and Bangladesh. A study from 2007 found that over 137 million people in more than 70 countries are probably threatened by arsenic poisoning of drinking water, most likely in collaboration with antimony poisoning (Smith et al, 2011).

Beside a wide contact of ground water with ore deposits this is due to increased solubility, dependent from the pH-value. First experiences on increased risks of antimony in acidic milieu are from antimonial color or Sb-abundanced dishes, when they came in contact with acid drinks instead of water (Merian, 1984).

Even more complex and of broad reach are export and transportation of antimony into the *atmospheric environment* depending from a wide spectra of possibly sources of emission and their specific transportation characteristic in the terrestrial atmosphere. In the ordinary natural case, aerosols are raised dust from the ground, with a composition similar to the upper continental crust. This kind of emission is the dominant source of almost half of the elements (e.g. REE, Sc, Zr, Th). Their relative abundances remain in the ratios of the upper continental crust. Their absolute concentration in pending dust [μg/m^3] can vary according to weather conditions up to one magnitude and more.

In contrast to these refractory lithophile elements the relative abundances of most metals and chalcophile elements like antimony are enriched in comparison to the upper continental crust (Jochum and Hofmann, 1997), signifying that additional components from metal- and sulfide-rich sources are contained in the aerosol. They can be also emitted from natural processes, like volcanoes, naturally generated forest fires or sea spray. How far these natural emissions are intensified by anthropogenic activities can be detected from sediments at least of elements which are retained sufficiently at their places in the temporal succession.

Shotyk et al (1996) could show that age controlled peat cores from bogs taken at several European highlands have to a large extend correlated patterns. This allows on one hand to normalize Europe-wide variations in the elemental atmospheric flux and dustfall and on the other hand to find far-reaching anthropogenic emissions. In specific they found 2000 year old peaks of Sc-normalized fluxes of As, Sb and Pb, most probably emitted from the Romans, who had already extracted and utilized these chalcophile elements. Further, they could take elemental fluxes and enrichment factors from pre-anthropogenic times (5300-8000 years BP). They concluded Sb-enrichment factors of natural dustfall in the area of three.

In modern peat samples Shotyk et al. (2004) found enrichment factors in the order of 30-80, with a strong decrease in the last 30 years. A similar pattern for the last 50 years was found in cores from sediments out of dammed lakes (DFG 419, A2).

Although, their abundances in the dustfall decreases a factor of 3 relative to 1970 antimony and zinc remained on an enrichment factor around 12 almost twice as high than other metals. Since this enrichment of Sb and Zn is typical for dustfalls in the surrounding of towns especially in the vicinity of traffic, it implies an origin from anthropogenic activities (→6.). Also the decrease after strong limitations of German industrial emissions got obligatory around 1980, proves the anthropogenic origin of the formerly higher amounts of dust-fall.

At least after this reduction in the eighties the absolute concentration of antimony in fine dust less 10μm is up to a factor of 5 higher than that of arsenic, but antimony is furthermore not environmentally controlled like this. The corresponding mean enrichment factor (EF) of antimony in PM10-aerosols lies with more than 1000 together with that of selenium ahead all other ecological relevant metals or metalloids (Weckwerth, 2010).

In comparison with much higher concentrations of heavy metals 40 years ago one could think, that the risk of heavy metals in aerosols has been reduced in Germany to an acceptable level. Nevertheless, big studies in the last 20 years have shown that already a small enrichment of 10μg/m^3 has significant influence on the health of bigger population groups (Katsouyanni et al., 2009). So far the main causers of diseases from aerosols are almost unknown e.g. whether the amount of particles of specific sizes or defined toxic ingredients are more important.

Beside diesel soot the sulfates are suspected to be of stronger relevance and intensively studied, therefore. Their occurrence is indirect connected with the high enrichment of the chalcophile elements, which are known to be unhealthy. May be they are effective in sum or high concentrations of only a few of these toxic elements in aerosols are primarily harmful to health. Antimony is according to its leading enrichment in aerosols one likely candidate. Dietl et al (1998) found Sb-concentrations in Munich which may result up to 56 cases in a group of 1 million people which is in a similar range as those from diesel soot.

The ecological and medical risks assumed to antimony in aerosols are based on animal experiments and on longtime studies of the health from people, who worked continuously with antimony (smelters) breathing high amounts of Sb-enriched fine dust before effective inhalation protections were obligatory. Although, exposures are often estimated and the statistical base is weak, there are concordant hints on carcinogenicity primarily on antimony trioxide (causing a classification as "most probably carcinogenic"). The risk increases to smaller sizes (< 4μm) in accordance to a better incorporation capacity and is comparable with risks from related compounds from arsenic and lead (Dietl. et al, 1997)

The high enrichment of antimony in aerosols give reason for further research on both, the real effects of so far only estimated risks to the health as well as effective ways to reduce locally and globally generated emission of antimony.

4. ANTHROPOGENIC UTILIZATION AND EMISSION OF ANTIMONY

Known most probably from early processed ores, the element antimony and some of its compounds have been utilized since ancient times. Early usage for medical applications and cosmetics have been later strongly limited according to increasing experiences about its toxic impacts. Nevertheless, a utilization in many further areas has been developed and sustained up to our times. Beside the application exploitation and removal can be sources of emission as well.

The actual extraction of antimony from ores depends from its concentrations in the ore, which is usually a sulfide. Antimony can be isolated from its ore by the reduction with iron:

$$Sb_2S_3 + 3Fe \rightarrow 2Sb + 3FeS$$

More frequently the sulfide is oxidized and separated by taken advantage of the volatility of antimony(III) oxide, which is recovered from an antimony specific roasting (Antimony, 2011).

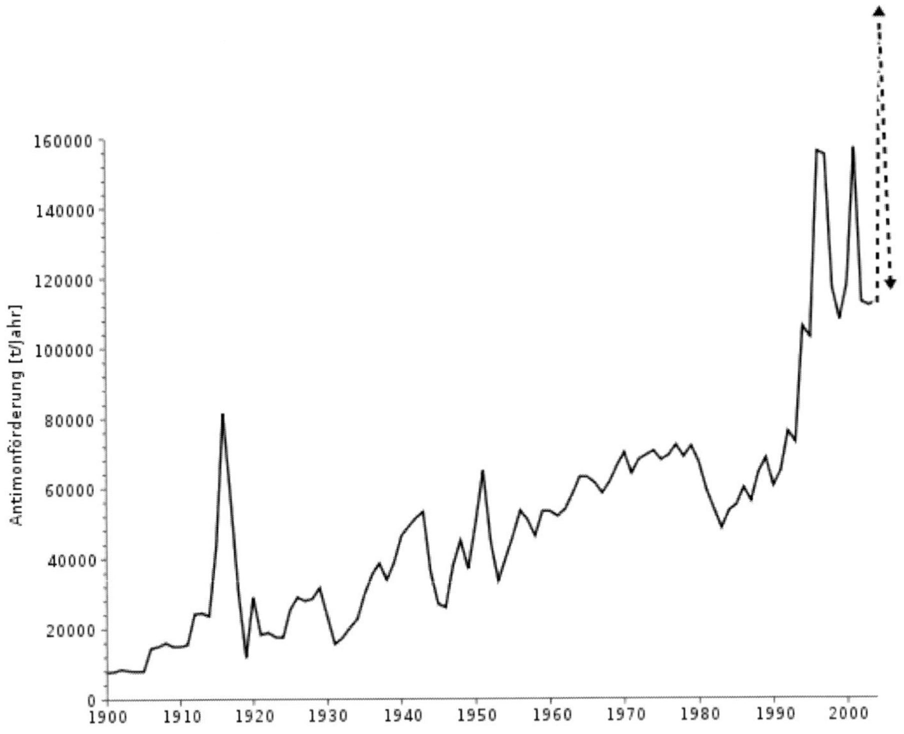

Figure 1. Development of Sb-production in the 20th century and trends up to 2010 (Antimony, 2011), (World Mining Data, 2010).

The main producer (~90%) of antimony is the people's republic of China, with 187000 metric tons in 2009. (The remaining 10% is distributed between South Africa, Bolivia, Tajikistan and Russia).

The Xikuangshan mine in the Chinese province Hunan has with 2.1 million metric tons the largest deposits. The online-magazine Physorg.com reports in April 2010 "Heaps of ore tailings are exposed directly to the open air at Xikuangshan. Therefore, Chen Zhu the world's largest antimony mine has become the world's largest laboratory for studying the environmental consequences of escaped antimony, an element whose environmental and biological properties are still largely a mystery".

Environmental issues have, however, in recent years, forced the closure of several of China's polluting mines and plants. As a result, the price of antimony reached new highs in 2010, almost $11,000 per ton, more than 150% higher than in 2009.

Who are the main consumers of the delivered raw Sb-products and for what kind of applications?

The annual consumption of the most important Sb-compound, *antimony trioxide,* in the United States and Europe is approximately 10,000 and 25000 tons, respectively in 2005. Since more than 90% of these amounts are produced outside, the toxic and most probably carcinogenic raw material has to be transported and imported with some safety requirements into the consuming countries.

The main application (~50%) of antimony trioxide is for flame retardants in combination with halogenated materials. This combination with halides is the key to the flame-retardant action for polymers, helping to form less flammable products. Such flame retardants are found in electrical apparatus, textiles, leather and coatings. Included are children's clothing, toys, aircraft and automobile seat covers.

The second main application is as a catalyst for the production of the polymer polyethyleneterephthalate (PET plastic) and the vulcanization of rubber. Since PET is used as recycled raw material most products from recycled plastics have between 100 and 200 ppm of antimony (Weckwerth, 2009). Both applications as flame retardants and catalyst are tested to be of low risk in emission to water (Shotyk and Krachler, 2007). A risk for atmospheric emission on lower level may be the long-term behavior of the incorporated antimony during corrosion and combustion of the plastic material.

Further applications like fines or opacifying agens for glasses or ceramics are of much lower range or they are hopefully in medical control when used in contact or incorporated in the human body (e.g. emetics, drugs, cosmetics, parasitizide).

The very brittle native antimony has no remarkable application, but it forms rather useful alloys especially with lead, increasing its hardness and mechanical strength. The Sb-Pb alloy is used in conventional lead-acid batteries, but it is increasingly replaced by calcium in modern maintenance-free batteries. A further application is in ammunition of small arms. Therefore, some antimony contamination can be found on shooting ranges. Antimony is found in various alloys especially with iron and tin, to optimize metal characteristics (e.g. solders, organ pipes) and in antifriction alloys such as Babitt metal.

Since antimony is also a siderophile element and present in ores, there could be also undesired enriched portions of antimony in iron or steel according to reducing conditions, selected during the extraction in furnaces. Analyses of wheels and parts of trains (Weckwerth, 2001) showed that the concentrations are significant lower than that of arsenic, which should be more in the focus of metal sources.

From earlier times smelters have suffered from a high contamination of the air at their working places. Several studies from the 50s and 60s analyzed increased risks of smelters to get acute toxic or late stochastic work-related diseases. Improved environmental controls of smeltery and steel works at least in Germany have limited emission and risks to the surrounded living population. Remarkable contributions to the fine dust in more distant areas are imaginable only from bigger accidents or unrecognized gaseous emissions.

Antimony trisulfide (stibnite) can be melted out at comparatively low temperatures from ores so that larger amounts are easily to obtain, but with some risk to loose volatilized portions. Beside still some medical and cosmetic applications stibnite is used in various pyrotechnic materials as glitter in fireworks, airbags and safety matches. For very fast flashes antimony trisulfide is mixed with 50% potassium chlorate. There are limitations and new developed replacements to avoid handling and emission risks.

Furtherly, antimony trisulfide can be used like other sulfides (Mo, Zn) as anti-friction or fine lubricant, but it became widely accepted only in traffic techniques, namely in most brake-linings (1-5%) and very rare in tires. In brake-linings the use as fine lubricant avoids that the brake shoes get stuck or produce too much ablation. The antimony trisulfide achieves this by stabilizing a thin oxidized layer which is generated between the heated brake shoes (Weckwerth, 2005).

In the 90s it was estimated that more than 70% of the brake-linings of German motor vehicles used antimony. The remaining cars used among others sulfides of molybdenum, tin or lead. Lead was ten years after asbestos in 2003 the second material in Germany, whose utilization in emitting materials like brake-linings was completely forbidden.

Since antimony was taken by the European Union on a list of necessary replacements, producers have to explain every year, why they have still antimony in use. In the year 2002 several of German brake-lining suppliers stated in a poll of the AvD (German Automobile Society) that they still can not avoid antimony in brake-linings because of its optimal characteristics in heavy lorries and its minimal risk to squeak. Nevertheless, since the year 1998 a competing company is on the German market with tin sulfide in their brake-linings with a confirmed share in 2008 of ~30% (Federal Mogul, 2008).

Since the fine lubricant in brake-linings is destined to get rubbed off with their original grain sizes (around 5μm), and grains are not retained in the disc brakes nor remarkably in the surrounded wheelhouse, it is certainly emitted to the pending dust. The concentration should depend mainly from its residence times, which are interdependent from the grain sizes and weather conditions. Almost independent from that details should be the contribution of brake-linings to the dust fall close to streets and wherever brake-linings of this kind are in use e.g. in trains.

Certainly, there are so far unmentioned fields of Sb-utilizations (e.g. in nuclear reactors or in the semi-conductor industry) but with most probably very low potentials of emission and low mentionable contribution to the fine-dust, perhaps with exceptions at very limited localities. More detailed descriptions of Sb-utilizations and their potential emissions can be found in the European diantimony risk assessment study (EURAR, 2008).

5. Non Sb-specific Sources of Emission (UCC, Fossil Fuels, Ores)

The basic source of any element at the terrestrial surface is the upper continental crust (UCC). Its mean composition corresponds with most ground materials, especially with that originated only from sedimentary processes. Since there is erosion and wind, an emission of ground material to the air as pending dust [$\mu g/m^3$] or dustfall [$mg/d*m^2$] is in the long term unavoidable. This UCC-related component from the ground has a very constant composition

(Taylor and McLennan, 1985) with time and differs only very little between most places around the world. Therefore, comparable enrichment factors (EF) of the elements have been usually calculated relative to this component. Mostly it is sufficient to calculate ratios only to scandium or an other Element, that is mainly controlled by the ground component (e.g. Al, Ti, Si, Ca, Zr, U).

High enrichment factors of antimony calculated in this way relative to UCC have caused also the question of this chapter: "Origin of the highly enriched Sb-abundances in fine dust". For a comparison of enrichment factors (EF) it is important to observe equal areas of grain sizes. While the EF of antimony measured in dustfall are between 10 and 50, the EFs in PM10 are around 100 times and in PM2.5 at least more than 1000 times higher (→Ch.6.)

An other non Sb-specific source is the combustion of fossil fuels like coal, oil and gas, and according to their content of 0.1 to 20% of sulfide antimony is in average 5 to 10 times enriched relative to the upper continental crust (Merian, 1984; EURAR, 2008). Although, modern furnaces have filters, which retain flue-ash and sulfur, antimony is mostly emitted by gaseous compounds, which later condenses as a fine grained aerosol (<2.5μm).

An estimation of the global atmospheric load from fossil fuels add up to 18 000 tons of antimony (Lantzy and Mackenzie, 1979). With the increase of the global combustion in the meantime by almost 70% (World-mining-data, 2010) the actual load of antimony should be close to 30 000 tons. These estimation summarizes the total load of the fossil fuels, but not the real emission rates, which may be much lower, mainly by the use of modern filters (<5%). Data of As-concentrations in coals differ from 1.5 ppm - 15ppm, suggesting a smaller enrichment relative to UCC and by this a depletion relative to antimony of 2-3. Nevertheless, that means arsenic is still significant more abundant than antimony, what matches with the use of arsenic as guiding element of flu-ash. For the global load of fossil fuels Lantzy and Mackenzie (1979) gave even a ten percent lower mass for arsenic than antimony, what is an agreement with an As-depletion of a factor 8 relative to antimony.

Only in first approximation a similar composed emission can be expected from roasting and smelting of sulfidic ores. To purify or to separate different kinds of metal this processing implicates also an oxidation and volatilization of unwanted metals and sulfides, which are mostly emitted as gaseous compounds. To what extend antimony is globally emitted by ore processing depends from:

1. The Sb-concentrations of the disturbing sulfides, which are between 100 and almost 1 million times enriched relative to UCC.
2. The temperature, which is necessary to separate a specific metal or metal group
3. Abundance and delivery rates of the extracted metals relative to the Sb-content
4. Whether emission of antimony is totally or only partly avoided.

Since antimony behaves very volatile, in most cases a closed system with cooling down and condensation of antimony, as well as filtration of condensed aerosols would be necessary, what needs an additional effort in technique and energy.

Assuming that this is more difficult for antimony compared to arsenic and other chalcophile elements, it may cause higher relative emission rates of antimony. But this is a theoretical assumption, which has to be rechecked on the really utilized ore composition, extracting procedures and emission controls. Whether higher Sb-emission rates from the mining of sulfidic ores can explain the high enrichment of antimony in the global fine grained aerosols can be estimated from the totally, with this method produced metals.

World-mining-data (2010) reports for 2008 a production of around 33 million tons of mostly from sulfidic ores extracted metals (copper, zinc, lead a.o.). The total amount of antimony in the extracted ores reaches from 10ppm to more than 10%. An emission of 0.1% (probably already 0.01%) antimony in average is enough to get ahead the load of antimony from fossil fuels respectively its estimated Sb-emission.

In specific, the Sb-production itself of around 200 000 tons, needs to be very clean. A loss by volatilization of 15% (probably 1.5%) is enough to exceed alone the emission of the total fossil fuels. Additionally, in Cu-mining the contribution of Sb-rich ores increases, according to decreasing availability of purer ores. All together it seems to be very likely that the Sb-emission rate from ores is significant higher than by combustion of fossil fuels. Since there are no real emission rates (e.g. from China, the leader in Zn- and Sb-production), a real proof of the main responsibilities are so far not possible.

A further non Sb-specific source of emission is the waste disposal or incineration. Since one can expect, that products, which use antimony (100ppm -100%) are more relevant than other wastes, with much lower Sb-content, this sources of emission can be discussed, if necessary, together with the respective Sb-utilization. Nevertheless, the total emission should be only a small fraction of the global annual Sb-production of 200 000 tons.

6. Emissions from Sb-Specific Uses

From the mentioned Sb-utilizations in section 4 we can consider and characterize, all uses of antimony, which have a chance to make a noticeable contribution to the fine-dust.

More than half of the production of antimony trioxide is used for flame retardants, the production of PET plastics and vulcanization. The required amount is not high and concentrated enough (<0.1%), to bring higher risks for the users of such products. They mostly consist out of different kind of plastics, which can be recycled and used several times. Exception may result in connection with certain acids, which increase the decomposition and solubility in water, but mainly from the waste incineration. An estimation of the proportions, which are emitted globally to the atmosphere may be difficult, especially in countries, where the environmental standards are low or not under control.

In specific in China, where a lot of these with antimony contaminated plastics are produced, a remarkable contribution to their anyway high emissions from ores could be possible. Since both emissions are condensates with mean grain sizes less 1 μm, the total northern hemisphere may be Sb-enriched from both sources. To distinguish both kinds of sources in the fine-dust of Europe or in longer distance from their origins might be difficult. After residence times of weeks in the atmosphere local and temporal separation seem to be hopeless. Conclusions from accompanying elements are in a global interfering competition also with the combustion of fossil fuels. At least a much better knowledge about the mean global composition of all kinds of contributions is necessary.

Further kind of uses of antimony, originating recognizable components in the fine-grained area of the pending dust in Germany, should be significant lower. They are overlapped by at least a factor of ten and even more difficult to distinguish, therefore. Far better are the chances to notice and to identify components with aerosols in the area of the coarse grained fine-dust, usually from mechanically processing like friction, erosion, fly-ash, rust bloom or the artificial pulverization.

A most significant source in this area is the use and the emission of antimony from brakes. Since the emitted particles of brake-linings in a town get mixed with that of mostly several 1000 other cars, the brake-lining component has an almost constant composition. It can be used as a chemical fingerprint of brake-linings, which changes only little with the locally used vehicle type and its development with time. Weckwerth (2001) succeeded to find in the coarse grained pending dust (2.5-10μm) a chemical fingerprint of averaged brake-linings out of Cu-, Sb- and Mo-abundances, although the total dust from brake-linings contributed less than 1% to the fine-dust. In principle the fingerprint-method to identify and to distinguish the component of brake-linings from fine-dust was confirmed in towns like Tokyo (Ozaki et al, 2004), Buenos Aires, Madrid and several others around the world.

Ten years later Weckwerth (2010) reconfirmed the values of his fingerprint with a 2-days measurement in a tunnel below the town-house of Cologne, with a record of 4% brake dust in the coarse grained aerosol. Since he founds for the first time a predicted amount of tin according to the new provider, with 30% market-share and tin sulfide in his brake-linings (Federal Mogul, 2007), the first changes of a fingerprint could be received directly from a coarse grained fine-dust collection.

The recovery of element ratios between Cu/Sb/Mo in almost 30 collections, proves that these 3 elements in the coarse grained filter are extensively controlled by the abrasion of brake-linings. Measurements in different distances to a railway close to a station show additionally, that also regional trains use similar brake-linings especially with similar amounts of antimony trisulfide (Weckwerth, 2010).

A simple estimation on the amount of antimony considers globally 100 million new cars per year, which need in the mean of all vehicles and all wheels roughly 1 kg of brake-lining in their lifetimes. If every second car has 2% Sb in its brake-linings, we can calculate with 0.5 * 20g * 100 million /year = 1000 tons/year. May be trains and other heavy vehicles double this value to only 1% of the global Sb-production per year?

An estimation of the totally emitted amount of antimony from brake linings using 50μg/km leads to a maximum of 100 tons or low estimations of less 10 tons (EURAR, 2008). This would signify that only 1% of the total brake-lining comes to an emission, what seems to be rather low in the light of high Sb-enrichments in the fine-dust. While other emissions of antimony are distributed regularly over the total atmosphere, the dust from brakes may be concentrated in the vicinity of humans and where they drive or go with their vehicles. So, high enrichments may be produced from apparently low amounts of emission.

Perhaps up to 20% of the global Sb-production one can still assume for a use of antimony in different metal applications, mostly in alloys with iron or lead, to improve metal characteristics. The emission out of such metals is very rare, with the exception of shooting ranges and rust e. g. from rails, mostly very local contamination without significance to an area-wide enrichment of antimony in fine-dust.

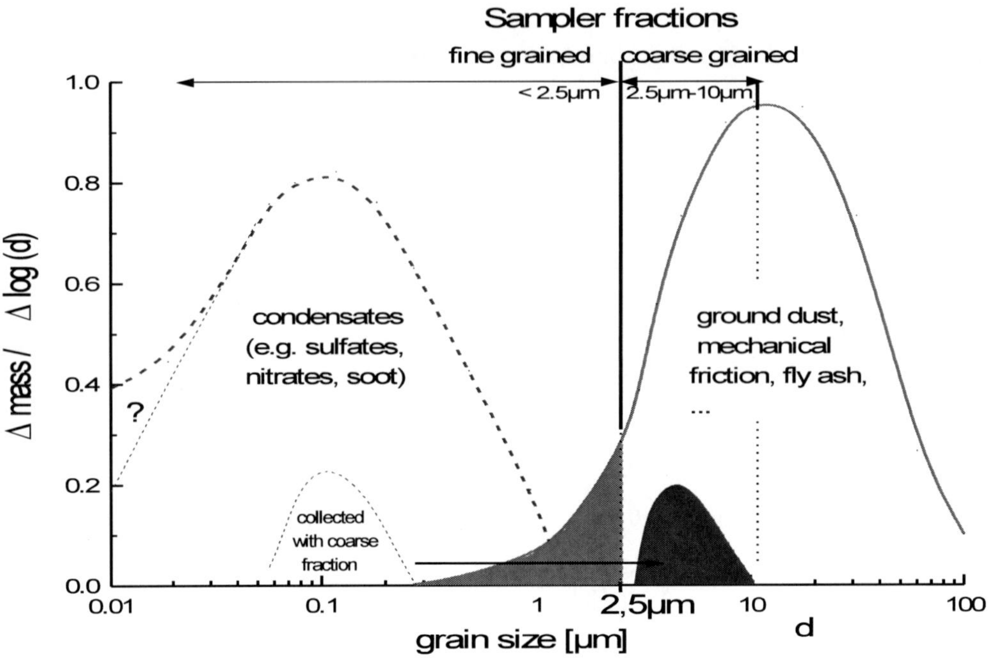

Figure 2. Size spectra of aerosols and the related fractions of the used sampler (Graseby-Anderson Instruments Inc.,1990)(Weckwerth, 2001).

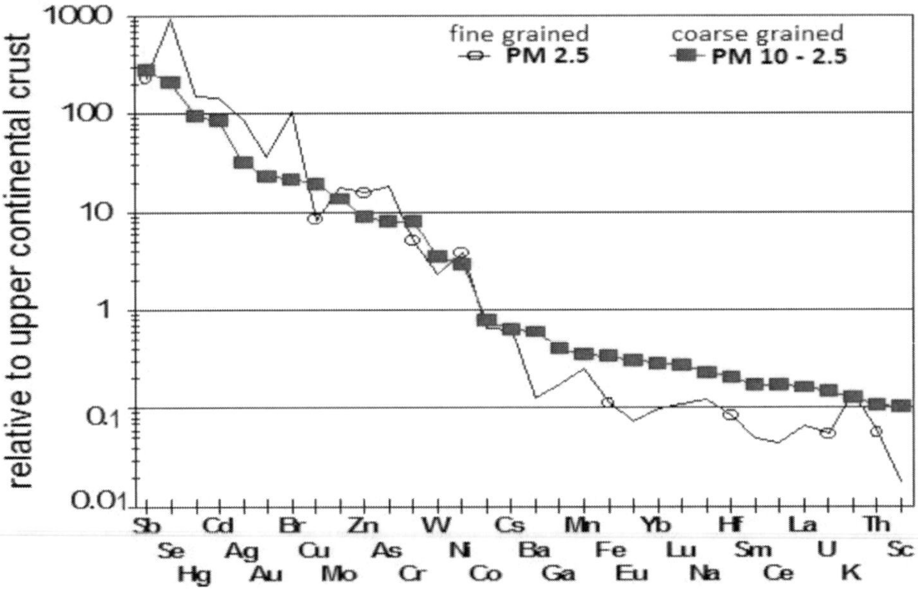

Figure 3. Mean normalized concentration of 32 elements in fine- and coarse-grained fraction of Dichotomous collections (Weckwerth, 2001).

7. Dichotomous Measurements and the Sources of Sb-Emission

An effective technique for the allocation of polluters including Sb-emissions are chemical fingerprints from dichotomous-collections and respective multi-element analysis. Figure 2 shows the grain-size fractions of such a two-channel sampler and the typical distributions of various components in the collected fine-dust.

The condensation of aerosols from the gas phase produces grain-sizes mainly less 2.5 μm and the reduction of grain-sizes by erosion, friction and other mechanical splittings gets unlikely below 2.5μm. Therefore, an additional grain-size separation at 2.5μm of PM10 fine-dust-collectors helps very effective to separate aerosols from different kinds of formation. On the other hand, there are still 10% - 20% of aerosols, which end up in the wrong channel according to their origin (condensation, mechanics). In specific, the used impactor technique can not avoid to collect 10-20% of the fine-grains together with the coarse fraction.

This can be seen in typical results of coarse and fine sizes in figure 3, which are never separated more than a factor of 6. Weckwerth (2001) developed a method of correction, using geochemical assumptions of the best separated elements to improve the splittings. While selenium and bromine disappear from the list of highest enriched elements in the coarse-grained fraction, this is certainly not true for antimony and cadmium, which are almost not separated between their fine- and coarse-grained values in all single measurements made in the agglomeration of Cologne.

In Figure 3. The mean concentrations of 12 measurements in and around of Cologne from 1997 normalized to the upper continental crust and sorted with decreasing abundances in the coarse-grained fraction are shown. The small differences of all elements between Eu and Sc from 0.1 to 0.2 is in good agreement with an UCC-like ground component of 15% in the mean in the coarse grained fraction and 10% in the mean of the total PM10 collections. The related PM10-enrichment of antimony to the UCC-like ground component is around 2000 and compared to the best analyzed element Sc more than 10000 in PM2.5. The Sb-enrichment is in contrast to decreased lead and selenium-values on the same level as in the seventies (Salomons and Foerstner, 1984).

Very striking is the strong difference between antimony (PM10: EF=2000) and arsenic (PM10: EF=100) enrichment, which seem to be significant higher than expected from the combustion of fossil fuels alone, where abundance ratios of Sb/As >1 (7.5x in UCC) seemed to be rare. But, perhaps this is a first hint on other, more Sb-specific sources with stronger differences between the emission of antimony and arsenic e.g. in sulfidic ores.

In the case of the coarse grained fraction (PM10-2.5) the problem of the high Sb-enrichment is almost completely solved. The method can be taken from figure 4.

After the described corrections, which eliminate still high Se- and Br-data in the coarse grained fraction, antimony is always significantly more enriched relative to UCC than all other elements, available by INAA (with some isolated exceptions of Au and Ir). For a comparison with the mean values of possibly sources from traffic by friction of brake-lining and tires, in Figure 4 mean normalized compositions from up to 20 standard samples of brake-linings and 8 samples of tires have been added (Weckwerth, 2001).

The vertical arrows with equal length to antimony in brakes-linings have been also drawn to the mean concentrations of copper and molybdenum in brake-linings, which end up close to the average of the 12 samples of coarse-grained fine-dust, showing that these concentrations are necessarily almost completely produced from the brake-linings, too. In the case of iron it is around 20%.

Therefore, antimony, copper and molybdenum in its mean abundances ratios of around 1: 4.6: 0.25 can be taken as a chemical fingerprint of the brake-lining component. In the case of tires a similar safe method is not available. Assuming that zinc in the coarse grained fraction is completely produced from tires, it enables at least to calculate the total content of the friction component from tires in any coarse-grained fine-dust. The concentrations are significantly lower than earlier calculations from PM10-measurements by Heinrichts and Brumsack (1997). This can be explained from even higher PM2.5 Zn-concentrations, which are most probably due to condensed trace gases of zinc, produced during the globally burning of sulfide-rich materials. By the same reasons also the concentration of copper and antimony from brake-linings is overlapped by condensed components, and need separate coarse-grained measurements for a reliable calculation.

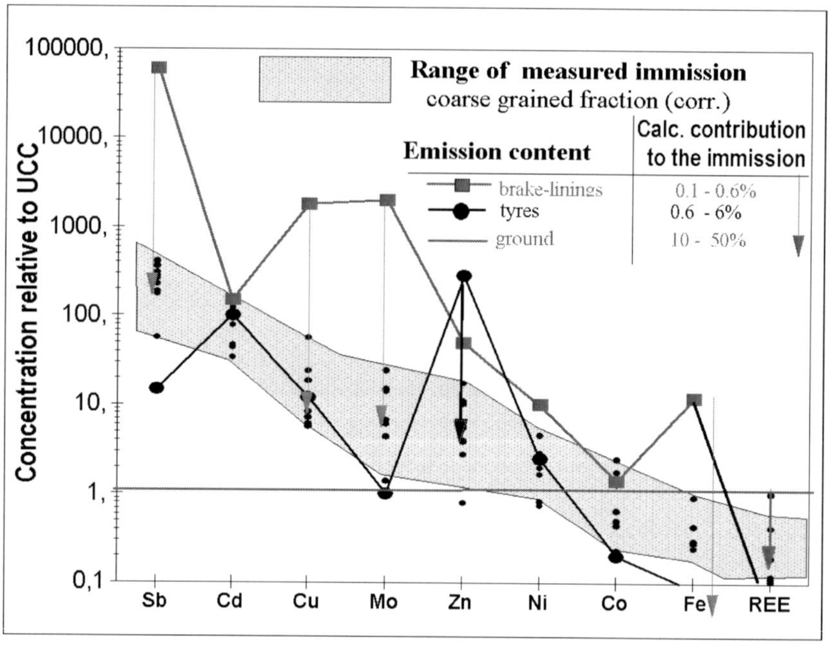

Figure 4. Normalized mean concentrations of coarse-grained aerosols in Cologne compared with the mean composition of emission from brakes and tyres (Weckwerth, 2001).

In the meantime the chemical fingerprint from brake-linings has been confirmed by two further series of measurements in Cologne (Weckwerth, 2010). Especially, the 2-days measurement of a mean concentration of 4%-brake-lining in a tunnel under the town-house of Cologne improved the exact chemical fingerprint.

Further analysis is planned to confirm expected changing of the mean German brake-lining content (Sn replaces Sb) from the concentration in fine-dust. One application of chemical fingerprints are certainly assessment studies, which need a detailed allocation of sources to estimate risks. Since both risk assessment studies on antimony trioxide (EURAR,

2008; Environment and Health Canada, 2010) did not make use of it, a new task would be, to analyze possibly improvements.

Summary

There is still some uncertainty, why small amounts of aerosols and especially people, which are living close to streets, have in spite of environmental control significantly increased health risks. The toxic and carcinogenic trace element antimony has world wide the highest enrichment factors in fine dusts and is therefore a candidate to make more unhealthy contributions than formerly thought. Collection of fine-dust (PM10) with two grain-size assigned channels (Dichotomous sampler) combined with INAA are very successful tools to identify and ascertain the main sources of Sb-enrichment. Almost everywhere the described technique can be used to verify the exact amounts of highly Sb-contaminated dust from brake-linings, which is, after asbestos and lead were banned, one of the next important contaminants for effective reduction in Europe.

Secondly, there are still some open questions about the origin of the high Sb-enrichment in fine-grained aerosols (PM2.5). It is suggested that abundance ratios of calcophile elements (e.g. Sb/As) including some continental variations have a chance to verify typical contributions, especially, if possibly negligent or disregarding emission rates from the processing of sulfidic ores would be responsible. The described technique seems to be most promising to find fingerprint relations as likely implications on the kind and position of sources. If necessary, this might be the best push to install countermeasures on Sb-emissions, especially when the remaining ores of copper or zinc are getting more Sb-abundant.

References

Antimony, http://de.wikipedia.org/wiki/antimony.

CAS# 7440-36-0 http://www.webelements.com/antimony/ (accessed March 2011) http://webmineral.com/chem/Chem-Sb.shtml.

Antimony trioxide, http://de.wikipedia.org/wiki/Antimonoxide.

CAS# 1309-64-4 http://en.wikipedia.org/wiki/Antimony_trioxide (accessed March 2011) http://www.inchem.org/documents/icsc/icsc/eics0012.htm.

Antimony trisulfide http://www.mindat.org/min-3782.html.

http://ntp.niehs.nih.gov/ntp/htdocs/Chem_Background/ExSumPdf/antimony_trisulfide.pdf.

CAS# 1345-04-6 http://www.chemicalbook.com/Search_DE.aspx?keyword=1345-04-6 or Stibnite http://webmineral.com/data/Stibnite.shtml (accessed March 2011) http://en.wikipedia.org/wiki/Stibnite.

Clement II, T. P.; Taie L.; Hager, J. P. (Golden, CO). Process for the recovery and separation of arsenic from antimony. United States Patent 4808221, 1989.

http://www.freepatentsonline.com/4808221.html (accessed March 2011).

Dietl, C.; Reifenhäuser, W. and Peichl, L. Association of antimony with traffic - occurrence in airborne dust, deposition and accumulation in standardized grass cultures. *Science of the Total Environment* 1997, 205, 235-244.

Dietl, C.; Faus-Keßler, T.; Wegenke, M.; Peichl, L. Verkehrsbezogene Immissionen von Antimon und anderen Metallen. *Bayerisches Landesamt für Umweltschutz, Schriftenreihe* 1998, 153.

EURAR (European Union Risk Assessment Report): *DIANTIMONY TRIOXIDE - RISK ASSESSMENT;* Swedish Chemicals Inspectorate: Sundbyberg, 2008, Sweden.

Environment Canada; Health Canada. *Screening Assessment for the Challenge: Antimony trioxide,* Canadian Ministeries of Environment and Health, Sept. 2010.

http://www.ec.gc.ca/ese-ees/9889ABB5-3396-435B-8428-F270074EA2A7/batch9_1309-64-4_en.pdf.

Federal Mogul, private communication with Paul, H. G., 2007.

Graseby-Anderson Instruments Inc., *Operators and Instruction Manual - Dichotomous Sampler Model* 241, Miami, 1990.

Heinrichts H. and Brumsack H.-J. *Anreicherung von umweltrelevanten Metallen in atmosphärisch transportierten Schwebstäuben aus Ballungszentren.* in Matschullat J.; Tobschall H.-J.; Voigt H.-J. (Hg.) *Geochemie und Umwelt* Springer Verlag: Berlin, Heidelberg, 1997.

Jochum, K. P. ; Hofmann, A. W. Constraints on earth evolution from antimony in mantle-derived rocks. *Chemical Geology* 1997, Vol. 139, Issues 1-4, 39-49.

Katsouyanni K.; Samet J.M.; Anderson H.R.; Atkinson R.; Le Tertre A.; Medina S.; Samoli E.; Touloumi G.; Burnett R.T.; Krewski D.; Ramsay T.; Dominici F.; Peng R.D.; Schwartz J.; Zanobetti A.; HEI Health Review Committee. *Air pollution and health: a European and North American approach (APHENA).* Research report (Health Effects Institute) 2009, 142, 5-90.

Lantzy, R. L.; Mackenzie, F. T. Atmospheric trace metals: Global cycles and assessment of man's impact. *Geochimica. Cosmochimica Acta* 1979, 43, 511–525.

Mason B.; Moore C.B. *Grundzüge der* Geochemie; Ferdinand Enke Verlag: Stuttgart, 1985

Merian E. (Ed.) *Metalle in der Umwelt;* Verlag Chemie: Weinheim, 1984.

Ozaki, H.; Watanabe, I.; Kuno, K. As, Sb and Hg distribution and pollution sources in the roadside soil and dust around Kamikochi, Chubu Sangaku National Park, Japan. *Geochemical Journal* 2004, 38, 473-484.

Physorg.com(2010). *A different kind of mine disaster* http://www.physorg.com/news190288453.html.

Salomons, W.; Förstner, U. *Metals in the Hydrocycle;* Springer Verlag: Berlin, Heidelberg, 1984.

Sonderforschungsbereich 419; *Ergebnisbericht 1999-2001, Teilprojekt A2;* Universität zu Köln, August 2001.

Shotyk, W.; Cheburkin, A.K.; Appleby, P.G.; Fankhauser, A. und Kramers, J.D. Two thousand years of atmospheric arsenic, antimony, and lead deposition recorded in a peat bog profile, Jura Mountains, Switzerland. *Earth Planetary Science Letters* 1996, 145, E1-E7.

Shotyk, W.; Krachler, M.; and Chen, B. Antimony in recent peat from Switzerland and Scotland: comparison with natural background values (5,320 to 8,020 ^{14}C yr BP), correlation with Pb, and implications for the global atmospheric Sb cycle. Global Biogeochemical Cycles 2004, 18, GB1016, 13 pp.

Shotyk, W.; and Krachler, M. Contamination of bottled waters with antimony leaching from PET increases with storage. *Environmental Science and Technology* 2007, 41, 1560-1563

Smith A.H.; Lingas E.O.; Rahman M. Contamination of drinking-water by arsenic in Bangladesh: a public health emergency. http://www.sos-arsenic.net/english/contamin/smith.html (accessed March 2011).

Taylor S.R.; McLennan S.M. *The Continental Crust: its Composition and Evolution;* Blackwell Scientific Publications: Oxford, Boston, Palo Alto, Victoria, 1985.

Weckwerth, G. Verification of traffic emitted aerosol components in the ambient air of Cologne (Germany). *Atmospheric Environment* 2001, 35, 5525-5536.

Weckwerth, G. Von Asbest zu Antimon - das weiterhin bestehende Krebsrisiko von Bremsbelägen. Immissionsschutz 2005, 1, 15-20.

Weckwerth, G. Origin of fine dust in urban environmental zones - Evidence from element patterns received by dichotomous collection and INAA. *Applied Radiation and Isotopes* 2010, Vol. 68, Issue 10, 1878-1883.

Weckwerth, G. *Die Einsatzmöglichkeiten von β–γ-Koinzidenzmessungen für instrumentelle Neutronenaktivierungsanalyse im Bereich der Geo- und Kosmochemie;* Thesis at the Max-Planck-Institute for Chemistry, cosmochemistry division: Mainz, 1989.

Weckwerth, G. *Unpublished results from INAA of PET and recycled PET-samples*, 2009.

Wedepohl, K. H. The composition of the continental crust. *Geochim. Cosmochim. Acta* 1995, 59, 1217–1232.

WORLD-MINING-DATA (L. Weber, G. Zsak, C. Reichl, M. Schatz) Vol. 25, April 2010, Wien auf http://www.bmwfj.gv.at/energieundbergbau/weltbergbaudaten/Seiten/default.asp.

In: Antimony: Characteristics, Compounds and Applications ISBN 978-1-62100-598-8
Editor: Manijeh Razeghi

Chapter 2

BRIDGING ANTIMONY KNOWLEDGE GAPS BY ASSESSING ITS BEHAVIOR IN A SEMI ARID CREEK

***Jessica L. Ruiz*[1], *Jianhong Ren*[1], *Thomas C. Vaughan*[2], *G. Christopher Shank*[3] *and Alfred Addo-Mensah*[2]**

[1]Department of Environmental Engineering, Texas A&M University-Kingsville, Kingsville, TX, US

[2]Department of Biology and Chemistry, Texas A&M International University, Laredo, TX, US

[3]University of Texas Marine Science Institute, Port Aransas, TX, US

ABSTRACT

Antimony (Sb) contamination in aquatic systems, due in large part to mining and smelting activities, is both an ecological and environmental concern. As will be discussed in the first section of this chapter, several knowledge gaps are hindering our complete understanding of antimony behavior in aquatic systems including a variety of issues relating to transport, sorption/desorption, speciation, and bioavailability and uptake of antimony. The second part of this chapter will summarize key results from recent Sb studies in a contaminated creek in a semi-arid environment and evaluate the contribution of these findings in filling some of the existing knowledge gaps in the understanding of antimony behavior. Results presented in section 2 include: an assessment of antimony concentrations in the water and sediment of this contaminated creek, an evaluation of the potential direct and indirect effects of rainfall on Sb cycling, chemical binding forms of antimony on streambed sediments and the implications for antimony mobility and bioavailability, and the potential effects of stream water chemistry on antimony release from contaminated sediment. Finally, preliminary bioavailability results of antimony uptake by the freshwater bivalve mollusk, *Corbicula fluminea*, will also be presented.

1. Current Information and Important Knowledge Gaps of Antimony Behavior in Aquatic Systems

Antimony (Sb) contamination in aquatic systems, due in large part to mining and smelting activities, is both an ecological and environmental concern. In the last several years, although more articles regarding Sb behaviors have been added in the current literature, numerous knowledge gaps are still hindering our complete understanding of Sb behavior. The purpose of this section is to identify and discuss some of these knowledge gaps, and to explain why the knowledge gaps exist, or what the challenge is in filling some of these gaps. Of particular interest are those knowledge gaps which are directly related to the environmental impacts of Sb. To this end, knowledge gaps discussed in this section include those related to transport, sorption/desorption, speciation, transformation, and bioavailability and uptake of antimony.

1.1. Overview of Sb Fate and Transport

Antimony in surface water and pore water may be present in the dissolved or particulate phases [Filella et al. 2002b] which are operationally defined as follows: dissolved phase metals are those which can pass through a nominal pore size filter (usually 0.2 μm or 0.45 μm); particulate phase metals are those which are associated with solid particles or colloids and are retained by the filter during processing [Schemel et al. 2000]. The fate and transport, or the movement and ultimate destination, of both dissolved and particulate Sb in aquatic systems such as rivers and streams are affected by a combination of physical, chemical, and biological processes [Filella et al. 2002a, b, 2007]. Physical processes include advection, dispersion, and hyporheic exchange. Advective transport, the dominant process in turbulent flow fields [Johansson et al. 2001], is the transport of contaminants in the direction of the bulk movement of the fluid containing the contaminant. Dispersive transport is responsible for the spread of a contaminant in three dimensions, and in turbulent flow fields, it is due to velocity fluctuations in the aquatic system. Hyporheic exchange is the exchange of contaminants between overlying surface water and the subsurface environment due to advective pumping and turnover mechanisms [Savant et al. 1987; Thibodeaux and Boyle 1987; Elliott and Brooks 1997a, b]. It is expected that the effect of these three physical processes on Sb transport in aquatic systems is similar to other divalent metal contaminants [Ren and Packman 2004; 2005; Huettel et al. 1998; Wörman 1998; Wörman et al. 1998; 2002]. In addition to these physical processes, the fate and transport of particulate Sb is also controlled by colloidal processes such as settling, filtration, and resuspension. Filtration refers to the attachment of transported colloidal particles to bed sediment grains as a result of chemical or electrostatic interactions [Ren and Packman 2004; Stumm and Morgan, 1996]. Just as for the physical processes mentioned above, colloidal processes affect Sb transport similarly as for other particulate metal contaminants [e.g., Ren and Packman 2004, 2005; Packman et al. 2000 a, b].

Chemical processes that could impact Sb dynamics are numerous and include complexation, redox, precipitation/dissolution, volatilization, photochemical reactions, and adsorption/desorption. For example, precipitation reactions cause the formation of metal oxides in stream water and sediment. Antimony readily sorbs to metal oxides, thus, as the colloidal or particulate metal oxides are transported in the stream water or subsurface environment, Sb transport is either facilitated or impeded [Ren and Packman 2004; 2005; Corapcioglu and Jiang, 1993; Grolimund et al. 1996; Saiers and Hornberger, 1999]. In another example, if Sb was sorbed to metal oxides in the subsurface, and those metal oxides dissolved due to changes in redox conditions, Sb could be released to the aqueous phase. The importance of adsorption/desorption reactions will be discussed in detail later and will be a focus of this paper due to its key role in regulating Sb distribution, and thus its fate and transport, in aquatic systems.

Biological processes also contribute to the fate and transport of Sb in aquatic systems and include bioturbation and irrigation, uptake, and transformation. Bioturbation is the random agitation of sediment fluids and solids by macroinvertebrates in the benthos [Buffle and De Vitre 1994]. Bioirrigation is a somewhat similar process, which occurs when solutes are transported into sediment burrows or tubes as a result of the ventilation of these spaces by burrowing or tube-dwelling benthic organisms [Buffle and De Vitre 1994]. During biotransformation, Sb transitions between oxidation states (III) and (V) or between methylated and nonmethylated species because of biotic activity. Depending on the particular aquatic system, biological processes may be as important as chemical and physical processes in controlling the transport of Sb.

As described above, there are numerous processes that govern the fate and transport of Sb, many of which can control both the dissolved phase and particulate phase Sb in aquatic systems. Biologically, both Sb phases may be utilized or transformed by biota [Filella et al. 2007]. Dissolved Sb may be carried downstream through advection/dispersion, undergo sorption/desorption to bed sediments, or be advectively pumped into the subsurface [Ren and Packman 2004]. In streambed pore waters, dissolved Sb may undergo sorption/desorption to bed sediments, or be advectively pumped through the streambed via hyporheic exchange and re-released to the overlying stream water [Savant et al. 1987; Thibodeaux and Boyle 1987; Elliott and Brooks 1997a, b]. For particulate phase Sb, depending on the particle size, they may settle onto the streambed surface, be pumped into and filtered by streambed during hyporheic exchange, or be carried downstream via advection/dispersion [Ren and Packman 2002; 2004]. Antimony sorbed to bed sediment may be desorbed and travel deeper into the streambed, or be re-released to overlying stream water. Because Sb commonly occurs in both dissolved and particulate phases, a major hindrance to a complete understanding of Sb cycling in natural aquatic systems is that very few studies have actually reported the dynamics of both the dissolved and particulate phase Sb [e.g., Mignon and Mori 1999; Filella et al. 2007; Landrum et al. 2009; Masson et al. 2009; Baeza et al. 2010]. It can be speculated that the reason for a lack of more detailed Sb characterization studies involving the determination of both dissolved and particulate Sb is the requirement for the filtration and preservation of an additional water sample, which incurs higher costs and is more labor intensive. Further, a holistic system view of Sb in an aquatic environment should include Sb in the pore water compartment, but these are rarely reported. Pore water sampling techniques are especially challenging to conduct and are also both time consuming and expensive. Thus, whenever feasible, future studies focusing on environmental assessments of Sb should be designed to determine the concentrations of both dissolved and particulate Sb and their complete distribution in aquatic systems which includes the water column, pore water, and sediment.

1.2. Sorption/Desorption of Sb

Partitioning of Sb in rivers and streams, defined as the distribution among overlying stream water, sediment pore water, and streambed sediment, is primarily the result of a combination of sorption/desorption and physical transport processes. A key knowledge gap in our understanding of Sb transport behavior in these aquatic systems is our limited understanding of Sb sorption/desorption which can largely control its mobility and bioavailability [Buschmann and Sigg 2004]. Few studies of Sb sorption on whole soils have been conducted, but more information is available regarding Sb sorption on pure phases such as humic acids or metal oxides [Wilson et al. 2010; Pilarski et al. 1995; Buschman and Sigg 2004; Tighe et al. 2005; Steely et al 2007; Thanalbalsingham and Pickering 1989; Leuz et al. 2006]. The challenge in understanding Sb sorption/desorption is rooted in the complex dependency of these processes on water and sediment chemistry and composition, especially pH, redox conditions, ionic strength, organic matter content, and metal oxide content. Further, a complete understanding of Sb sorption/desorption requires considerably more information than is available on binding mechanisms of the various Sb moieties.

1.2.1. Effects of Chemistry Conditions on Sorption/Desorption

As will be discussed in greater detail in Section 1.3, water chemistry conditions such as pH and redox play an important role in the speciation of Sb in aquatic systems which includes the two primary Sb(V) and Sb(III) species. Sb(V) is expected to be the dominant species under oxic conditions, while Sb(III) is expected to be the dominant species under anoxic conditions. It has also been observed that Sb(III) and Sb(V) demonstrate different sorption/desorption behaviors in contaminated sediment. Furthermore, sorption/desorption mechanisms of Sb to different sediment phases (i.e., organic matter or metal oxides) are uniquely influenced by different water chemistry conditions [Mok and Wai 1990; Johnson et al. 2005].

Associations between Sb and organic matter in soils and sediments can be significant [Crecelius et al. 1975; Pilarski et al. 1995; Leleyter and Probst 1998; Buschmann and Sigg 2004; Buschmann et al. 2005; Tighe et al. 2005; Steely et al. 2007]. Sb(III) and Sb(V) both exhibit significant sorption or binding to humic acids [Pilarski et al. 1995; Buschman and Sigg 2004; Tighe et al. 2005; Steely et al 2007], though sorption of Sb(V) as $Sb(OH)_6^-$ may be limited due to the stability and structure of the hexa-hydroxy species when the initial Sb concentration is low [Pilarski et al. 1995]. Sorption of both Sb(III) and Sb(V) to sedimentary organic matter is largely a function of pH. Sorption of Sb(V) to humic acid appears to decrease as pH increases in the pH range of 3–7, with maximum Sb(V) sorption (up to 60%) occurring at acidic pH values, generally between 3-3.5 [Tighe et al. 2005]. The decrease in binding affinity with increase in pH can be attributed to two possible explanations: 1) specific adsorption, or inner sphere complexing, the dominant sorption mechanism, experiences a decrease in sorption energy with the decrease in positive surface charge as pH increases; and 2) occurrence of some dissolution of humic acid when pH reaches 6-6.5 leads to less sorption sites available [Tighe et al. 2005]. For Sb(III) species, the effect of pH on adsorption to humic acid was found to be related to the type of Sb(III) compounds tested. For example, sorption of Sb(III) from potassium antimonyl tartrate ($C_8H_4K_2Sb_2O_{12}$), generally increased in the pH range of 3-5.5, with maximum sorption at pH 5-5.5 due to the enhanced access to sorption sites through de-protonation of some functional groups and/or via dispersion of colloidal particle aggregates at high pH [Pilarski et al. 1995]. However, for a Sb_2O_3 solution, sorption of Sb(III) was found high at pH 3, decreased and then increased to a maximum when pH increased to 3.8, then decreased between pH 3.8-5.5. This interesting Sb(III) sorption pattern for Sb_2O_3 was attributed to changes in the chemical nature of the Sb species caused by the pH changes [Pilarski et al. 1995]. In addition, the pH effect on Sb sorption/desorption also depends on the specific humic acid used. For example, among the three types of humic acids (terrestrial, coal, and aquatic) tested, Buschmann and Sigg [2004] found that maximum Sb(III) binding to aquatic and terrestrial humic acids was largest at pH 6.1, while for the coal humic acid binding was not dependent on pH between pH 3-6, but decreased significantly at pH above 6. The difference in sorption behavior was attributed to unique characteristics of the different humic acids, such as the involvement of different functional groups and/or varying degrees of stabilization by H-bridges or chelation.

Just as the sorption of Sb to organics is influenced by pH, Sb exhibits an affinity for metal oxides that is also regulated by pH. Sb(V), prevalent under oxic conditions as soluble $Sb(OH)_6^-$ [Filella et al. 2002b; Scheinost et al. 2006] will sorb to metal oxides at low pH above 2.5. However, it will not sorb to metal oxides above their point of zero charge (PZC) at neutral or basic pHs, and will therefore be mobilized under such conditions [Leuz et al. 2006; Scheinost et al. 2006]. In contrast, the most stable form of Sb(III) under reducing conditions is $Sb(OH)_3^0$, which will sorb to metal oxides across a wide range of pH values, and therefore is largely immobile [Scheinost et al. 2006; Leuz et al. 2006]. Further, dissolved Sb(III) can sorb to goethite, be oxidized to Sb(V), and then released as Sb(V) back to solution [Leuz et al. 2006].

In addition to the effects of pH on sorption, ionic strength of the surrounding waters can also influence Sb sorption/desorption characteristics in a unique way for different substrates. For example, sorption of Sb(III) to sedimentary organic matter has been found to increase with increasing NaCl concentrations because the addition of high levels of salt displace hydrogen ions from the organic moiety surface and create more sites for Sb sorption [Pilarski et al. 1995]. Yet, Sb(III) sorption to metal (Mn, Al, and Fe) hydrous oxides can decrease at increased ionic strength due to the addition of acetate buffer solution because the addition of the acetate species minimizes the retention of weakly bound Sb (III) [Thanalbalsingham and Pickering 1990]. However, in another study, ionic strength had no effect on Sb(III) sorption to goethite, which was indicative of the formation of Sb(III) inner-sphere surface complexes at the surface [Leuz et al. 2006]. Synergistic influences of pH and ionic strength have also been noted. For Sb(V), Leuz et al. [2006] found that sorption under ionic strength of 0.01 and 0.1 M $KClO_4$ was similar at pH 2-6, but the higher ionic strength caused lower adsorption at pH 6-12, which was attributed to the formation of ion pairs of $KSb(OH)_6^0$. In general, the effect of ionic strength on Sb sorption/desorption is closely related to the governing sorption mechanisms. If the sorption occurs through a specific (inner-sphere complexing), ionic strength usually has less effect [Tighe et al. 2005; Sparks 2003; Wilson et al. 2010]; however, if it is non-specific adsorption mechanism (outer-sphere complexing), increased ionic strength decreases negative surface charge at high pH thus increasing adsorption of Sb anionic species such as $Sb(OH)_6^-$. At pHs below the point of zero charge, non-specific adsorption decreases when ionic strength increases because increased ionic strength decreases electrostatic potential at the solid surface [Wilson et al. 2010].

It is clear that progress toward a fundamental understanding of the effects of water chemistry conditions on Sb adsorption/desorption has been made. However, substantial additional research is needed to further elucidate Sb sorption/desorption behaviors in aquatic systems with different environmental characteristics. One area of special importance is the effect of ionic strength on Sb adsorption/desorption. Although a few studies have demonstrated an effect of ionic strength of Sb sorption to humic acid or metal oxides as mentioned above, most of these studies were not specifically designed for investigating the effect of ionic strength. Another needed area for research is the evaluation of Sb sorption/desorption to whole soils or sediments. Most studies have examined Sb sorption to pure phases such as metal oxides or humic acid, but it is often not clear how the findings obtained from these pure phase studies are applicable to the whole soils/sediments. In addition, studies on how the synergistic effects of pH, ionic strength, and redox conditions on the sorption/desorption of Sb are related to the intrinsic properties of suspended or bed sediments such as sediment surface physicochemical conditions will always be important. The key to answering these questions is to generate more data in a systematic matter using well characterized systems. In particular, recent advances in applying sophisticated instrumentation such as X-ray absorption spectroscopy (XAS) in analyzing Sb in natural sediments/soils samples [Scheinost et al 2006; Mitsunobu et al. 2006] will provide a valuable tool in improving the overall understating of the underlying mechanisms of Sb mobility and its transport in aquatic systems.

1.2.2. Antimony Phase Associations in Sediment

Like other trace metals, Sb concentrations found in streambed sediment may be further partitioned into specific phase associations in the sediment matrix. Antimony mobility and availability vary depending on its associated soil fraction. Determination of the partitioning of Sb in different phases of the sediment matrix is performed through sequential extraction techniques [Crecelius et al. 1975; Brannon and Patrick 1985; McLaren et al. 1998; Leleyter and Probst 1999; Hou et al. 2006; He 2007; Tighe and Lockwood 2007; Baeza et al. 2010; Denys et al. 2008; Wang et al. 2011]. The general procedure of sequential extraction is to take a small sample of dry, contaminated sediment, and sequentially expose it to various chemical solutions that target specific phases in the soil matrix. Phases commonly targeted include dissolved, soluble, or labile concentrations, fractions associated with Fe, Al, or Mn oxides, fractions associated with organic matter, and recalcitrant metalloid fraction. The recalcitrant metalloid fraction, the fraction that is unextractable, usually is determined with a strong acid digest, and generally represents the portion of unavailable metalloid in a soil [Wilson et al. 2010]. After the sediment is mixed with a particular chemical solution, it is allowed to equilibrate, sometimes under specific environmental conditions (i.e. temperature, light, and pH) and then centrifuged to separate the supernatant solution for analysis. Limitations of sequential extraction techniques include the lack of selectivity and efficiency of the various steps in the procedures, and the possible readsorption of extracted elements [Leleyter and Probst 1998; Ahnstrom and Parker 1999]. However, despite these limitations, sequential extraction techniques are widely used and developed due to their economic feasibility and phase association information [Wilson et al. 2010]. Several studies have shown that sediment Sb found in the surface-adsorbed fraction can be associated with organic matter, and this association is affected by water chemistry parameters such as pH and ionic strength as mentioned earlier [Buschmann and Sigg 2004; Tella and Pokrovski 2009; Baeza et al. 2010]. Sb can also be strongly associated with metal oxides in sedimentary envionments [Leuz et al. 2006; Scheinost et al. 2006; Casado et al. 2007; Murciego et al. 2007], but this association is also strongly dependent upon pH and/or redox conditions [Neaman et al. 2004; Johnson et al. 2005; Tan et al. 2005; Scheinost et al. 2006; Baeza et al. 2010; Wilson et al. 2010]. Sb mobility is additionally limited by its association with recalcitrant metalloid fractions (see definition above), which are generally considered to be immobile [McLaren et al 1998; Tighe and Lockwood 2007; Wilson et al. 2010].

Sequential extractions have been frequently used to examine Sb phase association in especially contaminated sediment. Crecelius et al. [1975] compared contaminated and noncontaminated (Sb concentrations less than 4 ppm dry weight) soil from the Puget Sound estuary looking specifically at Sb bound to readily oxidizable organic matter and to amorphous iron and aluminum compounds. Results indicated that for the "noncontaminated" soils less than 10% of total Sb was bound to readily oxidizable organic matter with about 50% bound to amorphous iron and aluminum compounds. For contaminated soils, less than 10% of total Sb was bound to readily oxidizable organic matter but less than 20% was bound to amorphous iron and aluminum compounds. Lower associations of Sb bound to amorphous iron and aluminum compounds in contaminated sediments was attributed to Sb present in chemically stable forms, most likely slag particles.

In another case, Brannon and Patrick [1985] determined Sb concentrations in several chemically extractable phases (i.e. water soluble, exchangeable, easily reducible, and moderately reducible sediment phases) in unamended and amended (samples spiked with Sb to achieve a concentration of 75 μg/g dry weight) sediment samples collected from ten locations across the United States. For unamended soil samples, all Sb extracted was in the moderately reducible phase, of which Al and Fe were found to be the major constituents. In the amended soils, Brannon and Patrick [1985] found that Sb was also predominantly present in the moderately reducible phase, ranging from 46 to 81% of total Sb, but a significant amount of Sb may also be in the easily reducible phase (up to 37% of total Sb), with only small amounts present in the exchangeable (1 to 2 % of total Sb) and water soluble phases (1 to 4% of total Sb).

In a study of sediments from the Garonne River (France), Leleyter and Probst [1998] developed a sequential extraction procedure to determine Sb concentrations in several different chemically extractable fractions including that dissolved in water, readily exchangeable (cations adsorbed to solids due to their structural charges), and fractions bound to carbonates, manganese oxides, amorphous iron oxides, crystalline iron oxides, and organic matter. Results indicated that 53.4% of the Sb leached was bound to crystalline Fe oxides, while 19.8% was bound to amorphous iron. Less than 1% was water soluble, 8.8% was really exchangeable, 6.6% was bound to carbonates, 2.0% was bound manganese oxides, and 8.0% was bound to organic matter [Leleyter and Probst 1998].

Although Sb is consistently found to be associated with iron oxides, the results above also show that Sb partitioning data from sequential extractions can vary due to highly site specific nature of phase associations which are a function of where sediment samples were collected, sediment composition (i.e. metal content, organic matter content, phyllosilicate minerals, carbonates, and sulfides), and water and sediment chemistry parameters (i.e. pH, ionic strength, redox conditions). Also, methods of sequential extraction vary in chemical extractants based on the soil phases targeted. For these reasons, applying knowledge of Sb phase associations across study sites is a significant challenge. Consistent experimental designs would be helpful in providing phase association data that could systematically guide future investigations of binding mechanisms, mobility, and transport of Sb in aquatic systems.

1.2.3. Mechanisms for Sb Adsorption/Desorption and Phase Associations

Various mechanisms for the adsorption of Sb onto Fe-oxides [Ambe 1987; Leuz et al. 2006] have been proposed. Ambe [1987] proposed four individual mechanisms for the adsorption of Sb onto α-Fe_2O_3. One mechanism is the second order adsorption of two Sb(V) monomers at α-Fe_2O_3 surface sites in close proximity. In the second mechanism, also based on monomer adsorption, the formation of a non-stable transient adsorption species, M_{ads}, would occur, and it would remain in equilibrium with the aqueous monomers. When two α-Fe_2O_3 particles with adsorbed monomers (M_{ads}) collided, a stable adsorbed species, P, would form. This second mechanism proposed by Ambe [1987] also conforms to second-order adsorption. In the third and fourth proposed mechanisms, the formation of dimers from monomers and the adsorption of dimer onto α-Fe_2O_3 are considered. The formation of the dimer is assumed very fast, and the dimer adsorption onto α-Fe_2O_3 is assumed as the rate limiting step in the third mechanism. In the fourth proposed mechanism, the formation of the dimer is assumed slow, thus rate limiting, and the dimer adsorption onto α-Fe_2O_3 process is assumed fast. In another study using goethite as the particle surface, Leuz et al. [2006] found evidence of inner-sphere surface complex formation for both Sb(III) and Sb(V). Still another mechanism was described by Scheinost et al. [2006] who examined Sb-contaminated shooting range soils using X-ray absorption spectroscopy (XAS) and determined that Sb(V) was sorbed to Fe in soils by the formation of edge-sharing and corner-sharing inner-sphere complexes.

When considering Sb binding with natural organic matter, few binding mechanisms have been identified because the complex polyfunctional structures of natural organic matter makes determination of sorption mechanisms and identification of chemical binding sites difficult [Tella and Pokrovski 2009]. However, several studies have reported binding mechanisms for Sb on pure phase humic acids [Buschmann and Sigg 2004; Leuz et al. 2006]. For Sb(III), Buschmann and Sigg [2004] proposed two Sb-humic acid complexation mechanisms including ligand exchange at the Sb center, and the subsequent release of one or two hydroxides followed by the formation of a negatively charged complex that can bind Sb. Previously, Pilarski et al. [1995] indicated that humic acid was able to retain various chemical forms of Sb, including Sb cations (e.g. SbO^+), anions (e.g. SbT^- and $Sb(OH)^{-}_{6}$), and uncharged molecules (e.g. $Sb(OH)_3$). Cationic species retention was attributed to interactions with carboxylate functional groups and anionic species retention was attributed to possible interactions with amine functional groups or chemical transformation of the adsorbed Sb species [Pilarski et al. 1995]. The mechanism proposed for the retention of neutral Sb species was the sorption of the hydroxyl species on hydrogen-form functional groups [Pilarski et al. 1995]. Steely et al. [2007] confirmed these previous studies using size exclusion chromatography-inductively coupled plasma-mass spectrometry (SEC-ICP-MS) to show that Sb can strongly bind to humic acids (HA) via forming Sb-HA complexes in contaminated soils. However, further investigations are clearly needed to better explain binding mechanisms to organic matter as well as iron oxides since these phases appear to play a major role in controlling Sb mobility and transport.

1.3. Speciation of Antimony

As mentioned above, Sb exists in the (III) and (V) oxidation states with Sb(V) dominating in oxic aqueous environments [Filella et al. 2002b; Johnson et al. 2005]. Understanding Sb speciation dynamics is important because the oxidation state has a large influence on its reactivity, toxicity, mobility, and transport [Scheinost et al. 2006] and antimony speciation is very sensitive to redox conditions [Filella et al. 2002b; Johnson et al. 2005; Chen et al. 2003; Scheinost et al. 2006; Vink 1996]. As such, redox conditions in the hyporheic zone (the portion of the streambed where surface water mixes with shallow groundwater) are expected to have a significant effect on Sb mobility in river systems. The most dominant form of Sb(V) at environmentally relevant pH values (above 3) is $Sb(OH)_6^-$ [Wilson et al. 2010], which means that $Sb(OH)_6^-$ dominates Sb speciation under nearly all oxic conditions [Scheinost et al. 2006]. In the reducing conditions of anoxic waters and pore waters, the most stable species is Sb(III) which is dominated primarily by its $Sb(OH)_3^0$ configuration [Crecelius et al. 1975; Chen et al. 2003; Scheinost et al. 2006; Wilson et al 2010]. The study of Sb speciation can be carried out via thermodynamic analysis or experimental methods. Eh(pe)-pH diagrams for Sb have been developed by numerous researchers to give insight into species existence, predominance, and solubility at varying conditions of redox and pH [Pitman et al. 1957; Vink 1996; Filella et al. 2002b, Wilson et al. 2010]. As such, a discussion of some general conclusions researchers have made based on Eh(pe)-pH diagrams is worthwhile. But, it is also important to keep in mind that these diagrams are often limited by the availability of thermodynamic data, and the number of elements and species included [Wilson et al. 2010].

For a Sb-O-H-S system, Vink [1996] showed that the common Sb(V) species $SbO_3^-(aq)$ occupied a broad field in the Eh-pH diagram suggesting a high mobility for Sb under oxidizing conditions across the pH scale. Others have also used Eh-pH diagrams to show that antimonate ($Sb(OH)_6^-$ sometimes represented as SbO_3^-) occupies a large field in the Eh-pH space [Pitman et al. 1957 ; Vink 1996; Filella et al. 2002b]. Additionally, work by Filella et al. [2002b] using a Sb-S-H_2O system at environmentally relevant concentrations of Sb and S confirms the likely presence of soluble $Sb(OH)_6^-$ under oxic conditions, but also the soluble Sb(III) oxide $Sb(OH)_3$ under anoxic conditions and circumneutral pH. Other common Sb(III) species senarmontite and valentinite (Sb(III) oxides found as Sb_4O_6) also exhibit good stability across the pH scale in environments which are moderately reducing to moderately oxidizing [Vink 1996]. Valentinite is an oxidation product of stibnite while senarmontite is a weathering product with low solubility commonly associated with contamination from mining activities [Ettler et al. 2007; Murciego et al. 2007]. Solid stibnite (Sb_2S_3) is the predominant Sb(III) ore that forms under reducing and low to intermediate pH, but it is susceptible to oxidation through weathering during which Sb is released providing a potentially large source to soil or biota [Gál et al. 2007; Wilson et al. 2010].

Though the thermodynamic predictions above provide some generic information of Sb species which may be present under various conditions of pH and redox, real world determination of Sb speciation in environmental systems is more convoluted. Due to a variety of challenges, experimental studies on Sb speciation and speciation related Sb mobility processes have been limited [Bufflap and Allen 1994]. One of the challenges is preserving true redox conditions during experiments which can require experimental procedures to be conducted in completely inert atmospheres using inert gasses, commonly of nitrogen or argon

[Bufflap and Allen 1994]. Another challenge is the absence of reliable analytical techniques for determination of Sb species. To overcome this challenge, researchers have attempted to determine Sb (III) and Sb (V) in environmental samples by combining separation techniques, such as high performance liquid chromatography (HPLC) with highly sensitive elemental detection equipment like inductively coupled plasma-mass spectrometry or inductively coupled plasma-optical emission spectrometry [Steely et al. 2007]. In addition, state-of-the-art instrumentation such as XAS has also recently been applied successfully to determine Sb oxidation states [Scheinost et al. 2006; Mitsunobu et al. 2006]. These recent advances in Sb speciation analysis have proven valuable in facilitating Sb detection in natural samples and assisting Sb speciation environmental impact studies.

1.4. Sb Transformation

Antimony transformation refers to Sb movement between oxidation states (III) and (V) [Buschmann and Sigg 2004; Buschmann et al. 2005; Steely et al. 2007], or between methylated and nonmethylated species [Filella et al. 2007; Duester et al. 2008]. Information on antimony transformation in the environment is very important because of the consequences of speciation for transport, availability, and toxicity. Although the number of studies in this area is limited, transformations of Sb species have been shown to occur in the presence of organic matter and/or microorganisms. For example, humic acid has been shown to oxidize Sb(III) to Sb(V) [Buschmann and Sigg 2004; Buschmann et al. 2005; Steely et al. 2007]. Further, Buschmann and Sigg [2004] showed that after Sb(III) was oxidized to Sb(V) by humic acids and released to solution, a binding site on the humic acid structure was freed for another Sb(III) to bind. Photochemical transformations are also possible. Buschmann et al. [2005] reported that not only was Sb(III) photooxidized in the presence of humic acid, but also that the rate of photooxidation was linearly correlated with dissolved organic carbon concentration. Further, these authors showed that the rate of Sb(III) photooxidation increased sharply with increasing pH and ascribed the photooxidation of Sb species to the likely production of oxidizing phenoxyl radicals.

Another transformation process involves biomethylation, which has important consequences for metalloid toxicity and biogeochemical cycling [Andrewes et al. 2000]. For Sb, it has recently been shown that microorganisms, most likely bacteria or fungi, form methylated antimony [Filella et al. 2007; Duester et al. 2008]. Because methylation increases metal volatility, methylated antimony may be especially mobile in natural systems [Duester et al. 2008]. Further, methylated metals have been shown to bioaccumulate in organisms [Stumm and Morgan 1996; Duester et al. 2008]. Thus, more information concerning formation of methylantimony and its toxicity, bioaccumulation, and potential effect on Sb mobility in natural environment is clearly needed [Duester et al. 2008; Filella et al. 2007]. In addition, very limited information is available concerning the uptake of Sb species or the oxidation and reduction of Sb by biota [Filella et al. 2007]. In one study [Kantin 1983], Sb(V) was the predominant Sb species found in three marine multi-cellular algae collected from San Diego Bay including *Ulva* sp., *Enteromorpha* sp., and *Sargassum* sp., However, *Sargassum* sp. has also been shown to contain significant proportions of (up to 30%) of Sb(III) which may indicate an ability of *Sargassum* sp. to transform Sb species internally (Kantin 1983). In a separate study, another species of algae (*Chlorella vulgaris*) excreted 40% Sb(V) and 60%

Sb(III) after being initially exposed only to Sb(III), demonstrating the ability for this algae species to oxidize Sb(III) into a less toxic Sb(V). Some species of bacteria (*Acidithiobacillus ferroxidans, Stibiobater senarmontii*) have also exhibited the ability to oxidize Sb [Fillela et al. 2007]. In any case, preliminary data suggests organisms play an important role in Sb species transformations, although there is a considerable need for future studies.

1.5. Sb Bioavailablility and Uptake

Research regarding the bioavailability and uptake of Sb is very limited, but there is some information on the bioavailability of Sb in soils and uptake by flora and fauna [Ainsworth et al. 1990a; Hammel et al. 2000; Flynn et al. 2003; Casado et al. 2007; Gál et al. 2007; Heier et al. 2009]. Several studies have confirmed that even in Sb contaminated sediments, Sb availability to plants can often be low [Hammel et al. 2000; Casado et al. 2007; Gál et al. 2007]. In these studies, despite the fact that total sediment Sb concentrations ranged from 10 to 1,200 mg/kg, plant available fractions were typically less than 1% of total Sb concentrations [Hammel et al. 2000; Casado et al. 2007]. Casado et al. [2007] reported that Sb concentrations in plants, which included a variety of grass and pasture herbs, were low due to low solubility of Sb in the studied area. In a different study, Hammel et al. [2000] found that Sb levels in numerous vegetable and crop species grown in contaminated areas were comparable to concentrations reported in uncontaminated soils. However, the determination of plant available Sb concentrations is typically achieved by single extraction techniques in these studies, where water soluble (extracted with Milli-Q or deionized water) or readily soluble (extracted with NH_4NO_3) fractions were considered as the plant available fractions [Hammel et al. 2000; Flynn et al. 2003; Casado et al. 2007; Gál et al. 2007]. In contrast, He [2007] reported Sb concentrations in radish plants of 3-10.5 mg/kg in roots and 1.5-121.4 mg/kg in leaves, with a significant correlation between leaf and soil Sb concentrations. Additionally, Murciego et al. [2007] reported occasionally high Sb concentrations >1000 mg/kg in three species of autochthonous plants, including *Cytisus striatus* (Portuguese broom), *Cistus ladanifer* (Gum Rockrose), and *Dittrichia viscose* (Aromatic Inula). To better quantify Sb accumulation and account for environmental variability, Murceigo et al. [2007] presented plant accumulation coefficients (PAC) data for a number of flora (PAC is the ratio of Sb concentrations in plants to those in soil). Results showed considerable variability as PAC ranged from 0.03 up to a maximum of 18.79. But, based on the PAC data, these researchers were able to determine that the *Cytisus striatus* species could be a Sb excluder due to low PACs. Alternatively, *Dittrichia viscose* species appears to hyperaccumulate Sb and was suggested as a potential for use in phytoremediation schemes.

Sb accumulation studies for fauna are very limited [Gál et al. 2007; Heier et al. 2009]. In one example, Heier et al. [2009] studied time-dependant metal accumulation in brown trout for stream waters that exhibited elevated concentrations (2 to 3 μg/L) of several metals due to runoff from a shooting range. After brown trout were exposed to these waters for 23 days, Sb accumulation was identified on the gills and in livers of the fish. Antimony concentrations on fish gills ranged from 0.01 to 0.08 μg/L, while in fish livers, concentrations ranged from 0.0 to 0.07 μg/L. The data indicates that Sb accumulation is not large in trout nor is it tissue specific. In another study, Gál et al. [2007] studied the bioavailability of Sb in earthworms near a former Sb mining and smelting site where total Sb sediment concentrations were very high ranging from 10 to 1,200 mg/kg. In the two species of earthworms studied, antimony concentrations ranged from 0.72 to 26 mg/kg and bioconcentration factors (ratio of Sb concentration in biota to that in soil) were less than 1. These studies have shown that despite high Sb concentrations in surrounding environments, relatively low to moderate concentrations may only be available to certain biota. However, for meaningful comparisons of Sb bioavailability, it would be beneficial to form a consensus on the extractable concentration of Sb for defining the "available" Sb concentration in soil/sediment. Also, additional research on the uptake of Sb by various plant species would allow for insight into why some species serve as Sb excluders and why others accumulate Sb. To date, studies focusing on Sb uptake in vegetation are more abundant than those on uptake by aquatic organisms, invertebrates, and mammals. Directions for future studies should include comparisons of Sb uptake in contaminated and uncontaminated environments, Sb uptake in test species (i.e. *Daphnia magna*, *Hyalella azteca*, *Chironomus tentans*, and *Pimephales promelas*), and Sb uptake by organisms under varying water and sediment chemistry conditions (i.e. pH, ionic strength, redox conditions).

2. Insight into Sb Dynamics from a Case Study in a Semi-arid Creek

In this section, information on Sb dynamics from previous studies conducted in Manadas Creek near Laredo, TX, will be presented. The results include Sb concentrations in water and sediment, potential effects of rainfall on Sb cycling, laboratory analysis of chemical binding forms of Sb on streambed sediments, effects of stream water chemistry on Sb release from contaminated sediment, and Sb uptake in freshwater bivalve mollusks. Contributions of these results to filling the knowledge gaps of Sb behavior in river systems discussed in Section 1 will be evaluated.

2.1. Overview

Manadas Creek is a semi-arid creek and urban tributary of the Rio Grande located in Laredo, Texas. This creek has a 10,122-acre total drainage area, is a major watershed within Laredo, and is located in an industrialized area surrounded by warehouses, major railway lines, and a cement ready-mix plant. Manadas Creek also carries wastewater effluent from the North Laredo Wastewater Plant to the Rio Grande. In addition, an antimony smelter that operated from 1928 until being decommissioned in 1999 [USEPA 1997] is located on the creek bank approximately 2 km upstream from its confluence with the Rio Grande. Studies conducted on the smelter site found contamination of both sediment and groundwater [USEPA 1997]. Near the plant operations section of the site, dissolved antimony concentrations in ground water samples reached 2.5 mg/l [USEPA 1997]. Soils collected in undeveloped portions of the site were found to have antimony concentrations of up to 231 mg/kg at depths from zero to two feet below the surface [USEPA 1997]. Presently, two slag sites, containing antimony byproducts remain on the site for remediation [TNRCC 2002].

Table 2.1. Summary of major findings from previous studies conducted in Manadas Creek

Major Findings	References
• Manadas Creek identified as a site of high concern for potential effects from toxic substances • High Sb concentrations in stream water and sediment	**USIBWC [1994]; TNRCC [1998]** ***Location:*** Rio Grande and its tributaries; ***Time:*** March 1993 and June 1995
• Sb contamination of sediment and groundwater • Elevated dissolved Sb concentrations in ground water near plant operations section of site • Very high Sb concentrations in soils found in undeveloped portions	**USEPA [1997]** ***Location:*** Smelter Site
• Elevated Sb concentrations in stream water • Very high sediment exchangeable Sb concentrations • Temporal variation observed with high concentrations found from September through November	**Flores [2007]; Flores et al. [2011]** ***Location:*** half-way between MC-3 and MC-4; ***Time:*** July – Dec. 2006
• Sb concentrations were very low at MC-1 and MC-2, very high at MC-3 and MC-4, and decreased downstream at MC-5 and MC-6 • Low particulate-phase Sb concentrations across all sites in stream water • Insignificant vertical variation at MC-1 and MC-2, heterogeneous vertical distribution at MC-3 and MC-4, and low degree of vertical variations at MC-5	**Baeza et al. [2010]** ***Location:*** MC-1 – MC-6; ***Time:*** Feb., Apr., May 2008
• Lower dissolved in-stream Sb concentrations during rainy season than dry season • Similar sediment concentrations in rainy and dry seasons • Decrease in Sb concentration in water after a major rain event and slow increase in the following dry months	**Baeza et al. [2011b]** ***Location:*** MC-1– MC-6; ***Time:*** Aug., Sept., Nov. 2008
• A very fast initial Sb release followed by a slow release from contaminated sediment at all pH and ionic strength values tested • The initial Sb release was a function of pH, with highest Sb release occurring at pH 6 • Similar extent of the slow release after the initial stage for different pH and ionic strength conditions • Substantially increased total Sb release under high flow and reduced Sb release under high ionic strength in flume experiments	**Baeza [2009]; Baeza et al. [2011b]** ***Location:*** sediments collected from MC-3 and MC-4; ***Time:*** 2009

Several studies have been conducted at Texas A&M University-Kingsville since 2007 to further examine Sb contamination and distribution in Manadas Creek [Flores 2007; Flores et al. 2011; Baeza 2009; Baeza et al. 2010; Baeza et al. 2011]. A summary of these studies and their findings is provided in Table 2.1. Flores [2007] and Flores et al. [2011] conducted a preliminary study and examined the metal contamination levels in water and sediment at one location (near

MC-3 shown in Figure 2.1) from July - December 2006 and gained preliminary perspective of the temporal variation of levels of heavy metals at the study site. Baeza [2009] and Baeza et al. [2010, 2011a] conducted extensive studies on spatial and temporal variations of Sb in water and sediment at locations MC-1 through MC-6 from February - November, 2008 (Figure 2.1). Baeza [2009] and Baeza et al. [2011b] carried out preliminary studies to examine the effects of pH and ionic strength on Sb release from contaminated sediment using laboratory batch and flume experiments. A recent study has also been conducted by Vaughan and Addo-Mensah to examine bioaccumulation of Sb in *Corbicula fluminea*, a bivalve fresh water mollusk, in Manadas Creek (Figure 2.1). Results from these studies are summarized below.

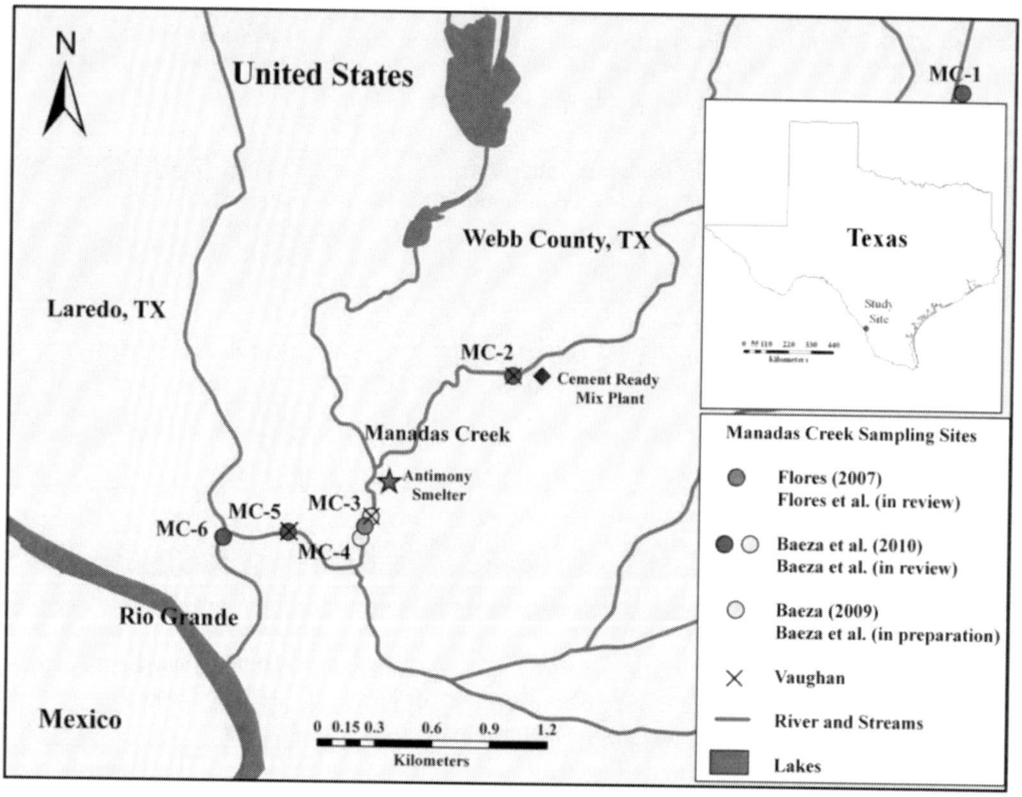

Figure 2.1. Manadas Creek sampling sites established in various studies to date.

2.2. Sb Contamination in a Highly Contaminated Field Site

2.2.1. Antimony Concentrations in the Water and Sediment

In the mid 1990's, two bi-national studies were conducted examining the presence of toxic substances in the Rio Grande/Rio Bravo and its tributaries [USIBWC 1994; TNRCC 1998]. Sampling of Manadas Creek was conducted in March 1993 and June 1995 at a site located 0.8 km upstream from its confluence with the Rio Grande, approximately half-way between MC-5 and MC-6 (Figure 2.1) [USIBWC 1994, TNRCC 1998]. In both studies, Manadas Creek was identified as being a site of high concern for potential effects from toxic substances [USIBWC 1994, TNRCC 1998]. In USIBWC [1994], dissolved Sb measured 78 μg/L in stream waters, a concentration that exceeded human health criteria by a factor of 5.6 indicating a risk if fish and or untreated water taken from the creek were consumed on a long-term basis [USIBWC 1994]. In addition, Sb concentrations exceeded the national 85th percentile by a factor of 1.4 [USIBWC 1994]. In TNRCC [1998], Sb concentrations in stream water and sediment were also high measuring 36.4 μg/L and 15.4 mg/kg, respectively.

In more recent studies (June to December 2006), elevated concentrations of Sb were also observed in stream water samples with Sb concentrations ranging from ~10 to 80 μg/L [Flores 2007; Flores et al. 2011]. In this work, aqueous Sb concentrations exceeded the USEPA limit of 5.6 μg/L of Sb for human health consumption of water and organisms [USEPA 2006] for all months [Flores et al. 2011]. In 2008, Sb concentrations in the stream water were found as high as 446 μg/L in April and 479 μg/L in August at locations of MC-3 and MC-4 [Baeza et al. 2010; 2011]. Sediment exchangeable concentrations of Sb, the fraction of the Sb extracted using 0.1 M $CaCl_2$ according to the sequential extraction method of Silveira et al. [2006], ranged from 1.3 to 4.3 mg/kg of dry sediment in 2006, exceeding the 1.1 mg/kg of dry sediment of Ecological Soil Screening Levels (Eco-SSLs) for Sb developed by USEPA [2005] for the state of Texas [Flores et al. 2011]. In 2008, the highest total sediment concentrations at locations MC-3 and MC-4 were found in the range of 957.1 – 1103.2 mg/kg of dry sediment [Baeza et al. 2010; 2011a]. In this same study, very high Sb enrichment factors (EF) of 146.6 – 390.7 were found at sites MC-3 and MC-4 in May and 1379.1 – 1792.4 in November (Baeza et al. 2010; 2011a]. These high sediment EF values clearly show that the streambed sediment is serving as a holding reservoir for Sb at sites MC-3 and MC-4, which may then serve as a potential continual source of Sb to the overlying stream water and negatively impact water quality downstream including in the Rio Grande [Baeza et al. 2010].

2.2.2. Spatial and Temporal Variations of Sb and Effect of Rainfall

The recent studies in Manadas Creek show that antimony concentrations in both stream water and sediment are low at sites MC-1 and MC-2 above the smelter site, and have never exceeded the soil screening level of 1.1 mg/kg of dry sediment [USEPA 2005; Baeza et al. 2010]. The highest Sb concentrations (both dissolved and total Sb) in stream water and sediment have always been found at MC-3 and MC-4, with decreasing levels downstream at sites MC-5 and MC-6 generally ranging from 45 to 135 μg/L (total Sb) in stream water and less than 60 mg/kg in sediment. The particulate phase Sb (the difference between the total and dissolved phase Sb) also varied significantly among sampling sites of MC-1 through MC-6 with the maximum level occurring at MC-4 [Baeza et al. 2010]. Vertical distributions of Sb in streambed sediment (total Sb associated with sediments) at sites MC-1 and MC-2 are insignificant, but at sites MC-3 and MC-4, sediment Sb concentrations range from 50 to 900 mg/kg dry sediment with a maximum of 1103.2 mg/kg dry sediment at depth of 33 cm [Baeza et al. 2010; 2011a]. While a heterogeneous vertical distribution has been observed at the highly contaminated sites, a low degree of vertical variation of Sb has been observed at MC-5 [Baeza et al. 2010; 2011a].

Sb levels in Manadas Creek vary temporally as well as spatially. Initially, the temporal variation of Sb in contaminated Manadas Creek waters appeared to be associated with rainfall events. Flores et al. [2011] found that dissolved Sb concentrations were highest during the autumn rainy season. A strong positive correlation was also found between stream flow and sediment associated Sb in the top layer of the streambed, which was attributed to the replenishment of particulate Sb via recent deposition of Sb contaminated fine sediments caused by surface runoff [Flores et al. 2011]. This seasonal Sb distribution pattern in Manadas Creek was further studied by Baeza et al. [2010; 2011a] who found, in contrast to Flores et al., that dissolved Sb concentrations in stream water were generally lower during the rainy season than the dry season [Baeza et al. 2011a]. Particulate phase Sb concentrations in the stream water were less than 1μg/L for all sites in May; however, they ranged from 4.7 to 12.0 μg/L in February and from 2.7 to 32.2 μg/L in April [Baeza et al. 2010]. In the autumn rainy months, the absence of particulate Sb was notable [Baeza et al. 2011a]. In this same work, Baeza et al. [2011a] found that after a major rain event, the Sb concentration in the stream water decreased significantly followed by a slow increase in subsequent dry months [Baeza et al. 2011a]. Also in contrast to Flores et al. [2011], Baeaza et al. [2011a] found that sediment concentrations appear similar during the rainy and dry seasons in Manadas Creek, implying little to no mobilization of Sb from streambed sediment following rain events. However, it is important to note that the sediment Sb levels detected in May by Baeza et al. [2010] were much lower than that observed in February and April and this difference was likely attributable to significant sediment disturbance caused by extremely high stream flow following especially heavy rainfall events.

2.2.3. Pending Research Questions for Future Studies

Recognizing the knowledge gaps identified in Part 1, a key facet of recent work in Manadas Creek is that both dissolved and particulate phase Sb have been examined [Baeza et al. 2010; 2011a]. The addition of the particulate phase Sb concentration data in the studies provided valuable information on Sb dynamics and also raised a set of research questions for future work in Manadas Creek. Example research questions include: Is the absence of particulate phase Sb in May due to specifically to a lack of resuspension of contaminated sediments? Since the sampling date was 10 days after the rainfall event in May, is it possible that some of the fine sediment resuspended by the rainfall flows might have settled to the bed surface prior to sample collection? Can the maximum particulate Sb observed at MC-4 be directly traced to the high sediment Sb level at this site? If yes, why was particulate phase Sb not observed at MC-3 which has similar high sediment Sb concentration? These questions can be addressed in future studies by collecting dissolved and particulate phase Sb concentrations at a high spatial and temporal resolution.

The longitudinal distribution of Sb in water and sediment at sites MC-3 through MC-6 indicates the downstream transport of Sb from the highly contaminated sites MC-3 and MC-4 to the confluence with the Rio Grande. As mentioned above, the effect of rainfall events on sedimentary Sb levels was inconsistent possibly because the dynamics of Sb associated fine sediments (particulate phase Sb) are controlled primarily by colloidal processes such as particle settling, filtration and resuspension (see Section 1). To completely understand the dynamics of Sb in Manadas Creek, the effect of these colloidal processes on controlling the Sb associated fine sediments transport under different stream flow and physicochemical conditions must be investigated in future studies.

2.3. Mechanistic Level Understanding of Sb Behavior

2.3.1. Chemical Binding Forms of Antimony on Streambed Sediments

Preliminary results on chemical distribution of Sb in different Manadas Creek sediment forms indicates that Sb is predominantly associated with the residual fraction of the sediment (65.7% to 76.9% of total sediment Sb), which includes Mn oxides, amorphous/crystalline Fe oxides, among other recalcitrant fractions. Sb associated with sediment organic matter has been measured at 15% to 23%, while surface-adsorbed Sb ranged from 5.2% to 10.2% and Sb in the soluble exchangeable fraction ranged from 3.0% to 6.7% [Baeza et al. 2010]. The low fraction of Sb present in the soluble exchangeable form implies a low bioavailability of the Sb in Manadas Creek, consistent with results reported in the literature [Ettler et al. 2007; Filella et al. 2002b; Murciego et al. 2007]. Although only three sediment samples have been analyzed to date, the results above provided good preliminary data for Sb phase associations in contaminated sediments of Manadas Creek. As mentioned in Section 1, one difficulty encountered in applying phase association information is their site-specific nature. Since a suite of studies have been/will be carried out at Manadas Creek, it is expected that Sb phase association information will be very useful because of the diverse studies conducted at the same study site. The results of a comprehensive study like that being performed at Manadas Creek should help address some of the key research gaps identified in Section 1 including identifying the primary factors controlling Sb mobility in aquatic environment.

2.3.2. Effects of Stream Water Chemistry on Antimony Release from Contaminated Sediment

Preliminary batch and flume experiments to examine the effects of pH, ionic strength, and various stream flow conditions on the release of Sb from contaminated streambed sediments have been conducted [Baeza 2009]. Batch sediment suspension experiments have been performed for pH 6-8 and ionic strengths of 10 and 100 mM NaCl or $CaCl_2$ during which the release of Sb was monitored over a 24-hr period and quantified [Baeza 2009]. The kinetics of Sb release included a very fast initial release followed by a slow release from contaminated sediment at all pH and ionic strengths [Baeza et al. 2011b]. The initial Sb release from contaminated sediment was a function of pH, with highest Sb release occurring at pH 6; however, the extent of the slow release after the initial stage was similar for different pH and ionic strength conditions [Baeza et al. 2011b]. Flume experiments have also been conducted at salinities of 1 ppt NaCl and 5 ppt $CaCl_2$ and two different flow rates with results indicating a substantially increased total Sb release under high flow conditions and reduced Sb release under high ionic strength conditions [Baeza et al. 2011b]. This data provides preliminary information on the extent of Sb release under practical field water chemistry conditions. Well-controlled laboratory experiments such as these also help address the knowledge gaps in understanding the mechanisms controlling the release of Sb from contaminated sediment (i.e. pH, ionic strength, redox conditions, composition) as mentioned in Section 1.

2.4. Antimony Uptake by the Freshwater Bivalve Mollusk *Corbicula Fluminea*

As stated earlier in this chapter, a significant knowledge gap exists for Sb bioavailability and uptake in flora and fauna. Whereas most studies report very insignificant bioaccumulation of Sb, statistically significant bioaccumulation of Sb in plant and animal tissues have also been reported [Besser et al; 2001; Duran et al; 2007; Fu et al; 2011; Qi et al; 2011]. Until recently, Sb bioaccumulation in the biota of Manadas Creek has not been investigated so a preliminary study was undertaken to estimate Sb uptake by the aquatic species *Corbicula fluminea*, commonly known as the Asiatic Clam.

Corbicula fluminea were collected from the mainstem of the Rio Grande upstream of the confluence with Manadas Creek. Clams were placed in stainless steel wire mesh cages with gravel and rock substrate suitable for clam survival. Each of the mesh cages was stocked with 125 to150 *C. fluminea.* Cages were placed at three locations, one upstream of the antimony smelter and processing plant (MC-2), a second just downstream from the smelter site (MC-3), and a third at MC-5 (see Figure 2.1). The MC-5 cage was lost as a result of creek flooding and was later replaced with a new cage and clams. Cages were initially placed in early September 2009. The MC-5 cage was replaced in early November 2009. The cages were checked weekly to ensure that they were still in place and underwater. The cages were retrieved monthly and checked, in the field, for clam survival; after which they were returned to the creek. The three cages were recovered after approximately 4-6 months. The clams were placed in fresh water for 24 hours after which they were held in sterile plastic bags at -80°C until analysis.

The soft tissue of the mussels was removed, cleaned and dried in an oven at 80°C for 24 hours prior to analysis. Approximately 200 mg of dried sample was weighed into a digestion liner and digested using 7 mL of HNO_3 and 3 mL HF, following EPA method 3051A [USEPA 2007]. The samples were prepared in triplicates, and blank process and spiking were used. Samples were analyzed by inductively coupled plasma optical emission spectrometry (ICP-OES). Known standards were run to determine accuracy of the ICP-OES. The results obtained from the preliminary study are summarized in Table 2.2.

Table 2.2. Preliminary results of Sb uptake by *C. fluminea* at three sites along Manadas Creek

	MC-2	MC-3	MC-5
Sb (mg/kg) in *C. fluminea*	7.08	8.66	5.26
Sb* (µg/L) in water	22.80	22.71	3.64

* Sb concentration in water at the time clams was removed from the creek.

The range of total Sb measured in *Corbicula* from the three sites was 5.26-8.66 mg/kg of dry sample. There was no significant correlation between the bioaccumulation of Sb in *Corbicula* and the sediment concentrations of Sb measured by Baeza et al. [2010; 2011a]. Sb concentrations in the dissolved phase measured at the end of the experiment on site MC-2 (22.80 µg/L) was slightly higher than previously reported (9.26 µg/L); whereas sites MC-3 and MC-5 were lower than previously reported [Baeza et al. 2010]. MC-3 which is in proximity to the decommissioned smelter recorded the highest Sb uptake with MC-5 recording the lowest and MC-2 which is upstream of the Sb smelter plant recorded Sb accumulation similar to MC-3. This was surprising because site MC-3 is downstream of the smelter where high Sb levels were reported in both water and sediment by Baeza et al. [2010], whereas MC-2 is upstream where very low level of Sb was reported by Baeza et al. [2010].

The preliminary data obtained addresses three key questions in our effort to bridge the knowledge gap in Sb bioavailability and bioaccumulation: Is the Sb in the Manadas Creek bioavailable to biota in the Creek? Does Sb bioaccumulate even though bioavailable concentration is low? Is *Corbicula* a good biomarker to determine the bioaccumulation of Sb in Manadas Creek? This data adds to the growing evidence that substantial Sb bioaccumulation does not correlate with Sb site pollution as other similar trace elements such as As. Ha et al. [2009] reported that only about a tenth of the available Sb was actually taken up by dwarf hairgrass *Eleocharis acicularis*, signifying that bioaccumulation was not only dependant on high bioavailable concentration of Sb. Bioaccumulation of Sb is tissue and species dependant rather than bioavailable concentration of Sb [Ainsworth 1990; Besser et al; 2001; Duran et al; 2007; Fu et al; 2011; Qi et al; 2011]. Fu et al. [2011] also reported a Sb concentration ranging of 0.7 to 51 mg kg^{-1} for animal species including crabs, fish and earthworms. *C. fluminea* which was used in this experiment is a filter feeder which feeds on algae, bacteria and fine detritus which are known to bioaccumulate to a greater extent than predators and omnivores [Duran et al 2007, Besser 2001]. Also, Duran et al. [2007], Besser et al. [2001], Qi et al. [2011], and Ainsworth [1990b] concluded that in general, bioaccumulation of metals in aquatic ecosystems relates to feeding groups with detritivores accumulating higher levels than herbivores and predatory groups. However, the length of time (4-6 months) that these clams were exposed to Sb in the creek could also explain the high Sb bioaccumulations observed in this preliminary study. Bioaccumulation of Sb has been shown to vary significantly with the length of exposure to the Sb rather than bioavailable concentration [Ha et al 2009].

As the knowledge gaps in bioaccumulation are being filled, it is becoming evident that there are more unanswered questions. Sb should be studied carefully in environments such as the Manadas Creek, where changes in hydrogeochemical conditions, such as changes in flowrate and disturbance of sediments, may make recalcitrant Sb readily available. Further experiments will be needed to address the knowledge gaps on the bioaccumulation and bioavailability of Sb in Manadas Creek. The initial questions set out in this preliminary study have been partially addressed by these findings: *Corbicula* is a good biomarker to study the bioaccumulation of Sb in Manadas Creek. The bioaccumulation observed could be due to length of exposure of the species to the Sb source, changes in geochemical conditions during the experiment that increased Sb bioavailability and the Sb tolerance of *Corbicula*.

SUMMARY

This chapter discussed some of the important knowledge gaps currently hindering a complete understanding of Sb behavior in aquatic systems. The knowledge gaps discussed involve several aspects of Sb dynamics including partitioning, phase associations, binding mechanisms, transformation, and Sb speciation. Results from recent studies of Sb distribution in a semi-arid creek have also been presented, including Sb concentrations in stream water and sediment, potential effects of rainfall on Sb cycling, laboratory analysis of chemical binding forms of antimony on streambed sediments, effects of stream water chemistry on Sb release from contaminated sediment, and Sb uptake in freshwater bivalve mollusks. These results address some of the knowledge gaps identified in Section 1; however, substantial work will still be needed in order to improve our overall understanding of Sb behavior in aquatic systems. The on-going and planned research work will include for example: in-depth understanding of the factors controlling the Sb release from contaminated sediment (i.e. water and sediment chemistry and rainfall); the role of Sb speciation in regulating Sb release from streambed sediment; differentiation of Sb distributions among all sediment phases (soluble fraction, fractions associated with Fe, Al, or Mn oxides, fractions associated with organic matter, and residual or recalcitrant fractions) and identification of factors affecting this distribution. All of these future studies will be designed based on the current findings presented in Section 2. Since these research work will all be conducted at the same study site as illustrated in Figure 2.1, the site specific nature of the some of the research findings will be minimized and cross interpretation of the results from different types of studies will be possible, thus some of the knowledge gaps discussed in Section 1 will be addressed in a better means.

Acknowledgements

This article was based upon work supported by the Center of Research Excellence in Science and Technology–Research on Environmental Sustainability of Semi-Arid Coastal Areas (CREST–RESSACA) at Texas A&M University–Kingsville (TAMUK) through a Cooperative Agreement (HRD-0734850 and HRD-0206259) from the National Science Foundation.

References

Ahnstrom, Z. S.; Parker, D. R. *Soil Sci. Soc. Am. J.* 1999, 63, 1650–1658.

Ainsworth, N.; Cooke, J. A.; Johnson, M. S. *Environ. Pollut.* 1990, 65, 65-77.

Ainsworth, N.; Cooke, J. A.; Johnson, M. S. *Environ. Pollut.* 1990, 65, 79-87.

Ambe, S. *Langmuir.* 1987, 3, 489-493.

Andrewes, P.; Cullen, W. R.; Polishchuk, E. *Environ. Sci. Technol.* 2000, 34, 2249-2253.

Baeza, M. Texas A&M University – Kingsville, (2009).

Baeza, M.; Ren, J.; Krishnamurthy, S.; Vaughan, T. *Arch. Environ. Contam. Toxicol.* 2010, 58, 299-314.

Baeza, M.; Ren, J.; Krishnamurthy, S.; Vaughan, T. Environmental Sustainability: South Texas-Mexico Border Region; Springer, 2011a (in review).

Baeza, M.; Ren, J.; Shank, G. C. 2011b (in preparation).

Besser, J. M.; Brumbaugh, W. G.; May, T. W.; Church, S. E.; Kimball, B. A. *Arch. Environ. Contam. Toxicol.* 2001, 40, 48-59.

Brannon, J. M.; Patrick, J., W.H. *Envionmental Pollution*. 1985, 9, 107-126.

Bufflap, S. E.; Allen, H. E. *Water Res.* 1995, 29, 165-177.

Buffle, J., De Vitre, R. R. *Chemical and biological regulation of aquatic systems*, CRC Press, Inc.: Boca Raton, FL, 1994; pp 1-385.
Buschmann, J.; Canonica, S.; Sigg, L. *Environ. Sci. Technol.* 2005, 39, 5335-5341.
Buschmann, J.; Sigg, L. *Environ. Sci. Technol.* 2004, 38, 4535-4541.
Casado, M.; Anawar, H. M.; Garcia-Sanchez, A.; Regina, I. S. *Commun. Soil Sci. Plant Anal.*
2007, 38, 1255-1275.
Chen, Y.; Deng, T.; Filella, M.; Belzile, N. *Environmental and Science Technology*. 2003, 37, 1163-1168.
Corapcioglu, M. Y.; Jiang, S. *Water Resour. Res.* 1993, 29, 2215-2226.
Crecelius, E.; Bothner, M. H.; Carpenter, R. *Environ. Sci. Technol.* 1975, 9, 325-333.
Denys, S.; Tack, K.; Caboche, J.; Delalain, P. *Chemosphere*. 2008, 74, 711-716.
Duester, L.; Vink, J. P.; Hirner, A. V. *Environ. Sci. Technol.* 2008, 42, 5866-5871.
Duran, M.; Kara, Y.; Akyildiz, G.; Ozdemir, A. *Bull. Environ. Contam. Toxicol.* 2007, 78, 395-399.
Elliot, A. H.; Brooks, N. H. *Water Resour. Res.* 1997, 33, 123-136.
Elliot, A. H.; Brooks, N. H. *Water Resour. Res.* 1997, 33, 137-151.
Ettler, V.; Mihaljevic, M.; Sebek, O.; Nechutný, Z. *Chemosphere*. 2007, 68, 455-463.
Filella, M.; Belzile, N.; Chen, Y.-W. *Earth Sci. Rev.* 2002, 59, 265-285.
Filella, M.; Belzile, N.; Chen, Y.-W. *Earth Sci. Rev.* 2002, 57, 125-176.
Filella, M.; Belzile, N.; Lett, M.-C. *Earth Sci. Rev.* 2007, 80, 195-217.
Flores, B. K. Texas A&M University – Kingsville; (2007).
Flores, B. K.; Ren, J.; Krishnamurthy, S.; Belzer, W. Environmental Sustainability: South Texas-Mexico Border Region; Springer, 2011 (in review).
Flynn, H. C.; Meharg, A. A.; Bowyer, P. K.; Paton, G. I. *Environ. Pollut.* 2003, 124, 93-100.
Fu, Z.; Wu, F.; Mo, C.; Liu, B.; Zhu, J.; Deng, Q.; Liao, H.; Zhang, Y. *Microchem. J.* 2011, 97, 12-19.
Gal, J.; Hursthouse, A.; Cuthbert, S. *J. Environ. Sci. Health, Part A: Toxic/Hazard. Subst.* 2007, 42, 1263-1274.
Grolimund, D.; Borkovec, M.; Barmettler, K.; Sticher, H. *Environ. Sci. Technol.* 1996, 30, 3118-3123.
Ha, N. T. H.; Sakakibara, M.; Sano, S. *CLEAN*. 2009, 37, 720-725.
Hammel, W.; Debus, R.; Steubing, L. *Chemosphere*. 2000, 41, 1791-1798.
He. *Environ. Geochem. Health*. 2007, 29, 209-219.
Heier, L. S.; Lien, I. B.; Strømseng, A. E.; Ljønes, M.; Rosseland, B. O.; Tollefsen, K.-E.; Salbu, B. *Sci. Total Environ.* 2009, 407, 4047-4055.
Hou, H.; Takamatsu, T.; Koshikawa, K.; Hosomi, M. *Eur. J. Soil Sci.* 2006, 57, 214-227.
Huettel, M.; Ziebis, W.; Forster, S.; Luther Iii, G. W. *Geochim. Cosmochim. Acta*. 1998, 62, 613-631.
Johansson, H.; Jonsson, K.; Forsman, K. J.; Wörman, A. *Sci. Total Environ.* 2001, 266, 229-238.
Johnson, C. A.; Moench, H.; Wersin, P.; Kugler, P.; Wenger, C. *J. Environ. Qual.* 2005, 34, Kantin, R. *Limnol. Oceanogr.* 1983, 28, 165-168.
Landrum, J. T.; Bennett, P. C.; Engel, A. S.; Alsina, M. A.; Pastén, P. A.; Milliken, K. *Appl. Geochem.* 2009, 24, 664-676.
Leleyter, L.; Probst, J.-L. *Int. J. Environ. Anal. Chem.* 1999, 73, 109 - 128.

Leuz, A.-K.; Mönch, H.; Johnson, C. A. *Environ. Sci. Technol.* 2006, 40, 7277-7282.
Masson, M.; Schäfer, J.; Blanc, G.; Dabrin, A.; Castelle, S.; Lavaux, G. *Appl. Geochem.* 2009, 24, 1747-1756.
McLaren, R. G.; Naidu, R.; Smith, J.; Tiller, K. G. *J. Environ. Qual.* 1998, 27, 348-354.
Mignon, C.; Mori, C. *Hydrobiologia.* 1999, 392, 81-88.
Mitsunobu, S.; Harada, T.; Takahashi, Y. *Environ. Sci. Technol.* 2006, 40, 7270-7276.
Mok, W. M.; Wai, C. M. *Environ. Sci. Technol.* 1990, 24, 102-108.
Murciego, A. M.; Sánchez, A. G.; González, M. A. R.; Gil, E. P.; Gordillo, C. T.; Fernández, J. C.; Triguero, T. B. *Environ. Pollut.* 2007, 145, 15-21.
Neaman, A.; Mouélé, F.; Trolard, F.; Bourrié, G. *Appl. Geochem.* 2004, 19, 973-979.
Packman, A. I.; Brooks, N. H.; Morgan, J. J. *Water Resour. Res.* 2000, 36, 2363-2372.
Packman, A. I.; Brooks, N. H.; Morgan, J. J. *Water Resour. Res.* 2000, 36, 2351-2361.
Pilarski, J.; Waller, P.; Pickering, W. *Water, Air, Soil Pollut.* 1995, 84, 51-59.
Pitman, A. L.; Pourbaix, M.; de Zoubov, N. *J. Electrochem. Soc.* 1957, 104, 594-600.
Qi, C.; Wu, F.; Deng, Q.; Liu, G.; Mo, C.; Liu, B.; Zhu, J. *Microchem. J.* 2011, 97, 44-51.
Ren, J.; Packman, A. I. *J. Environ. Eng.* 2002, 128, 624.
Ren, J.; Packman, A. I. *Environ. Sci. Technol.* 2004, 38, 6571-6581.
Ren, J.; Packman, A. I. *Environ. Sci. Technol.* 2005, 39, 6387-6394.
Saiers, J. E.; Hornberger, G. M. *Water Resour. Res.* 1999, 35, 1713-1727.
Savant, S. A.; Reible, D. D.; Thibodeaux, L. J. *Water Resour. Res.* 1987, 23, 1763-1768.
Scheinost, A. C.; Rossberg, A.; Vantelon, D.; Xifra, I.; Kretzschmar, R.; Leuz, A.-K.; Funke, H.; Johnson, C. A. *Geochim. Cosmochim. Acta.* 2006, 70, 3299-3312.
Schemel, L. E.; Kimball, B. A.; Bencala, K. E. *Appl. Geochem.* 2000, 15, 1003-1018.
Silveira, M. L.; Alleoni, L. R. F.; O'Connor, G. A.; Chang, A. C. *Chemosphere.* 2006, 64, 1929-1938.
Sparks, D. L.; *Environmental Soil Chemistry*; 2nd Ed.; Academic Press: New York, NY, 2003; pp 1-352.
Steely, S.; Amarasiriwardena, D.; Xing, B. *Environ. Pollut.* 2007, 148, 590-598.
Stumm, W.; Morgan, J. J. Aquatic *Chemistry: Chemical Equilibria and Rates in Natural Waters*; John Wiley & Sons, Inc.: New York, NY, 1996; 1-1022.
Tan, W.; Liu, F.; Feng, X.; Huang, Q.; Li, X. *J. Colloid Interface Sci.* 2005, 284, 600-605.
Tella, M.; Pokrovski, G. S. *Geochim. Cosmochim. Acta.* 2009, 73, 268-290.
Thanabalasingam, P.; Pickering, W. F. *Water, Air, Soil Pollut.* 1990, 49, 175-185.
Thibodeaux, L. J.; Boyle, J. D. *Nature.* 1987, 325, 341-343.
Tighe, M.; Ashley, P.; Lockwood, P.; Wilson, S. *Sci. Total Environ.* 2005, 347, 175-186.
Tighe, M.; Lockwood, P. *Commun. Soil Sci. Plant Anal.* 2007, 38, 1487-1501.
TNRCC (Texas Natural Resource Conservation Commission). (1998).
TNRCC (Texas Natural Resource Conservation Commission). (2002).
USEPA (United States Environmental Protection Agency). (1997).
USEPA (United States Environmental Protection Agency). (2005).
USEPA (United States Environmental Protection Agency). (2006).
USEPA (United States Environmental Protection Agency). (2007).
USIBWC (United States International Boundary and Water Commission). (1994).
Vink, B. W. *Chem. Geol.* 1996, 130, 21-30.
Wang, X.; He, M.; Xi, J.; Lu, X. *Microchem. J.* 2011, 97, 4-11.

Wilson, S. C.; Lockwood, P. V.; Ashley, P. M.; Tighe, M. *Environ. Pollut.* 2010, 158, 1169-1181.

Wörman, A. *Water Resour. Res.* 1998, 34, 2703-2716.

Worman, A.; Forsman, J.; Johansson, H. *J. Environ. Eng.* 1998, 124, 122-130.

Wörman, A.; Packman, A. I.; Johansson, H.; Jonsson, K. *Water Resour. Res.* 2002, 38, 1001.

In: Antimony: Characteristics, Compounds and Applications
Editor: Manijeh Razeghi
ISBN 978-1-62100-598-8

Chapter 3

LEISHMANIA ANTIMONY RESISTANCE: CHALLENGES AND OPPORTUNITIES

Denis Sereno*
MIVEGEC (UM1-CNRS 5290-IRD 224), Département Santé,
Représentation IRD au Maroc, 15 rue Abou Derr, 10000, Rabat, Agdal, Morocco

ABSTRACT

Leishmania are the causative protozoal agents of various forms of leishmaniasis, a disease that is a significant cause of morbidity and mortality in more than 88 countries. The clinical manifestations of the disease range from self-healing cutaneous and mucocutaneous skin ulcers to a fatal visceral form, called visceral leishmaniasis or kala-azar. In the absence of any effective vaccines, the only feasible way to treat and control leishmaniasis is through the use of affordable medications. The current chemotherapeutic arsenal is still composed of molecules that were developed in the 1950s, including pentavalent antimony compounds (i.e., Pentostam, Glucantime) Pentamidine, various formulations of the antibiotic Amphotericin B and, more recently, miltefosine. The choice of treatment used is frequently dictated by economic considerations; therefore, in a large majority of countries, chemotherapeutic approaches rely solely on the use of cheaper pentavalent antimonial compounds. Recent instances of antimonial therapy failures in India were shown to be linked to parasite resistance. However, in most parts of the world, the frequency of parasite antimony resistance linked to treatment failure is unknown. This information is crucial for addressing the risk of selection and transmission of drug-resistant parasites. Poor knowledge about factors that favor selection of resistant parasites, the multiplicity of the agents that play a role in the *in vivo* antileishmanial activity of antimony, and the lack of a standard protocol to diagnose and survey parasite resistance all contribute to insufficient monitoring of antimony resistance. In this review we discuss on the factors, including environmental ones, involved in the selection of antimony resistance in the field and discuss on the methods available for its diagnosis and survey in natural *Leishmania* populations.

---

* E-mail address: denis.sereno@ird.fr.

1. Introduction

Leishmania are the causative protozoal agents of various forms of leishmaniasis, which is a significant cause of morbidity and mortality in more than 88 countries. The clinical manifestations of leishmaniasis range from self-healing cutaneous and mucocutaneous skin ulcers to a fatal visceral form (i.e., visceral leishmaniasis or kala-azar). The clinical manifestations of leishmaniasis range from various cutaneous forms to a fatal visceral form. In the Mediterranean Basin, anthroponotic cutaneous leishmaniasis (ACL) caused by Leishmania. tropica, zoonotic cutaneous leishmaniasis (ZCL) caused by L. major or, less frequently L. infantum and zoonotic visceral leishmaniasis (ZVL) caused by L. infantum are all endemic. Visceral leishmaniasis is widely distributed around the Mediterranean Basin, unlike ZCL and ACL, which are more restricted to meridional and oriental regions. In South America, ZCL caused by various Leishmania species (e.g., L. mexicana, L. (Vianna) peruviana, L. (V.) guyanensis, L. (V.) panamensis and others), mucocutaneous leishmaniasis caused by L. (V.) braziliensis, diffuse cutaneous leishmaniasis caused by L. amazonensis and ZVL caused by L. chagasi are all highly endemic. In the Indian subcontinent, anthroponotic visceral leishmaniasis (AVL) caused by *L. donovani* poses a major health problem. In several regions of the world, the incidence of leishmaniasis outbreaks has been associated with urbanization, travel, climatic change and social conflict (Croft et al., 2006a; Cardenas *et al.*, 2006; Dereure *et al.*, 2009).

In the absence of effective vaccines, the only feasible way to treat and control leishmaniasis is through the use of affordable medications. The current chemotherapeutic arsenal consists of molecules that were developed in the 1950s, including pentavalent antimony (SbV) compounds (e.g., pentamidine), various formulations of the antibiotic Amphotericin B and, more recently, miltefosine (Santos *et al.*, 2008; Mondal *et al.*, 2010; Murray, 2010). Instead of determining treatment based on rational therapeutic indications, treatment choice is frequently dictated by economic considerations, pentavalent antimonial compounds being one of the most affordable (Vanlerberghe *et al.*, 2007). Therefore, in a large majority of countries, chemotherapeutic approaches for all forms of leishmaniasis rely solely on the use of pentavalent antimonial compounds. Therapeutic failure during antimony treatment remains a well know problem, but antimony resistance is not the only factor that leads to therapeutic failure. Indeed, factors that are linked to the host (e.g., immunosuppression due to HIV or malnutrition), the drug itself (e.g., drug batch or counterfeit drugs), the Leishmania species, or the practitioner (e.g., incomplete treatment follow-ups) will also play a role in parasite drug resistance and treatment failure (Croft 2001; Croft et al., 2006a, 2006b; Llanos-Cuentas *et al.*, 2008). In most parts of the world, the frequency of parasite antimony resistance linked to treatment failure is unknown. This information is crucial for addressing the risk of selection and transmission of drug-resistant parasites, particularly in areas where antimony is the only chemotherapeutic alternative. Poor knowledge about factors that favor selection of resistant parasites, the multiplicity of the agents that play a role in the *in vivo* antileishmanial activity of SbV, and the lack of a standard protocol to diagnose and survey parasite resistance all contribute to insufficient monitoring of antimony resistance.

A

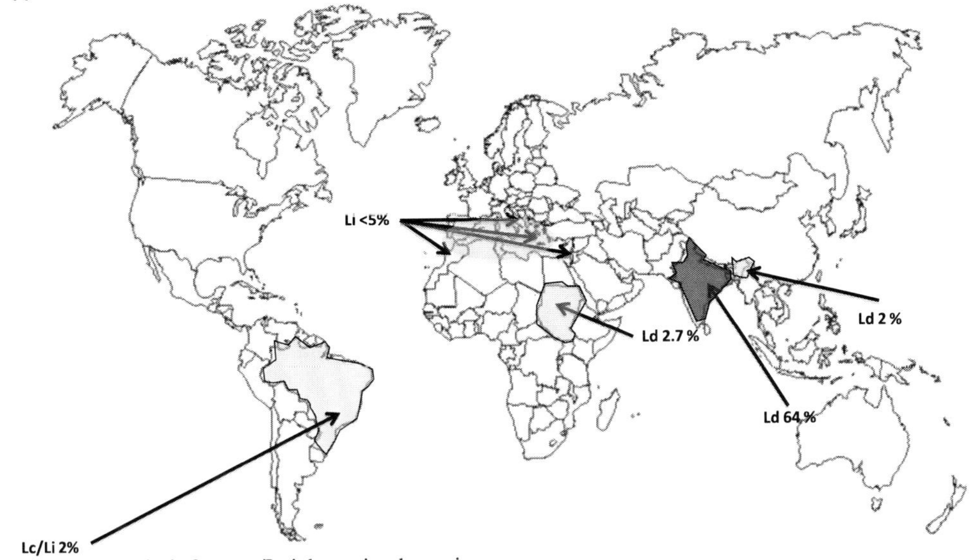

LI/LC Leishmania infantum/Leishmania chagasi
LD Leishmania donovani

B

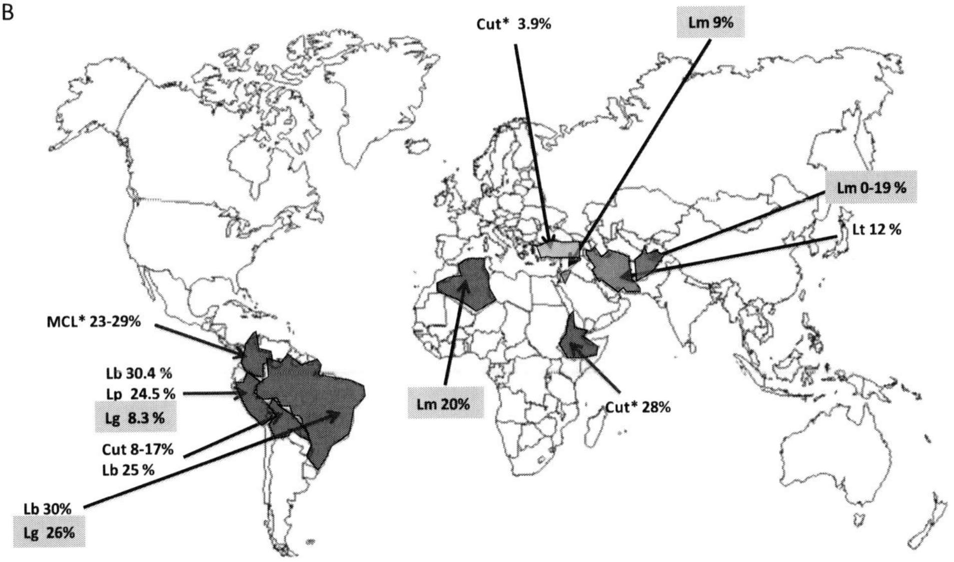

Lm Leishmania major.
Lt Leishmania tropica.
Lb Leishmania braziliensis.
Lp Leishmania panamensis.
Lg Leishmania guyanensis.
*MCL Mucocutaneous Leishmaniosis isolates not typed, *Cut Cutaneous Leishmaniosis: isolates not typed.

Figure 1. Treatment failure rate. Treatment failures in visceral (A) and cutaneous, muco-cutaneous (B) leishmaniasis. Treatment of new world cutaneaous leishmaniasis caused by *L. mexicana* is considered to be efficient and is therefore not quoted in the map.

2. Antimony Clinical Resistance

Antimonials have been in clinical use for all forms of leishmaniasis, for more than 60 years and high rate of therapeutic failure, associated with antimonial begin now to emerged (see Figure 1A & B). However, the link between parasite resistance and therapeutic failure is currently poorly documented particularly in zoonotic context of leishmaniasis.

High rate of therapeutic failure linked to drug resistant parasites are reported for *L. donovani* the ethiological agent of AVL in India but not in Sudan. In 1997, the first indications of therapeutic failure linked to *Leishmania* drug resistance originated in northern Bihar province of India (Sundar *et al.*, 1997). Subsequently, an explicit link between therapeutic failure and parasite antimony-resistance was demonstrated (Lira *et al.*, 1999). In endemic foci of *L. donovani* in Sudan, a longitudinal survey of 820 AVL patients reported that treatment failures occurred in 22 (a failure rate of roughly 3%), among whom only 6 were infected with antimony-resistant parasites (Abdo *et al.*, 2003). In Bangladesh a 2% treatment failure during AVL treatment has been reported no information is available on the susceptibility of *L. donovani* isolates that are transmitted this area (Bern & Chowdhury, 2005). Concerning ACL caused by *L. tropica*, in Iran, therapeutic failures linked to parasite resistance were reported in 12% of patients (Hadighi *et al.*, 2006). In the Mediterranean Basin, even though ACL is endemic, no information on the efficiency of antimony treatments is available.

A failure rate of less than 5% was reported in cases of ZVL therapy (Gradoni *et al.*, 2008, Gradoni *et al.*, 2003). Among New World countries, treatment failure of visceral leishmaniasis caused by *L. infantum/L. chagasi* was recently reported in Columbia (Velez *et al.*, 2009). Despite the studies cited here, reports relating to the antimonial susceptibility of *L. infantum* isolates are scarce, even though selection for antimony resistance during therapy readily occurs (Farault-Gambarelli *et al.*, 1997, Carrio *et al.*, 2001, Carrio and Portus, 2002).

During treatment of ZCL caused by *L. major* a high rate of treatment failure (20%) has been found in Algeria (Belazzoug and Neal, 1981), but not in Israel (9%) or Turkey (3%) (Solomon *et al.*, 2009; Uzun *et al.*, 2004). Surprisingly treatment failure between 0 and 19% has been reported in Pakistan (Firdous *et al.*, 2009).

In Ethiopia, unresponsiveness to antimonial treatment is reported in 28% of patients bearing cutaneous lesions; however, no information either on the *Leishmania* species involved or on their antimony susceptibility is provided (Padovese *et al.*, 2009). A case of relapse after Pentostam treatment of a cutaneous case of leishmaniasis has been reported in 1979 in Senegal, though the strain was not typed (Neva *et al.*, 1979).

In South America, antimony complications in cases of mucocutaneous leishmaniasis have long been recognized (Rocha *et al.*, 1980, Palacios *et al.*, 2001). Mean cure rates of 71.3% and 74% were reported in Brazilian patients infected with *L. braziliensis* or *L. guyanensis*, respectively (Rojas *et al.*, 2006, Romero *et al.*, 2001, Tuon *et al.*, 2008). In both Brazil and Peru, differences in treatment outcome during *L. braziliensis* or *L. guyanensis* therapy have been documented. These two species differ in their intrinsic susceptibility to antimonials, but susceptible and resistant strains of both species have been isolated. It may be assumed that the proportion of SbV-tolerant *L. guyanensis* isolates may vary geographically for unidentified reasons (Arevalo *et al.*, 2006, Yardley *et al.*, 2006).

3. *In vivo* Antileishmanial Activity of Antimony

The multiplicity of the factors that shape the intracellular antimony susceptibility of *Leishmania* (see Figure 2) can explain *a posteriori* why a *Leishmania* susceptibility profile does not always correlate with clinical outcome. Indeed, the nature of the active compound(s) *in vivo* has been debated since 1945 when Goodwin and coworkers (Goodwin, 1995) proposed that trivalent antimonials are the active forms of pentavalent compounds. Today, scientists assume that trivalent antimonials are the leishmanicidal form, which makes the bioconversion of SbV by the host or the parasite a key activation event. Interestingly, studies have identified two genes encoding for proteins that are able to convert SbV to the trivalent form (Denton *et al.*, 2004; Zhou *et al.*, 2004).

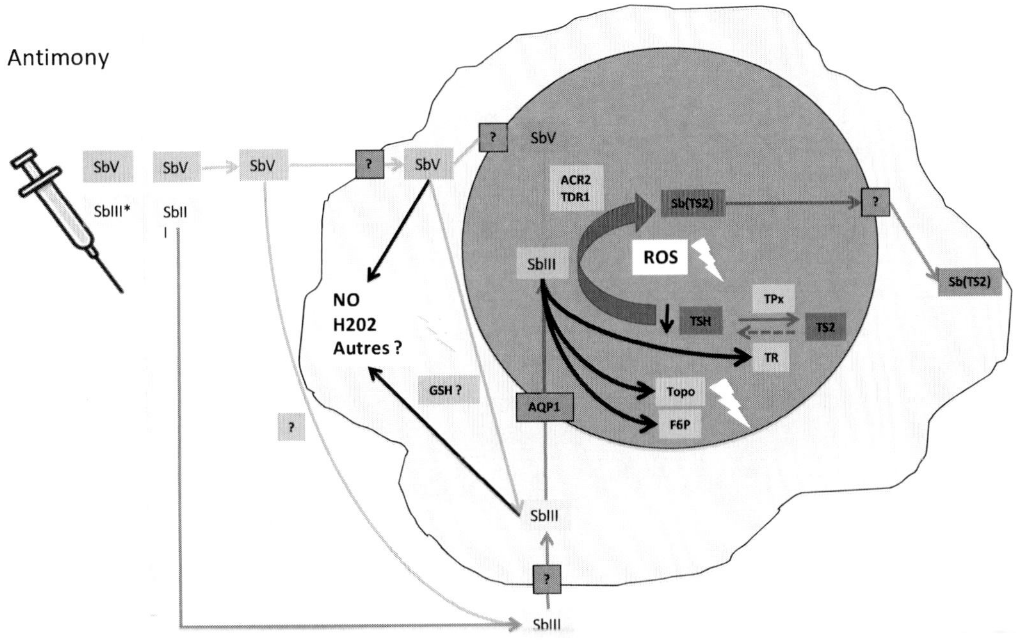

Figure 2. Schematic representation of the proposed mode of action of SbV containing drugs. Even if trivalent form of antimony is considered as the active form of commercially available pentavalent antimony compounds but hosts-derived molecules also play a role in the antileishmanial activity of SbV *in vivo*. The entry of SbV and SbIII into the host occurs yhrpugh unidentified transporter. Both species of antimony are able to induce the generation of NO and H202 that are harmful for intracellular parasites. The entry of Sb(V) into the parasite occurs by an unknown transporter, and entry of Sb(III) through AQP1. Once Inside leishmania Sb vis reduced into SbIII by the action of réductases. Such réduction can also take place in the host cell (Glutathione can play role in such réduction). SbIII antimony can inhibit various enzymes (Topo: Topoisomerase, F6P: fructose 6 phosphatase and TR: trypanothione reductase). SbIII interferes with the antioxydant defence of the parasite through inhibition of the TR and the formation of Sb(III) TSH complex that can are actively exported. These action led to a drastic reduction in TSH and cocomitantly to an accumulation of Reactive oxydant species (ROS) toxic for leishmania. Other abbreviations: TSH, reduced trypanothine; TS2, oxidized trypanothione; Sb(TS)2, conjugate of Sb(III) with trypanothione. *Trace of SbIII can be found in pentavalent antimony formulations (Łukaszczyk & Zyrnicki, 2010; Santos *et al.*, 2009).

The first clue on the mode of action (see Figure 2) of SbIII was provided by Bueding and Mansour (1957) when demonstrating that SbIII inhibit *Shistosoma mansoni* phosphofructo-kinase enzymatic activity. Antimony drastically affects glucose metabolism, fatty acid beta-oxidation and ATP formation (Berman *et al.*, 1987; 1989). Besides these pathways, specific targets were also identified *in vitro*, like Topoisomerase or Trypanothione reductase (Walker and Saravia., 2004; Cunningham & Fairlamb, 1995). SbIII can also competes with Zn(II) for its binding to the CCHC of the zinc finger domain, indicating that zinc finger proteins may be among target of antimony (Demichelli *et al.*, 2008). The exposure of *Leishmania* to SbIII led to the accumulation of damaging reactive oxygen species within *Leishmania* parasites (Mandal *et al.*, 2007; Wyllie *et al.*, 2004). Thiol metabolism plays central role in the in the maintenance of an intracellular environment having a role in the defence against oxidative damage caused by oxydative stress. SbIII also decreases the intracellular thiol buffering capacity of *Leihmania* by promoting loss of trypanothione, a glutathione (GSH)–spermidine compound, and by inhiting Trypanothione reductase activity (Cunningham and Fairlamb, 1995). All these thereby profoundly perturb the thiol redox potential of the parasite and therefore its antioxydant defence.

Microbicidal compounds generated by the host during antimony treatment (e.g., H_2O_2 and NO) play also a role in the *in vivo* leishmanicidal activity of antimony formulations (Carter *et al.*, 2005; Mookerjee Basu *et al.*, 2006; Mandal *et al.*, 2007; Sarkar *et al.*, 2001). Therefore, the *in vivo* chemotherapeutic outcome of antimonial therapy depends on the susceptibility of the parasite to not only antimonial compound but also to other host-derived microbicidal factors. Likewise, during treatment of *L. donovani*, the outcome of pentavalent antimonial therapy *in vivo* correlates with tolerance towards macrophage-derived microbicidal factors (Carter *et al.*, 2005; Carter *et al.*, 2006). The acquisition of drug-resistance markedly alters the response of the host immune system following parasitic infection (Haldar *et al.*, 2010).

3. Selective Pressures for *In vivo* Antimonial Resistance

The link between drug pressure and selection of drug-resistant microorganisms has been extensively documented in various infectious models (Blower *et al.*, 1996; Arestrup 1999); however, this link may not be obvious for leishmaniasis and many other vector-transmitted diseases, because visceral and cutaneous leishmaniasis are transmitted in foci with zoonotic or anthroponotic life-cycles (See figure 3). In anthroponotic form of leishmaniasis, the main known mammalian reservoirs are humans that are subject to treatment. Obviously, in a *Leishmania* population, the probability for a parasite of being in contact with antimony of chemotherapeutic origin will be greater in anthroponotic forms of leishmaniasis than in zoonotic ones. We can therefore consider that the contact time between *Leishmania* and antimony is greater in anthroponotic forms of leishmaniasis than in zoonotic ones. As a consequence, antimony of chemotherapeutic origin will certainly represent the main selective agent involved in the selection of drug-resistant parasites in anthroponotic forms of leishmaniasis. Indeed, the extensive use of antimony compounds to treat anthroponotic visceral leishmaniasis caused by *L. donovani* in India has certainly favored the selection of antimony-resistant parasites (Croft *et al.*, 2006b). In zoonotic forms of leishmaniasis, where wild or feral mammals are the reservoir, the contact time of parasites with antimony is dramatically reduced or even absent. *Leishmania* isolates that naturally have poor susceptibility to antimony, however, have been characterized in patients prior to treatment (Faraut-Gambarelli *et al.*, 1997; Gambarelli *et al.*, 1987; Robledo *et al.*, 1999). Substantial differences in antimony treatment outcome have also been reported in both Brazil and Peru during treatment of *L. guyanensis* and *L. braziliensis* (See figure 1B). Researchers have proposed that these differences in treatment outcome are related to geographic variations in the proportion of SbV-tolerant isolates. Although clear explanations of differences in treatment outcomes have not been determined, but treatment differences between *Leishmania* species highlight that antimony chemotherapy is not the only selective pressure that shapes *Leishmania* antimony susceptibility in the field (Yardley *et al.*, 2006). In zoonotic leishmaniasis, environmental sources of antimony or other selective agents might contribute to *in vivo* selection of antimony-resistant parasites (See figure 3). Although antimony concentrations in unpolluted environmental samples are often very low (0.2 μg/l in water to 1.0 μg/g in soil), concentrations up to 3 mg/l have been reported in spring sources (Krachler *et al.*, 2005; Shotyk *et al.*, 2005; Shotyk *et al.*, 2005).

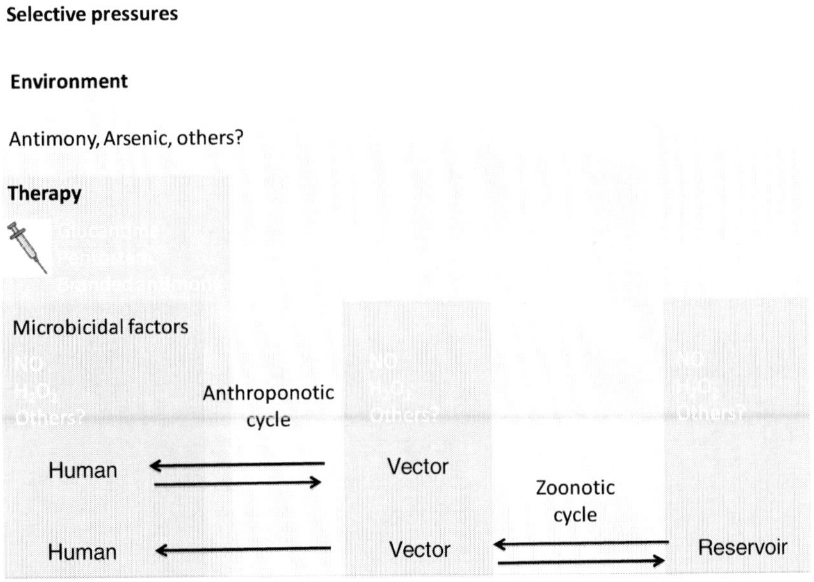

Figure 3. Shematic representation of selective factors shaping antimony susceptibility of *Leishmania* population. Antimony is applied to treat all form of human leishmaniasis in most part of the world, with the exception of the occidental part the Mediterranean basin, Israel and India. SbV antimony is also extensively used to treat the canine reservoir of *Leishmania* in Spain, Portugal, France, Italy or Greece. Beside antimony delivered during Leishmaniasis chemotherapy, the one present in the environment of the various hosts (human or vector) and reservoir (wild or domestics mammals) might also act as selective factors. Because resistance to arsenic and/or to other heavy metals could also select for cross-resistance to antimony, it can be considered as a potential selective factor. Immune derived host microbicidal factors, like NO, can also be involved in the selection of antimony resistant parasites. All these factors can reach *Leishmania* during the whole life cycle of the parasite, including its insect vector: Immune response, at the level of the epithelium of the sandfly lumen where *Leishmania* accomplish its intravectorial development.

Aside from antimony, arsenic (As) concentrations can reach up to 500 µg/l in particularly polluted locations (Baghure *et al.*, 2011). Interestingly, *in vitro* studies have shown that common mechanisms are involved in the occurrence of arsenite and antimony resistance (Rosen 1995; Ouellette *et al.*, 1998). Therefore, cross-resistance between environmental sources of Sb and As may play a role in shaping the Sb-tolerance of natural populations of *Leishmania.* In heavily polluted areas, such as mining sites, swamps and urban environments, sandflies feed on plants that could be contaminated by local atmospheric dust, and wild or feral animal reservoirs can be exposed to Sb and As by drinking polluted water. The impact of such theoretical contamination can be important if it contributes to shaping the behavior of the parasite offspring after genetic exchanges that may take place in the invertebrate host (Akyopants *et al.*, 2009).

Nitric oxide (NO) is present in various *Leishmania* hosts, including the macrophages of its mammalian hosts and the epithelium of the sandfly vector. Tolerance to NO involves trypanothione metabolism, which is also crucial for antimony resistance (Bocedi *et al.*, 2010). Experimental evidence supports the notion that NO may be a crucial factor that can modify the antimony response of *Leishmania* populations. *Leishmania* that were selected *in vitro* for SbIII resistance expressed cross-resistance to NO (Holzmuller *et al.*, 2005). Although studies have not addressed the effect that NO resistance has on antimony cross-resistance, *L. braziliensis* isolated from antimony-unresponsive patients were generally less susceptible to NO (Souza *et al.*, 2010). The extent to which lifetime exposure to NO influences the antimony susceptibility of *Leishmania* populations in endemic areas remains to be investigated.

4. Monitoring *Leishmania* Antimony Resistance

Various hosts and drug-derived molecules play a role in the antileishmanial activity of SbV *in vivo*. Therefore, to reflect the *in vivo* activity of antimonials, the *in vitro* tests must monitor the antileishmanial activity of all the microbicidal compounds involved in the killing pathway of SbV, including SbIII, H_2O_2 and NO. During these tests the origin of the antimonial batch has also to be taken into account because commercially available SbV solutions are not pure (Santos *et al.*, 2009; Lukaszczyk and Zyrnicki, 2010). Therefore various factors are supposed to effect affect the outcome of the test.

Cellular Models for *In vitro* Tests

Obviously, the intracellular forms of the parasites (amastigotes) represent the ideal form to assay because both indirect activity through the host cell and direct activity on the parasite can be assessed. Unfortunately, methods that involve intracellular amastigotes are labor intensive, difficult to standardize, and the outcome is strongly dependent upon the nature of the host cell (Berman 1984; Seifert *et al.*, 2010; Gebre-Hiwot et al., 1992). The development of reporter gene technologies has enabled the quantification of *Leishmania* parasites in host cells and whole mammalian hosts (Sereno *et al.*, 2001; Sereno et al., 2007; De La Llave *et al.*, 2011; Lang *et al.*, 2009), and the capacity of these technologies to determine the drug susceptibility of *Leishmania* field isolates has been tested (Hadighi *et al.*, 2006; Lachaud *et al.*, 2009). The time required for transfection and the selective pressure that is required for the selection of recombinant parasites, however, will certainly affect the composition of the isolates in cases of mixed infections (Antoniou *et al.*, 2004); therefore, its utility for *Leishmania* field isolates is limited. The nature of the host cell also greatly influences the outcome of the test (Seifert *et al.*, 2010). The lack of consensus in the standardization of this methodology makes interpretations of potential intra and interspecific variations in antimony susceptibility difficult (See table 1). In fact, even if this system takes into account all the factors that are involved in intracellular SbV antileishmanial activity, its versatility and cost make its application very impractical for large-scale analysis of parasite antimony susceptibility and resistance surveys.

Table 1. Sensitivities to pentavalent antimonials of intracellular amastigote stages of *Leishmania* species

Host cell	Parasite strain/species	Incubation (days)	ED50	References
THP-1	*L. infantum*	NA	272.0 μg/ml	Antoniou*et al.*, 2004
	L. infantum	NA	13.0μg/ml	
	L.infantum	5	35.0 μg/ml	Sereno*et al.*, 2001
	L. tropica	5	2- 40 μg/ml	Hadighi*et al.*, 2006
	L. donovani	5	85.0μg/ml	Vergnes *et al.*, 2007
Mouse peritoneal macrophages	*L. tropica*	5	2-50 μg/ml	Hadighi*et al.*, 2006
	L. major	5	6-17 μg/ml	
	L. amazonensis	2	30.4 μg/ml	Ayres*et al.*, 2008
	L.amazonensis	2	49-99 μg/ml	Zauli-Nascimento*et al.*, 2010 Zauli-Nascimento*et al.*, 2010
	L. baziliensis	2	35-77 μg/ml	Farraut –Gambarelli*et al.*, 1997
	L. infantum	7	40-70 μg/ml	
	L. infantum	7	7.5-45.0 μg/ml	Farraut-Gambarelli*et al.*, 1987 Farraut-Gambarelli*et al.*, 1987
	L. Topica	7	7.5-30.0 μg/ml	Lira *et al.*, 1999
	L. donovani	3	1.8-28.0 μg/ml	Kumar*et al.*, 2009
	L. donovani	3	14.0-18.0 μg/ml	Vouldoukis*et al.*, 2006
	L. infantum	2	40 -100 μg/ml	
RAW264.7				Carrio*et al.*, 2001
	L. infantum	2	10-50 μg/ml	Carrio*et al.*, 2000
	L. infantum	2	7.4-7.9 μg/ml	
J774				Callahan*et al.*, 1997
	L. mexicana	3	29.0 μg/ml	
U937				Rojas *et al.*, 2006
	L. panamensis	3	1.6 to 10μg/ml	Goyeneche-Patino*et al.*, 2008
	L. panamensis	3	7.0-14 μg/ml	

Various *Leishmania* species can be grown *in vitro* as amastigotes under axenic conditions (Gupta *et al.*, 2001; Sereno *et al.*, 2007). The determination of drug activity is simple, typically inexpensive and does not require host cells, which makes standardization easier (Sereno and Lemesre, 1997). Not all *Leishmania* species, however, are amenable to being cultured in axenic conditions. Currently, no extensive studies using axenic amastigote systems have been performed to evaluate the antimony susceptibility of *Leishmania* field isolates.

The usefulness of promastigotes to ascertain the antimony susceptibility status of *Leishmania* field isolates has been debated. Because promastigote forms of *Leishmania* are not very susceptible to pentavalent antimony, trivalent antimonial forms are used. Unfortunately, the susceptibility of *Leishmania* promastigotes to trivalent forms of antimony does not always reflect the sensitivity of intramacrophagic amastigotes to pentavalent antimonial formulations (Yardley et al., 2006; Sundar et al., 2001; Vermeersh et al., 2009; Lira et al., 1999). Interestingly, various experimental reports that used *L. tropica* (Hadighi *et al.*, 2006), *L. infantum* (Aït Oudhia *et al.*, unpublished results) or *L. donovani* (Vergnes *et al.*, 2007) support the notion that promastigote trivalent antimony susceptibility partially reflects the intrinsic SbV susceptibility of intramacrophagic amastigotes. Because the host cell is omitted in tests that use promastigote forms, we hypothesized that the discrepancies observed between experiments might reflect the differential role of the host-derived microbicidal factors involved in the SbV-induced intracellular killing. Most of these factors (either drug or host-derived) are currently identified, but tests aimed at simultaneously ascertaining the intrinsic susceptibility of *Leishmania* promastigotes to the molecules known to play a role in the antileishmanial action of antimonials (i.e., SbIII, NO, H_2O_2) might be more representative. Because *Leishmania* has the capacity to reduce SbV antimony to trivalent forms, differences in the expression level or the capacity of *Leishmania* antimony reductase to readily activate antimony between isolates can also determine its susceptibility to SbV. In such cases, tests that use the promastigote form of the parasite and trivalent antimony formulations will never accurately reflect the antimony susceptibility of the isolates.

Biomarker for Antimony Resistance

The search for resistance biomarker is based on the in-depth understanding of the antileishmanial mode of action of antimony and resistance associated with *in vitro* selection (Ashutosh *et al.*, 2007). Studies on the intracellular mode of action of SbIII demonstrates that thiol buffer capacity of *leishmania* is deeply affected by SbIII, through the inhibition of the Trypanothione reductase (Cunningham & Fairlamb, 1995) and via a diminution in the intracellular thiol content through an efflux mechanism yet unidentified (Wyllie *et al.*, 2004). These two mechanisms combine to profoundly compromise the thiol redox potential in drug-sensitive parasites and lead ultimately to the accumulation of reactive oxygene species ROS (Wyllie *et al.*, 2004; Mandal *et al.*, 2007). *Leishmania* parasites can act at different levels to induce resistance to antimony, and many of the specific genes associated have been discovered. These mechanisms generally involve:

- limiting antimony entry to the cell. The entry of SbIII occurs through an AQP1 transporter (Marquis *et al.*, 2005; Gourbal *et al.*, 2004), but the route of entry of SbV is currently not identified.

- increasing the expression of the target. As mentioned above SbIII inhibit trypanothione reductase activity leading to an accumulation of reduced form of trypanothione. Overexpression of the Trypanothione reductase has been characterized in some antimony resistant field isolates (Mittal et al., 2007) with some exception (Wyllie *et al.*, 2010).
- decreasing the activation of the prodrug form of Sb (i.e., SbV) to the active form SbIII. This function is assumed by two reductases, TDR1 a thiol dependent reductase belonging to Glutathione S-tranferase that share homology with the *T. cruzy* Tc52 protein (Denton *et al.*, 2010). The second reductase characterized, LmACR2, share homology with arsenate reductases (Zhou *et al.*, 2004). Although, currently variations in reductase capacity of field isolates of *Leishmania* resistant to antimony has not been investigated.
- conjugation with thiol containing molecules and sequestration or efflux of the thiol conjugated drug. In *Leishmania*, the major low molecular mass thiol is trypanothione in contrast with to most other organisms wich utilize glutathione. Trypanothione is a glutathione–spermidine conjugate (Fairlamb and Cerami, 1992) which is formed following several enzymatic steps, via the synthesis of gluthatione (GSH) and its subsequent conjugation to spermidine. Steps in the synthesis of spermidine involved Ornithine decarboxylase (ODC) and spermidine synthase. The first step in the biosynthesis of GSH is catalyzed by γ-glutamylcysteine synthetase (γ-GCS) and Gluthatione synthase. The conjugation of GSH to spermidine in the following step, is then catalyzed by trypanosome specific enzymes (glutathionyl-spermidine synthetase and Trypanothione synthase) (Oza *et al.*, 2002). Therefore the overexpression of some of the genes in this pathway (γ-GCS or Ornithine decarboxylase ODC) that lead to an increase in the intracellular level of $T(SH)_2$ induces resistance towards SbIII (Grondin *et al.*, 1997; Haimeur *et al.*, 1999). The Sb–Trypanothione conjugate that is formed inside the parasite cell is then sequestered within a vacuole by the intracellular ABC transporter MRPA (Légaré *et al.*, 2001; Leprohon *et al.*, 2009).

Additionnaly, a set of genes able to confer antimony resistance have also been characterized *in vitro*. These enclose genes coding for proteins that belong to the superfamilly of the Leucine Rich Repeat Proteins, Histone H2A, and a large hypothetical polypeptide (Genest *et al.*, 2008; Singh *et al.*, 2010; Choudhury *et al.*, 2008). The mechanism by which the over-expression of these genes confers antimony resistance has yet not been addressed.

The identification of genetic mutations associated with resistance represents molecular tools largely applied for drug resistance surveillance. Genetic mutations associated with resistance can limit potential population selection during *in vitro* culture. In plasmodium for example, the prevalence of some single nucleotide polymorphisms (SNPs) associated with parasite susceptibility is often a good indicator of the level of clinical resistance in a population (Anderson et al., 2010; Wilson et al., 2005). Unfortunately, currently no SNPs in genes involved in antimony resistance have been proved to be indicative of antimony resistance in *Leishmania* field isolates.

In fact in *Leishmania*, upregulation of resistance genes is frequently associated with genomic rearrangement, which leads to gene amplification through homologous recombination between repeated sequences (Leprohon *et al.*, 2009b). Therefore, either quantification of copy number or expression of genes known to be involved in antimony susceptibility should represent good biomarkers for addressing antimony resistance (Guimond et al., 2003; Ubeda et al., 2008). If some of the above-mentioned mechanisms are effectively encountered in resistant lines from the field, however, then no rules could be applied (Adaui *et al.*, 2010; Mandal *et al.*, 2010; Murkherjee *et al.*, 2007; Torres *et al.*, 2010; Maharjan *et al.*, 2008), because of the presence of additional unrelated and less specific mechanisms (Kumar et al., 2010; Wyllie et al., 2008; Vergnes et al., 2007), or because of other yet unidentified mechanisms of resistance (Goyeneche-Patino *et al.*, 2008). Finally, a comparative proteome mapping, carried out on drug susceptible and resistant parasites, have characterized additional protein over-expressed in field antimony-resistant isolates of *L. donovani*, their capacity to confer antimony resistance have now to be investigated (Kumar *et al.*, 2010).

Recent metabolomic approaches have been used to investigate parasites with different susceptibilities to antimony. Interestingly, a hierarchical clustering approach (antimony resistant and susceptible) revealed that differences in the metabolite abundance profiles of the drug-resistant and -sensitive clones could clearly be distinguished. The sample size of this experiment, however, was not large enough to conclusively detect drug-resistant parasites (t'Kindt *et al.*, 2010). The extensive procedures involved in culturing the parasites, the high cost, the large amount of time required and the need for highly skilled individuals to perform the metabolomic analysis limit its usefulness for antimony susceptibility surveys, but metabolomic studies might provide information on new metabolic markers for monitoring *Leishmania* resistance.

Overall, because drug-resistant isolates might be composed of a heterogeneous population of parasites that express various drug resistance mechanisms, no accurate and simple molecular marker is currently available to efficiently diagnose the antimony susceptibility of field isolates.

CONCLUDING REMARKS

During the past 10 years, the emergence and spread of drug-resistant parasites have led to a high rate of antimonial therapy failure in India and have raised questions about the selection and propagation risk of chemoresistant parasites. In terms of population structure, *Leishmania* parasites appear to be essentially clonal, with genetic exchange possibly taking place in invertebrate hosts (Akyopants *et al.*, 2009). In endemic areas, the probability of such an event between drug susceptible and drug resistant parasites is currently unknown but is likely to be low. Thus, the spread of a chemoresistant phenotype in the field likely depends more on the capacity of resistant parasites to be successfully transmitted between hosts rather than on genetic exchange. The selection *in vitro* for antimony-resistant parasites impacts the fitness of the parasites (Sereno *et al.*, 2001), and such selection is associated with a competitive cost (Agnew *et al.*, 2001). Likewise, in laboratory-derived *Mycobacterium tuberculosis* mutants, rifampicin resistance is universally associated with a competitive fitness cost that is caused by the specific resistance mutation and the genetic background of the strain (Gagneux *et al.*, 2006). In contrast, prolonged patient treatment can result in the emergence of resistant strains of *M. tuberculosis* with no fitness defects (Gagneux *et al.*, 2006); furthermore, strains with low- or no-cost resistance mutations are the most frequent types found among clinical isolates. A report describing the fitness advantages conferred by the selection for antimony resistance in Nepalese *L. donovani* strains has recently been published (Vanaershot *et al.*, 2010). The results of that study will likely result in the prioritization of studies aimed to understand better the antimony susceptibility of *Leishmania* parasites in the field. Particularly, it is crucial to: (i) gather more information on the frequency of therapy failure linked to proven parasite resistance, (ii) search for cofactors present in the natural environments of *Leishmania* parasites that could play a role in shaping antimony susceptibility and (iii) further delineate the fitness cost associated with resistance to antimonials and other associated cofactors. In particular, the capacity of drug-resistant parasites to compete with susceptible parasites during intravectorial development will further clarify the fitness of drug-resistant parasites and the potential of these strains to be transmitted in endemic areas.

REFERENCES

Aarestrup FM (1999). Association between the consumption of antimicrobial agents in animal husbandry and the occurrence of resistant bacteria among food animals. *Int J Antimicrob Agents* 12:279-85.

Abdo MG, Elamin WM, Khalil EAG, Mukhtar MM (2003) Antimony-resistant *Leishmania donovani* in eastern Sudan: incidence and in vitro correlation. *East Mediterr Health J* 9: 837-43.

Adaui V, Schnorbusch K, Zimic M, Gutierrez A, Decuypere S, Vanaerschot M, De-Doncker S, Maes L, Llanos-Cuentas A, Chappuis F, Arevalo J, Dujardin JC (2010) Comparison of gene expression patterns among *Leishmania braziliensis* clinical isolates showing a different *in vitro* susceptibility to pentavalent antimony. *Parasitology* 3:1-11.

Agnew P, Holzmuller P, Michalakis Y, Sereno D, Lemesre JL, Renaud F (2001) *In vitro* growth of *Leishmania amazonensis* promastigotes resistant to pentamidine is dependent on interactions among strains. *Antimicrob Agents & Chemother* 45:1928-1929.

Akopyants NS, Kimblin N, Secundino N, Patrick R, Peters N, Lawyer P, Dobson DE, Beverley SM, Sacks DL (2009). Demonstration of genetic exchange during cyclical development of *Leishmania* in the sand fly vector. *Science* 324:265-268.

Anderson T, Nkhoma N, Ecker A, Fidock D (2010) How can we identify parasite genes that underlie antimalarial drug resistance? *Pharmacogenomic* 12:59-85.

Antoniou M, Doulgerakis C, Pratlong F, Dedet JP, Tselentis Y (2004) Treatment failure due to mixed infection by different strains of the parasite *Leishmania infantum*. *Am J Trop Med Hyg* 71:71-72.

Arevalo J, Ramirez L, Adaui V, Zimic M, Tulliano G, Miranda-Verestegui C, Lazo M, Loayza-Muro R, De Doncker S, Maurer A, Chappuis F, Dujardin JC, Llanos-Cuentas A (2007) Influence of Leishmania (Viannia) species on the response to antimonial treatment in patients with American tegumentary leishmaniasis. *J Infect Dis* 195:1846-1851.

Ashutosh S, Sundar N, Goyal N (2007) Molecular mechanisms of antimony resistance in *Leishmania*. *J Med Microbiol* 56:143-53.

Ayres DC, Pinto LA, Giorgio S (2008) Efficacy of pentavalent antimony, amphotericin B, and miltefosine in *Leishmania amazonensis*-infected macrophages under normoxic and hypoxic conditions. *J Parasitol* 94: 1415-1417.

Bellazoug S, Neal RA (1981) Failure of meglumine antimoniate to cure cutaneous lesions due to *Leishmania major* in Algeria. *Trans Roy Soc Trop Med & Hyg* 80:670-671.

Berman JD, Gallalee JV, Best JM (1987) Sodium stibogluconate (Pentostam) inhibition of glucose catabolism via the glycolytic pathway, and fatty acid beta-oxidation in *Leishmania mexicana* amastigotes. *Biochem Pharmacol* 3:197-201.

Berman JD (1984) *Leishmania tropica*: quantitation of *in vitro* activity of antileishmanial agents by Giemsa staining, viability, and 3H-formycin B incorporation. *J Parasitol* 70: 561–562.

Berman JD, Edwards N, King M, Grogl M (1989) Biochemistry of Pentostam resistant *Leishmania*. *Am J Trop Med Hyg* 40:159-64.

Bern C, Chowdhury R (2005) The epidemiology of visceral leishmanisis in Bangladesh: prospects for improved control. *Indian J Med Res* 123:275-288.

Bhagure GR, Mirgane SR (2011) Heavy metal concentrations in groundwaters and soils of Thane Region of Maharashtra, India. *Environ Monit Assess* 173:643-52.

Blower SM, Small PM, Hopewell PC (1996) Control strategies for tuberculosis epidemics: new models for old problems. *Science* 273:497-500.

Bocedi A, Dawood KF, Fabrini R, Federici G, Gradoni L, Pedersen JZ, Ricci G (2010) Trypanothione efficiently intercepts nitric oxide as a harmless iron complex in trypanosomatid parasites. *FASEB J* 24:1035-1042.

Bueding E, Mansour JM (1957) The relationship between inhibition of phosphofructokinase activity and the mode of action of trivalent organic antimonials on *Schistosoma mansoni*. *Br J Pharmacol Chemother* 12:159-65.

Cardenas CM, Sandoval AJ, Rodriguez-Morales P, (2008) Zoonoses and climate variability. *Ann N Y Acad Sci* 1149:326-330.

Carrio J, Portus M (2002) *In vitro* susceptibility to pentavalent antimony in *Leishmania infantum* strains is not modified during *in vitro* or *in vivo* passages but is modified after host treatment with meglumine antimoniate. *BMC Pharmacol* 2:11.

Carrio J, Riera C, Gallego M, Portus M (2001) *In vitro* activity of pentavalent antimony derivatives on promastigotes and intracellular amastigotes of *Leishmania infantum* strains from humans and dogs in Spain. *Acta Trop* 79:179-183.

Carter KC, Hutchison S, Boitelle A, Murray HW, Sundar S, Mullen AB (2005) Sodium stibogluconate resistance in *Leishmania donovani* correlates with greater tolerance to macrophage antileishmanial responses and trivalent antimony therapy. *Parasitology* 131: 747-757.

Carter KC, Hutchison S, Henriquez FL, Légaré D, Ouellette M, Roberts CW, Mullen AB (2006) Resistance of *Leishmania donovani* to sodium stibogluconate is related to the expression of host and parasite gamma-glutamylcysteine synthetase. *Antimicrob Agents Chemother* 50:88-95.

Choudhury K, Zander D, Kube M, Reinhardt R, Clos J (2008) Identification of a *Leishmania infantum* gene mediating resistance to miltefosine and SbIII. *Int J Parasitol* 38:1411-1423.

Croft SL (2001) Monitoring drug resistance in leishmaniasis. *Trop Med Int Health* 6:899-905.

Croft SL, Seifert K, Yardley V (2006a) Current scenario of drug development for leishmaniasis. *Indian J Med Res* 123:399-410.

Croft SL, Sundar S, Fairlamb AH (2006b) Drug resistance in leishmaniasis. *Clin Microbiol Rev* 19:111-126.

Cunningham ML, Fairlamb AH (1995) Trypanothione reductase from *Leishmania donovani.* Purification, characterisation and inhibition by trivalent antimonials. *Eur J Biochem* 230:460-468.

De La Llave E, Lecoeur H, Besse A, Milon G, Prina E, Lang T (2011) A combined luciferase imaging and reverse transcription polymerase chain reaction assay for the study of *Leishmania* amastigote burden and correlated mouse tissue transcript fluctuations. *Cell Microbiol* 13:81-91.

Demicheli C, Frézard F, Mangrum JB, Farrell NP (2008) Interaction of trivalent antimony with a CCHC zinc finger domain: potential relevance to the mechanism of action of antimonial drugs. *Chem Commun (Camb)* 39:4828-4830.

Denton H, McGregor JC, Coombs GH (2004) Reduction of anti-leishmanial pentavalent antimonial drugs by a parasite-specific thiol-dependent reductase, TDR1. *Biochem* 381: 405-412.

Dereure J, Vanwambeke SO, Malé P, Martinez S, Pratlong F, Balard Y, Dedet JP (2009) The potential effects of global warming on changes in canine Leishmaniasis in a focus outside the classical area of the disease in southern france. *Vect Born & Zoonotic Dis* 9: 687-694.

Fairlamb AH, Cerami A (1992) Metabolism and functions of trypanothione in the Kinetoplastida. *Annu Rev Microbiol* 46:695-729.

Faraut-Gambarelli F, Piarroux R, Deniau M, Giusiano B, Marty P, Michel G, Faugère B, Dumon H (1997) *In vitro* and *in vivo* resistance of *Leishmania infantum* to meglumine antimoniate: a study of 37 strains collected from patients with visceral leishmaniasis. *Antimicrob Agents Chemother* 41:827-830.

Firdous R, Yasinzai M, Ranja K (2009) Efficacy of glucantime in the treatment of Old World cutaneous leishmaniasis. *Int J Dermatol* 48:758-762.

Gagneux S, Long CD, Small PM, Van T, Schoolnik GK, Bohannan BJ (2006) The competitive cost of antibiotic resistance in *Mycobacterium tuberculosis*. *Science* 312: 1944-1946.

Gambarelli F, Franck J, Dumon H (1987) Sensibilité des Leishmania spp aux derives stibiès. Utilisation de macrophages péritonéaux de souris. *Ann Soc Belge Med Trop* 67:149-155.

Gebre-Hiwot A, Tadesse G, Croft SL, Frommel D (1992) An *in vitro* model for screening antileishmanial drugs: the human leukaemia monocyte cell line, THP-1. *Acta Trop* 51: 237-245.

Genest PA, Haimeur A, Légaré D, Sereno D, Roy G, Messier N, Papadopoulou B, Ouellette M (2008) A protein of the leucine-rich repeats (LRRs) superfamily is implicated in antimony resistance in *Leishmania infantum* amastigotes. *Mol Biochem Parasitol* 158: 95-99.

Goodwin LG (1995) Pentostam (sodium stibogluconate); a 50-year personal réminiscence. *Trans R Soc Trop Med Hyg* 89:39-341.

Gourbal B, Sonuc N, Bhattacharjee H, Légaré D, Sundar S, Ouellette M, Rosen BP, Mukhopadhyay R (2004) Drug uptake and modulation of drug resistance in *Leishmania* by an aquaglyceroporin. *J Biol Chem* 279:31010-31017.

Goyeneche-Patino DA, Valderrama L, Walker J, Saravia NG (2008) Antimony resistance and trypanothione in experimentally selected and clinical strains of *Leishmania panamensis*. *Antimicrob Agents Chemother* 52:4503-4506.

Gradoni L, Gramiccia M, Scalone A (2003) Visceral leishmaniasis treatment, Italy. *Emerg Infect Dis* 9:1617-1620.

Gradoni L, Soteriadou K, Louzir H, Dakkak A, Toz SO, Jaffe C, Dedet JP, Campino L, Canavate C, Dujardin JC (2008) Drug regimens for visceral leishmaniasis in Mediterranean countries. *Trop Med Int Health* 13:1272-1276.

Grondin K, Haimeur A, Mukhopadhyay R, Rosen BP, Ouellette M (1997) Co-amplification of the gamma-glutamylcysteine synthetase gene gsh1 and of the ABC transporter gene pgpA in arsenite-resistant *Leishmania tarentolae. EMBO J* 16:3057-3065.

Guimond C, Trudel N, Brochu C, Marquis N, El-Fadili A, Peytavi R, Briand G, Richard D, Messier N, Papadopoulou B, Corbeil J, Bergeron MG, Légaré D, Ouellette M (2003) Modulation of gene expression in *Leishmania* drug resistant mutants as determined by targeted DNA microarrays. *Nucleic Acids Res* 31:5886-5896.

Gupta N, Goyal N, Rastogi AK (2001) In vitro cultivation and characterization of axenic amastigotes of Leishmania. *Trends Parasitol* 17:150-153.

Hadighi R, Mohebali M, Boucher P, Hajjaran H, Khamesipour A, Ouellette M (2006) Unresponsiveness to Glucantime treatment in Iranian cutaneous leishmaniasis due to drug-resistant *Leishmania tropica* parasites. *PLoS Med* 3:e162.

Haimeur A, Guimond C, Pilote S, Mukhopadhyay R, Rosen BP, Poulin R, Ouellette M (1999) Elevated levels of polyamines and trypanothione resulting from overexpression of the ornithine decarboxylase gene in arsenite-resistant *Leishmania. Mol Microbiol* 34: 726-735.

Haldar AK, Yadav V, Singhal E, Bisht KK, Singh A, Bhaumik S, Basu R, Sen P, Roy S (2010) *Leishmania donovani* isolates with antimony-resistant but not sensitive phenotype inhibit sodium antimony gluconate-induced dendritic cell activation PLoS *Pathog* 6:e1000907.

Holzmuller P, Sereno D, Lemesre JL (2005) Lower nitric oxide susceptibility of trivalent antimony-resistant amastigotes of *Leishmania infantum. Antimicrob Agents Chemother* 49:4406-4409.

Krachler M, Zheng J, Koerner R, Zdanowicz C, Fisher D, Shotyk W (2005) Increasing atmospheric antimony contamination in the northern hemisphere: snow and ice evidence from Devon Island, Arctic Canada. *J Environ Monit* 7:169-1176.

Kumar A, Sisodia B, Misra P, Shasany AK, Dube A (2010) Proteome mapping of overexpressed membrane-enriched and cytosolic proteins in sodium antimony gluconate (SAG) resistant clinical isolate of Leishmania donovani. *Br J Clin Pharmacol* 70:609-617.

Kumar D, Kulshrestha A, Singh R, Salotra P (2009) In vitro susceptibility of field isolates of *Leishmania donovani* to Miltefosine and amphotericin B: correlation with sodium antimony gluconate susceptibility and implications for treatment in areas of endemicity. *Antimicrob Agents Chemother* 53:835-838.

Lachaud L, Bourgeois N, Plourde M, Leprohon P, Bastien P, Ouellette M (2009) Parasite susceptibility to amphotericin B in failures of treatment for visceral leishmaniasis in patients coinfected with HIV type 1 and *Leishmania infantum. Clin Infect Dis* 48:e16-22.

Lang T, Lecoeur H, Prina E (2009) Imaging *Leishmania* development in their host cells. *Trends Parasitol* 25:464-473.

Légaré D, Richard D, Mukhopadhyay R, Stierhof YD, Rosen BP, Haimeur A, Papadopoulou B, Ouellette M (2001) The *Leishmania* ATP-binding cassette protein PGPA is an intracellular metal-thiol transporter ATPase. *J Biol Chem* 276:26301-26307.

Leprohon P, Légaré D, Ouellette M (2009) Intracellular localization of the ABCC proteins of *Leishmania* and their role in resistance to antimonials. *Antimicrob Agents Chemother* 53: 2646-26499.

Leprohon P, Légaré D, Raymond F, Madore E, Hardiman G, Corbeil J, Ouellette M (2009b) Gene expression modulation is associated with gene amplification, supernumerary chromosomes and chromosome loss in antimony-resistant *Leishmania infantum. Nucleic Acids Res* 37:1387-1399.

Lira R, Sundar S, Makharia A, Kenney R, Gam A, Saraiva E, Sacks D (1999) Evidence that the high incidence of treatment failures in Indian kala-azar is due to the emergence of antimony-resistant strains of *Leishmania donovani. J Infect Dis* 180:564-567.

Llanos-Cuentas A, Tulliano G, Araujo-Castillo R, Miranda-Verastegui C, Santamaria-Castrellon G, Ramirez L, Lazo M, De-Doncker S, Boelaert M, Robays J, Dujardin JC, Arevalo J, Chappuis F (2008) Clinical and parasite species risk factors for pentavalent antimonial treatment failure in cutaneous leishmaniasis in Peru. *Clin Infect Dis* 46: 223-231.

Lukaszczyk L, Zyrnicki W (2010) Speciation analysis of Sb(III) and Sb(V) in antileishmaniotic drug using Dowex 1 x 4 resin from hydrochloric acid solution. *J Pharm Biomed Anal* 2010 52:747-51.

Maharjan M, Singh S, Chatterjee M, Madhubala R (2008) Role of Aquaglycoprotein (AQP-1) gene and drug uptake in antimony-resistant isolates of *Leishmania donovani. Am J Trop Med Hyg* 79:69-75.

Mandal G, Wyllie S, Singh N, Sundar S, Fairlamb AH, Chatterjee M (2007) Increased levels of thiols protect antimony unresponsive *Leishmania donovani* field isolates against reactive oxygen species generated by trivalent antimony. *Parasitology* 134:1679-1687.

Mandal S, Maharjan M, Singh S, Chatterjee M, Madhubala R (2010) Assessing aquaglyceroporin gene status and expression profile in antimony-susceptible and resistant clinical isolates of *Leishmania donovani* from India. *J Antimicrob Chemother* 65:496-507.

Marquis N, Gourbal B, Rosen BP, Mukhopadhyay R, Ouellette M (2005) Modulation in aquaglyceroporin AQP1 gene transcript levels in drug-resistant *Leishmania. Mol Microbiol* 57:1690-1699.

Mittal MK, Rai S, Ravinder A, Gupta S, Sundar S, Goyal N (2007) Characterization of natural antimony résistance in *Leishmania donovani* isolates. *Am J Trop Med Hyg* 76: 681–688.

Mondal S, Bhattacharya P, Ali N (2010) Current diagnosis and treatment of visceral Leishmaniasis. *Expert Rev Anti Infect Ther* 8:919-944.

Mookerjee-Basu J, Mookerjee A, Sen P, Bhaumik S, Sen P, Banerjee S, Naskar K, Choudhuri SK, Saha B, Raha S, Roy S (2006) Sodium antimony gluconate induces generation of reactive oxygen species and nitric oxide via phosphoinositide 3-kinase and mitogen-activated protein kinase activation in *Leishmania donovani*-infected macrophages *Antimicrob Agents Chemother* 50:1788-1797.

Mukherjee A, Padmanabhan PK, Singh S, Roy G, Girard I, Chatterjee M, Ouellette M, Madhubala R (2007) Role of ABC transporter MRPA, gamma-glutamylcysteine synthetase and ornithine decarboxylase in natural antimony-resistant isolates of *Leishmania donovani. J Antimicrob Chemother* 59: 204-211.

Murray MW (2010) Treatment of visceral leishmaniasis in 2010: direction from bihar state, India. *Future Microbiol* 5:1301-1303.

Neva FA, Wyler D, Nash T (1979) Cutaneous leishmaniasis--a case with persistent organisms after treatment in presence of normal immune response. *Am J Trop Med Hyg* 28:467-471.

Ouellette M, Légaré D, Haimeur A, Grondin K, Roy G, Brochu C, Papadopoulou B (1998) ABC transporters in *Leishmania* and their role in drug resistance. *Drug Resist Update* 1: 43-48.

Oza SL, Tetaud E, Ariyanayagam MR, Warnon SS, Fairlamb AH (2002) A single enzyme catalyses formation of Trypanothione from glutathione and spermidine in Trypanosoma cruzi. *J Biol Chem* 277:35853-35861.

Padovese V, Terranova M, Toma L, Barnabas GA, Morrone A (2009) Cutaneous and mucocutaneous leishmaniasis in Tigray, northern Ethiopia: clinical aspects and therapeutic concerns. *Trans R Soc Trop Med Hyg* 103:707-711.

Palacios R,Osorio LE,Grajalew LF,Ochoa MT (2001) Treatment failure in children in a randomized clinical trial with 10 and 20 days of meglumine antimonate for cutaneous leishmaniasis due to Leishmania viannia species. Am J Trop Med Hyg 64:187-193.

Robledo SM, Valencia AZ, Saravia NG (1999) Sensitivity to Glucantime of *Leishmania viannia* isolated from patients prior to treatment. *J Parasitol* 85:360-366.

Rocha RA, Sampaio RN, Guerra M, Magalhaes A, Cuba CC, Barreto AC, Marsden PD (1980) Apparent Glucantime failure in five patients with mucocutaneous leishmaniasis. *J Trop Med Hyg* 83:131-139.

Romero GA, Guerra MV, Paes MG, Maccedo VO (2001) Comparison of cutaneous leishmaniasis due *to Leishmania (Viannia) braziliensis* and *L. (V.) guyanensis* in Brazil: therapeutic response to meglumine antimoniate. *Am J Trop Med Hyg* 65:456-465.

Rojas R, Valderrama L, Valderrama M, Varona MX, Ouellette M, Saravia NG (2006) Resistance to antimony and treatment failure in human *Leishmania (Viannia)* infection. *J Infect Dis* 193:1375-1383.

Rosen BP (1995) Resistance mechanisms to arsenicals and antimonials. *J Basic Clin Physio Pharmacol* 6:251-263.

Santos DO, Coutinho CE, Madeira MF, Bottino CG, Vieira RT, Nascimento SB, Bernardino A, Bourguignon SC, Corte-Real S, Pinho RT, Rodrigues CR, Castro HC (2008) Leishmaniasis treatment--a challenge that remains: a review. *Parasitol Res* 103:1-10.

Santos VS, Kubota LT, Tarley CR (2009) Speciation of Sb(III) and Sb(V) in meglumine antimoniate pharmaceutical formulations by PSA using carbon nanotube electrode. *J Pharm Biomed Anal* 50:151-157.

Sarkar A, Saha P, Mandal G, Mukhopadhyay D, Roy S, Singh SK, Das S, Goswami RP, Saha B, Kumar D, Das P, Chatterjee M (2011) Monitoring of intracellular nitric oxide in leishmaniasis: its applicability in patients with visceral leishmaniasis. *Cytometry* 79:35-45.

Seifert K, Escobar P, Croft SL (2010) In vitro activity of anti-leishmanial drugs against *Leishmania donovani* is host cell dependent. *J Antimicrob Chemother* 65:508-511.

Sereno D, Cordeiro da Silva A, Mathieu-Daude F, Ouaissi A (2007) Advances and perspectives in *Leishmania* cell based drug-screening procedures. *Parasitol Int* 56:3-7.

Sereno D, Guilvard E, Maquaire S, Cavaleyra M, Holzmuller P, Ouaissi A, Lemesre JL (2001) Experimental studies on the evolution of antimony-resistant phenotype during the *in vitro* life cycle of *Leishmania infantum*: implications for the spread of chemoresistance in endemic areas. *Acta Trop* 80:195-205.

Sereno D, Lemesre JL (1997) Use of an enzymatic micromethod to quantify amastigote stage of *Leishmania amazonensis* in vitro. *Parasitol Res* 83:401-403.

Sereno D, Roy G, Lemesre JL, Papadopoulou B, Ouellette M. DNA transformation of Leishmania infantum axenic amastigotes and their use in drug screening. *Antimicrob Agents Chemother*. 2001 Apr;45(4):1168-73.

Shotyk W, Krachler M, Chen B (2005) Antimony: global environmental contaminant. *J Environ Monit* 7:1135-1136.

Shotyk W, Krachler M, Chen B, Zheng J (2005) Natural abundance of Sb and Sc in pristine groundwaters, Springwater Township, Ontario, Canada, and implications for tracing contamination from landfill leachates. *J Environ Monit* 7:1238-1244.

Singh R, Kumar D, Duncan RC, Nakhasi HL, Salotra P (2010) Overexpression of histone H2A modulates drug susceptibility in *Leishmania* parasites. *Int J Antimicrob Agents* 36: 50-57.

Solomon M, Baum S, Barzilai A, Pavlotsky F, Trau H, Schwartz E (2009) Treatment of cutaneous leishmaniasis with intralesional sodium stibogluconate. *J Eur Acad Dermatol Venereol* 23:1189-1192.

Souza AS, Giudice A, Pereira JM, Guimaraes LM, De-Jesus AR, De-Moura TR, Wilson ME, Carvalho EM, Almeida RP (2010) Resistance of *Leishmania (Viannia) braziliensis* to nitric oxide: correlation with antimony therapy and TNF-alpha production. *BMC Infect Dis* 10:209.

Sundar S, Pai K, Kumar R, Pathak-Tripathi K, Gam AA, Ray M, Kenney RT (2001) Resistance to treatment in Kala-azar: speciation of isolates from northeast India. *Am J Trop Med Hyg* 65:193-196.

t'Kindt R, Scheltema RA, Jankevics A, Brunker K, Rijal S, Dujardin JC, Breitling R, Watson DG, Coombs GH, Decuyper S (2010) Metabolomics to unveil and understand phenotypic diversity between pathogen populations. *PLoS Negl Trop Dis* 4:e904.

Torres DC, Adaui V, Ribeiro-Alves M, Romero GA, Arévalo J, Cupolillo E, Dujardin JC (2010) Targeted gene expression profiling in *Leishmania braziliensis* and *Leishmania guyanensis* parasites isolated from Brazilian patients with different antimonial treatment outcomes. *Infect Genet Evol* 10:727-733.

Tuon FF, Amato VS, Graf ME, Siqueira AM, Nicodemo AC, Neto VA (2008) Treatment of new world cutaneous leishmaniasis-a systematic review with a meta-analysis. *Int J Dermatol* 47:109-124.

Ubeda JM, Légaré D, Raymond F, Ouameur AA, Boisvert S, Rigault P, Corbeil J, Tremblay MJ, Olivier M, Papadopoulou B, Ouellette M (2008) Modulation of gene expression in drug resistant *Leishmania* is associated with gene amplification, gene deletion and chromosome aneuploidy. *Genome Biol* 9:R115.

Uzun S, Durdu M, Culha G, Allahverdiyev AM, Memisoglu HR (2004) Clinical features, epidemiology, and efficacy and safety of intralesional antimony treatment of cutaneous leishmaniasis: recent experience in Turkey. *J Parasitol* 90:853-859.

Vanlerberghe V, Diap G, Guerin JP, Meheus F, Gerstl S, Van der Stuyft P, Boelaert M (2007) Drug policy for visceral leishmaniasis: a cost-effectiveness analysis. *Trop Med Int Health* 12:274-283.

Vélez ID, Colmenares LM, Munoz CA (2009) Two cases of visceral leishmaniasis in Colombia resistant to meglumine antimoniate treatment. *Rev Inst Med Trop S Paulo* 51: 231-236.

Vergnes B, Gourbal B, Girard I, Sundar S, Drummelsmith J, Ouellette M (2007) A proteomics screen implicates HSP83 and a small kinetoplastid calpain-related protein in drug resistance in *Leishmania donovani* clinical field isolates by modulating drug-induced programmed cell death. *Mol Cell Proteomics* 6:88-101.

Vanaerschot M, Maes I, Ouakad M, Adaui V, Maes L, De Doncker S, Rijal S, Chappuis F, Dujardin JC, Decuypere S (2010) Linking *in vitro* and *in vivo* survival of clinical *Leishmania donovani* strains. *PLoS One* 5:e12211.

Vermeersch M, da Luz RI, Toté K, Timmermans JP, Cos P, Maes L (2009) *In vitro* susceptibilities of Leishmania donovani promastigote and amastigote stages to antileishmanial reference drugs: practical relevance of stage-specific differences *Antimicrob Agents Chemother* 53:3855-3859.

Vouldoukis I, Rougier S, Dugas B, Pino P, Mazier D, Woehrlo F (2006) Canine visceral leishmaniasis: comparison of *in vitro* leishmanicidal activity of marbofloxacin, meglumine antimoniate and sodium stibogluconate. *Vet Parasitol* 135:137-146.

Walker J, Saravia NG (2004) Inhibition of Leishmania *donovani* promastigote DNA topoisomerase I and human monocyte DNA topoisomerases I and II by antimonial drugs and classical antitopoisomerase agents. *J Parasitol* 90:1155-1162.

Wilson PE, Alker AP, Meshnick SR (2005) Real-time PCR methods for monitoring antimalarial drug resistance. *Trends Parasitol* 21:278-283.

Wyllie S, Cunningham ML, Fairlamb AH (2004) Dual action of antimonial drugs on thiol redox metabolism in the human pathogen *Leishmania donovani. J Biol Chem* 279: 39925-39932.

Wyllie S, Mandal G, Singh N, Sundar S, Fairlamb AH, Chatterjee M (2010) Elevated levels of tryparedoxin peroxidase in antimony unresponsive *Leishmania donovani* field isolates. *Mol Biochem Parasitol* 173:162-164.

Wyllie S, Vickers TJ, Fairlamb, AH (2008) Roles of trypanothione S-transferase and tryparedoxin peroxidase in resistance to antimonials. *Antimicrob Agents Chemother* 52: 359-365.

Yardley V, Ortuno N, Llanos-Cuentas A, Chappuis F, Doncker SD, Ramirez L, Croft S, Arevalo J, Adaui V, Bermudez H, Decuypere S, Dujardin JC (2006) American tegumentary leishmaniasis: Is antimonial treatment outcome related to parasite drug susceptibility?. *J Infect Dis* 194:1168-1175.

Zauli-Nascimento RC, Miguel DC, Yokoyama-Yasunaka JK, Pereira LI, Pelli de Oliveira MA, Ribeiro-Dias F, Dorta ML, Uliana SR (2010) *In vitro* sensitivity of *Leishmania (Viannia) braziliensis* and *Leishmania (Leishmania) amazonensis* Brazilian isolates to meglumine antimoniate and amphotericin B. *Trop Med Int Health* 15:68-76.

Zhou Y, Messier N, Ouellette M, Rosen BP, Mukhopadhyay R (2004) *Leishmania major* LmACR2 is a pentavalent antimony reductase that confer sensitivity to the drug pentostam. *J Biol Chem* 279:37445-37451.

In: Antimony: Characteristics, Compounds and Applications
Editor: Manijeh Razeghi
ISBN 978-1-62100-598-8

Chapter 4

STRUCTURE OF LIQUID SB AND SB-BASED ALLOYS

A. Yakymovych and S. Mudry
Department of Physics of Metals, Ivan Franko National University
79005, Kyrylo I Mephodiy str. 8, Lviv, Ukraine

ABSTRACT

The structure of liquid Sb in liquid state is analyzed using own results and literature data.

For structure analysis the data on diffraction measurements as well as structure sensitive properties are analyzed.

Binary Sb-based melts are considered also from the standpoint of structure change versus temperature and concentration. We have analyzed the structure of binary Sb-based molten alloys of different kind: eutectics, chemical compounds, solid solutions. Structure in liquid state is compared with solid one using the quasicrystalline models. Among binary Sb-based alloys, whose structure is planned to be considered are Sn-Sb, Cd-Sb, Bi-Sb, Ge-Sb and other binaries. It is shown that inhomogenous atomic distribution of liquid Sb is the characteristic of structure in many binary molten alloys, especially in melts of eutectic and neareutectic concentration.

Systems, where Sb forms the intermetallic phases with other elements (Sn-Sb, Cd-Sb, Bi-Sb, Ge-Sb etc.) structure is analyzed from the viewpoint of another kind cluster-associates, where atoms are chemically ordered. Namely, the model of associated and self-associated clusters is used in order to interpret the data of X-ray diffraction and viscosity measurements.

Structure of chemically ordered clusters is compared with one for crystalline chemical compounds at room temperature. Unfortunately, the literature data on high temperature X-ray investigation of solid crystalline Sb-based are unavailable.

Taking into account the thermodynamic data (enthalpy of mixing) we have analyzed our data and literature own in order to find the correlation between structure parameters (most probable interatomic distance, number of neighbors) with enthalpy of mixing and viscosity.

These and other available data on structure, thermodynamic data and results on structure-sensitive properties allowed us to conclude that structure of Sb-based molten alloys is inhomogeneous before solidification. Random atomic distribution can be reached at overheating only. Features of structure changes before solidification are just determining factor at control of solidificated structure that is related with properties, responsible for application of Sb-based materials in various areas (lead-free solders, semiconductors for electronic devices, heterojunction bipolar transistors etc.).

INTRODUCTION

In solid state Sb shows the structure, whose any atom has three neighbors. Upon melting such structure transforms into more dense atomic distribution with about somewhat higher than 3 neighbor atoms. At heating this parameter slightly decreases. In [1] it is shown that structure factors (SFs) obtained at 933 and 1073 K are relatively temperature insensitive.

The neutron diffraction studies of liquid antimony show also the untypical shape of structure factor [2]. Right hand side of principal peak reveals a shoulder, which the characteristic of liquid semiconductors and semimetals. It is difficult to use the hard sphere model to fit the experimental structure factors. Analysis of diffraction data and physical properties measurement allowed to conclude that structure of liquid Sb is inhomogeneous. It consists of two kind structural units: clusters with metallic bonds of higher atomic density and clusters covalent bonds of less density. Such structural inhomogeineities are important at alloying of Sb with other elements and can significantly change the atomic arrangement not only in liquid state but also upon solidification. On that reason data on structure of binary melts with Sb are needed in order to control new materials fabricating including nanomaterials. For example the neutron diffraction studies of $K_{100-x}Sb_x$ (x=15; 25; 40; 50 at.%) molten alloys show a tendency for segregation at x_{Sb}=15 at.%. It is also shown the existence of salt-like order at the octet composition, changing continuously to a polyanionic order with chain-like Sb clusters at equiatomic composition [3].

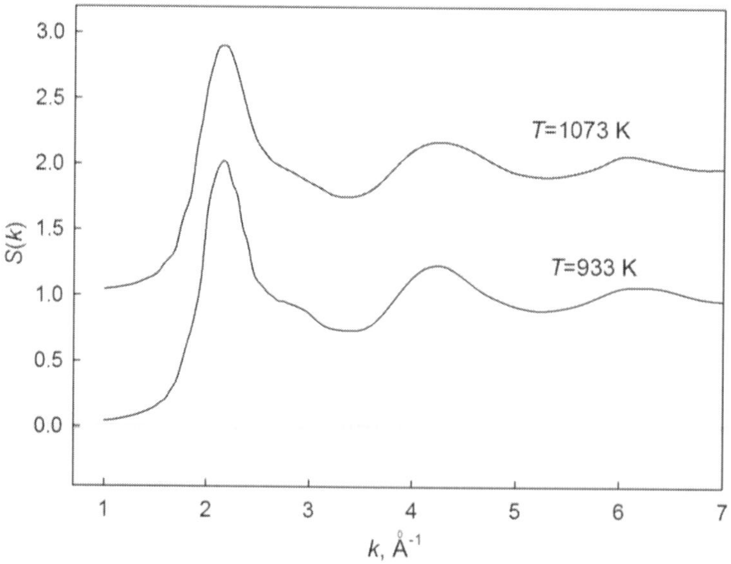

Figure 1. Structure factors of liquid Sb.

The width of the peaks in the radial distribution function was not found to increase and even decreased within a certain temperature range. This behavior correlates with anomalous temperature dependence of sound velocity [4]. This correlation may be accounted for by increased rigidity on the liquid structure with temperature [5].

PHASE DIAGRAM AND STRUCTURE FEATURES

Before analyzing of binary Sb-based melts it is reasonably to consider the binary systems of Sb with other elements. Antimony forms only one binary system with unlimited solubility in solid state (Bi-Sb) [6] (fig. 2).

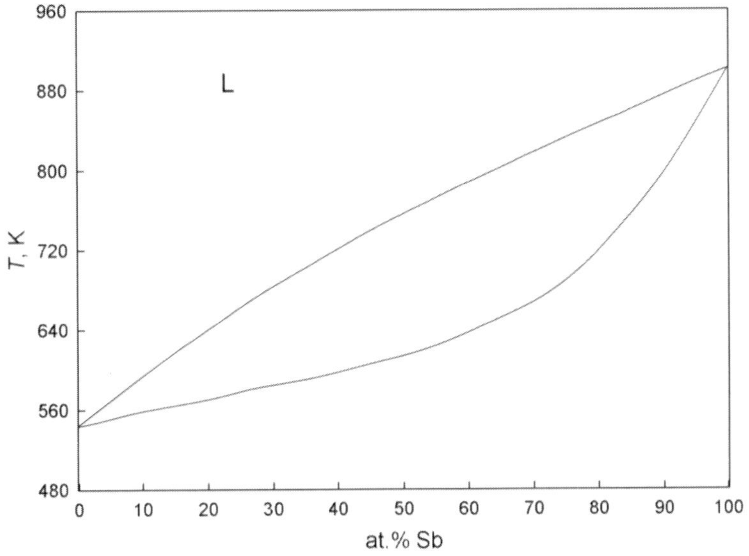

Figure 2. Phase diagram of Bi-Sb system.

Similar to those is the system As-Sb (fig. 3), whose phase diagram reveals minimum point liquidus curve and at lower temperature there is two-phase region [7]. As a group-V element Sb attempts to form the chemical compounds with some metals and chalcogen elements the chemical compounds. Such compounds with IIIb-elements (AlSb, InSb, GaSb) reveal semiconducting properties due to which they are used widely in electronics. The limited solubility of Sb with some elements is the reason of formation of systems with eutectic points in phase diagram. Most of them antimony creates with IVb-elements, whose structure in liquid state, excepting Pb, is similar to structure of Sb.

It should be noted here that antimony does not form the binaries with miscibility gap in liquid phase. Nevertheless the eutectic alloys, which are formed due to limited solubility in solid state, are the characteristic for many binaries with antimony. Various kind eutectics are formed: from simple to complex eutectics, whose one component is chemical compound. Between them are also eutectic points corresponding very small content of one component (zero-eutectic). Therefore, Sb reveals the different tendency to interaction with other elements dependending on atomic size factor, electronegativity difference and other characteristics.

From analysis of phase diagrams follows that limited solubility in solid state for eutectic alloys can be precursor of Sb-based cluster existence in liquid state at temperatures near solidification point. Other kind clusters – associates with Sb, are supposed to be formed in binaries with chemical compounds. Such clusters exist as dynamical structural units and their transformation by means of thermodynamic condition changes and other factors can affect the structure of Sb-alloys after solidification. On that reason the main trends in structure formation features in liquid state before transition to solid state should be investigated.

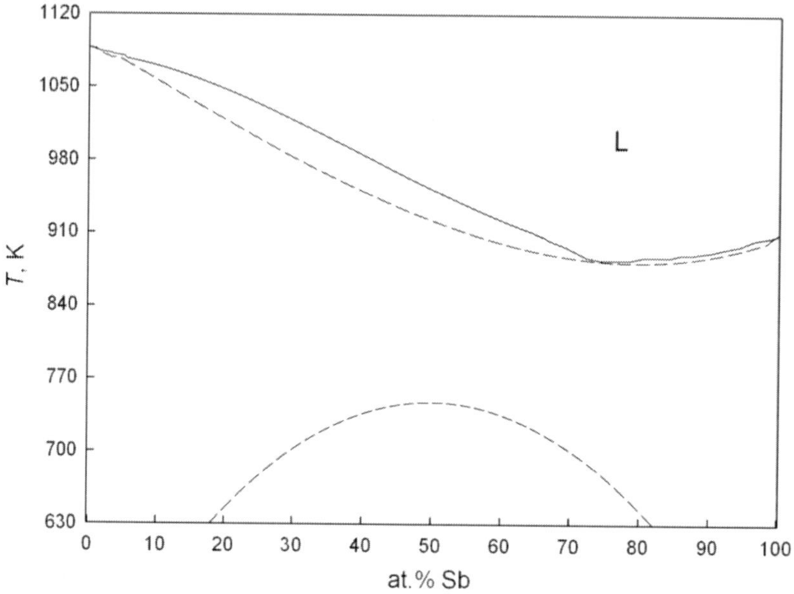

Figure 3. Phase diagram of As-Sb system.

Unlimited Solubility Systems

Bi and Sb are neighbor elements in Periodic Table and have the similar structure in solid state. The structure factors show the shoulder on right hand side of principal peak. Analyzing the structure parameters for Bi and Sb one can see that most probable interatomic distances for them are close (r_1^{Bi}=3.38 Å, r_1^{Sb}=3.33 Å). Number of nearest neighbor atoms are also in fact the same (Z_1^{Bi}=8.8; Z_1^{Sb}=8.7). Therefore there are good conditions for formation of random atomic distribution at mixing of these elements.

Structure of liquid $Bi_{100-x}Sb_x$ alloys has been investigated by means of X-ray diffraction method in [8]. The structure factors show the features, which are the characteristic of liquid Bi and Sb. It is seen (fig. 4) that SF for equiatomic melt shows the stability of shoulder when temperature is increased to 845 K and thermal disordering attempts to break the surviving after melting covalent bounds, due to which the shoulder is pronounced.

This allows to suppose that atoms of Bi are diluted with similar atoms of Sb, forming the random atomic distribution.

In assumption of this kind atomic distribution the main structure parameters including the principal peak position in SF should depend on concentration as linear function. In case of Bi-

Sb system (fig. 5) this parameter reveals the significant deviation from linear dependence. Positive deviation shows the maximum values within concentration region 20÷50 at.% Sb. As is known the positive deviation of $k_1(x)$ linear dependence corresponds to opposite deviation of most probable interatomic distance $r_1(x)$. This fact allows us to state that within above mentioned concentration range exists the tendency to preferred interaction of unlike kind atoms, which results in formation of clusters with heterocoordinated atomic distribution.

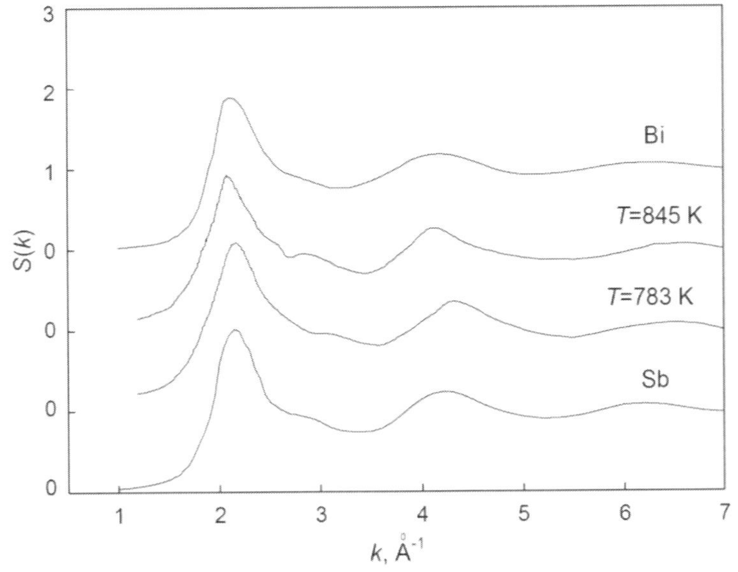

Figure 4. Structure factor for $Bi_{50}Sb_{50}$ melt.

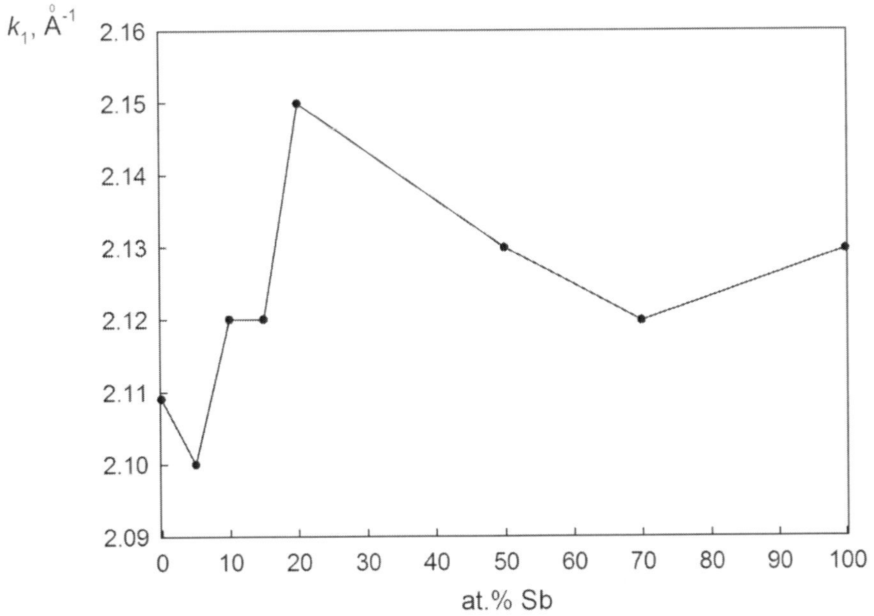

Figure 5. Principal peak position for Bi-Sb melts.

It is shown also that addition of Bi-atoms to Sb does not affect the structure of Sb as at addition of Sb to Bi. May be a slightly less value of most probable interatomic distance of Sb comparing to one of Bi is the main reason of such behavior in concentration dependence $k_1(x)$. When a little content of Sb is added to Bi a rapid increase of k_1 (decrease of r_1) is observed. Such behavior is supposed to be caused by formation of more dense structure due to diluting of Sb-atoms and Sb_n small clusters in vacant places of inhomogeneous (two kind clusters) structure of Bi. In case of Sb-enriched melts Sb-like structure persists at addition of Bi-atoms that results in substitution of Sb-atoms by Bi ones in clusters.

Therefore, an inhomogeneous structure of Sb, whose main elements are Sb_n clusters, does not disappear at alloying with Bi. Cluster solution changes the atomic structure as well as inrterclusters distribution with concentration variation. At heating the main feature of such kind structure persist within some temperature range but the cluster solution attempts to be rearranged into random atomic distribution.

Eutectic Systems

Now we consider the structure of binary molten alloys, whose second component is element with more inhomogeneous structure than Bi. Ge-Sb binary is just such a system (fig. 6) [9].

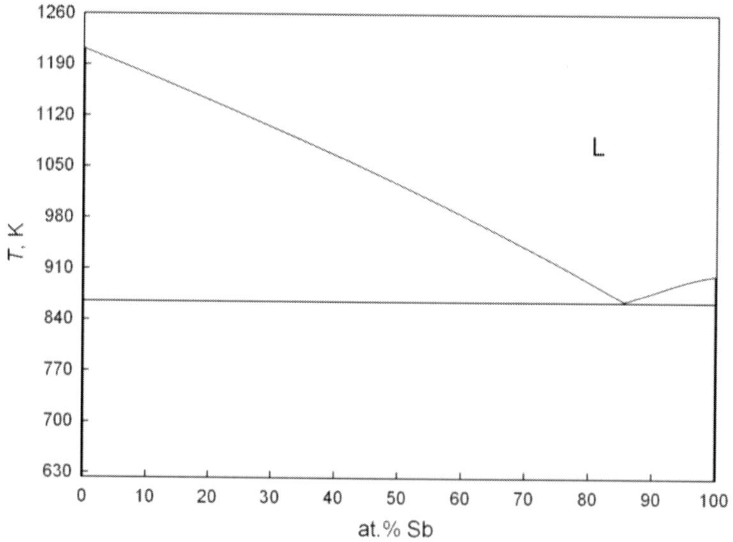

Figure 6. Phase diagram of Ge-Sb system.

Liquid Ge reveals the SF with low height of principal peak and large its width with pronounced shoulder. As follows from X-ray diffraction data, Ge at melting also consists of some part of unbroken bonds, fraction of which is maximum at temperature close to melting point and is decreasing with heating.

Another important feature of this system is that such elements as Ge and Sb are in fact insoluble each other. On that reason it is interesting, what is the stability of Sb-clusters and structural inhomogeneities when Sb is alloyed with Ge.

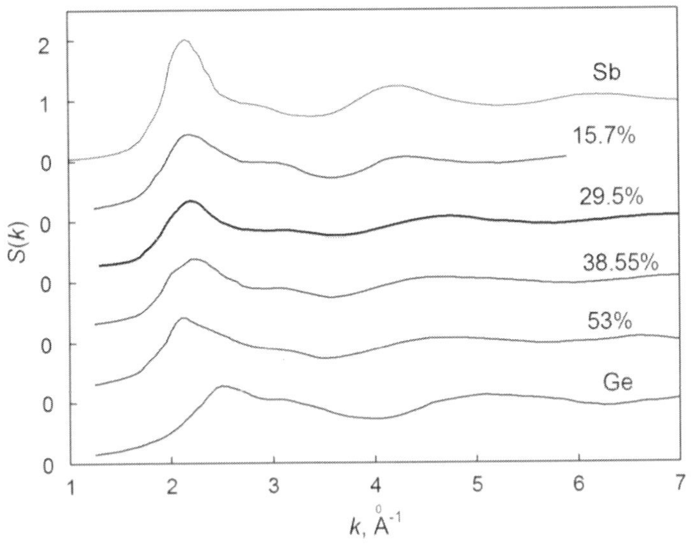

Figure 7. Structure factor for Ge_xSb_{100-x} melts.

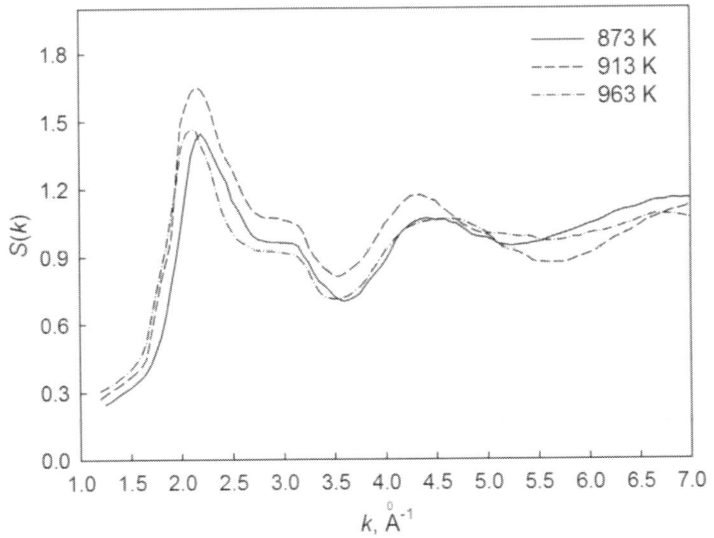

Figure 8. Structure factor for $Ge_{15.7}Sb_{84.3}$ melt.

Structure factors for alloys with different content of Ge are more sensitive to temperature than in case of Bi-Sb system (fig. 7). It is seen that addition of Sb significantly transforms the atomic distribution of this semiconducting element. Structure factors for molten alloys, containing 53 at.% and less are more close to SF of Sb than for Ge. Unfortunately, it should be noted that there no investigated alloys with concentrations between liquid Ge and 53 at.% Ge. Nevertheless the concentration dependence of SF implies that in this concentration range Sb significantly affect the structure of liquid Ge.

Another important feature of experimental SFs is the transformed shoulder on right hand side of principal peak. If compare positions of main peaks for Ge and Sb, which equal 2.11 and 2.50 Å^{-1} respectively showing significant difference, one can conclude that SF is the weighted additive corresponding to structural units Ge_m and Sb_n. This kind structure on the base of self-associated atoms has a high thermal stability that follows from temperature dependence of SF (fig. 8).

It is seen that heating of eutectic melt (15.7 at.% Ge) from 873 to 963 K is not accompanied by significant changes of SF. Height of peaks slightly shifts to low wave vector value, that corresponds to increasing of interatomic distance. Increase of half height width is related with size reduction of clusters. The shoulder does not change its profile with heating within this temperature range. Thus, one can conclude that such thermal stability of structure without clusters existence is impossible.

Therefore, the results on structural studies of Ge-Sb molten alloys shown that most probable structural units are the self-associated atomic groups. Atoms of antimony occupy the vacant sites in structure of Ge, increasing in such way its thermal stability. Certainly that at more higher temperatures thermal disordering should destroy the clusters that results in formation of more microhomogeneous atomic distribution.

Similar to Sb is its neighbor in Periodic Table – tin. This element also belongs to group of semimetal, showing the reduction of most probable interatomic distance due to formation of more dense short range order than in solid state.

As was mentioned above binary phase diagram Sn-Sb implies that in liquid state atoms of Sn and Sb will attempt to be as neighbors (fig. 9) [10].

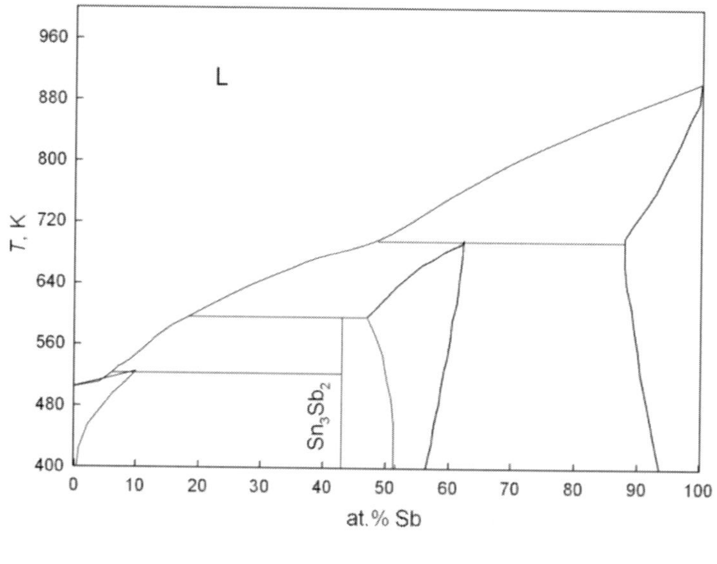

Figure 9. Phase diagram of Sn-Sb system.

This suggestion was confirmed by X-ray diffraction studies [11]. Structure factors for molten alloys, containing 10; 30; 50; 70 and 90 at.% Sb, show the high sensitivity to concentration at low content of Sb (≤10 at.% Sb) (fig. 10). As is seen the addition of 10 at.% Sb leads to essential shift of principal peak position of Sn to low wave vector value. It is untypical that strong covalent bonds of Sn, unbroken at melting upon melting can not stabilize the Sn-like structure in Sn-enriched melts. This fact that these bonds in Sn are stronger than in Sb follows from changes of enthalpy at melting [12]. Besides, the ratio of electroconductivity in solid state to liquid state one ρ_S/ρ_L for Sn is also larger than for Sb [13].

With further increase of antimony content SF is almost unchangeable, that allows us to state that Sb-like atomic arrangement exists within wide concentration range. May be one of main reason of such Sb-influence on structure of melts.

Another melts, whose phase diagram is similar to one of Ge-Sb system and reveals the eutectic point are Pb-Sb molten alloys [14].

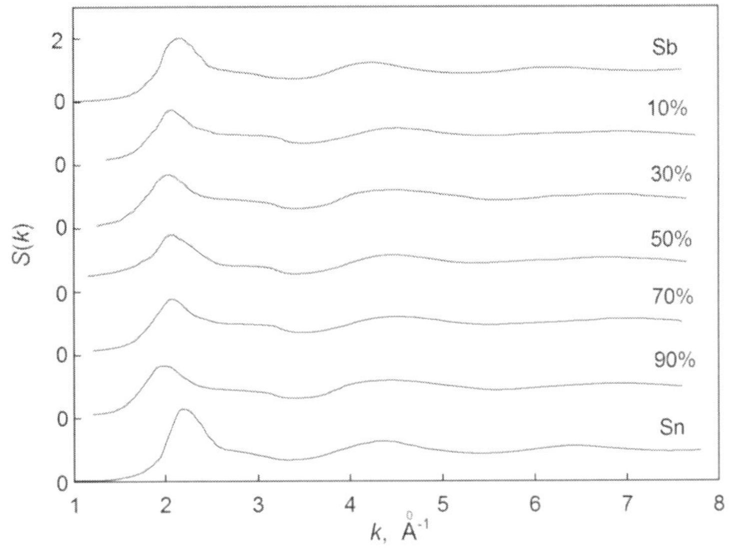

Figure 10. Structure factor for Sn_xSb_{100-x} melts.

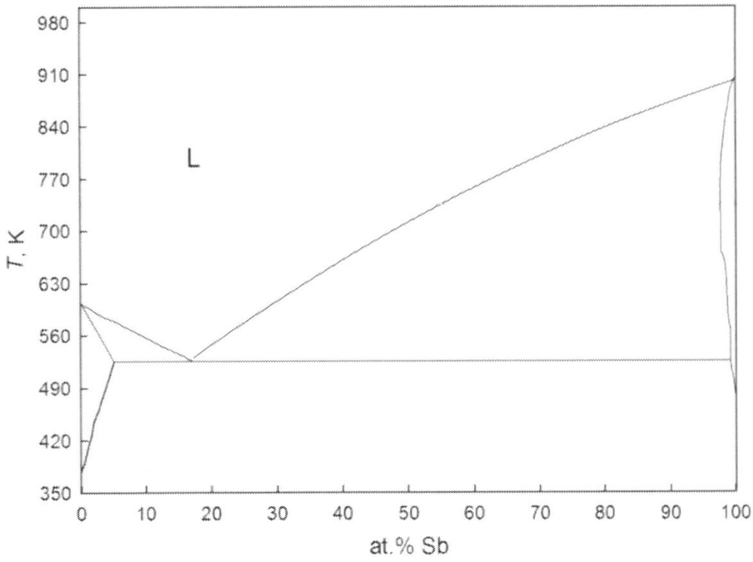

Figure 11. Phase diagram of Pb-Sb system.

In case of Ge-Sb system antimony is alloyed with semiconductor element, whereas in Pb-Sb eutectic system Sb is alloyed with typical metal – Pb (fig. 11).

Binary Pb-Sb shows a negative enthalpy of mixing (-80 cal/g atom) with no known electron or intermetallic compounds or order/disorder change. Structure of Pb-Sb molten alloys has been studied in [15] with help of Kumar-Samarin technique of centrifuging liquid metals.

Alloys, containing 10; 17; 45 and 87 at.% of Sb have been investigated in the temperature range from 300 to 700 ^{0}C. Logarithm of atomic concentration of Sb in liquid Pb-Sb alloys versus distance from farther end of the crucible is analyzed. It is shown that the size of either lead or antimony clusters decreases as the eutectic composition is approached in the hypo- and hyper-eutectic alloys respectively. Once the minimum size is reached at the eutectic composition there is no appreciable change thereafter. Authors suggest that within concentration ranges 0÷35 at.% Sb and 65÷100 at.% Sb molten alloys can be considered as colloidal dispersions of lead- and antimony-rich clusters in monotectic matrices. Besides, clusters size is minimum at eutectic composition. It was also suggested that the solutions in the intermediate composition range, corresponding to inflection, which early regarded as homogeneous, are in fact incipiently immiscible and can separate into the conjugate lead-rich and antimony-rich liquids on centrifuging.

Compound Forming Systems

Few chemical compounds are formed in binary system of Sb with Cd. Besides, this system has also two eutectic points, created by CdSb chemical compound and pure Cd and Sb [16]. In solid state can be crystallized into two metastable chemical compounds – Cd_4Sb_3 and Cd_3Sb_2 (fig. 12). It is clear that such different phase formed upon solidification should in some respects inherit the structure in liquid state.

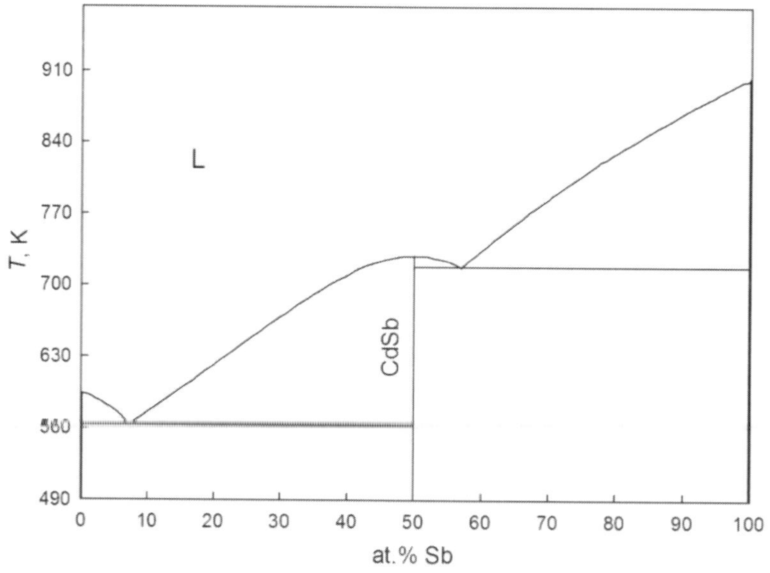

Figure 12. Phase diagram of Cd-Sb system.

In [17] the structure of liquid Cd_xSb_{100-x} containing 96; 93; 60; 57.1; 50; 42 and 30 at.% Sb has been studied by means of X-ray diffraction at temperatures 10 K above liquidus curve. Scattered intensities as functions of wave vector are shown in fig. 13 and the main structure parameters, obtained from them – in table 1.

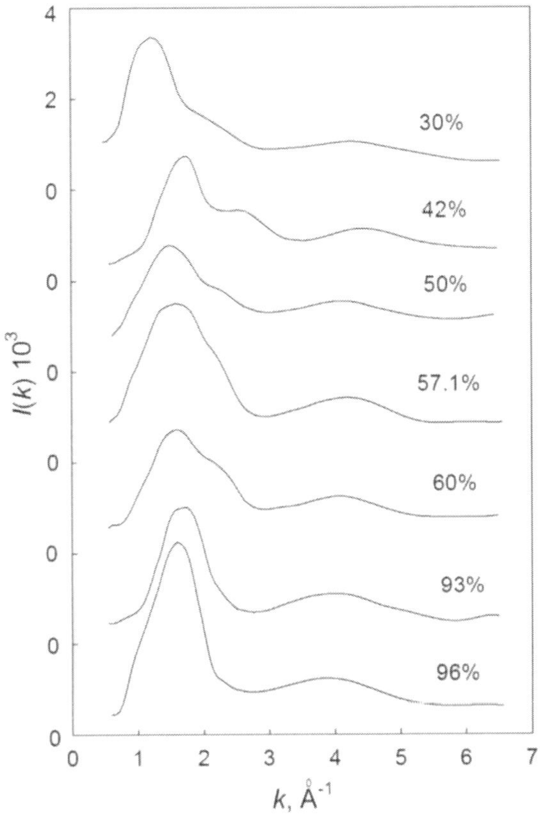

Figure 13. Scattered intensities for Cd_xSb_{100-x} melts.

It is seen that intensity curves strongly depend on concentration. With addition atoms of Sb-atoms to Cd principal peaks shifts to low values of wave vector, but this shift is not linear function and deviates it in positive direction that notes a departure of structure from random atomic distribution. Increase of Sb-content is accompanied by transformation of right hand side of principal peak and appearance of shoulder, which is the characteristic of SF for liquid Sb.

Peak positions were compared with positions of related diffraction reflexes for crystalline chemical compounds and it was shown that Cd_4Sb_3-like structure is formed before solidification of melt, containing 42.9 at.% Sb. Intensity curves for molten alloys with 40 and 50 at.% Sb are similar and show the shoulders at S=2.7-2.9Å^{-1}. Chemical ordering in these melts is based on CdSb-like short range order. Liquid alloys with 7and 70 at.% Sb have a structure close to the structure of Cd and Sb respectively. Clusters of these elements are stable at near melting temperature region and they coexist with other kind clusters, which are characterized by preferred interaction of unlike atoms with some degree of chemical ordering. Contrary Pb-Sb eutectic system, where tendency to separation of like-kind atoms exists within concentration range around equiatomic concentration, in binary Cd-Sb system such tendency is observed for Cd- and Sb-rich melts. Intermediate concentration region corresponds to another kind clusters than self-associated ones, namely, there the formation of chemically ordered clusters (associates) occurs.

Table 1. Structure features of $Cd_{100-x}Sb_x$ melts

Cd, at. %	T, K	Maxima positions			r, Å	Z
		S_1, Å^{-1}	S_{Sub}, Å^{-1}	S_2, Å^{-1}		
96	575	2.4	3.1	4.3	3.10	8.2
93	595	2.5	2.7-3.1	4.5	3.00	8.6
60	695	2.3	2.7	4.5	3.15	6.5
57.1	705	2.2	-	4.1	3.10	7.0
50	735	2.3	2.9	4.5	3.15	6.2
42	740	2.4	3.1	4.7	2.90	6.1
30	795	2.0	2.7	4.1	3.00	6.8

Therefore, one can conclude that in case of liquid Cd-Sb alloys the formation of chemical compounds begins still before crystallization. On other hand, melting of these chemical compounds is accompanied by process, when chemical bonds are not entirely destroyed. Clusters, formed due to the strong preferred interaction of unlike kind atoms, persist significantly within some temperature interval at heating after melting.

It should be noted here that structure state of molten alloys of Cd and Sb is responsible for the formation of metastable phases (Cd_3Sb_2) in this system. Besides, this process depends also on cooling conditions.

Existence of chemically ordered clusters is confirmed by results on electroconductivity measurements, which show two maxima on the resistivity -composition curve at concentrations, corresponding to CdSb and Cd_3Sb_2 [18].

The structure of molten Cd-Sb alloys, as in case of Pb-Sb melts, has been studied with using of Kumar-Samarin technique of centrifuging liquid metals [19]. In result it was shown that the strong Cd-Sb interaction in solid intermetallic compound retains in liquid state and that liquid alloys of compositions other that around the equiatomic, consist of a colloidal dispersion of pseudo-compound clusters in matrix of monoatomic atoms. The size of the clusters is a function of composition and temperature.

Antimony and copper form the alloy of different kind, depending significantly on concentration (fig. 14). Atoms of Sb show some limited solubility in Cu, whereas Cu is unsolvable in Sb. Flat maximum in liquidus curve of phase diagram implies that chemical compounds, existing in solid state can dissociate significantly upon melting [20].

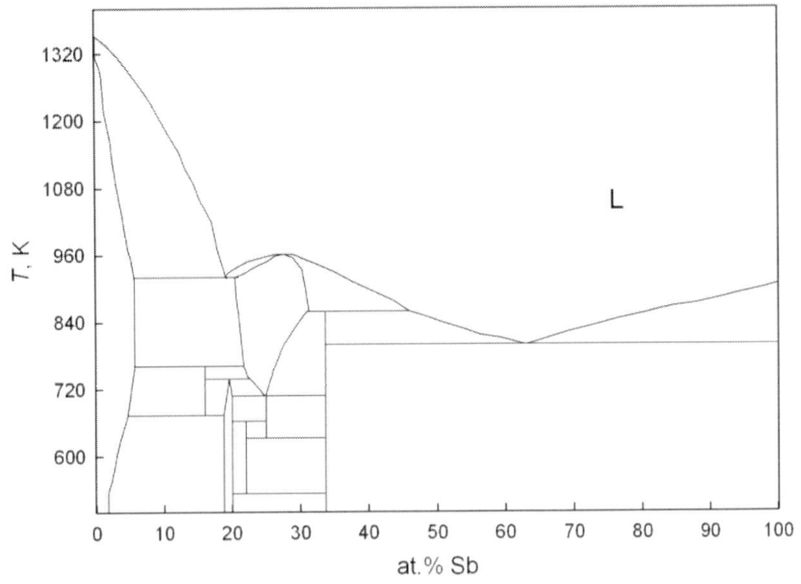

Figure 14. Phase diagram of Cu-Sb system.

In [21] molten alloys of this system have been studied by means both X-ray diffraction and neutron diffraction method. Structure factors for Cu-Sb melts of different content at temperatures 10 K and 200 K above liquidus have been obtained by means of neutron diffraction. As follows from this paper, the addition of copper to antimony up to equiatomic concentration display the influence of Sb-like short range order on atomic structure of molten alloys. On other hand, molten alloys containing about 30 at.% of Sb or less show the intensity curves, which are very close to one for pure liquid cooper. Therefore first maximum of intensity curves shows the splitting between 52 and 90 at.% of Sb that implies the existence of two kind clusters. Authors concluded that within this concentration range molten alloys consist of structural units on the base Sb and Cu. It is also possible that some regions of melts are chemically ordered with Cu_2Sb-like short range order.

According to [22], the pair correlation function of Cu-Sb melts haven the appearance represented on figure 15.

In order to estimate the degree of atomic segregation within this concentration range the total intensity curves were calculated by weighted superposition of the intensity curves for liquid Cu and Sb. These calculated curves were in good agreement with experimental ones.

Most probable interatomic distances r_1 and number of neighbors have been also analyzed. Number of neighbors as function of Sb content shows the deviation in opposite directions from linear dependence, which occurs at random atomic distribution. It is seen that for Cu- and Sb-rich melts there no deviations.

Such deviation is negative for melts within concentration range 20÷60 at.% Sb, whereas it is positive in the range 60÷90 at.% Sb. Consequently Cu- and Sb-rich melts can be regarded as structural units with random atomic distribution; liquid alloys, whose concentration lies between 20 and 60 at.% Sb attempt to be chemically ordered; melts, containing Sb from 60 to 90 at.% Sb show the tendency to microsegregation. In this work author state that chemically ordered clusters can be Cu_2Sb and Cu_3Sb, while in [21] it is concluded Cu_2Sb-agglomerates are structure units of Cu-Sb melts.

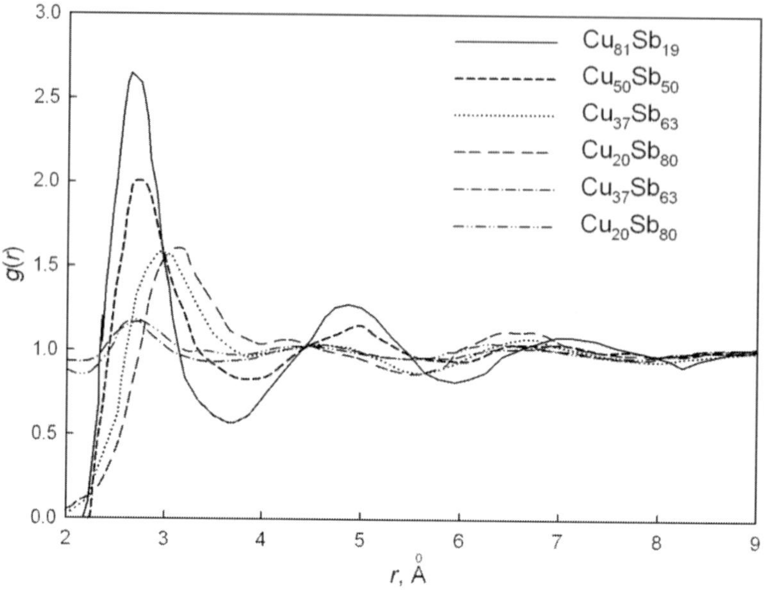

Figure 15. Pair correlation function of Cu-Sb melts.

Analyzing the results of these two papers we have suggested that in the concentration region, where tendency to preferred interaction on unlike kind atoms is manifested of Cu_2Sb and Cu_3Sb – compounds formation, negative deviation of structure parameters (r_1, N) and positive deviation of enthalpy of mixing, the chemical ordering in liquid state is formed on the base of Cu_2Sb clusters with wide homogeneity range. Certainly that such homogeneity range is not the same as in solid state and it supposes more disordered atomic distribution. Therefore binary Cu-Sb system, which shows significant dependence of interatomic interaction versus concentration, reveals the similar trend in liquid state, especially before crystallization. At near crystallization temperatures cluster structure is in many respects responsible for the formation of crystalline phases.

Authors concluded that within this concentration range molten alloys consist of structural units on the base Sb and Cu. It is also possible that some regions of melts are chemically ordered with Cu_2Sb-like short range order.

Among binary systems of group III elements with Sb, the In-Sb and Ga-Sb molten alloys are most studied. Recently, interest of researchers is also focused to Al-Sb binary system. Motivation of this interest is due to their use in electronics and moreover in nanotelectronics, nanooptics and other new fields of advanced materials application.

On the figure 16 the phase diagram of In-Sb system is represented [23]. Structure of liquid InSb as most interest alloy of this system, which represents III-V compounds has been studied by X-ray diffraction in [2]. The pair correlation function (PCF) (fig. 17) show the profile typical for liquid semiconductors such as Si, Ge or GaAs. The interatomic distance is 3.05Å, whereas this parameter for crystalline compound is 2.81Å. At the same time the number of neighbors for liquid InSb is the same as for liquid Sb (6.8). Authors suppose that liquid InSb has a structure analogous to the sixfold coordinated simple cubic or β–tin structure.

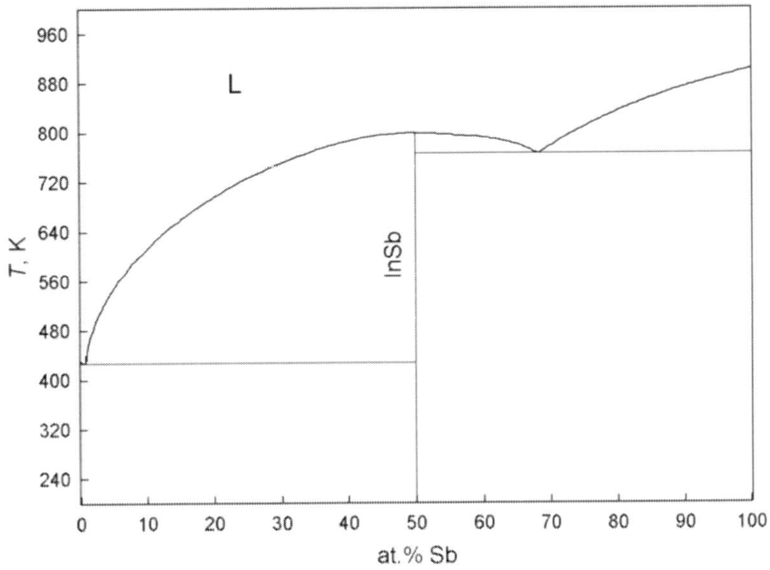

Figure 16. Phase diagram of In-Sb system.

This conclusion is in agreement with entropy of melting (7.1 cal/g atom) and the relative volume change during melting $\Delta V/V$=-13.7% [24].

Structure of molten InSb compound was also studied by means XAFS method [25]. It was shown that melting destroys the tetrahedral covalent network. Upon transition to liquid state the average coordination number Z and the average bond length r_1 of the first nearest neighbors around In (or Sb) increases by 1.5~1.8 and 0.1Å for liquid InSb and a few percent In-In (or Sb-Sb) coordination appears in the first shell of liquid InSb.

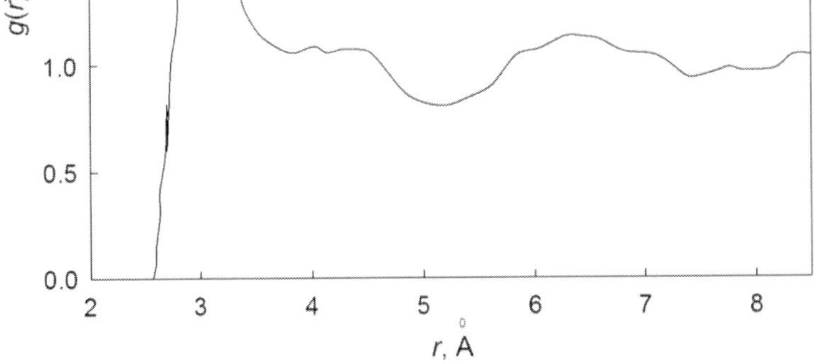

Figure 17. The pair correlation function of liquid InSb at 923 K.

Liquid InSb and GaSb have been investigated early in [26]. Temperature dependence of coordination number for InSb show the rapid increase just upon melting and then a slight decreasing at further heating. Such behavior shows that semiconductor-metal transition occurs not just at the melting but within some temperature range including liquid phase.

Intensity curves for liquid GaSb show the similar tendency. There no agreement between diffraction maxima of liquid phase and crystalline one. This allowed to suppose that structure of liquid GaSb significantly changes upon melting. Existence of shoulder on right hand side of principal peak is the evidence that covalent bonds are not completely destroyed after melting and exist within some temperature range. The coordination number at temperature 10 K above melting point is about 6 than allowed authors to sate that short range order structure in liquid state is similar to octahedral coordination of atoms.

Unfortunately the data on structure in liquid state of another system of this III-V group – AlSb are unavailable.

VISCOSITY COEFFICIENT FEUTURES

The clusters are dynamical structural units simultaneously change their size and other parameters. On that reason the liquid state is characterized by fluctuations in the atomic distribution in near-neighbor regions. The viscosity, as a structure sensitive property of liquids playing the most important role in study of the flow and reaction mechanism, strictly reflects their structure. In the present capital section the viscosity data are represented in correlation with phase diagram. Moreover, the changes of activation energy obtained from temperature dependence of viscosities are considered from the viewpoint of structural transformations in liquids.

Viscosities of liquid Sb and Sb-M (M = Bi, Cd, Cu, Ge, In, Pb, Sn) alloys have been analyzed with accounting of phase diagram features.

For description of temperature dependence of viscosity for liquid metals commonly use an Arrhenius-type equation:

$$\eta(T)=\eta_0 exp(E_A/RT), \quad (1)$$

where η is the viscosity of liquid, η_0, is the pre-exponential factor, E_A, is the activation energy of viscous flow, R is the gas constant, T is the absolute temperature.

Relevant data for pure Sb are collected in Table 2.

Table. 2. Values of parameters for (1) of liquid antimony

η_0, mPa s	E_A, kJ mol^{-1}	References
0.337	7.817	exp

Theoretical calculation of viscosity at melting point in [27] according to hard-sphere description shows disagreement with experiment ($\eta_{exp}(T_m)/\eta_{calc}(T_m)$ = 1.22/2.44). Besides, existence of associates in liquid Sb due to deviation of temperature dependence of viscosity from Arrhenius type (Fig. 18). Such deviation is in accordance with structure data: antimony is a semi-metal and is characterized by existence of covalent bonds in solid state. In this case not all bonds are breaking after melting and some part of them remains in liquid as was shown also by diffraction studies. Consequently, the experimental viscosity of liquid Sb in this temperature region is less relative to theoretical ones with maxima deviation at T_m.

To our knowledge, the relationship between viscosity and phase diagram is ambiguous [28]. Therefore, it is interesting to study the viscosity of different type systems in terms of correlation with the phase diagram view.

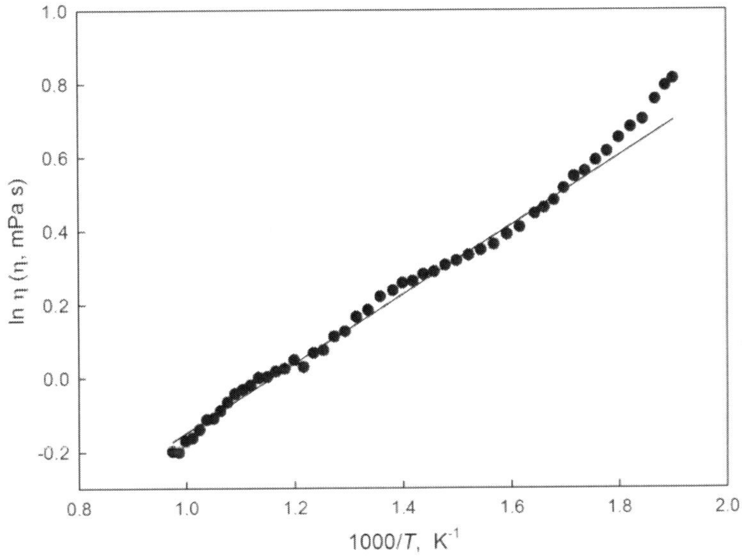

Figure 18. Viscosity as function of reciprocal temperature for the liquid Sb (• - our data, - Arrhenius type).

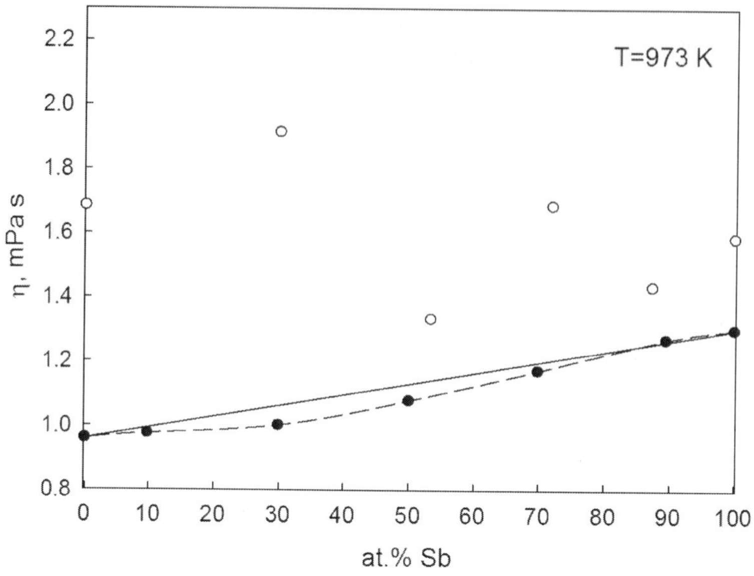

Figure 19. The concentration dependence of viscosity for Bi-Sb melts (○ – [29]; ● – [30]).

At first shows the viscosity changes for infinitely mutual solution system (Bi-Sb). Such type systems characterized no complex interactions, due to similar physical parameters of components on the viscosity influence only the interaction forces between neighbor atoms. Fig. 19 shows the concentration dependence of viscosity for Sb_xBi_{100-x} alloys. This system is interesting due to recent work of Sun Chunjing et al. [29] where the influence of magnetic field on the viscosity of liquid Bi-Sb alloys was studied.

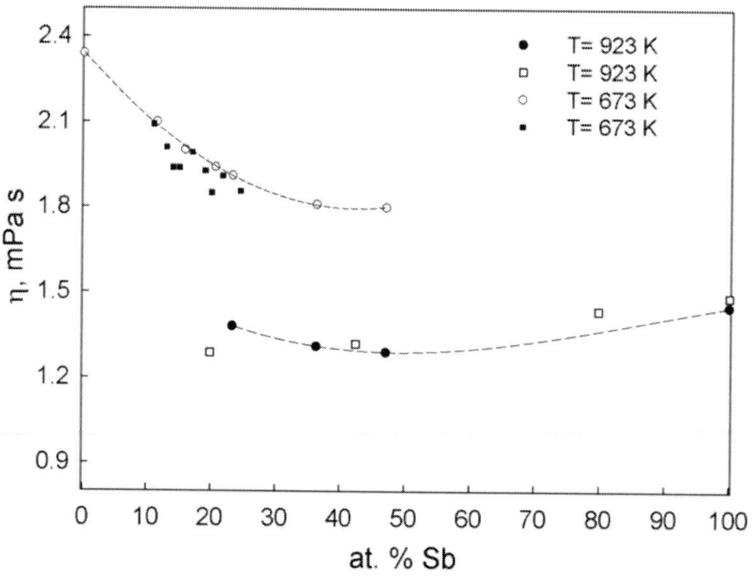

Figure 20. The viscosity of Pb-Sb melts versus content of Sb (●, ○ – [32]; □ – [33]; ■ – [34]).

According to data of [29] the relative excess viscosity without magnetic field reaches to -0.16 at equiatomic concentration by the temperature 973 K. Taking into account not explained atypical change of viscosity with concentration in [29], we suggest that such data are vague and consider the experimental viscosity-composition curves of [30]. In this case, Bi-Sb belongs to simple solid solution systems, whose viscosity isotherms are usually either lineal or slightly concave.

The most studying metal systems are eutectics. It is sometimes stated that such type systems shows the negative deviations in the viscosity isotherms, but there are exceptions to the rule for systems [31]. We suggest that the minimum point of viscosity and activation energy corresponds to eutectic point in case of a large difference between the viscosity values of pure compounds, otherwise should be a smooth running of concentration dependence with extreme value about equiatomic composition. To study we choose the eutectics Pb-Sb, Sn-Sb and Ge-Sb systems.

The literature data [32-34] on the concentration dependence of the viscosity for liquid Pb-Sb alloys are in general agreement (Fig. 20). The viscosity continuously changes with concentration, herewith the minimum locates at ~ 50 at.% Sb. It was found a slight negative deviation from additive rule in the concentration dependence of viscosity for liquid Sn-Sb alloys, as schematically shown in fig. 21.

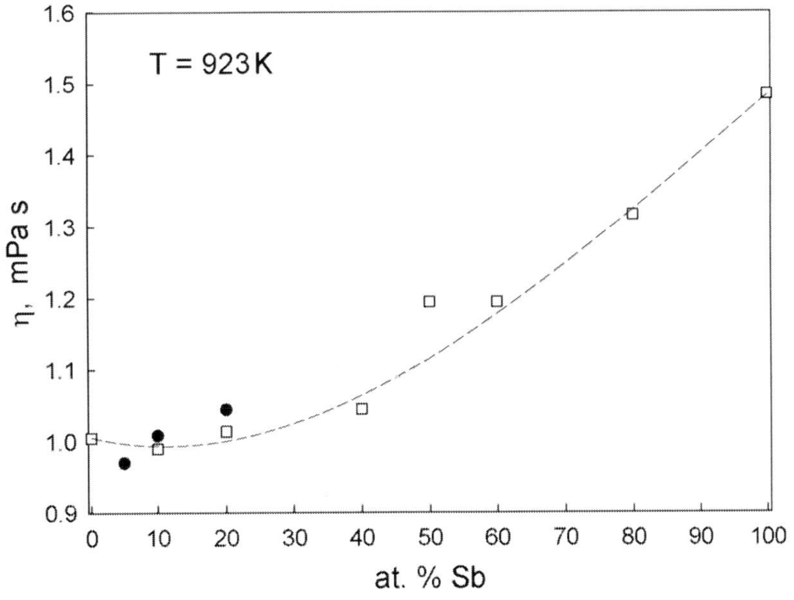

Figure 21. The concentration dependence of viscosity of Sn-Sb melts(□ – [33]; ● – [35]).

A similar viscosity behavior for Sn-Sb liquids to Pb-Sb was expecting, account similar structure and physicochemical properties of rare components and, the rather mutual solid solution for both systems. Another testimony similarity of these systems is the negative deviation of activation energy from additive rule (tab. 3).

Table 3. Activation energies for Pb-Sb [32] and Sn-Sb [33] liquid alloys

at.% Sb	E_A, kJ/mol for Pb-Sb	at.% Sb	E_A, kJ/mol for Sn-Sb
0.0	4.446	0.0	6.625
11.5	4.087	4.9	6.734
15.9	4.015	19.8	6.813
23.3	3.920	39.7	8.737
36.3	3.944	49.7	8.735
47.1	4.087	79.8	8.994
100.0	6.405	100.0	13.205

It can be seen that addition of second component on Sb-Pb melts leads to decrease of the activation energy with minimum point about 30 at. % Sb. In this case one can conclude that concentration at the minimum values of viscosity and activation energy are in the disagreement with eutectic composition 17.5 at. % Sb for Sb-Pb system. Due to almost linear concentration dependence E_A with slightly negative deviation for Sn-Sb we suggest that interatomic interaction with fraction of components is independent. The least value is in concentration range about 80 at.% of antimony. Unfortunately, we can not find the literature data on concentration dependence of viscosity for Ge-Sb system are unavailable.

The last reviewed type of systems is compound formation ones. In the main the viscosity isotherms do appear to pass through maxima in the concentration range corresponding roughly to the compounds in the solid state in such systems [36].

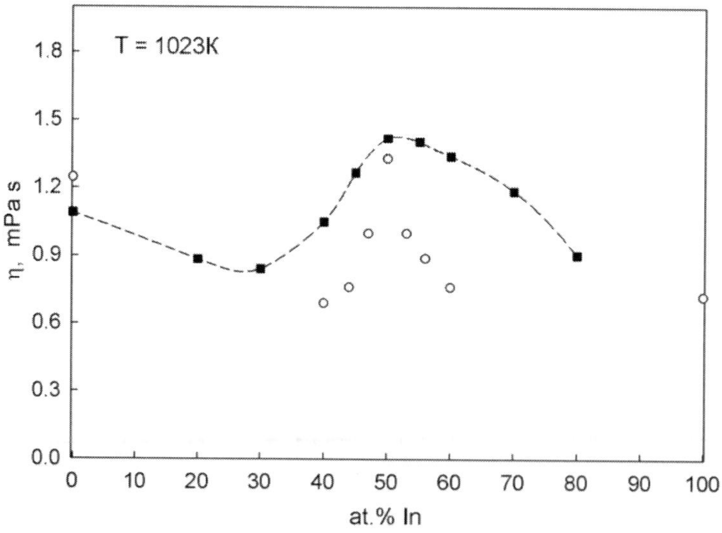

Figure 22. The concentration dependence of viscosity of In-Sb melts (■ – [37]; ○ – [38]).

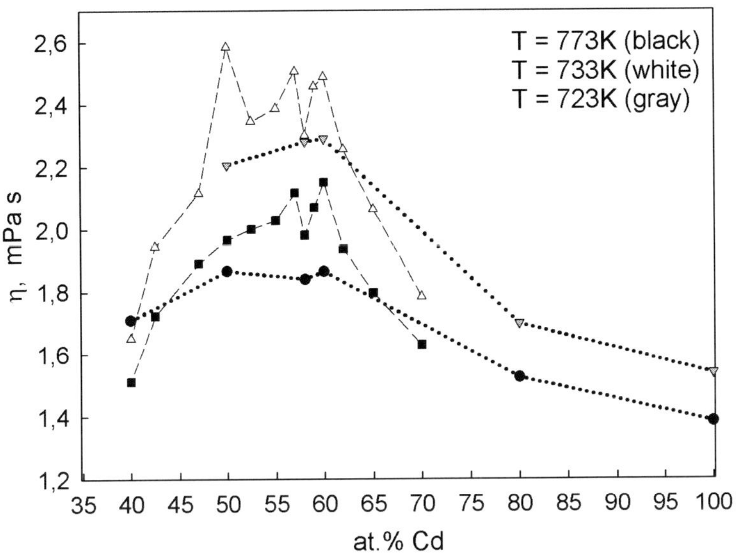

Figure 23. The concentration dependence of viscosity of Cd-Sb melts (■ Δ – [39]; , ▽ – [40]).

The maximum value of viscosity reveals at compounds stoichiometry, for such anomalous viscosity behavior are responsible the structural inhomogeneities (Fig. 22-24). One can note the deviation in literature data (discrepancies of absolute values etc.) can be induced by singularity of experimental investigations, but procession for all reviewed systems is similar. Rapid decrease of viscosity in InSb+Sb sybsystem is supposed to be quasieutectic, on the other case in InSb+In sybsystem is supposed to formation of In_3Sb associates. Consequently, taking into account the viscosity data, one can note that in liquid In-Sb alloys the InSb and In_3Sb associates exist. Fig. 23 shows the comparison viscosity data [39, 40] with phase diagram for Cd-Sb melts. Authors [40] measured the viscosity by heating and in [40] in process of cooling. According to results of structure investigations [19] in liquid state exists the stable CdSb compound and two metastable Cd_3Sb_2, Cd_4Sb_3 compounds. The confirmation of microheterogeneity structure is a large values of viscosity by heating and cooling with concentration in region 50-70 at.% Cd. Smaller viscosity by cooling as by heating for the same melt is attributed to quantity of clusters. After melting in melts exists the unlike atoms clusters, which volume fraction is largest as formed by cooling ones. While the metastable phases Cd_4Sb_3 and Cd_3Sb_2 forming by conversion Cd+CdSb, such type clusters existing in higher temperature region as CdSb, in another hand on the isotherm we don't see drastic decrease of viscosity for $Cd_{60}Sb_{40}$ and $Cd_{57}Sb_{43}$ melts in comparison with $Cd_{50}Sb_{50}$ one. Also the Cd_4Sb_3 and Cd_3Sb_2 clusters are more stable in liquid state as CdSb and crystallization in this system should be in accordance to metastable phase diagram [40]. In literature [41] investigated the viscosity and electroconductivity of Cd-Sb eutectic melt. It is stated that eutectic structure of this melts is not destroyed in some temperature range after melting and confirm the propose about existing of microheterogeneity structure in liquid state of Cd-Sb melts.

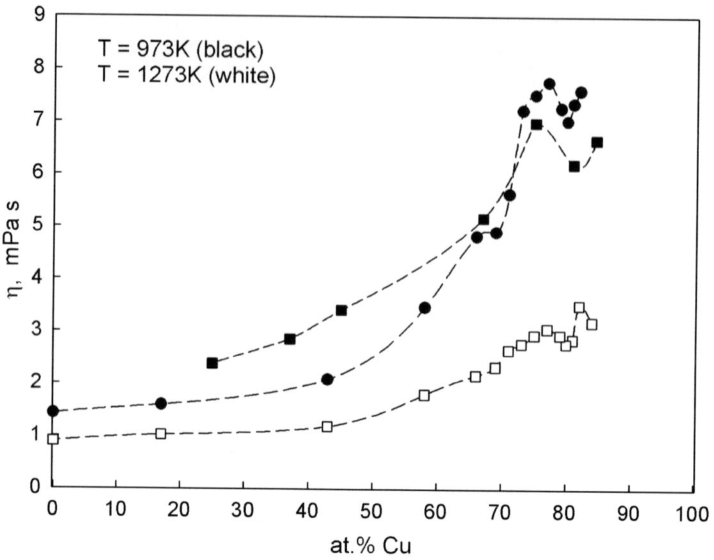

Figure 24. The concentration dependence of viscosity of Cu-Sb melts (■ – [42]; ○ – [43]).

Fig. 24 shows three maxima on concentration dependence of viscosity, which correspond to 67, 75 and 83 at. % Cu and a minimum at 80 at.% Cu, what is in agreement with phase diagram. It was concluded that in Cu-Sb melts exist an unlike-atoms based clusters with different composition. Most stable of them corresponds to Cu_3Sb. The clusters of type Cu_2Sb dissociate with temperature, on another hand the h-phase appears on the isotherm of viscosity starting from temperature 1273K. Although the Cu-Sb belongs to complex systems literature data [42] and [43] are in general agreement. In table 4 is shown the data of E_A for liquid In_xSb_{100-x} alloys (x= 0; 40; 44; 47; 50; 53; 56; 60; 100 at.% from [37] and x=31.3; 45.5; 99 at.% from [44]) and for liquid Cu_xSb_{x100-x} alloys from [43].

Table. 4. Values of activation energy for In-Sb and Cu-Sb systems

at.% In	**E_A, kJ/mol for In-Sb**	**at.% Cu**	**E_A, kJ/mol for Cu-Sb**
0.0	22.000	83.0	31.895
40.0	22.581	82.0	29.639
44.0	20.511	81.0	32.419
47.0	18.923	80.0	31.441
50.0	23.321	79.0	30.939
53.0	19.405	77.0	31.790
56.0	18.139	75.0	32.733
60.0	16.700	73.0	32.200
100.0	5.922	71.0	26.034
		69.0	25.654
at.% In	E_A, kJ/mol for In-Sb	67.0	27.254
		58.0	22.397

31.3	26.489	43.0	20.427
45.5	16.548	17.0	15.917
99.0	9.226	0.0	17.087

In [44] has been found an anomalous phenomenon in variation of viscosity of $In_{45.5}Sb_{54.5}$ and $In_{31.3}Sb_{68.7}$ melts with temperature at about 1073K. Therefore, the whole temperature range was divided into two segments: low temperature region LTR (from the liquidus to 1023K) and high temperature region HTR (from 1073K to 1223K). According to this assumption authors calculated the activation energy for low temperature region and high temperature region. In our opinion, these dependences are linear in whole temperature region; therefore by comparison our E_A agreement with data for LTR from literature [44]. Moreover, a maximum value of activation energy is in the composition range of the intermetallic phases and shows increase about the eutectic point. The maxima of activation energy for Cu-Sb system are in agreement with maxima on the concentration dependence of viscosity.

In literature [45] it is noted a number of systems, which show the anomalous behavior of specific electroresistivity at composition of components as 2:1 and 1:2. Cu (In, Cd)-Sb systems belong to such systems and it is interesting to analyze the viscosity feature to comparing. The anomalous behavior of viscosity near those concentrations commonly is related with existing of such composition clusters in liquid state (Cu_2Sb, Cd_3Sb_2, In_3Sb). Complicated situation is with InSb+Sb sybsystem, where the rapid decrease of viscosity about 1:2 composition with increase of activation energy of viscous flow is observed. At first observe what influence on viscosity changes:

- strength interatomic bounds;
- atom sizes of components;
- coordinational number.

It is suggested, that activation energy responds only for changes of interatomic interactions. In this case decrease of viscosity by small adding of Sb to InSb is attributed with structural transformation in melt.

For theoretical description of viscosity commonly use the models based on the correlation between interchange energy, which can be expressed through the enthalpy of mixing and viscosity.

The differences between melts with unlike type of phase diagram agree with suggestion about the possible similarity of viscosity behavior within each type. To indicate such characteristic we plots available temperature dependence of viscosity in the coordinates $\ln(\eta)$-T_m/T.

Fig. 25 contains only viscosities for melts with concentration of compounds for compound-forming systems, and melts with above equiatomic concentration for eutectic systems. For Sb-Cd and Ge-Sb systems available are the data of kinematic viscosity only, so to obtain the dynamic viscosity data the concentration and temperature dependences of density were calculated due to Wagner law. The densities of pure components were taken from [36]. As to Ge-Sb system, the literature data of viscosity we find [46] only for $Sb_{25}Ge_{75}$, therefore this melt is as an exception in fig. 25. Through pure literature data the Ge-Sb system requires more detailed investigation. Fig. 25 shows some similarity of same type melts with higher slope of the tangent to the curve for compound-forming systems alloys and lower slope for eutectic systems ones. Our results general agree with equal studies of glass-forming melts [47].

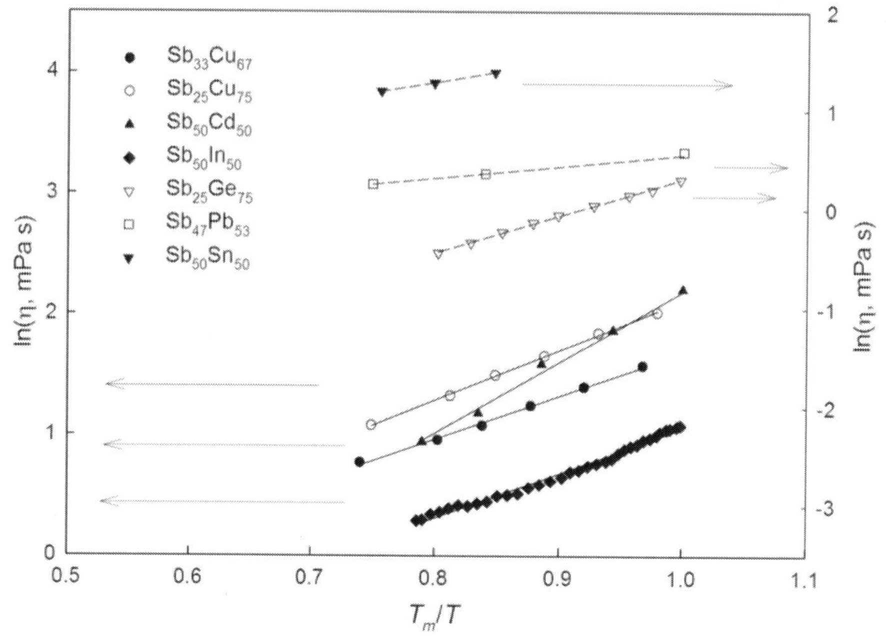

Figure 25. Viscosity as function of reduced temperature.

Concentration Fluctuations

The investigated binary melts shown the deviations from ideal mixture, that is connected with existence of clusters in liquid state. Describing these by an association model it is possible to reproduce quantitatively the concentration dependences of the thermodynamic properties or the kinetic structure-sensitive properties (viscosity, diffusion). In the light of this, on the basis of a simple association model the long wavelength correlation function have been calculated. Following this model in the melts the short range ordered regions (associates) with non-associates randomly distributed atoms in a steady dynamic equilibrium coexists.

Since the work [48], the correlation function in the long wavelength limit corresponding to the fluctuations in concentration-concentration function $S_{cc}(0)$ have been widely used to study of structural inhomogeneities of the binary mixtures.

This function is related with thermodynamic quantities,

$$S_{cc}(0) = RT\left(\frac{\partial^2 G}{\partial C}\right)_{T,p,N} = (1-C)a_A\left(\frac{\partial a_A}{\partial C}\right)^{-1}_{T,p,N} = Ca_B\left(\frac{\partial a_B}{\partial(1-C)}\right)^{-1}_{T,p,N}, \quad (2)$$

where C is contain of component A in binary A-B alloy and N=1, T is the absolute temperature, p is the damp, R is the gas constant, G_m is the free energy of mixing; a_A and a_B are the activity of elements A and B, respectively.

The calculation of $S_{cc}(0)$ can be also defined as [49]:

$$S_{cc}(0)=C(1-C)\eta_{ad}/\eta, \tag{3}$$

where η_{ad} is the additive viscosity. According to additive rule the η_{ad} is equal:

$$\eta_{ad}=C\eta_1+(1-C)\eta_2, \tag{4}$$

where η_1, η_2 are the viscosities of components A and B, respectively. For ideal solutions,

$$S_{cc}^{id}(0)=C(1-C).$$

The relationship between viscosity and enthalpy of mixing ΔH were obtained in [50]:

$$\frac{\Delta\eta}{\eta_{ad}}=-\frac{\Delta H}{RT}. \tag{5}$$

Using (3) in (2) we receive:

$$S_{cc}(0)=\frac{C(1-C)}{1-\frac{\Delta H}{RT}}.$$

The fluctuations in concentration-concentration function have been widely investigated for different categories of binary melts. This factor ascribed preferred interactions in liquid state: $S_{cc}(0)\sim S_{cc}^{id}(0)$ for simple liquid mixture, $S_{cc}(0)< S_{cc}^{id}(0)$ for compound forming liquid alloys, $S_{cc}(0)> S_{cc}^{id}(0)$ for segregating and immiscible liquid alloys.

The available experimental results on enthalpy of mixing for investigated systems [6, 9, 10, 14, 51, 52] was used for calculate of $S_{cc}(0)$ (we indicate such data as $S_{cc}(0)_H$). The calculated data on the fluctuations in concentration-concentration function in the long wavelength limit marks as $S_{cc}(0)_\eta$.

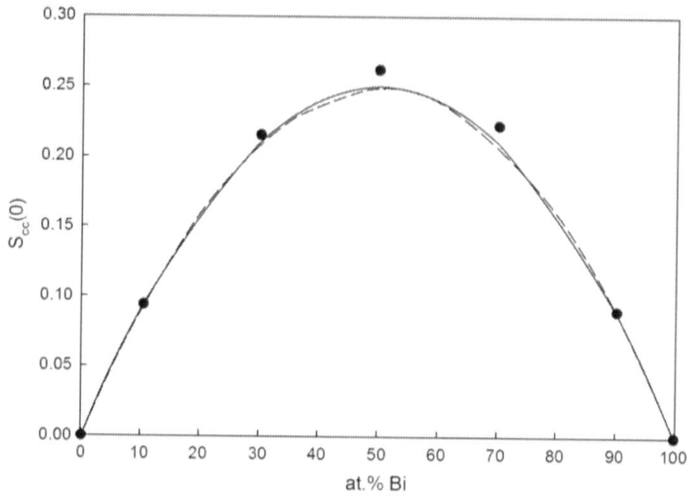

Figure 26. Concentration dependence of $S_{cc}(0)$ for Bi-Sb system by temperature 973K ($-S_{cc}^{id}(0)$; - - $S_{cc}(0)_H$; • $S_{cc}(0)_\eta$).

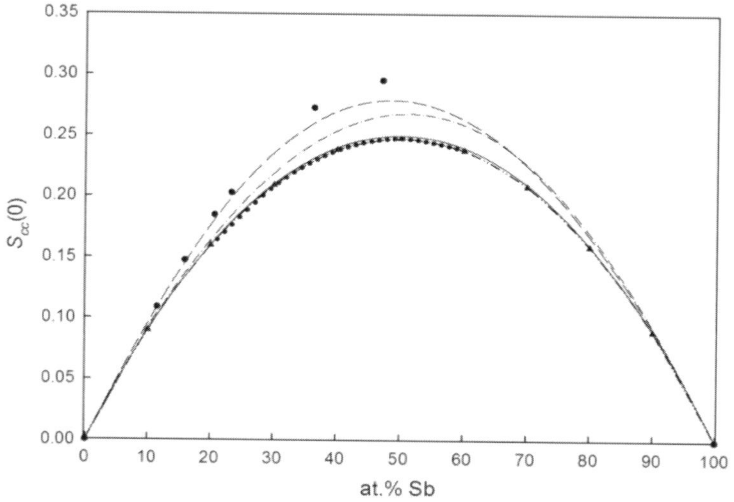

Figure 27. Concentration dependence of $S_{cc}^{id}(0)$ (–), $S_{cc}(0)$ for Pb-Sb ($S_{cc}(0)_H$; • $S_{cc}(0)_\eta$) and for Sn-Sb (-·-·- $S_{cc}(0)_H$; - - $S_{cc}(0)_\eta$) at 923 K, for Ge-Sb (-·-·- $S_{cc}(0)_H$) at 1273K.

Fig. 26 shows the congruence $S_{cc}(0)_H$ with ideal behavior slight deviation $S_{cc}(0)_\eta$. These results are in agreement with viscosity data and confirm suggestions about the random atomic distribution in melts of such type phase diagram.

The feature of investigated eutectic systems is that on the one hand positive divergence of $S_{cc}(0)_\eta$ is observed, on the other hand $S_{cc}(0)_H$ are in agreement with ideal case (fig. 27). In this case we suggest that the atom distribution and the size factor essential influence on $S_{cc}(0)_\eta$ and η, respectively. Fig. 27 shows deviations of $S_{cc}(0)_H$ from $S_{cc}^{id}(0)$ for Ge-Sb. It should be noted, that maximum value for all three eutectic systems is located about equiatomic composition.

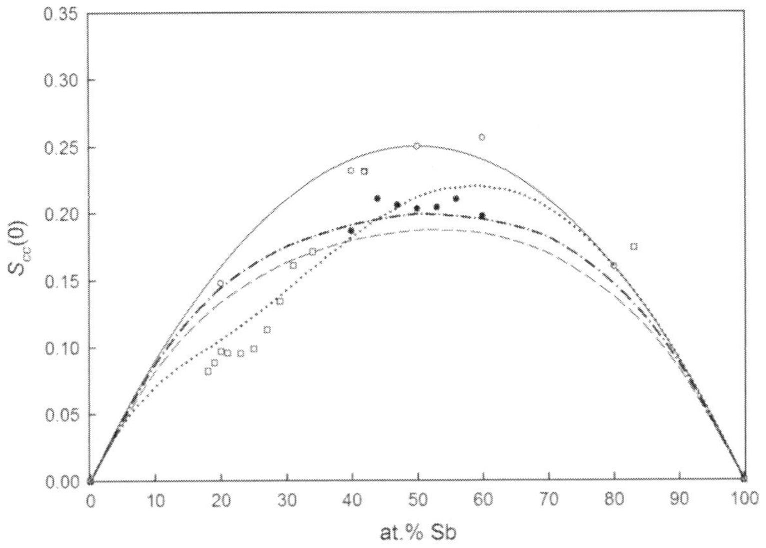

Figure 28. Concentration dependence of $S_{cc}^{id}(0)$ (–), $S_{cc}(0)$ for In-Sb (- - $S_{cc}(0)_H$; • $S_{cc}(0)_\eta$) at 900 K, for Cd-Sb (-·-·- $S_{cc}(0)_H$; ° $S_{cc}(0)_\eta$) at 923 K and for Cu-Sb ($S_{cc}(0)_H$ (1003K); □ $S_{cc}(0)_\eta$ (1023K)).

The calculated concentration fluctuations in long wavelength limit using mixing enthalpy data characterized by negative deviation on ideal one for investigated compound forming systems. This fact denotes the preferred interactions between unlike-kind of atoms. On the other hand the behavior of $S_{cc}(0)_\eta$ is more complicated and depends from the shape of phase diagram: $S_{cc}(0)_\eta < S_{cc}^{id}(0)$ in concentration regions of compounds existence (InSb compound for In-Sb system; Cd_4Sb_3 *and* Cd_3Sb_2 compounds for*;* Cu_3Sb and Cu_2Sb compounds for Cu-Sb system) and $S_{cc}(0)_\eta \geq S_{cc}^{id}(0)$ in concentration regions of eutectic point existence (CdSb+Cd sybsystem in Cd-Sb; no compounds and eutectic point in Cu-rich site of Cu-Sb system) (fig. 28). The negative deviation of $S_{cc}(0)_\eta$ from $S_{cc}^{id}(0)$ in InSb+Sb sybsystem is supposed to existence a large fraction of InSb clusters in the melts, that neglected the two another factors of viscosity effect.

CONCLUSION

Liquid Sb has a structure, which consists of two kinds: clusters with incompletely destroyed at melting covalent bonds and clusters with metallic bonds. Mixture of these clusters coexists within some temperature range and fraction of clusters with metallic bonds increases with heating.

Inhomogeneous structure of Sb persists at alloying with other elements. Behavior of Sbn clusters as well as Sb-atoms at mixing with other elements is different and suppose the formation of chemically ordered clusters and self-associated atomic groups. Tendency to formation of each kind clusters is in agreement with phase diagram. Clusters are dynamical structural units and their parameters are sensitive to temperature and composition being responsible for viscous properties.

Structural and thermodynamic features of cluster solutions are related with properties of solidificated alloys. Changing the structure and other parameters it is possible to control the solidification process and in result to improve the properties of materials, especially nanomaterials on the base GaSb, InSb and others.

REFERENCES

[1] Waseda Y. *The* Structure *of Non-Crystalline Materials: Liquids and Amorphous Solids*; Mcgraw-Hill: New York, USA, 1980; p. 326.

[2] Gaspard J.-P.; Bellissent R.; Menelle A.; Bergman C.; Ceolin R. *Il Nuovo Cimento*. 1990, vol 12D(4-5), 649-655.

[3] Bergman C.; Coulet M. V.; Bellisent R.; Seife-Lorenz K.; Hafner J. *J. Non-Crystalline Solids*. 1999, vol 250-252, 253-257.

[4] Greenberg Y.; Yahel E.; Caspi E. N.; Beuneu B.; Moshe P.; Dariel M. P.; Makov G. *J. Chem. Phys*. 2010, vol 133, 094506.

[5] Zou X. W.; Jin Z. Z.; Tang S. Q. *Phys status solidi B*. 1987, vol 142 (1), 9-14.

[6] Manasijevic D., Vrestal J., Minic D., Kroupa A., Zivkovic D., Zivkovic Z. *J. Alloys Comp*. 2007, vol 438, 150–157.

[7] Massalsky T. (ed.) Binary Alloy *Phase Diagrams* (2nd edition); ASM, Materials Park: Ohio, USA, 1990; p. 3589.

[8] Klym N. M.; Mudry S. I.; Halchak V. P.; Komarnizky M. S. Izv. *AN SSSR, ser Neorg. Mater*. 1990, vol 26 (10), 2123-2126.

[9] Wang J.; Leinenbach C.; Roth M. *J. Alloys Comp*. 2009, vol 485, 577–582.

[10] Chen S.-W., Chen C.-C., Gierlotka W., Zi A.-R., Chen P.-Y., Wu H.-J. *J. Electron. Mater*. 2008, vol 37(7), 992-1002.

[11] Klym N. M.; Mudry S. I.; Kozyrenko V. N. Izv. *AN SSSR, ser Neorg. Mater*. 1986, vol 22 (2), 201-203.

[12] Kubashevsky O.; Alcock S. Metallurgical termochemistry. *Metallurgy*: Moscow, Russia, 1982, p. 390 (in Russian).

[13] Belashchenko D. K. Transfer Phenomena in *Liquid Metals and Semiconductors*; Atomizdat: Moscow, Russia, 1970, p. 400 (in Russian).

[14] Hassam S.; Boa D.; Fouque Y.; Kotchi K. P.; Rogez J. *J. Alloys Comp*. 2009, vol 476, 74–78.

[15] Kumar R.; Sivaramakrishnan C. S. *J. Mater. Sci*. 1969, vol 4, 383-388.

[16] Zabdyr L.A. *Calphad* 1993, vol 17, 125–132.

[17] Dutchak Y. I.; Klym N. M.; Mudry S. I.; Kozyrenko V. N. *Metallofizika*. 1983, vol 5(3), 102-107.

[18] Bath A.; Gasser J. G.; Kleim R. *Phys. Letters A*. 1982, vol 91(7), 355-357.

[19] Kumar R.; Sivaramakrishnan C. S. *J. Mater. Sci*. 1971, vol 6, 48-53.

[20] Gierlotka W., Jendrzejczyk-Handzlik D. *J. Alloys Comp*. 2009, vol 484, 172-176.

[21] Knoll W.; Steeb S. *Phys. Chem Liquids*. 1973, vol 4(1), 39-60.

[22] Halm Th.; Neumann H.; and Hoyer W. Z. *Naturforsch*. 1994, vol 49a, 530-534.

[23] Manasijevic D., Minic D., Zivkovic D., Vrest'ál J., Aljilji A., Talijan N., Stajic-Trosic J., Marjanovic S., Todorovic R. *Calphad.* 2009, vol 33, 221-226.
[24] Nachtrieb N. H.; Clement N. *J. Phys. Chem.* 1958, vol 62(6), 747-750.
[25] Wei S.;Lu K.; Li C.; Zhong W.; Yan W.; Uruga T.; Chen W.; Niu X.;Wu Z.; Chen D.;Sun Z. *Physica Scripta.* 2005, vol T115, 399-401.
[26] Poltavtsev Yu. G. Structure of semiconductor metals; *Metallurgiya*: Moscow, Russia, 1984, p. 176 (in Russian).
[27] Yokoyama I. *Physica B.* 2000, vol 291, 145-151.
[28] Iida T.; Gurthrie R. I. L. *The physical properties of liquid metals*; Clarendon Press: Oxford, England, 1988; p. 312.
[29] Sun, C.; Geng H.; Zhang N.; Teng X.; Ji L. *Materials Letters.* 2008, vol 62, 73–76.
[30] Iida T.; Ueda M.; Morita Z. I. *Tetsu-to-Hagane.* 1976, vol 62(9), 1169-1178.
[31] Faber T. E. *Introduction to the theory of liquid metals*; Cambridge University Press: Cambridge, England, 1972; p. 587.
[32] Gebhardt E.; Köstlin K. Z. *Metallkde.* 1957, vol 48(12), 636-641.
[33] Sato T.; Munakata S. *Bull. Res. Inst. Min. Dress. Metall.*, Tohoku Univ. 1955, vol. 11(2), 183-187.
[34] Gvozdeva L. I.; Lyubimov A. P. *Izv. Vyssh. Uchebn.* Zaved., *Chern. Met.* 1965, vol 7, 16-19.
[35] Mudry S. I.; Sklyarchuk V. M.; Plevachuk Yu. O.; Yakymovych A. S. *Inorg. Mater.* vol 46(8), 837-839.
[36] Arsent`ev P. P.; Koledov L. A. Metallic melts and their properties; *Metallurgiya*: Moscow, Russia, 1976, p. 376 (in Russian).
[37] Glazov V. M.; Chizhevskaya S. N.; Glagoleva N. N. *Liquid Semiconductors*; Nauka: Moscow, Russia, 1967, p. 244 (in Russian).
[38] Mudry S.; Yakymovych A. *J. Mol. Liquids.* 2009, vol 149, 105-107.
[39] Chizhevskaya S. N.; Danilchenko V. A.; Kurbatov V. P. Izv. *AN SSSR, ser Neorg. Mater.* 1974, vol 10(8), 1397-1400.
[40] Belashchenko D. K. *Rus. J. Phys. Chem.* 1957, vol 31(10), 2269-2277.
[41] Dutchak Y. I.; Panasyuk I. V. *Fiz. Met. Metalloved.* 1964, vol 18, 155-156.
[42] Ning S.; Bian X. F.; Ren Z. F. *Physica B.* 2010, vol 405, p. 3633–3637.
[43] Eretnov K. I.; Lyubimov A. P. *Ukr. Fiz. Zh.* 1967, vol. 12(2), 214-220.
[44] Cheng S. J.; Bian X. F.; Qin X. B.; Si P. C. *Indian. Eng. Mater. Sci.* 2004, vol 11, 521-524.
[45] Wilson J. R. *Metall. Rev.* 1965, vol 10, 381-590.
[46] Glazov V. M. *Zh. Neorg. Khim.* 1961, vol 6(4), 933-936.
[47] Ubbelohde A. R. *Molten state of matter: melting and crystal structure*; John Wiley and Sons Ltd.: New York, USA, 1978, p. 154.
[48] Bhatia A. B.; Thornton D. E. *Phys. Rev. B.* 1979, vol 2, 3004–3012.
[49] Singh R. N.; Sommer F. *Phys. Chem. Liq.* 1998, vol 36(1), 17–28.
[50] Odusote Y. A.; Hussain L. A.; Awe O. E. *J. Non-Cryst. Solids.* 2007, vol 353, 1167–1171.
[51] Liu H.S.; Liu C. L.; Wang C.; Jin C. P.; Ishida K. *J. Electron. Mater.* 2003, vol 32, 81-88.
[52] Hultgren R.; Desai F. D.; Hawkins D. T.; Gleiser M.; Kelley K. K. *Selected values of the thermodynamic properties of binary alloys*; ASM, Materials Park: Ohio, USA, 1973, p. 1445.

In: Antimony: Characteristics, Compounds and Applications ISBN 978-1-62100-598-8
Editor: Manijeh Razeghi

Chapter 5

TYPE-II ANTIMONIDE-BASED SUPERLATTICE PHOTON DETECTORS AND FOCAL PLANE ARRAYS

Manijeh Razeghi and Binh-Minh Nguyen
Center for Quantum Devices, Northwestern University,
Evanston, IL, US

ABSTRACT

Vision is one of the most developed senses of human beings. Unfortunately, Nature allows human vision only a very small window in the wide, continuous spectrum of electromagnetic radiation. With the help of science and technology, people are now able to broaden their vision range, which facilitates the discovery of the world in different new perspectives. Since its discovery, the infrared regime has been an attractive window for advantageous imaging and sensing applications. Infrared imaging and thermography find applications in diverse aspects of life, from night vision and target tracking for homeland security and defense to non-destructive failure detection in industry, chemical sensing in medicine, and free-space communication. Currently, the dominant technologies of photodetectors based upon HgCdTe (MCT) and Quantum Well Photoconductors are experiencing many limitations. Such limitations include non-uniformity, difficulty in controlling the material growth, and high power consumption, which hamper the emergence of these technologies into many applications. In recent years, Type-II InAs/GaSb superlattices have been emerging as a viable alternative to current infrared detector technologies. This chapter aims to present this antimonide-based material system and discuss its development progress and contributions to the field of infrared detection.

A. INTRODUCTION

Antimonide based Type-II InAs/GaSb superlattice [1] is a very important material system for infrared detection. This is due to the Type-II band alignment between the closely lattice matched InAs and GaSb layers, where electrons are spatially confined in the InAs layer and holes in the GaSb layer. Among the currently developing infrared detection technologies, the only three that take advantages of low dimensional properties of quantum mechanics include: Type-II InAs/GaSb superlattice photodetectors, the quantum well intersubband photodetectors and the quantum dot infrared photodetector. Going by dimensionality, the Type-II superlattice is still a three dimensional system, the same as a bulk semiconductor, while the other two systems are so to speak "two and zero dimensional systems" respectively, meaning that the confinements of carriers are locally in one or more than one direction. However, compared with the quantum well and the quantum dot, Type-II superlattices share the same capability of spatially localizing electrons in a nanoscale space. This is due to the spatial separation of carriers in the constituent layers. Moreover, the physics that underlie the Type-II superlattice give us the unique ability to control the interaction between localized electrons by way of varying potential barrier thickness. This gives us, in this system, an additional degree of freedom for designing and creating new "materials". This is not unlike the way Nature created solid crystals from atoms. Having bulk-like properties, Type-II superlattices can be exploited using the considerable understanding acquired for bulk semiconductors in the past. This knowledge can be used to optimally design device architectures for photodetecting applications. The close match between theoretical work and experimental developments has allowed Type-II superlattice photodetectors to become one of the fastest growing technologies in infrared detection. During a decade of maturing, the material system has experienced significant progress that has lifted the detector performance to comparable and better levels than the state-of-the-art Mercury Cadmium Telluride (MCT) technology.

B. Material System

1. The 6.1. Angstrom Family

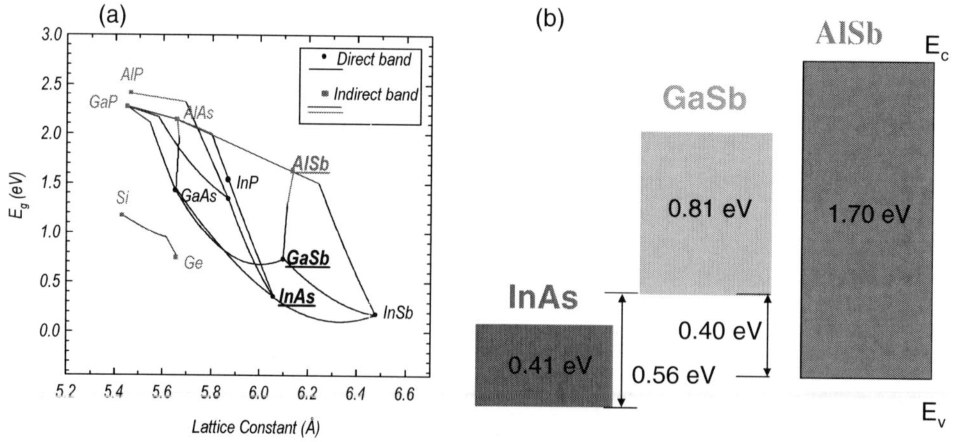

Figure 1. (a) The energy gap versus the lattice constant of InAs, GaSb, and AlSb compared with other semiconductors. (b) InAs, GaSb, and AlSb energy band lineups.

The 6.1 Angstrom family (Figure 1) is an important group in the semiconductor technology. It consists of three members that are closely lattice matched to each other: InAs (a= 6.0584 Å), GaSb (a= 6.0959 Å) and AlSb (a= 6.1355 Å). Since GaSb substrates became available, and the growth of III-antimonide semiconductors was feasible, the 6.1 Å and its compounds have been providing enormous flexibility in designing heterostructures for optical and electronic applications. One great advantage of the family is the small lattice mismatch of the family to GaSb substrates and similar growth windows for the three materials. This enables the growth of high quality materials with a low density of defects and dislocations. The energy gap of the family and related compounds ranging from 0.41 eV (for InAs) to 1.70 eV (for AlSb) is in particular interest of optoelectronic devices in the Short Wavelength Infrared (SWIR) and Mid-Wavelength Infrared (MWIR) regimes. Moreover, the heterojunctions between InAs and the other two members benefit of the unique features of the Type-II band alignment. On the one hand, the InAs/AlSb interface forms a Type-II staggered line up where the conduction band of InAs is slightly above the valence band of AlSb. The high energy gap of AlSb leads to an exceptionally large conduction band off set of about 1.45 eV, enabling the realization of very deep quantum wells and very high tunneling barriers. This heterostructure has been widely utilized in high frequency field effect transitors (FETs) and resonant interband tunneling diodes (RITDs). On the other hand, the heterojunction between InAs and GaSb leads to the exotic so called *broken gap* line up, where the conduction band of InAs is about 0.15 eV lower than the valence band of Gasb. This type of *misaligned* structure is the reason why Type-II superlattices have a flexible band gap engineering capability and this will now be discussed in the next sections.

2. Type-II InAs/GaSb Superlattice

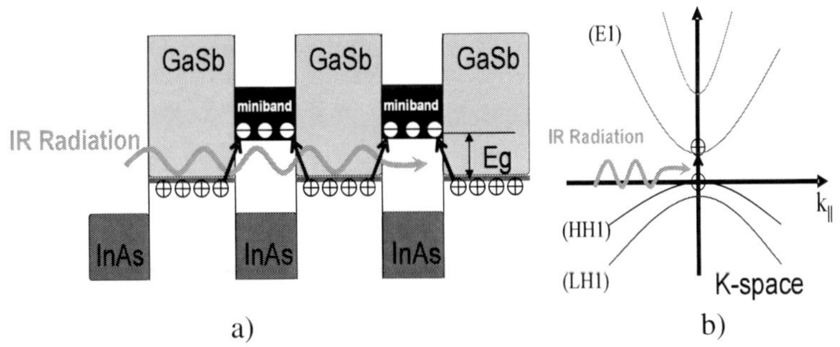

Figure 2. a) Spatial band alignment in Type-II superlattice (red stands for InAs forbidden gap and green stands for GaSb forbidden gap) and b) band structure with direct band gap and absorption process in k-space.

The idea of Type-II InAs/GaSb superlattice was first proposed by Sai-Halasz and Esaki in the 1970's. The superlattice is formed by alternating the InAs and GaSb layers over several periods. This creates a one dimensional periodic structure, like the periodic atomic chain in naturally occurring crystals (Figure 2). The Type-II broken gap alignment leads to the separation of electrons and holes into the InAs and GaSb layers, respectively. Electrons partition themselves in quantized energy levels of InAs quantum wells and construct conduction band structure of the superlattice while hole's energy level in GaSb wells form the superlattice's valence band. Moreover, the unique nature of Type-II misaligned structure allows electrons in InAs wells to tunnel through the GaSb layer via GaSb's valence band channel. Energy levels of electrons therefore are determined by the quantization process in

InAs well as well as tunneling process in GaSb barrier. These energy levels depend on the thickness of the quantum well layers, thus enabling a band gap engineering capability of Type-II superlattice. Large period superlattices behave like semimetals but if the superlattice period is shortened, the quantization effects are enhanced causing a transition from a semi-metal to a narrow gap semiconductor. Despite spatial separation of carriers, in reciprocal k-space, the superlattice is a direct bandgap material with large optical coupling of direct band-to-band transition as shown in Figure 2 (b).

C. Advantages of Type-II Superlattice

1. Band Gap Engineering

Infrared photodetectors based on bulk semiconductors, such as the state-of-the-art Mercury Cadmium Telluride (HgCdTe) or Indium Arsenide Antimonide (InAsSb) are often limited in the choice of the detection wavelength. For a binary semiconductor, the energy gap is unchangeable. In order to adjust the cut-off wavelength, one or more elements must be added to form ternary or quaternary compounds. However, in many cases, it is not practical for experimental growth. The compositional change in the material forces the material lattice constant to change. This can result in a strong lattice mismatch to the substrate and completely destroy the structural quality of the device.

In Type-II superlattices, the energy gap is determined by the thickness of the constituent layers. Without any change in growth parameters, superlattices with cut-off wavelength ranging from 3μm up to 32μm have been experimentally demonstrated [2].

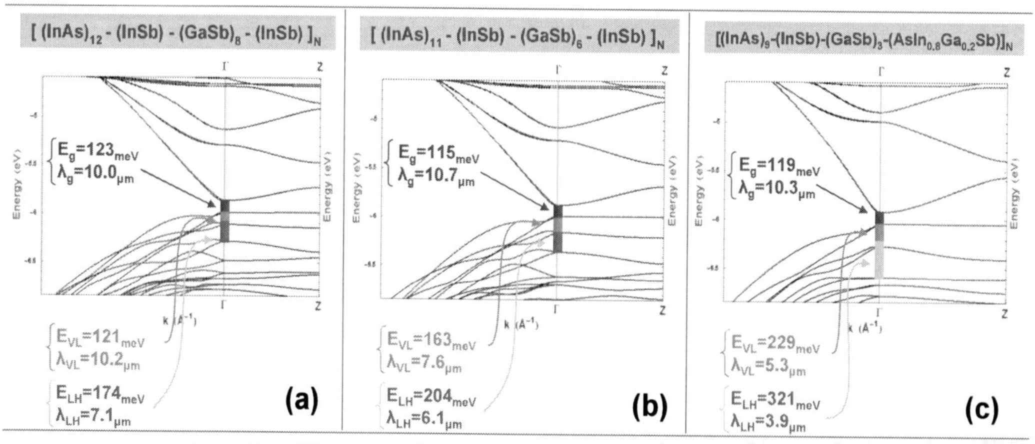

Figure 3. Example of bandgap engineering for the suppression of Auger recombination. Three designs which have the same energy gap but different inter-valence band gaps. Design C is the most promising design for Auger supression.

2. Auger Suppression

In addition to the flexibility in tailoring the energy gap, there are a large number of possible superlattice configurations which allow one to generate different superlattice designs for any given cut-off wavelength. Therefore, it is possible to select the superlattice structure that can minimize non-radiative recombinations, such as Auger recombination, or the Shockley-Read-Hall recombination which use the mid-gap trap centers. Figure 3 is an example of the bandgap engineering used to suppress Auger recombination. Three designs are selected for a cut-off wavelength around 10μm. The GaSb quantum well for holes is thinnest in Design C, and as a consequence of quantum mechanics, the difference of the energy levels in the valence band is most pronounced in Design C. The design therefore reduces the Auger recombination process, allowing for longer carrier lifetime and better device performance.

3. Large Effective Mass

Low bandgap semiconductors often produce a small effective mass. As a rule the effective mass is inversely proportional to the energy gap:

$$m^* = \left[\frac{1}{m_e} + \frac{E_P(E_g + 2/3\Delta)}{E_g(E_g + \Delta)} \right]^{-1} \quad (1)$$

where E_P is the Kane parameter, E_g is the energy gap and Δ is the spin-orbit interaction energy.

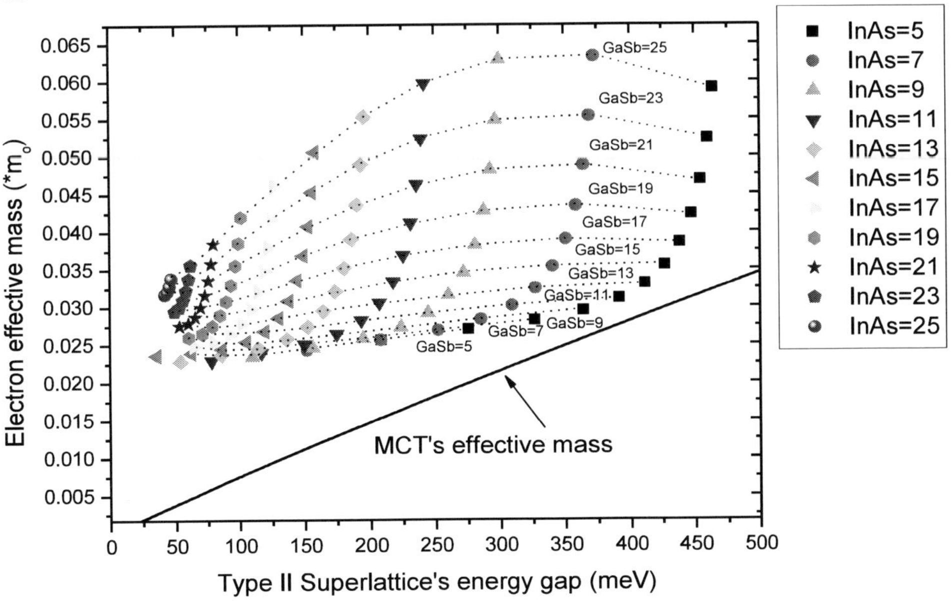

Figure 4. Electron effective mass for low bandgap materials: in bulk HgCdTe, the effective mass rapidly decreases with the cut-off wavelength, while in Type-II superlattice, the effective mass is similar to the effective mass of the InAs layer.

For example, the effective mass of the state-of-the-art Mecury Cadmium Telluride system is below 0.01 m_e when the cut-off wavelength is above 10μm, and reduced to 0.005 m_e with a 15μm cut-off (Figure 4). The problem arises for LWIR and VLWIR detection. Here, the electron can tunnel through the very small forbidden gap and give rise to tunneling leakage currents which depend exponentially on the square root of the effective mass.

Type-II superlattices, in contrast, are predicted to have stable values of effective mass, regardless of the cut-off wavelength. This is because the effective mass is mainly determined by the overlap between electron wavefunction in adjacent InAs wells, and not by the interaction between the conduction band and the valence band like in the case of bulk materials. The large effective mass allows Type-II superlattice devices to operate with lower tunneling current, and therefore, they can be optimized for higher performance.

4. Normal Incident, Broad Band Absorption

While being low dimensionality and benefiting from a large effective mass, the Type-II superlattice, keeps the advantage of the bulk as far as the optical properties are concerned. Unlike Quantum Wells or Quantum Dots, Type-II superlattices have a broad band absorption spectrum with a specific cut-off wavelength corresponding to the energy gap. Any photon with energy greater than the band gap can be absorbed and contributes to the integrated signal response. Moreover, the delocalization of the electron wavefunction allows for normal incident absorption without requiring a grating.

5. Good Uniformity

Another advantage of Type-II superlattices is the high quality, high uniformity and stable nature of the material. The energy gap and electronic properties of Type-II superlattices are determined by the layer thicknesses rather than the molar fraction as is the case for Mercury Cadmium Telluride. The growth of Type-II superlattices can be carried out with better control over the structure and with higher reproducibility. The spatial uniformity is also improved, since the effects of compositional change due to flux and temperature non-uniformity are not as important as they are in ternary/ quaternary bulk materials. We have demonstrated that Type-II superlattices can achieve excellent uniformity across a three inch in diameter wafer [3-4] where in contrast, with Mercury Cadmium Telluride, one experiences great difficulties in controlling the material growth (Figure 5).

D. Historic Development of Type-II Superlattice Photodetectors

1. Earlier Years: Proof of Concept for Type-II Superlattice Detectors

In 1975,Sai-Halasz and Esaki [6] first proposed the concept of the semiconductor superlattice using an alternating GaAs/GaAlAs heterostructure. They identified the quantum states and predicted how this would give rise to the phenomenon of negative differential conductance "NDR" in the device. Two years later, the same group predicted important effects for superlattices with Type-II alignment between two constituent layers in which, the conduction band of one layer lies below the valence band of the other [7]. Taking InAs/GaSb as a typical example for a Type-II superlattice, theoretical calculations were carried out to illustrate the energy gap shifting and the semiconductor/semimetal transition when the layer thicknesses are changed. The structure was then proposed by Smith and Mailhiot [8] for infrared detection applications, exploiting the fact that the cut-off wavelength could be adjusted in the infrared range, and that the effective mass was constant, regardless of the energy gap.

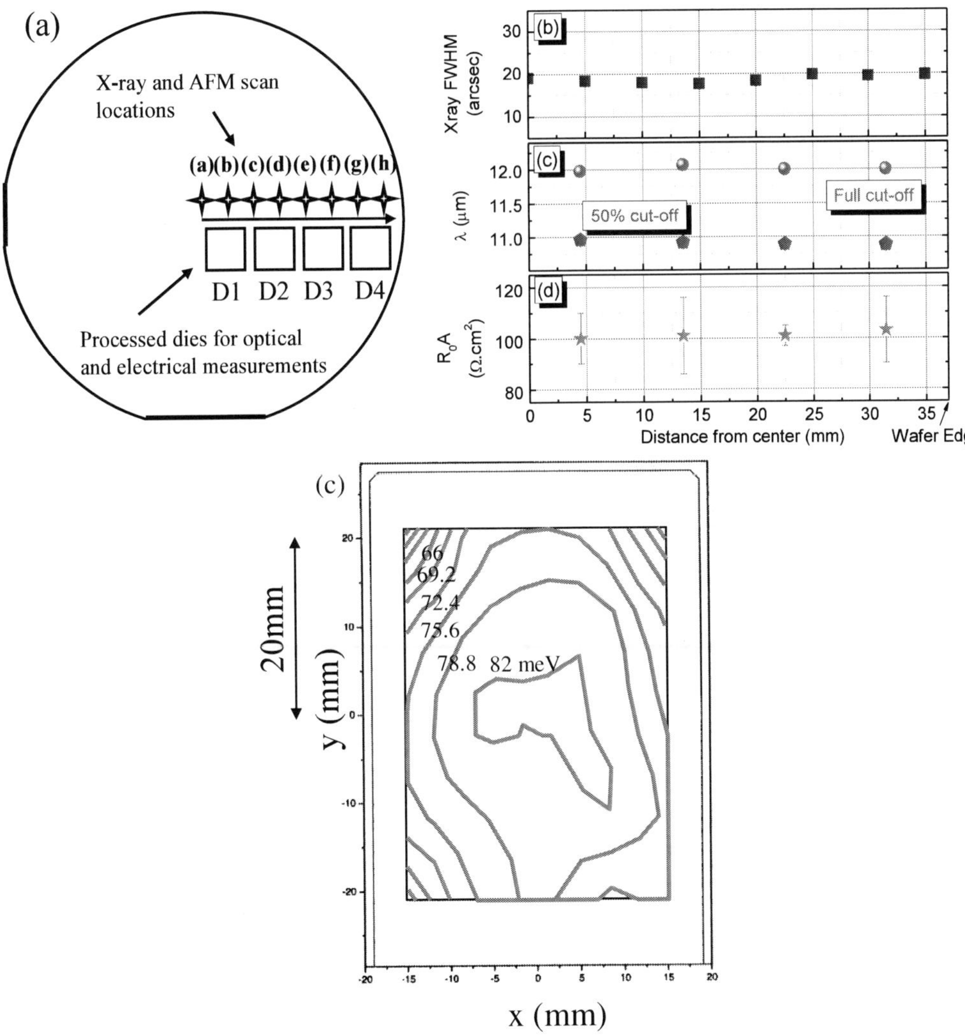

Figure 5. Comparison of growth uniformity over large wafer area between low band gap materials: Type-II superlattice (a, b) [4] and HgCdTe (c) [5].

Despite positive theoretical predictions, high-quality growth of this type of materials was not demonstrated until recent advances in the technology of the molecular-beam epitaxy (MBE) technique. The first experimental demonstration of photon detectors comprising superlattices was by Johnson et al [9]. The first photoconductors grown on GaAs substrate were demonstrated at the Center for Quantum Devices (CQD) at Northwestern University in 1997 [10]. Uncooled photoconductors grown on GaAs substrates have been demonstrated for the LWIR detection [11].

However, Type-II superlattices grown on GaAs substrate experienced many difficulties due to the 7% lattice mismatch between the epitaxial layer and the substrate. Since epi-ready GaSb substrates became commercially available in the late 1990s, the growth of Type-II superlattice on lattice matched GaSb have been significantly improved. First high performance photodiodes using ternary InAs/GaInSb superlattices were demonstrated in 1997 [12]. Uncooled photodiodes grown on GaSb substrates have been demonstrated for the detection in the LWIR regime [13] and MWIR regime [14]. Later on, effort expanding the detection range of Type-II superlattice photodetectors have been conducted by Wei, with the demonstration of cut-off wavelength ranging from 3.7μm [15], 5μm [14], 7μm [16], 9μm [17], 19μm and 32μm [18]. The suppression of Auger Recombination was also realized experimentally by interface engineering to reduce interface-induced defect centers [19].

Thanks to the growth on GaSb substrates, defect and dislocation free crystals enable the realization of photodiode detectors which require large differential resistance and low leakage. In the perpendicular p-i-n superlattice photodiode configuration, the signal current flows in the growth direction, the size of the detector element could thus be reduced, and multi-element devices, such as Focal Plane Arrays became feasible. Since then, efforts have been concentrated in the optimization of photodiode performance and the fabrication of FPAs. In 2003, the first demonstration of an FPA based on Type-II superlattice material systems was carried out at the CQD [20]. This provided the proof of concept of this technology. One year later, a high performance FPA with 5 μm cutoff was demonstrated in 2004 in Germany [21].

2. Recent Progress: Type-II Superlattices as a Viable Technology for the Third Generation Infrared Imagers

While the first decade of Type-II superlattice was focused on the demonstration of the advantages and detection concept of the material system, the second decade are experiencing unprecedented growth in the understanding of the materials and the emergence of Type-II superlattice in the field of infrared detection. Multi-facet research has been paid to optimize the detector performance, making Type-II superlattice a viable alternative to the state of the art HgCdTe technology.

a. Theoretical Modeling

The physics of Type-II superlattices has been extensively investigated and modeled. The Empirical Tight Binding frame work has shown excellent agreement with experiments, and become a powerful tool for the study of the material [2]. Not only standard information on the energy gap, band structure is precisely predicted but also the effective mass of superlattices is deeply analyzed, confirming the qualitative assessment on the higher effective mass of Type-II superlattice as compared to HgCdTe (FIGURE). The understanding of Type-II superlattices have stimulated the creation of novel material design, such as M-structure superlattice [22]. While being still a Type-II superlattice, the M-structure preserves all the benefits that a traditional design possesses, it exhibits numerous advantages that the predecessor does not have: larger effective mass; more flexible tunability of the band edges, including the valence band, etc. [23].

b. Material Growth and Characterization

With the guidance of theoretical modeling, material growth using the state-of-the-art molecular beam epitaxy has also been making significant progress [24]. High quality superlattice is often characterized by:

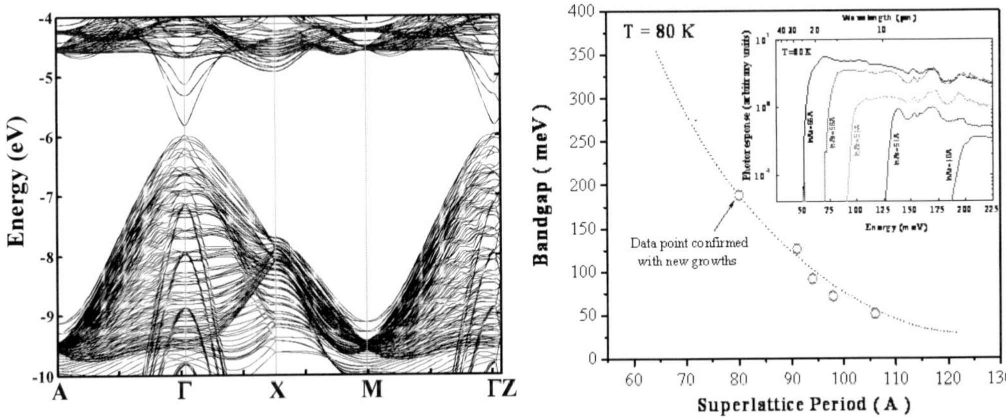

Figure 6. a) Band structure of a typical Type-II superalttice design and b) comparison between experimental energy gap (extracted from the optical responses in the inset) and theoretical prediction for several Type-II superlattice designs.

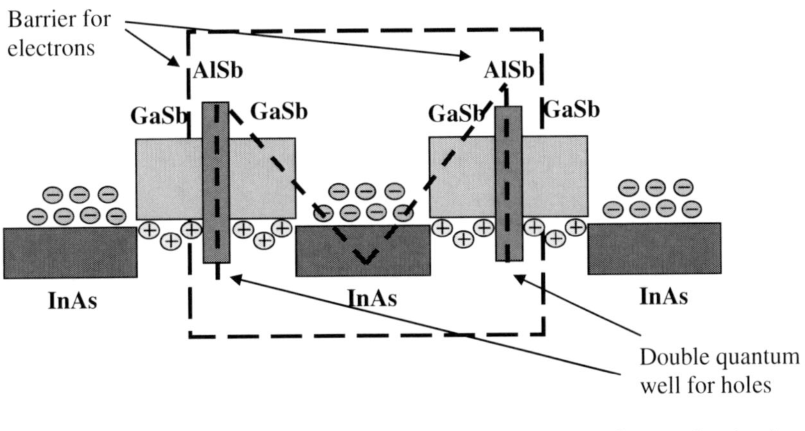

Figure 7. Schematic diagram of the M structure. The inserted AlSb layer forms a barrier for electrons in the conduction band and a double quantum well for holes in the valence band [22].

- Smooth surface with clear atomic steps under Atomic Force Microscopy
- Narrow peaks, High order diffraction patterns under High Resolution X-ray diffraction.
- Narrow Photoluminescence peaks at exactly the expected position.
- Sharp interface and precise layer thickness as revealed by Transmission Electron Microscopy or Scanning Tunneling Microscopy.
- Type-II superlattice.

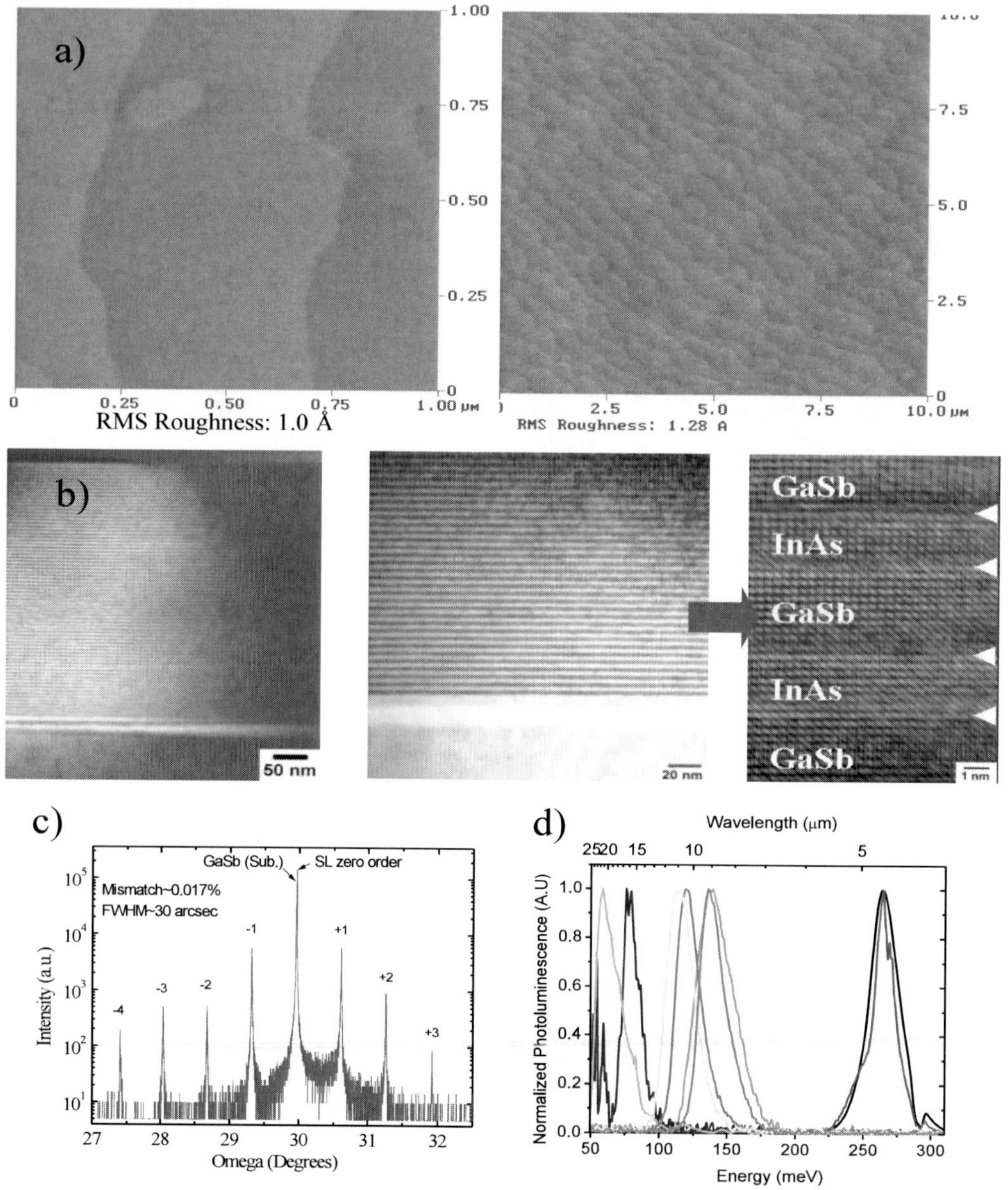

Figure 8. Routined structural characterization for Type-II superlattice using: a) Atomic step microscopy; b) Transmission Electron Microscopy, c) High Resolution X-ray Diffraction, d) Photoluminesence [25].

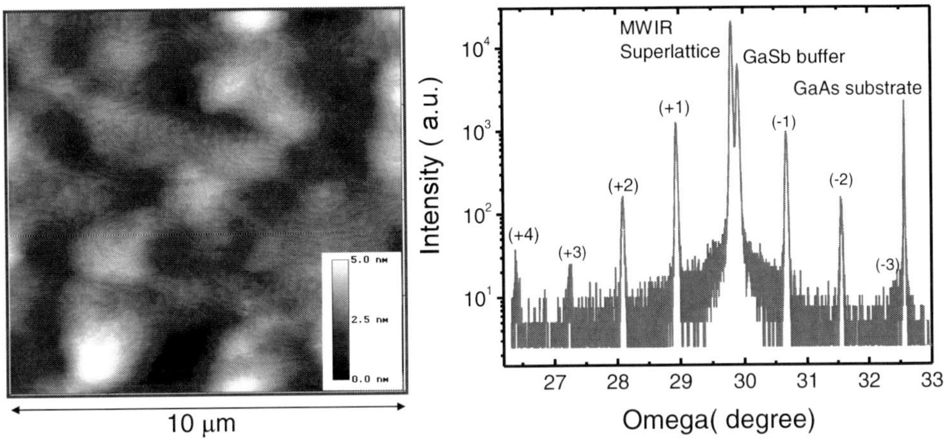

Figure 9. Structural characterization of MWIR Type-II superlattice grown on GaAs substrates: a) Atomic Force Micrograph of a 10x10mm^2 area with a RMS roughness of 8.1 Angstrom, b) High resolution X-ray diffraction showing higher order satellite peaks [30].

New growth conditions for many III-V compounds have been established for the growth of lattice mismatched layer on GaSb substrate, enabling the realization of numerous device architectures. For example, with the presence of an InAsSb layer, Type-II superlattice can be grown with a p-on-n polarity, which is the primordial requirement for the superlattice can be bonded with commercial Read Out Integrated Circuits [26]. The layer also acts as a highly selective etch stop layer for the removal of GaSb substrates, which enhance the optical efficiency and extend the lifetime of Type-II superlattice Focal Plane Arrays [27]. Optimal growth conditions also allows for the growth of high quality material with epitaxial thickness of up to 15μm [28] without any structural degradation.

In parallel, there has been a renew interest for the growth of Type-II superlattice on GaAs substrates, due to a great need for large format, cost reduction fabrication of Type-II superlattice focal plane arrays. Despite a 7.8% lattice mismatch between the superlattice and the GaAs substrates, high quality Type-II superlattices photodiodes have been demonstrated for the first time in the MWIR [29-30], and the LWIR [31]. Both detectors achieved background limited performance with a 300K background under a 2π field of view. Focal plane arrays based on Type-II superlattices on GaAs substrates were also demonstrated [32].

c. Device Fabrication

As narrow bandgap materials are very sensitive to the surface state, device fabrication and passivation are important steps to ensure high performance detectors. Thus far, all Type-II superlattice photodiodes are mesa-based devices, whose fabrication process involves an etching step to isolate the mesa. Isotropic chemical wet etching was found to be not compatible with the fabrication of focal plane arrays due to the slented sidewall and the reduction of array sizes. Dry etching using electron cyclotron resonant plasma (ECR) has been considered the state-of-the-art for mesa isolation for the last 10 years. Recently, inductively coupled plasma (ICP) etcher has been utilized and resulted in smooth, vertical sidewalls without need of post-wet etching. The etching has shown to increase the surface resistivity by one order of magnitude [33-34].

For narrow bandgap material and small size devices, the surface leakage becomes a limiting factor to the device performance and has been a known processing challenge for years. The surface leakage results from the abrupt termination of the periodic structure at the diodes' side walls. Due to the incomplete bond of surface atoms, a bending of the conduction and valence bands near the surface occurs, creating a surface conduction channel along the mesa sidewalls. Moreover, foreign adsorbents and process contaminants can further alter the surface potential and introduce trap levels within the energy gap, leading to more efficient trap-assisted tunneling currents. A good passivation layer is expected to fix the pinning of the Fermi level near the surface and to physically protect the mesa side walls from outside contaminants. Three main families of passivants used for Type-II superlattice photodetectors include: wet chemical passivation, dielectric passivation and polymer passivation.

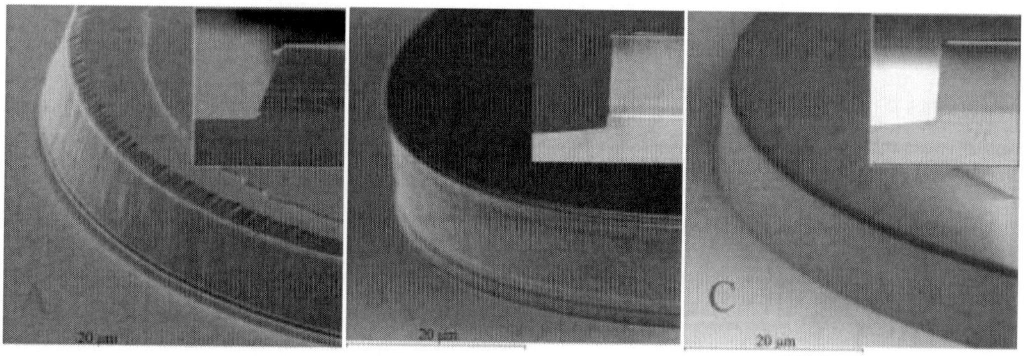

Figure 10. SEM images of 7.7μm deep photodiodes etched by (A) ECR and wet chemical etch, (B) ICP and wet chemical etch and (C) ICP only. The inset of each image shows the resulting sidewall angle to be (A) 70°, (B) 90° and (C) 90°. The samples followed by wet chemical etch show sidewall striations and undercut, whereas (C) is visually smooth [34].

Wet chemical passivation operates in two steps. First, the solution removes the native oxides at the surface of the semiconductor. Then, new atoms occupy the dangling bonds of the superlattice to prevent reoxidation of the material, surface contamination and to minimize band bending. The most common chemical passivations used for Type-II superlattices are ammonium sulfide and sodium sulfide based solutions [35]. The solution etches the surface of the material, removing the oxide and, chains of sulfur atoms occupy the dangling bonds. The differential resistance of type-II InAs/GaSb photodiodes improves by one order of magnitude after an exposition of one minute to the solution.

The most common passivation scheme used for electronic devices consists in depositing a thin layer of insulator to protect the surface of the semiconductor. The dielectric satisfies the dangling bonds and prevents contaminants from shorting the p-i-n junction. SiO_2 is often chosen for such applications as it is one of the best known insulators. While SiO_2 is a reliable passivation in the MWIR, the situation is more complex at longer wavelengths. Homojunctions passivated with SiO_2 deposited with Plasma Enhanced Chemical Vapor Deposition (PECVD) present differential resistances two orders of magnitude lower than unpassivated samples.

Recently, a double heterostructure with large bandgap p and n contacts was suggested to reduce the surface leakage current [36-38]. The contact regions with a cutoff wavelength in the MWIR are less sensitive to the presence of surface contaminants and fixed charges. The absorbing region is not changed in order to preserve the optical response of the devices. The high bandgap contacts are not inverted by the fixed charges of the dielectric layer. Thus, even if the LWIR absorbing superlattice is inverted, the electron or hole conduction channel is stopped at the interface with the higher bandgap superlattice (Figure 11). The device does not get shorted by the surface conduction channels and the I-V characteristic of the diodes is not affected by the passivation layer.

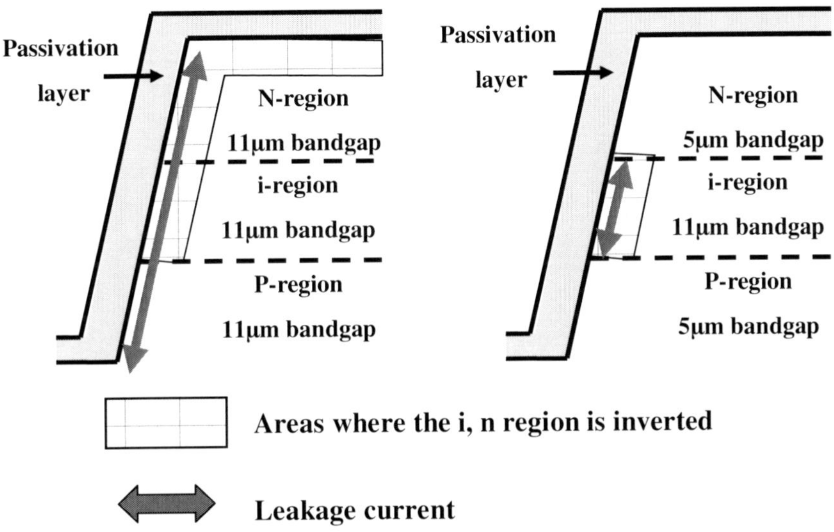

Figure 11. Sketch of the technique used to suppress the surface leakage in a double heterostructure. This figure represents the case when the n contact region is inverted [36].

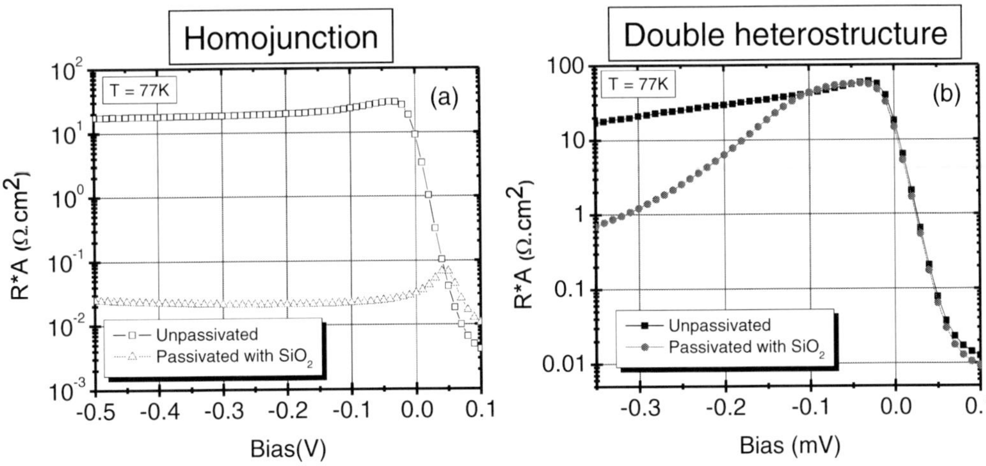

Figure 12. (a) R_0A of an homojunction before and after SiO_2 passivation. (b) R_0A of a double heterostructure before and after SiO_2 passivation [36].

Conformal polyimide physically passivates and protects the underlying semiconductor from the ambient environment while maintaining or even improving the electrical performance of the device measured prior to passivation. This technique demonstrated nearly bulk limited performance in type-II InAs/GaSb superlattice photodiodes [39]. The surface resistivity of polyimide passivated photodiodes is increased at least by one order of magnitude compared to as-processed devices. This process has been demonstrated to be compatible with small mesa dimensions and pitches found in typical FPA fabrication processes (Figure 13). The polyimide passivation was shown to be stable upon exposure to various ambient conditions as well as over time.

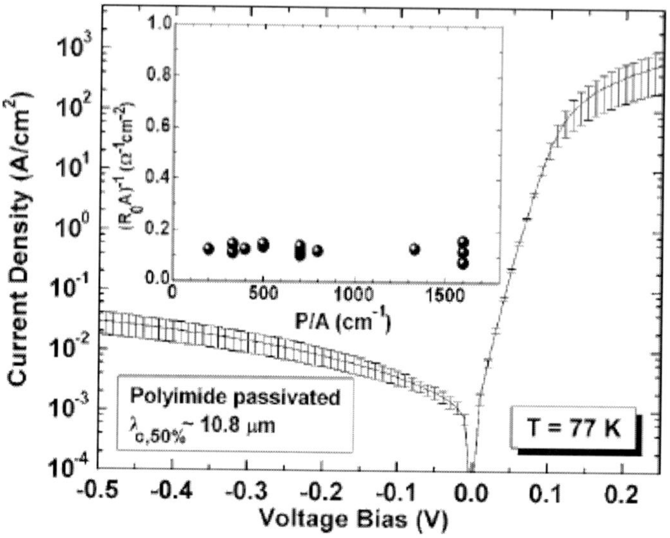

Figure 13. Average (solid line) and standard deviation (vertical bars) of ten representative *I-V* curves from polyimide passivated diodes ranging in size from 25 to 55 μm a side.

d. Optimization of Device Architectures

Device architecture of Type-II superlattice has recently undergone drastic upgrade and modification. No longer is a simple design of a p-i-n homojunction; photodetector devices nowadays are utilized in many configurations to maximize the effect of each device parameters for an overall performance optimization.

Dark current is the most important parameter to be reduced in order to optimize the performance of a photodetectors. Traditionally, dark current in a p-i-n photodetectors is dominated by the diffusion current, which is associated with a differential resistance area product given by:

$$(RoA)_D = \frac{kT}{q^2}\left\{\frac{qD_h n_i^2}{L_h N_d}\tanh(x_n / L_h) + \frac{qD_e n_i^2}{L_e N_a}\tanh(x_p / L_e)\right\}^{-1} \tag{2}$$

where k is Boltzmann constant, T is absolute temperature, q is electron charge, $D_{e,h}$, $L_{e,h}$ are diffusivity and diffusion length of electron, hole respectively. $N_{a,d}$ are acceptor and donor concentration, n_i is intrinsic concentration.

It has been shown that reducing the layer thicknesses of the devices (x_n, x_p) can reduce the diffusion current as the number of minority concentration at inside the devices is reduced [40]. However, a negative effect of shortening the dimension is the reduction of the quantum efficiency, which impairs with the electrical improvement. Another way to suppress the dark current is to increase the doping concentration. Devices with active region intentionally doped p-type have shown one order of magnitude of improvement while preserving the optical efficiency [41].

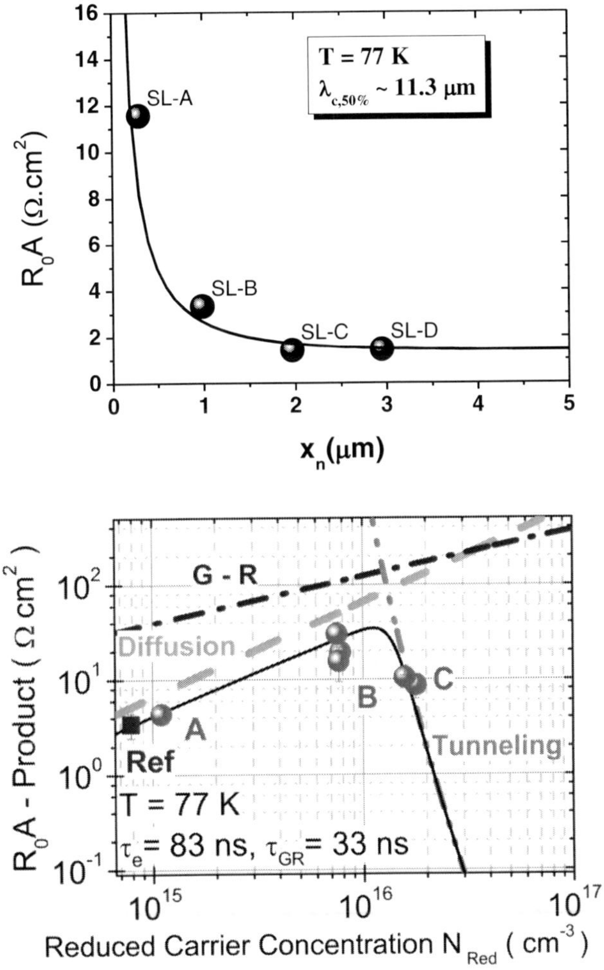

Figure 14. Illustration of methods for reducing diffusion current: a) by reducing the device thicknesses and b) by increasing the active region doping concentration [42].

However, reducing the diffusion current by increasing the doping concentration leads to higher built-in electric field and shorter depletion region. As a result, the devices become dominated by the tunneling mechanism. In order to suppress the tunneling component, the M-structure superlattice was utilized as a potential barrier at the junction, spatially separating the carriers and spreading out the electric field distribution for a smaller tunneling probability. The p-π-M-n device architecture (Figure 15) was experimentally realized and lead to an electrical enhancement by one order of magnitude when the M-structure barrier was thicker than 500 nm [43-45].

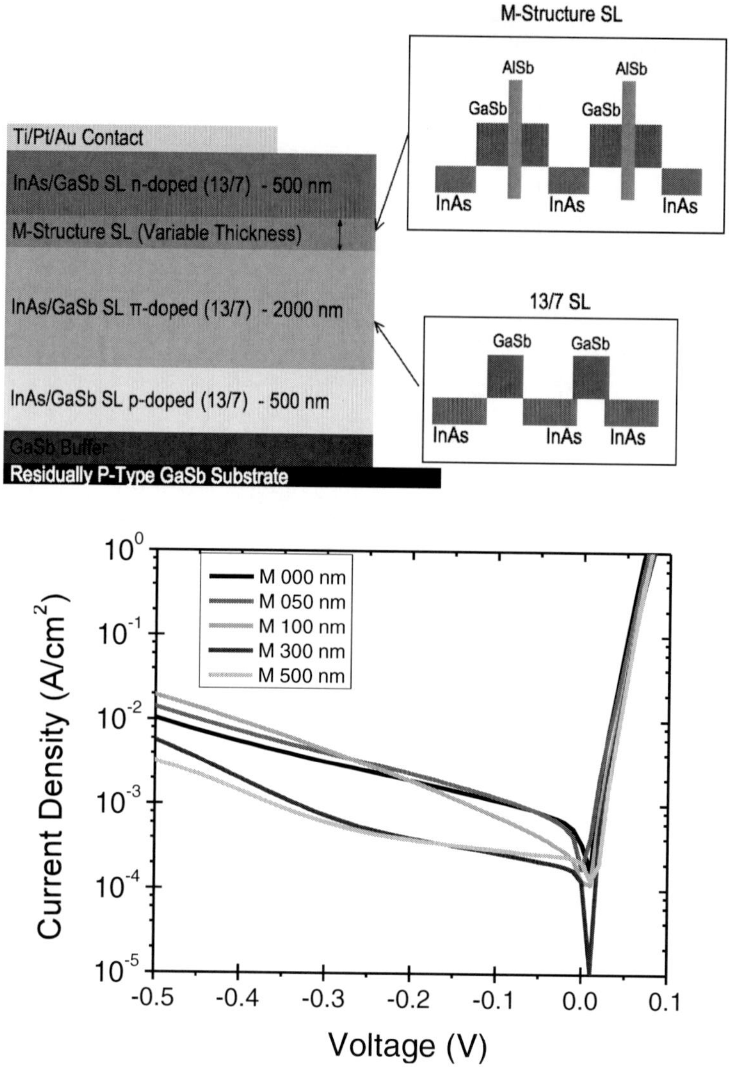

Figure 15. a) schematic diagram of the p-π-M-n device architecture for the suppression of tunneling current, b) I-V curves of devices with different M-barrier thicknesses, with thickness above 500nm, the dark current is reduced by one order of magnitude [45].

Recently, a novel device architecture called p-M-p proposed to combine the advantages of photoconductive and photovoltaic detectors. By using the unipolar mode of minority carrier devices, the device was expected to have low generation-recombination current, low tunneling current, and high quantum efficiency at very small applied bias. The proof of concept of the architecture has been demonstrated at the VLWIR regime, and exhibit comparable performance to the best VLWIR Type-II superlattice photodiodes. The architecture is also beneficial for two color device operating in bias selectable mode [46-47].

In parallel with electrical optimization, optical performance of Type-II superlattice photodiodes were also improved by assuring high quality material with long diffusion length and long diffusion lifetime. Devices with an active region thickness more than 6μm exhibit a quantum efficiency more than 50% in front side (Figure 17), single path configuration, and

75% in backside multiple path configuration [42, 48]. These values set Type-II superlattice apart form any other low dimensional quantum systems (such as quantum dots, quantum wells) which were known to have low quantum efficiency, and attain the level of the state-of-the-art HgCdTe. Figure 18 summarizes the development progress of LWIR Type-II superlattice photodiodes over the past few years. Similar progress was obtained at different wavelengths, which is shown in Figure 19.

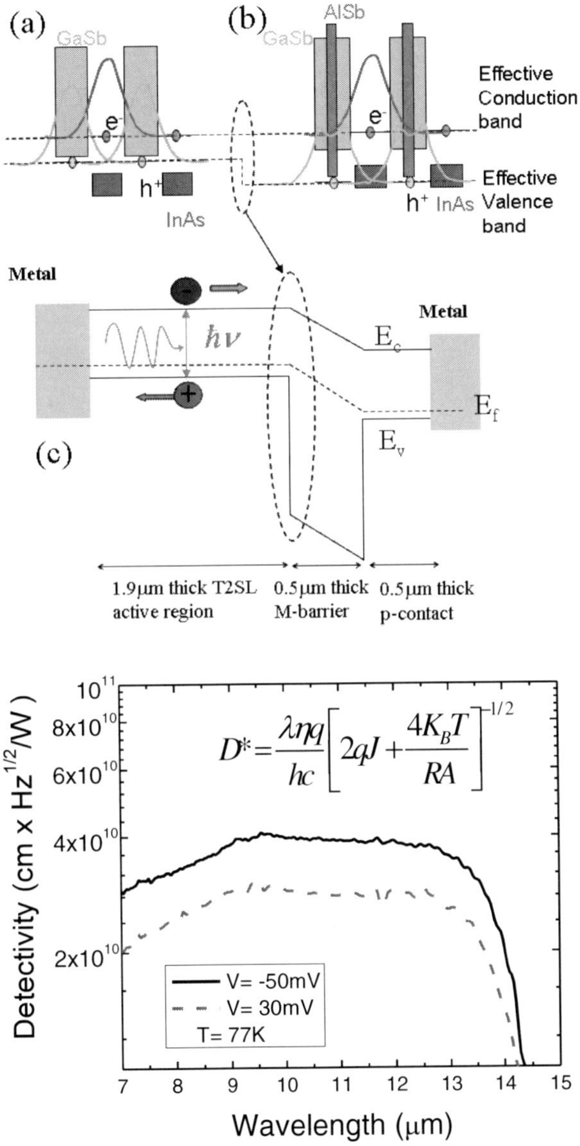

Figure 16. a) schematic diagram of the p-M-p device architecture b) Device detectivity at both negative and positive bias illustrates the capability for bias selectable two color detection [46].

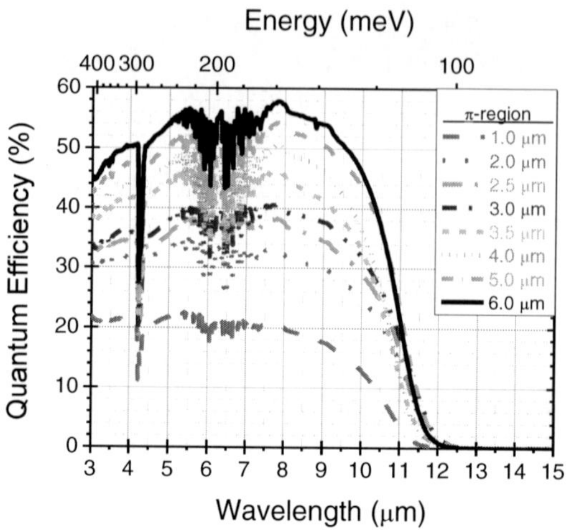

Figure 17. Demonstration of high quantum efficiency in LWIR Type-II superlattice photodiodes with thick active region [42].

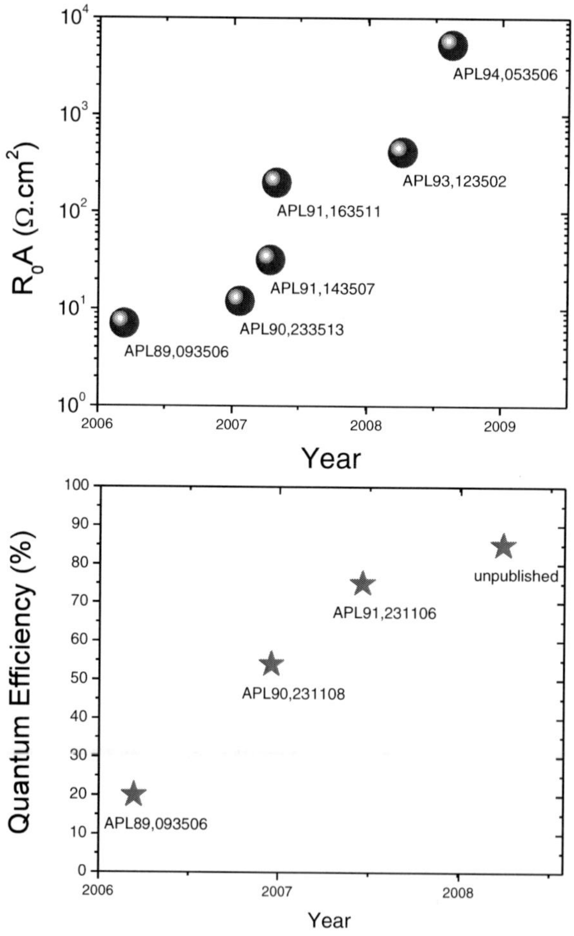

Figure 18. Development progress of LWIR Type-II superlattice over the past few years: a) electrical performance, b) Optical performance.

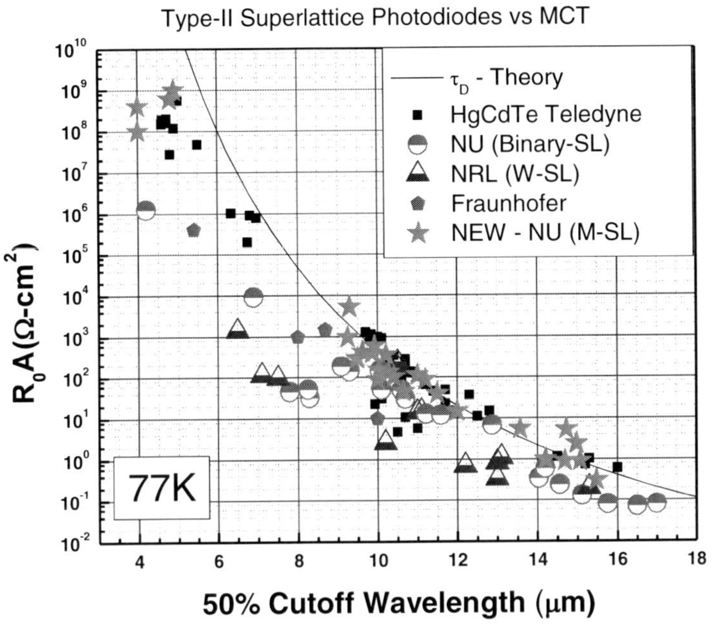

Figure 19. Performance of Type-II superlattice photodiodes vs. HgCdTe technology.

3. State-of-the-art FPAs Based on Superlattices

Together with the development of material and structural design, the fabrication of Type-II superlattice focal plane arrays has been significantly improved. All processing steps have been refined, optimized, resulting in the demonstration of high performance focal plane arrays.

High Operating Temperature MWIR FPAs [49-50]

Based on the development of material and device architecture, an M-barrier heterojunction was designed for high operating temperature (HOT) mid-infrared detection. The device exhibited a 50% cut-off wavelength of 4.9 μm at 150K and a quantum efficiency above 50% in front side illumination. A MWIR camera was fabricated and hybridized to an Indigo ISC9705 Read Out Integrated Circuit and characterized at different temperatures.

Noise equivalent temperature dependence (NEdT), a performance metric dictating the minimum temperature difference that a thermal camera can differentiate, was measured under f/2.3 optics (~16° field of view) and 300K background. Below 120K, the NEdT stayed constant at ~10mK for integration time of 10.02 ms, and can be reduced to 8 mK when longer integration time was set. This temperature insensitivity indicates that the camera was limited by either photon noise or system noise rather than shot noise since the later exhibited very strong dependency on temperature. At higher temperature, the dark current increased and became limiting factor to FPA operation. Higher operating temperature led to larger NEdT (Figure 20). Human imaging is recognizable when FPA is operated at 170K (Figure 21) while uncooled camera can image hot objects such as a soldering iron.

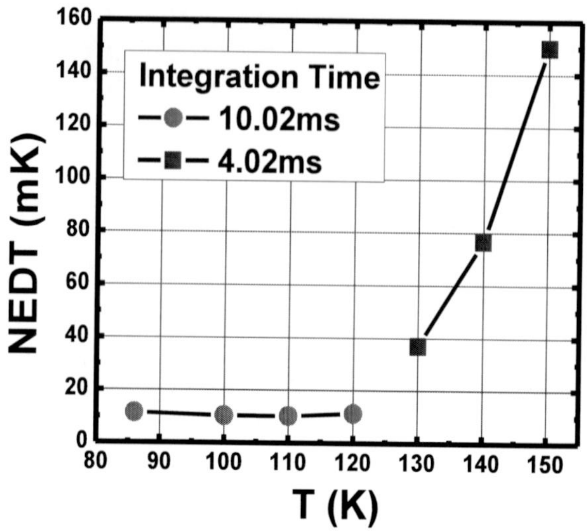

Figure 20. NEDT measured from 86K to 150K. An f/2.3 optics is used, the integration time is reduced above 120K to avoid capacitor saturation [50].

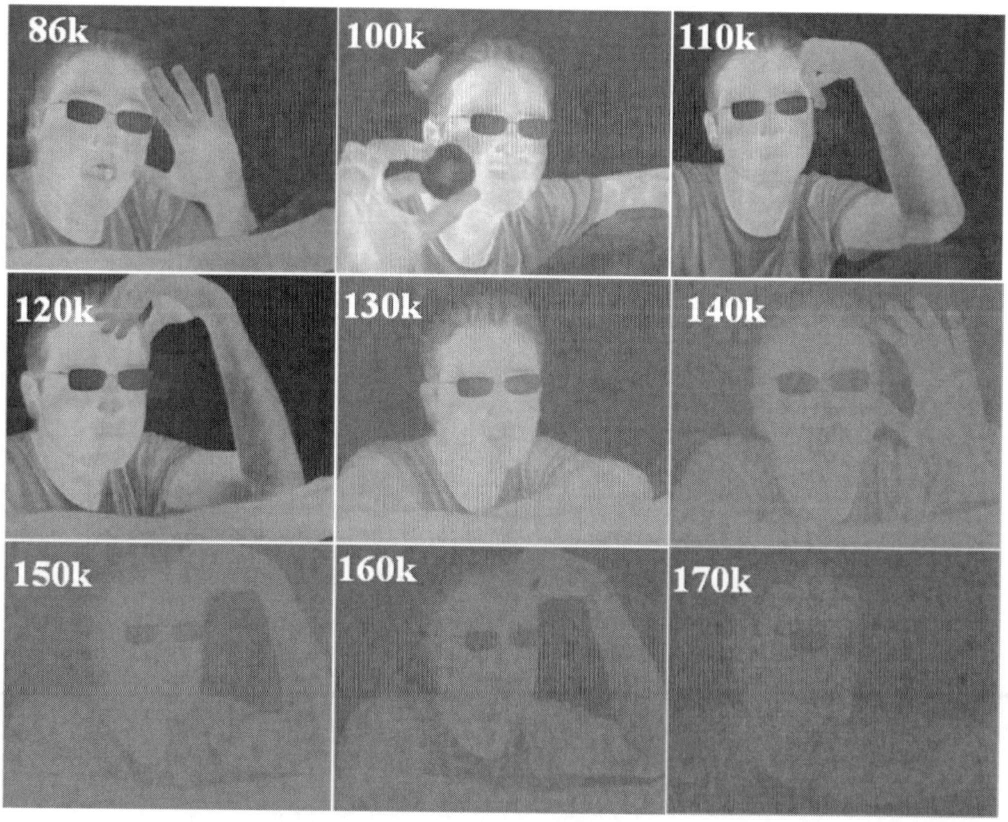

Figure 21. Pictures taken by HOT FPA at different temperatures [50].

LWIR FPAs [51-52]

Recent progress in growth techniques, structure design and processing has lifted the performances of LWIR superlattice photo-detectors to a level where it can compete with existing technologies such as MCT and quantum well infrared photo-detectors. The introduction of an M-structure barrier at the interface of the main p-n junction reduced the dark current by one order of magnitude [43]. The additional progress made in the FPA fabrication process led to the fabrication of the first BLIP FPA based on this material platform.

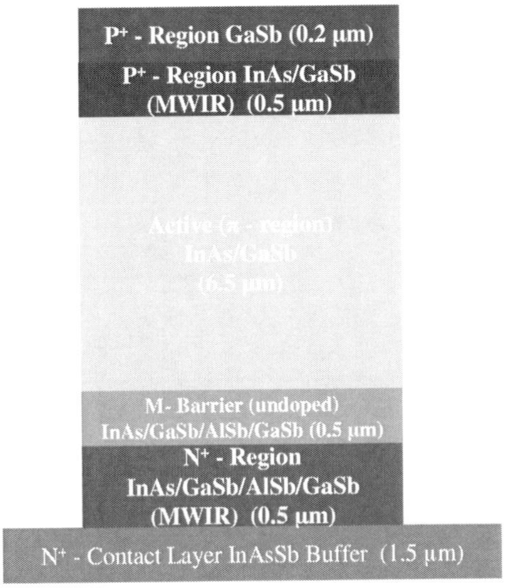

Figure 22. Structure of type-II superlattice devices.

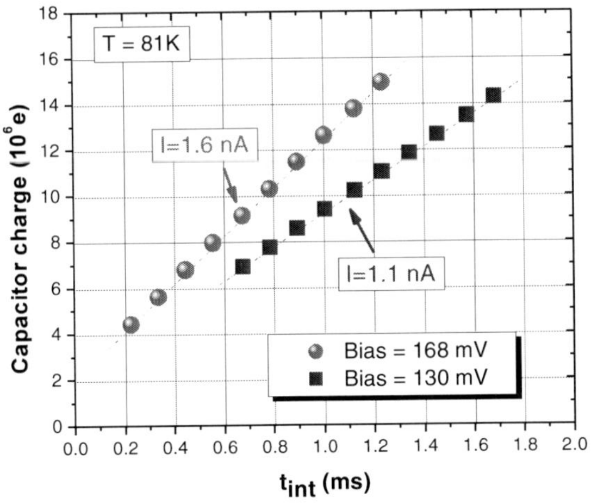

Figure 23. Variation of the capacitor charge with the integration time at 81K, under dark conditions.

The device structure presented in Figure 12 include different characteristics, each of them necessary to maximize the performance of the superlattice FPA. An imager was characterized at 80K with a 0.11 ms integration time. The median dark current measured with an 81K cold shield was between 1.1 and 1.6 nA, depending on the bias settings (Figure 13).

The quantum efficiency at 8 μm was 74%, with an operability of 98% (Figure 24). The uniformity of 3.6% was limited by the non-uniformity of the blackbody illumination. The uniformity improved to 0.19% after two point non-uniformity correction, similar to QWIP based FPAs. For integration times shorter than 0.5 ms, the noise of the imager was limited by the ROIC. The shot noise became dominant for longer integration times. The FPA approached BLIP at 9 10^{15} $ph.cm^{-2}.s^{-2}$.

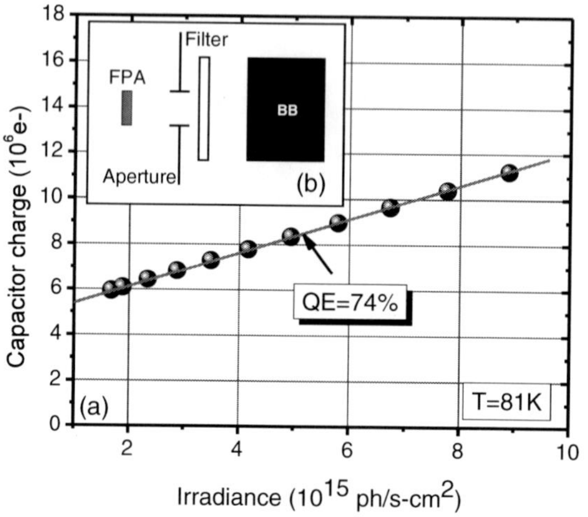

Figure 24. Variation of the capacitor charge versus the photon irradiance at 8 μm [53].

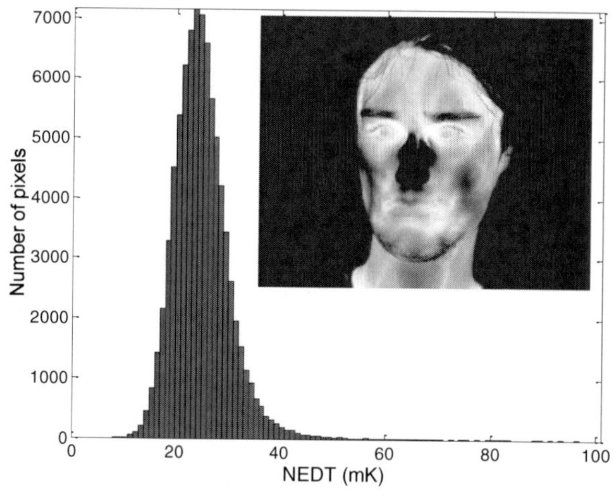

Figure 25. Histogram of the NEDT at 81K with a LWIR lens (F#=2.0). The inset is picture of a student taken with the type-II superlattice FPA [36].

The NEDT was measured as low as 23 mK, corresponding to an operability of 98% (Figure 25). In this configuration, the dark current still corresponds to 5-10% of the charges accumulated on the capacitors. This means that the dark current needs to be further reduced in order to use the full capacitor charge to store the photo excited carriers. This would also improve the performance of the imager at lower backgrounds where longer integration times are necessary.

Noise measurement has been performed on LWIR FPA. The noise of the imager was dominated by photon noise for photon fluxes higher than 1.8×10^{15} ph s−1 cm−2. At lower irradiance, the imager was limited by the shot noise generated by the dark current or the noise of the testing system. The superlattice detector did not create 1/f noise for frequencies above 4 mHz. As a result, the focal plane array did not require frequent calibrations [54]. The spatial noise of the array was dominated by the nonuniformity of the illumination through the circular aperture. A standard two-point nonuniformity correction improved the inhomogeneity equivalent temperature difference to 16 mK. The correctability just after calibration was 0.6. The noise behavior of the detectors was properly described by a model based on thermal, shot, read out integrated circuit, and photon noises. The long-term stability time was superior to 25 hours [55].

Large Format 1024x1024 LWIR FPA [56]

Advances in material development have enabled realization of very high quality material with excellent uniformity over large wafer area [3]. Technological challenges of focal plane arrays have been overcome, and resulted in demonstration of the first high performance, large format focal plane array based on Type-II superlattice. The 11 μm cut-off FPA showed excellent imaging with a 300 K background at an integration time of 0.13 ms using f/2 optics (Figure 26). A noise equivalent temperature difference (NEDT) of 23.6 mK, a dark current (J_{dark}) of 3.3×10^{-4} A/cm^2, a differential resistance area product at zero bias (R_0A) of 166 Ωcm^2 at 81 K and 22.5 mK, 5.1×10^{-5} A/cm^2, 1286 Ωcm^2 respectively at 68 K were observed. A quantum efficiency of 78% was obtained [56].

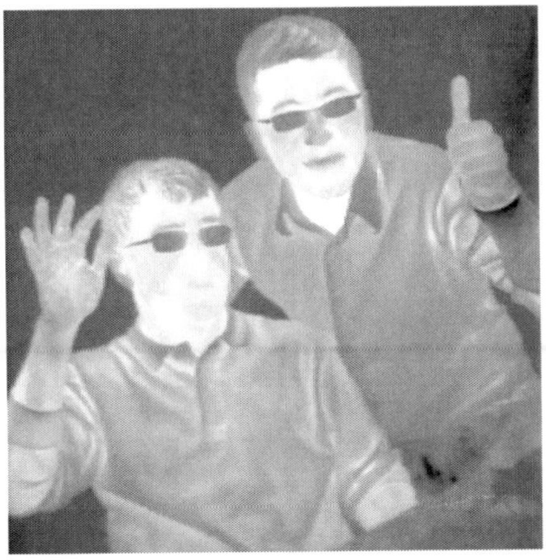

Figure 26. LWIR Imaging achieved at 81K with a high performance, high operability 1024x1024 FPA based on Type-II superlattice.

Focal Planes Arrays on GaAs Substrates [32, 57]

Realization of large format focal plane arrays raises a need for larger size substrates for mass production and cost reduction. While a conventional 2" GaSb substrate can only allocate one 1024x1024 FPA die, a 3" GaSb substrate can allow for a production batch of 4 FPAs. In order to even further reduce the fabrication cost, large size GaAs substrates are desirable since these wafers are commercially available with size upto 6" in diameter and at low cost. However, a challenge for realization of Type-II superlattice FPAs on GaAs substrates is the lattice mismatch of 7% between the superlattice and the substrate, which causes severe degradation of the material quality.

Efforts have been paid to overcome this challenge and achieved high quality Type-II superlattice on GaAs substrates. The materials were processed into 32-x256 focal plane arrays for both MWIR and LWIR regimes. Excellent imaging quality was demonstrated with cameras cooled to 67K (Figure 27).

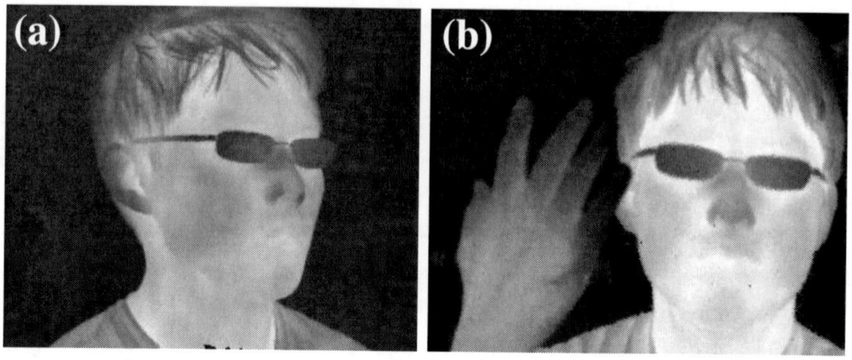

Figure 27. (a) MWIR and (b) LWIR imaging at 67K of Type-II superlattice focal plane arrays on GaAs substrates [32].

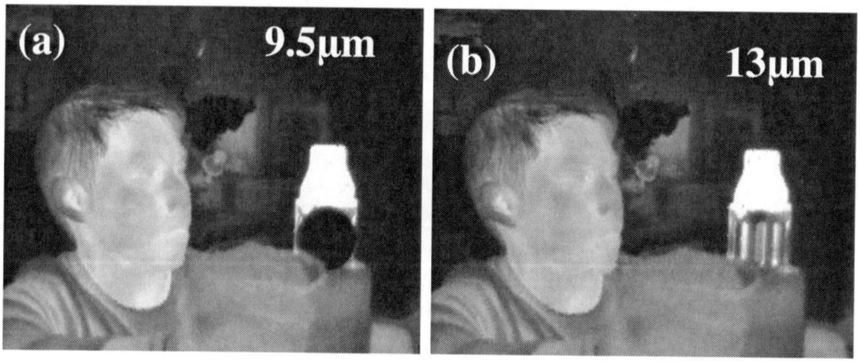

Figure 28. Dual band imaging from a T2SL FPA with 9.5μm (a) and 13μm (b) 100% cutoff wavelengths is shown at 77K. A narrowband filter centered at 11.3μm is placed in front of a heat gun [32].

Dual Band LWIR/LWIR Two Color FPA [32, 58]

High performance single color FPA have been already demonstrated for the LWIR regime. However, despite the high performance, single band FPAs are sometimes unable to discriminate a target from its background when they present similar infrared radiation at a given wavelength. This may occur even if the temperature of the object is very different from the background temperature. In such a case, two color FPAs offer an additional wavelength that allows easy identification of the target.

Development of material growth and structural designs have enabled the realization of dual band detection in coregistered pixels. Fabrication processes have matured, resulting in demonstration of a high performance 256x320 dualband T2SL focal plane array in the LWIR regime. Shown in Figure 28 a is an image captured from the shorter cut-off band (or blue channel with cut-off wavelength of 9.5μm) while image captured in Figure 28 b is from the longer cut-off band (or red channel with cut-off wavelength ~13μm). Due to the capability to detect longer wavelength, the red channel is capable to capture photons from a heat gun passing through a narrow band filter centered 11.3 μm placed in front while such photons are not detectable in the blue channel with shorter cut-off wavelength. At 77K, both channels exhibit excellent performance with the NEdT ~25 mK.

SUMMARY

We have reviewed the progress and performance of the state-of-the-art Type-II superlattice infrared photon detectors. Important timeline of the development is shown in Figure 29. Clearly Type-II superlattices have made tremendous achievement and have become a viable technology for the third generation infrared focal plane arrays. Review on progress of other competitive technology is out of the scope of this chapter; however, we still want to bring out in the appendix a general outlook for the future of infrared technology.

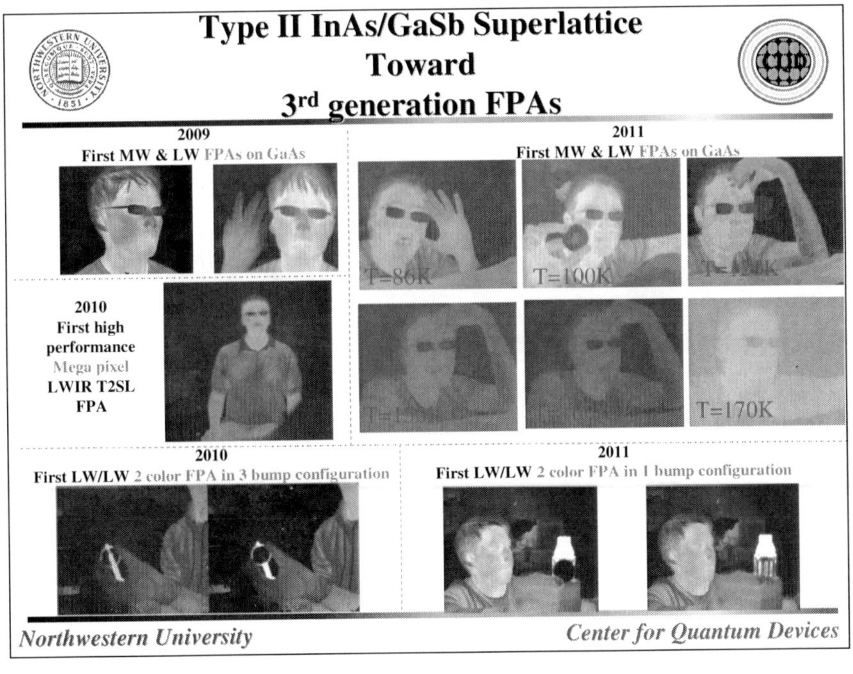

Figure 29. Historic development of Type-II superlattice focal plane arrays for the third generation infrared imagers.

ACKNOWLEDGEMENTS

The authors would like to thank past and present members of the Center for Quantum Devices who have been contributing to the development of Type II superlattice technology at the center. Funding and supports from various government agencies are also acknowledged.

APPENDIX: COMPARISON BETWEEN CURRENT INFRARED DETECTION TECHNOLOGIES

	Bolometer	InSb	HgCdTe	Type II SLs	QWIP	QDIP/QDWIP
Physics	- Low speed - Multicolor detection impossible	- Fixed cut-off or sensitive to the lattice parameters + Simple material, excellent performance at the MWIR	-Small effective mass. - Unstable II-VI materials - Energy gap sensitive to the molar fraction + Can cover the whole Infrared regime.	+ Flexible capability of bandstructure engineering and heterostructure designs. + Large effective mass + Robust III-V semiconductor + High uniformity	-High dark current due to majority carrier operation, requiring extensive cooling. -narrow band absorption. + No spectral cross-talk in multicolor detection	+ Interesting physics and promising capabilities + No spectral cross-talk in multicolor detection
Technological challenges	- Challenges for micro-machining small pixels and large format.	- Low melting point, risk of thermal degradation in high temperature applications.	-Low yield and poor uniformity. - Small size substrates and high fabrication cost + matured technology, high performance	-Young technology that still needs a lot of investment. - Surface effects at device sidewalls are still unknown.	-Requires grating for normal incident signal + Simple fabrication process	-Experimental realization is far behind the theory. - Fundamental challenges on the control of material quality, uniformity and reproducibility. + Simple fabrication process
Choice of future???	Low cost and uncooled applications	MWIR imaging	Will continue to be the state-of-the-art for ~3-5 years	Fast growing and promising technology. Will soon be the choice of the future, replacing the HgCdTe.	Low cost applications	If the gap between theory and experiments can be diminished, QDWIP will the choice for high operation temperature, polarization sensitive and multicolor applications. But it remains still an open question.

REFERENCES

[1] M. Razeghi, US Patent 6864552, Focal Plane Arrays in Type-II Superlattices, 2005.

[2] Y. Wei, and M. Razeghi, "Modeling of type-II InAs/GaSb superlattices using an empirical tight-binding method and interface engineering," *Physical Review B (Condensed Matter and Materials Physics)*, 69(8), 085316-7 (2004).

[3] B. M. Nguyen, G. Chen, M.-A. Hoang, and M. Razeghi, "Growth and Characterization of Long-Wavelength Infrared Type-II Superlattice Photodiodes on a 3-in GaSb Wafer," Quantum Electronics, *IEEE Journal of,* 47(5), 686-690 (2011).

[4] B.-M. Nguyen, G. Chen, M.-A. Hoang, and M. Razeghi, "Growth and characterization of long wavelength infrared type II superlattice photodiodes on a 3" GaSb wafer," *Proceeding of.* 7945, 79451O (2011).

[5] S. Tobin, M. Hutchins, and P. Norton, "Composition and thickness control of thin LPE HgCdTe layers using x-ray diffraction," *Journal of Electronic Materials,* 29(6), 781-791 (2000).

[6] R. Tsu, L. L. Chang, G. A. Sai-Halasz, and L. Esaki, "Effects of Quantum States on the Photocurrent in a "Superlattice"," *Physical Review Letters,* 34(24), 1509 (1975).

[7] G. A. Sai-Halasz, R. Tsu, and L. Esaki, "A new semiconductor superlattice," *Applied Physics Letters,* 30(12), 651-653 (1977).

[8] D. L. Smith, and C. Mailhiot, "Proposal for strained type II superlattice infrared detectors," *Journal of Applied Physics,* 62(6), 2545-2548 (1987).

[9] J. L. Johnson, L. A. Samoska, A. C. Gossard, J. L. Merz, M. D. Jack, G. R. Chapman, B. A. Baumgratz, K. Kosai, and S. M. Johnson, "Electrical and optical properties of infrared photodiodes using the InAs/Ga [sub 1 - x]In [sub x]Sb superlattice in heterojunctions with GaSb," *Journal of Applied Physics,* 80(2), 1116-1127 (1996).

[10] H. Mohseni, E. Michel, J. Sandoen, M. Razeghi, W. Mitchel, and G. Brown, "Growth and characterization of InAs/GaSb photoconductors for long wavelength infrared range," *Applied Physics Letters,* 71(10), 1403-1405 (1997).

[11] H. Mohseni, J. Wojkowski, M. Razeghi, G. Brown, and W. Mitchel, "Uncooled InAs-GaSb type-II infrared detectors grown on GaAs substrates for the 8-12-μm atmospheric window," Quantum Electronics, *IEEE Journal of,* 35(7), 1041-1044 (1999).

[12] F. Fuchs, U. Weimer, W. Pletschen, J. Schmitz, E. Ahlswede, M. Walther, J. Wagner, and P. Koidl, "High performance InAs/Ga [sub 1-x]In [sub x]Sb superlattice infrared photodiodes," *Applied Physics Letters,* 71(22), 3251-3253 (1997).

[13] H. Mohseni, and M. Razeghi, "Long-wavelength type-II photodiodes operating at room temperature," Photonics Technology Letters, *IEEE,* 13(5), 517-519 (2001).

[14] Y. Wei, A. Hood, H. Yau, A. Gin, M. Razeghi, M. Z. Tidrow, and V. Nathan, "Uncooled operation of type-II InAs/GaSb superlattice photodiodes in the midwavelength infrared range," *Applied Physics Letters,* 86(23), 233106-3 (2005).

[15] Y. Wei, J. Bae, A. Gin, A. Hood, M. Razeghi, G. J. Brown, and M. Tidrow, "High quality type II InAs/GaSb superlattices with cutoff wavelength ~ 3.7 mu m using interface engineering," *Journal of Applied Physics,* 94(7), 4720-4722 (2003).

[16] Y. Wei, A. Hood, H. Yau, V. Yazdanpanah, M. Razeghi, M. Z. Tidrow, and V. Nathan, "High-performance type-II InAs/GaSb superlattice photodiodes with cutoff wavelength around 7 mu m," *Applied Physics Letters,* 86(9), 091109-3 (2005).

[17] Y. Wei, A. Hood, A. Gin, V. Yazdanpanah, M. Razeghi, and M. Tidrow, "High performance LWIR type II InAs/GaSb superlattice photodetectors and infrared focal plane arrays," *Proceeding of Quantum Sensing and Nanophotonic Devices II.* 5732, 309-315 (2005).

[18] Y. Wei, A. Gin, M. Razeghi, and G. J. Brown, "Type II InAs/GaSb superlattice photovoltaic detectors with cutoff wavelength approaching 32 mu m," *Applied Physics Letters,* 81(19), 3675-3677 (2002).

[19] H. Mohseni, V. I. Litvinov, and M. Razeghi, "Interface-induced suppression of the Auger recombination in type-II InAs/GaSb superlattices," *Physical Review B (Condensed Matter and Materials Physics),* 58(23), 15378-15380 (1998).

[20] M. Razeghi, Y. Wei, J. Bae, A. Gin, A. Hood, J. Jiang, and J. Nah, "Type II InAs/GaSb superlattices for high-performance photodiodes and FPAs," *Proceeding of.* 5246, 501-511 (2003).

[21] W. A. Cabanski, K. Eberhardt, W. Rode, J. C. Wendler, J. Ziegler, J. Fleissner, F. Fuchs, R. H. Rehm, J. Schmitz, H. Schneider, and M. Walther, "Third-generation focal plane array IR detection modules and applications," *Proceeding of Infrared Technology and Applications XXX.* 5406, 184-192 (2004).

[22] B. M. Nguyen, M. Razeghi, V. Nathan, and G. J. Brown, "Type-II M structure photodiodes: an alternative material design for mid-wave to long wavelength infrared regimes," *Proceeding of Quantum Sensing and Nanophotonic Devices IV.* 6479, 64790S-10 (2007).

[23] B.-M. Nguyen, D. Hoffman, P.-Y. Delaunay, E. K.-W. Huang, M. Razeghi, and J. Pellegrino, "Band edge tunability of M-structure for heterojunction design in Sb based type II superlattice photodiodes," *Applied Physics Letters,* 93(16), 163502-3 (2008).

[24] M. Razeghi, B.-M. Nguyen, D. Hoffman, P.-Y. Delaunay, E. K.-w. Huang, M. Tidrow, and V. Nathan, "Development of material quality and structural design for high performance Type II InAs/GaSb superlattice photodiodes and focal plane arrays," *Proceeding of Infrared Spaceborne Remote Sensing and Instrumentation XVI.* 7082, 708204-10 (2008).

[25] M. Razeghi, Y. Wei, A. Hood, D. Hoffman, B. M. Nguyen, P. Y. Delaunay, E. Michel, and R. McClintock, "Type-II superlattice photodetectors for MWIR to VLWIR focal plane arrays," *Proceeding of Infrared Technology and Applications XXXII.* 6206, 62060N-7 (2006).

[26] B.-M. Nguyen, D. Hoffman, P.-Y. Delaunay, M. Razeghi, and V. Nathan, "Polarity inversion of type II InAs/GaSb superlattice photodiodes," *Applied Physics Letters,* 91(10), 103503-3 (2007).

[27] P.-Y. Delaunay, B. M. Nguyen, D. Hofman, and M. Razeghi, "Substrate removal for high quantum efficiency back side illuminated type-II InAs/GaSb photodetectors," *Applied Physics Letters,* 91(23), 231106-3 (2007).

[28] P.-Y. Delaunay, B.-M. Nguyen, D. Hoffman, A. Hood, E. K.-W. Huang, M. Razeghi, and M. Z. Tidrow, "High quantum efficiency two color type-II InAs/GaSb n-i-p-p-i-n photodiodes," *Applied Physics Letters,* 92(11), 111112-3 (2008).

[29] B.-M. Nguyen, D. Hoffman, E. K.-w. Huang, S. Bogdanov, P.-Y. Delaunay, M. Razeghi, and M. Z. Tidrow, "Demonstration of midinfrared type-II InAs/GaSb superlattice photodiodes grown on GaAs substrate," *Applied Physics Letters,* 94(22), 223506-3 (2009).

[30] B.-M. Nguyen, D. Hoffman, E. K. Huang, P.-Y. Delaunay, and M. Razeghi, "High performance antimony-based type-II superlattice photodiodes on GaAs substrate," *Proceeding of Infrared Technology and Applications XXXV.* 7298, 72981T-13 (2009).

[31] S. A. Pour, B. M. Nguyen, S. Bogdanov, E. K. Huang, and M. Razeghi, "Demonstration of high performance long wavelength infrared type II InAs/GaSb superlattice photodiode grown on GaAs substrate," *Applied Physics Letters,* 95(17), 173505-3 (2009).

[32] E. K. Huang, B.-M. Nguyen, S. Ramezani-Darvish, S. A. Pour, G. Chen, A. Haddadi, and M.-A. Hoang, "Recent advances in high-performance antimonide-based superlattice FPAs," *Proceeding of.* 8012, 80120T (2011).

[33] E. K.-w. Huang, D. Hoffman, B.-M. Nguyen, P.-Y. Delaunay, and M. Razeghi, "Surface leakage reduction in narrow band gap type-II antimonide-based superlattice photodiodes," *Applied Physics Letters,* 94(5), 053506-3 (2009).

[34] E. K. Huang, B. M. Nguyen, D. Hoffman, P. Y. Delaunay, and M. Razeghi, "Inductively coupled plasma etching and processing techniques for type-II InAs/GaSb superlattices infrared detectors toward high fill factor focal plane arrays," *Proceeding of Quantum Sensing and Nanophotonic Devices VI.* 7222, 72220Z-8 (2009).

[35] A. Gin, Y. Wei, A. Hood, A. Bajowala, V. Yazdanpanah, M. Razeghi, and M. Tidrow, "Ammonium sulfide passivation of Type-II InAs/GaSb superlattice photodiodes," *Applied Physics Letters,* 84(12), 2037-2039 (2004).

[36] P.-Y. Delaunay, and M. Razeghi, "High-performance focal plane array based on type-II InAs/GaSb superlattice heterostructures," *Proceeding of Quantum Sensing and Nanophotonic Devices V.* 6900, 69000M-10 (2008).

[37] P.-Y. Delaunay, A. Hood, B. M. Nguyen, D. Hoffman, Y. Wei, and M. Razeghi, "Passivation of type-II InAs/GaSb double heterostructure," *Applied Physics Letters,* 91(9), 091112-3 (2007).

[38] I. Vurgaftman, E. H. Aifer, C. L. Canedy, J. G. Tischler, J. R. Meyer, J. H. Warner, E. M. Jackson, G. Hildebrandt, and G. J. Sullivan, "Graded band gap for dark-current suppression in long-wave infrared W-structured type-II superlattice photodiodes," *Applied Physics Letters,* 89(12), 121114-3 (2006).

[39] A. Hood, P.-Y. Delaunay, D. Hoffman, B.-M. Nguyen, Y. Wei, M. Razeghi, and V. Nathan, "Near bulk-limited R [sub 0]A of long-wavelength infrared type-II InAs/GaSb superlattice photodiodes with polyimide surface passivation," *Applied Physics Letters,* 90(23), 233513-3 (2007).

[40] A. Hood, D. Hoffman, B.-M. Nguyen, P.-Y. Delaunay, E. Michel, and M. Razeghi, "High differential resistance type-II InAs/GaSb superlattice photodiodes for the long-wavelength infrared," *Applied Physics Letters,* 89(9), 093506-3 (2006).

[41] D. Hoffman, B.-M. Nguyen, P.-Y. Delaunay, A. Hood, M. Razeghi, and J. Pellegrino, "Beryllium compensation doping of InAs/GaSb infrared superlattice photodiodes," *Applied Physics Letters,* 91(14), 143507-3 (2007).

[42] M. Razeghi, D. Hoffman, B.-M. Nguyen, P.-Y. Delaunay, E. K. Huang, and M. Z. Tidrow, "Recent advances in LWIR Type-II InAs/GaSb superlattice photodetectors and focal plane arrays at the Center for Quantum Devices," *Proceeding of Infrared Technology and Applications XXXIV.* 6940, 694009-12 (2008).

[43] B.-M. Nguyen, D. Hoffman, P.-Y. Delaunay, and M. Razeghi, "Dark current suppression in type II InAs/GaSb superlattice long wavelength infrared photodiodes with M-structure barrier," *Applied Physics Letters,* 91(16), 163511-3 (2007).

[44] B.-M. Nguyen, D. Hoffman, E. K.-w. Huang, P.-Y. Delaunay, and M. Razeghi, "Background limited long wavelength infrared type-II InAs/GaSb superlattice photodiodes operating at 110 K," *Applied Physics Letters,* 93(12), 123502-3 (2008).

[45] B.-M. Nguyen, D. Hoffman, P.-Y. Delaunay, E. K.-w. Huang, and M. Razeghi, "Very high performance LWIR and VLWIR Type-II InAs/GaSb superlattice photodiodes with M-structure barrier," *Proceeding of Infrared Spaceborne Remote Sensing and Instrumentation XVI.* 7082, 708205-12 (2008).

[46] B.-M. Nguyen, S. A. Pour, S. Bogdanov, and M. Razeghi, "Minority electron unipolar photodetectors based on type II InAs/GaSb/AlSb superlattices for very long wavelength infrared detection," *Proceeding of.* 7608, 760825-9 (2010).

[47] B. M. Nguyen, S. Bogdanov, S. A. Pour, and M. Razeghi, "Minority electron unipolar photodetectors based on type II InAs/GaSb/AlSb superlattices for very long wavelength infrared detection," *Applied Physics Letters,* 95(18), 183502-3 (2009).

[48] B.-M. Nguyen, D. Hoffman, Y. Wei, P.-Y. Delaunay, A. Hood, and M. Razeghi, "Very high quantum efficiency in type-II InAs/GaSb superlattice photodiode with cutoff of 12 mu m," *Applied Physics Letters,* 90(23), 231108-3 (2007).

[49] S. A. Pour, E. K. Huang, G. Chen, A. Haddadi, B. M. Nguyen, and M. Razeghi, "High operating temperature midwave infrared photodiodes and focal plane arrays based on type-II InAs/GaSb superlattices," *Applied Physics Letters,* 98(14), 143501-3 (2011).

[50] M. Razeghi, S. A. Pour, E. Huang, G. Chen, A. Haddadi, and B.-M. Nguyen, "High-operating temperature MWIR photon detectors based on Type II InAs/GaSb superlattice," *Proceeding of.* 8012, 80122Q (2011).

[51] P. Y. Delaunay, N. Binh Minh, D. Hoffman, and M. Razeghi, "High-Performance Focal Plane Array Based on InAs–GaSb Superlattices With a 10-<formula formulatype="inline"><tex>$mu{hbox {m}}$</tex></formula> Cutoff Wavelength," *Quantum Electronics, IEEE Journal of,* 44(5), 462-467 (2008).

[52] P. Y. Delaunay, B.-M. Nguyen, D. Hoffman, E. K. W. Huang, and M. Razeghi, "Background Limited Performance of Long Wavelength Infrared Focal Plane Arrays Fabricated From M-Structure InAs–GaSb Superlattices," Quantum Electronics, *IEEE Journal of,* 45(2), 157-162 (2009).

[53] P.-Y. Delaunay, B. M. Nguyen, D. Hoffman, E. K.-w. Huang, P. Manurkar, S. Bogdanov, and M. Razeghi, "Background limited performance of long wavelength infrared focal plane arrays fabricated from M-structure InAs/GaSb superlattices," *Proceeding of Quantum Sensing and Nanophotonic Devices VI.* 7222, 72220W-10 (2009).

[54] P. Y. Delaunay, and M. Razeghi, "Noise analysis in type-II InAs/GaSb focal plane arrays," *Journal of Applied Physics,* 106(6), (2009).

[55] P. Y. Delaunay, and M. Razeghi, "Spatial Noise and Correctability of Type-II InAs/GaSb Focal Plane Arrays," *Quantum Electronics, IEEE Journal of,* 46(4), 584-588 (2010).

[56] P. Manurkar, S. Ramezani-Darvish, B.-M. Nguyen, M. Razeghi, and J. Hubbs, "High performance long wavelength infrared mega-pixel focal plane array based on type-II superlattices," *Applied Physics Letters,* 97(19), 193505 (2010).

[57] E. K. W. Huang, P. Y. Delaunay, N. Binh-Minh, S. A. Pour, and M. Razeghi, "Photovoltaic MWIR Type-II Superlattice Focal Plane Array on GaAs Substrate," *Quantum Electronics, IEEE Journal of,* 46(12), 1704-1708 (2010).

[58] E. K.-w. Huang, A. Haddadi, G. Chen, B.-M. Nguyen, M.-A. Hoang, R. McClintock, M. Stegall, and M. Razeghi, "Type-II superlattice dual-band LWIR imager with M-barrier and Fabry?Perot resonance," *Opt. Lett.*, 36(13), 2560-2562 (2011).

In: Antimony: Characteristics, Compounds and Applications ISBN 978-1-62100-598-8
Editor: Manijeh Razeghi

Chapter 6

ANTIMONY BASED CHALCOGENIDE THIN FILMS FOR PHOTOVOLTAIC APPLICATIONS

***B. Krishnan*[1,2,*], *G. Alan Castillo*[1], *T. K. Das Roy*[1] *and S. Shaji*[1,2]**

[1]Facultad de Ingeniería Mecánica y Eléctrica, Universidad Autónoma de Nuevo León, San Nicolás de los Garza, Nuevo León, México

[2]CIIDIT- Universidad Autónoma de Nuevo León, Apodaca, Nuevo León, México

ABSTRACT

Antimony based binary and ternary chalcogenides have attained much technological interest from research and development community due to their optoelectronic properties. Antimony sulfide (Sb_2S_3), Silver antimony selenide ($AgSbSe_2$), Copper antimony sulfide ($CuSbS_2$) are some of the antimony based materials having appropriate optoelectronic properties for various applications. Antimony sulphide (Sb_2S_3) thin films have gained much attention during the past two decades due to their special properties such as high refractive index, well defined quantum size effects and their photosensitive and thermoelectric properties. In this chapter, we mainly discuss the preparation and characterization of these materials as thin films describing their physical properties such as crystallinity, morphology, optical absorption and electrical conduction. Doping of Sb_2S_3 by elements such as carbon is included. Further, modification of various properties of this compound using postdeposition treatments such as heating and laser irradiation are discussed. Fabrication of different photovoltaic structures using these materials and their performance are also described. Some results obtained in our laboratory in the preparation and characterization of $AgSbSe_2$ and $CuSbS_2$ are also discussed. Incorporation of these materials in various photovoltaic structures is another highlight of this chapter.

* E-mail address: kbindu_k@yahoo.com; Tel: (office) (+52) 18114920380 (Corresponding author).

INTRODUCTION

Antimony based compounds such as antimony sulfide, antimony selenide, silver antimony selenide and copper antimony sulfide are newly investigated materials for various optoelectronic applications. In this chapter we will discuss preparation of each of these compound thin films in our laboratory and studies on their structural and physical properties. Results of photovoltaic devices fabricated using some of these materials are also included.

1. ANTIMONY SULFIDE THIN FILMS (SB_2S_3)

1.1. Introduction

Antimony sulphide (Sb_2S_3) belongs to V_2-VI_3 compound semiconductors. This compound in thin film form possess special properties such as high refractive index [1], well defined quantum size effects [2] and high photosensitivity [3-5]. For the last few years our group have been focusing on studying this material as it illustrated its potential in various photovoltaic structures [6, 7]. Here we discuss preparation of Sb_2S_3 thin films by chemical bath deposition, their characterization and the modification of their physical properties by thermal and laser annealing.

1.1.1. Chemical Bath Deposition (CBD) of Sb_2S_3 Thin Films

Chemical bath deposition is one of the low cost thin film deposition techniques which involves controlled precipitation of the compound on a substrate from a precursor solution containing the corresponding cations and the anions [8]. Chemical bath deposition of antimony sulphide thin films was developed by various research groups using different types of precursor solutions at low temperatures [1-9].

Here we explain a possible chemical reaction for the formation of Sb_2S_3 thin films from a chemical bath at room temperature [10,11]. The bath composition is as follows: a 100 ml aqueous solution (pH value of 3.5) consisting of 650 mg of $SbCl_3$, 2.5 ml of acetone, 25 ml of 1 M $Na_2S_2O_3$ (pre-cooled). Uniformly reflective, smooth and adherent orange thin films were formed from the bath. Sb_2S_3 thin films of thickness 180 nm and 550 nm were formed after 30 min and 5 h of deposition respectively.

Chemical Reaction

The formation of antimony sulphide in the thiosulphate bath may involve different mechanisms depending on the dissociation of thiosulphate ions [7]:

In acidic solutions:

$$S_2O_3^{2-} + H^+ \rightarrow \mathrm{S} + HSO_3^-$$

$$S_2O_3^{2-} + 2H^+ \rightarrow \mathrm{S} + \mathrm{SO}_2 + H_2O$$

and in alkaline solutions:

$$S_2O_3^{2-} + OH^- \rightarrow HS^- + SO_4^{2-}$$

Since the thiosulphate is a reducing agent, it may act as an electron donor and reduce the S to S^{2-} in both cases. In acidic (pH=3.5) bath, the reaction involving the formation of sulphur and their subsequent reduction into sulphide ions are possible. It has been suggested that thiosulphate forms very strong complex with metals ions such as Sb [10,11]. These metal-thiosulphate complexes undergo decomposition to form corresponding metal-sulphides. First antimony thiosulphate ($Sb_2(S_2O_3)_3$) is formed due to the reaction of $SbCl_3$ and $Na_2S_2O_3$ which in turn hydrolyzed to form Sb_2S_3 as discussed in ref.10. Specifically, the Sb^{3+} and S^{2-} ions produced through the decomposition of intermediate antomony-thiosulphate complex ($Sb_2(S_2O_3)_3$) and the subsequent dissociation of $(S_2O_3)^{2-}$ react on the glass substrates to form Sb_2S_3 as described by chemical reaction formula:

$$Sb_2(S_2O_3)_3 + 3H^+ + 6e^- \rightarrow Sb_2S_3 + 3HSO_3^-.$$

1.1.2. Crystalline Structure of Chemical Bath Deposited Sb_2S_3 Thin Films

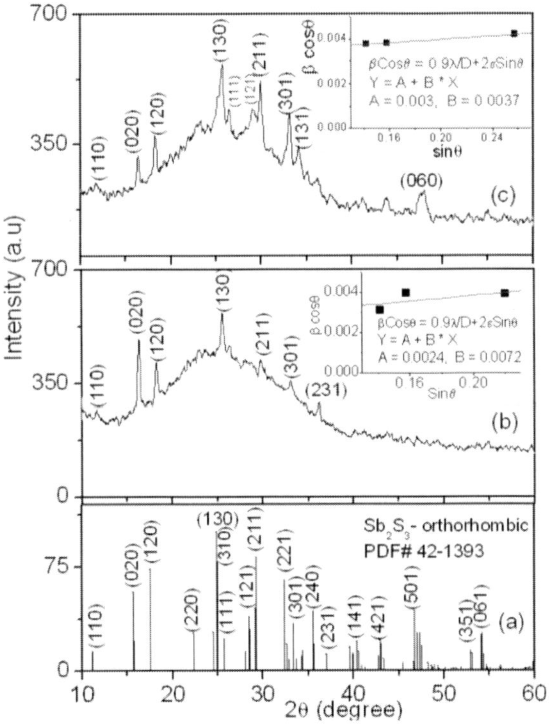

Figure 1. XRD patterns of Sb_2S_3 thin films (a) standard lines for orthorhombic Sb_2S_3 (stibnite) (b) and (c) Sb_2S_3 of 550 nm and 180 nm in thicknesses annealed in nitrogen at 300 °C for 30 min respectively. Inset: Williams-Hall plot corresponding to each pattern.

The chemical bath deposited Sb_2S_3 thin films showed nearly amorphous nature in X-ray diffraction analysis [10] or poor crystallinity in electron diffraction [11]. However, these thin films crystallized into orthorhombic structure (stibnite) after annealing at 300 °C in low vacuum as shown in figure 1. In this figure, the growth of different crystal planes after heating is illustrated using XRD patterns recorded for Sb_2S_3 thin films of different thicknesses. Average grain size of polycrystalline Sb_2S_3 thin films was evaluated using the line broadening analysis of the diffraction patterns in figure 1(b) and (c). The line broadening can occur due to small grain size as well as due to microstrains associated with the thin films. The size and the strain effects can be separated by the Williamson–Hall plot [12] in which $\beta\cos\theta$ vs $\sin\theta$ are plotted, β being the full width at half maximum (FWHM) of the diffraction peak located at 2θ. This plot can be fitted to a straight line. The inverse of the intercept of the line on $\beta\cos\theta$ axis gives the ratio of the crystalline grain diameter (D) to wavelength (1.54056 Å) from which the value of D can be extracted. Moreover, the slope of the line indicates the microstrain (ε). From the corresponding Williamson–Hall plots, the values obtained for the grain size and microstrain are 55 nm and 0.4×10^{-3} for Sb_2S_3 thin films of 550 nm in thickness and those for 180 nm thin films are 46 nm and 0.2×10^{-3} respectively.

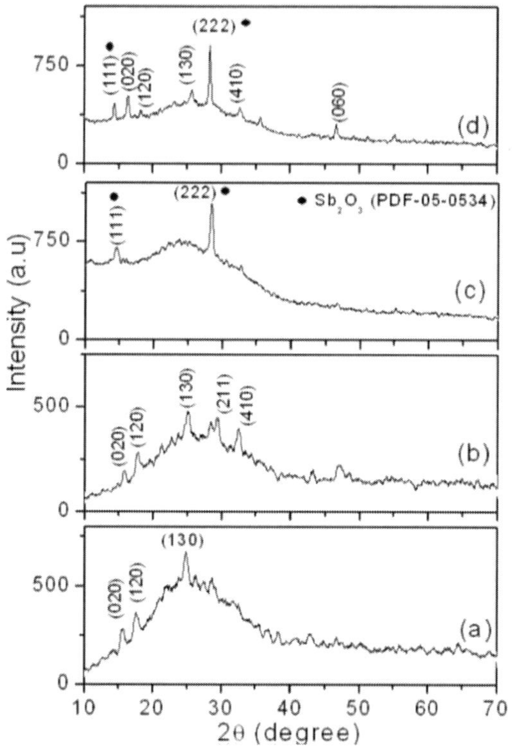

Figure 2. XRD patterns of Sb_2S_3 thin films annealed in air (a) and (b) 550 nm and 180 nm thick films for 15 minutes (c) and (d) 550 nm and 180 nm in thicknesses for 30 minutes.

After air annealing at 300 °C, amorphous Sb_2S_3 was converted to Sb_2O_3 with cubic structure (Senarmontite) or Sb_2S_3- Sb_2O_3 mixed phases depending on the duration of heating as shown in figure 2.

1.1.3. Morphology

The scanning electron micrographs of as-deposited and annealed Sb_2S_3 thin films at different conditions show compact spherical granular morphology structures as given in figure 3 (a) and (b). The grains are homogenous and well distributed indicating the formation of uniform thin films. After air annealing, Sb_2O_3 crystallites are dispersed on the glass substrates as shown in figure 4. The magnified image given in the inset illustrates the octahedron shaped crystallites of Senarmontite phase.

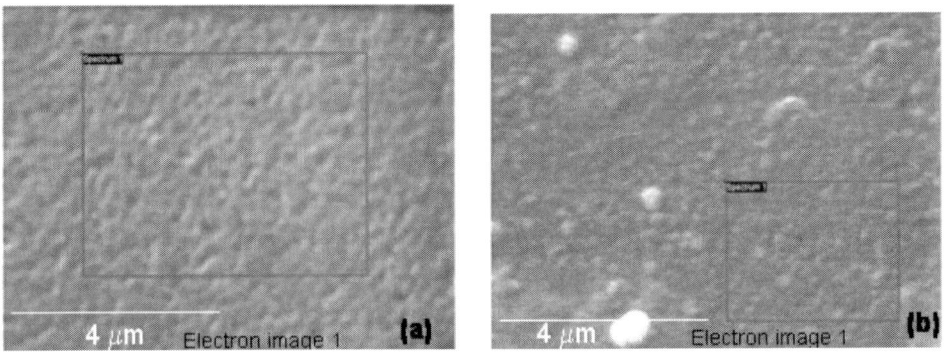

Figure 3. Scanning electron micrograph of as-deposited Sb_2S_3 thin films of thickness (a) 180 nm (b) 550 nm. The marked area corresponds to S/Sb ratio analysis ~ 1.3 and 1.1.

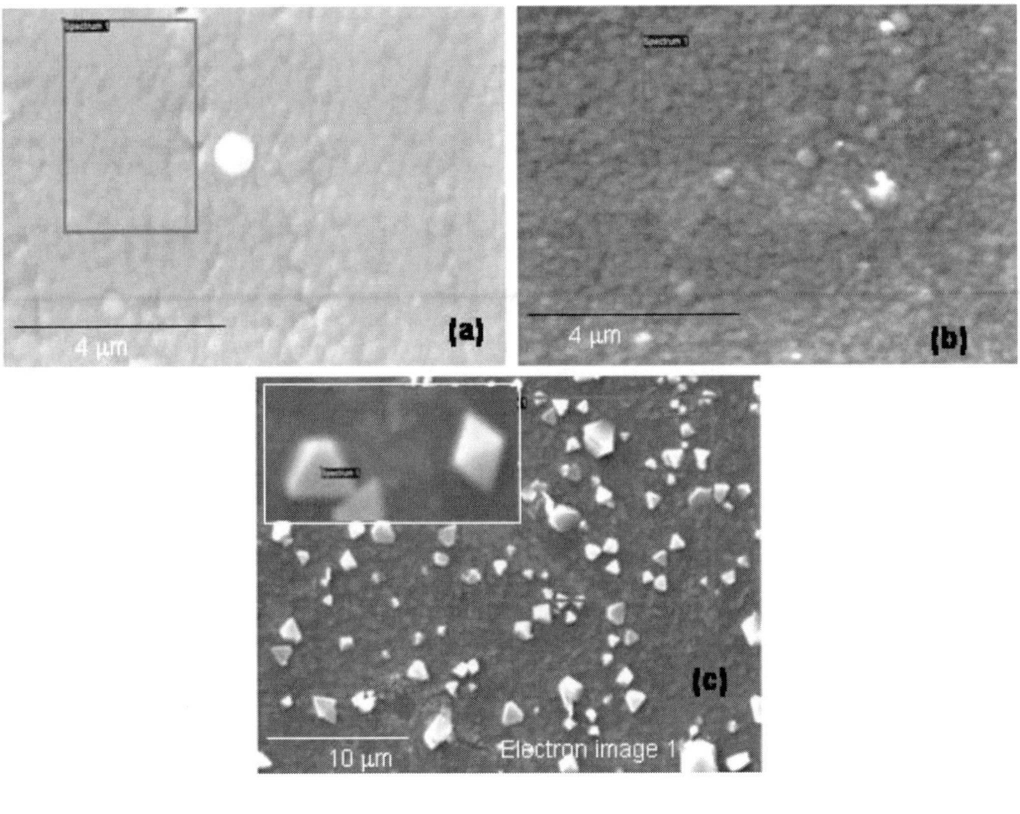

Figure 4. Scanning electron micrograph of Sb_2S_3 thin films of thickness (a) 550 nm annealed in nitrogen for 30 min (b) 550 nm annealed in air for 15 min (c) 180 nm annealed in air for 30 min. S/Sb ratio ~ 1.2 and 1.0.

1.1.4. Optical Properties

In this section, we give the results of evaluating optical absorption coefficient (α) and optical band gap (E_g) values of the thin films. Figure 5(a) shows the optical absorborption spectra of as prepared and annealed Sb_2S_3 thin films. From the spectra, we can see a red shift for the absorption edge for the annealed thin film. The optical band gap of the Sb_2S_3 thin films due to band-band transition is 2.57 eV and 1.73 eV for the as-deposited and the annealed films respectively. This was determined from the Tauc plots of $(\alpha h\nu)^2$ vs hν as shown in figure 6.

The absorption spectrum of Sb_2O_3 shows that Sb_2O_3 is a transparent semiconductor, figure 5(e).

1.1.5. Electrical Properties

Electrical resistivity of the as-prepared Sb_2S_3 thin films is high, ~10^{10} (Ω cm) which limited their use as photovoltaic absorber material. This value was reduced to more than two orders due to annealing, still high value for PV applications. Doping of this material using carbon resulted in a low value, 40 and 92×10^2 Ω cm for Sb_2S_3 thin films of thickness 180 nm and 550 nm respectively, p-type conduction in both cases [13]. Doping was done by heating graphite coated Sb_2S_3 deposited on Glass substrate (Glass/Sb_2S_3/C, carbon layer thickness constant) in vacuum at 300 °C.

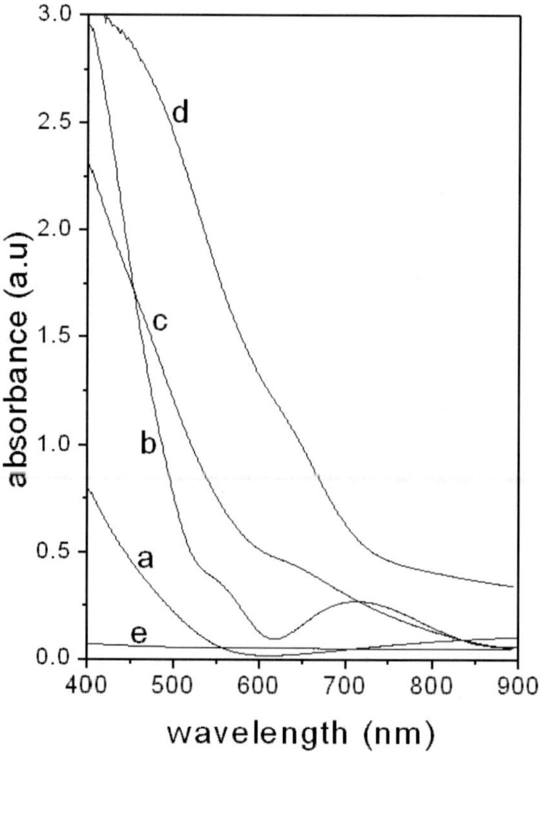

Figure 5. Optical absorption spectra of Sb_2S thin films (a) and (b) as-deposited 180 nm and 550 nm in thickness (c) and (d) annealed at 300 °C in nitrogen (e) Sb_2O_3 formed by heating Sb_2S_3 thin films for 30 min in air.

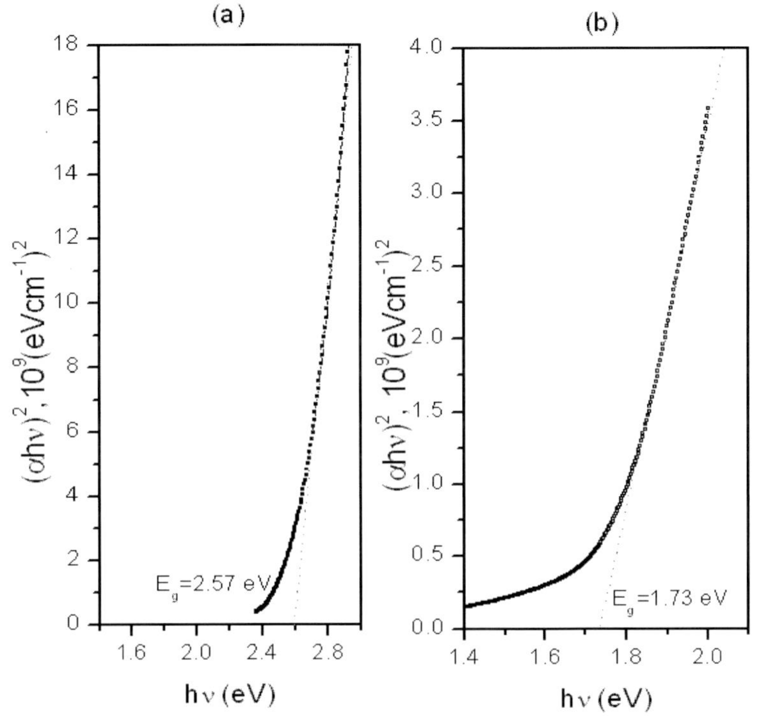

Figure 6. The plot of $(\alpha h\nu)^2$ vs hν for 550 nm Sb_2S_3 thin films (a) as-deposited (b) annealed at 300 °C in nitrogen for 30 min.

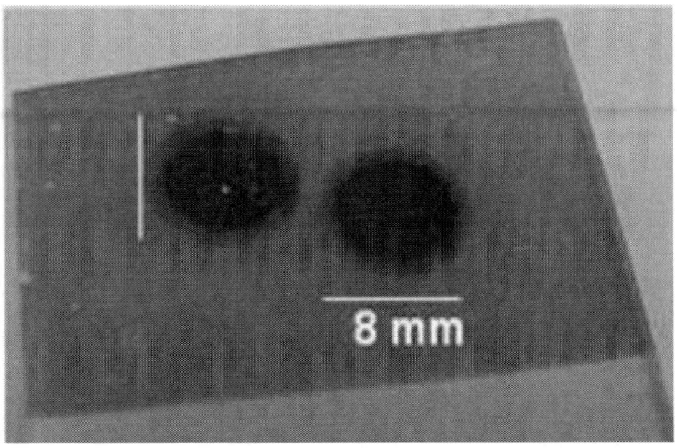

Figure 7. Photograph of Sb_2S_3 thin film after exposure to an expanded laser beam of power density 6 W/cm^2 for 2 min. Orange yellow color is the as-prepared form which was turned to dark brown after irradiation.

1.2. Modification of Physical Properties of Sb_2S_3 by Laser Irradiation [14]

Laser induced material processing or laser annealing has been widely used for crystallization [15], recovering crystal damage and electrically activating the dopants [16-18] in various materials in the wafer or the thin film form. Here we describe laser induced changes in the properties of Sb_2S_3 thin films prepared by chemical bath deposition as specified earlier. Sb_2S_3 thin films of 500 nm in thickness were irradiated by 532 nm continuous laser beam (both focused and 1 cm expanded, 5 W power, 532 nm diode pumped solid state (DPSS) laser) under different conditions at ambient atmosphere. Modifications in structure, morphology, optical and electrical properties are described here. Figure 7 shows the photograph of Sb2S3 thin film irradiated for 2 min using laser beam of power density 6 W/cm2. The dark brown spots correspond to the irradiated region using the expanded beam.

1.2.1. Structure and Morphology Change

As prepared Sb_2S_3 thin films are amorphous to XRD as mentioned earlier. Laser irradiation induced an amorphous to polycrystalline transformation in these thin films. This transformation into stibnite form under different conditions of irradiation is shown in figure 8.

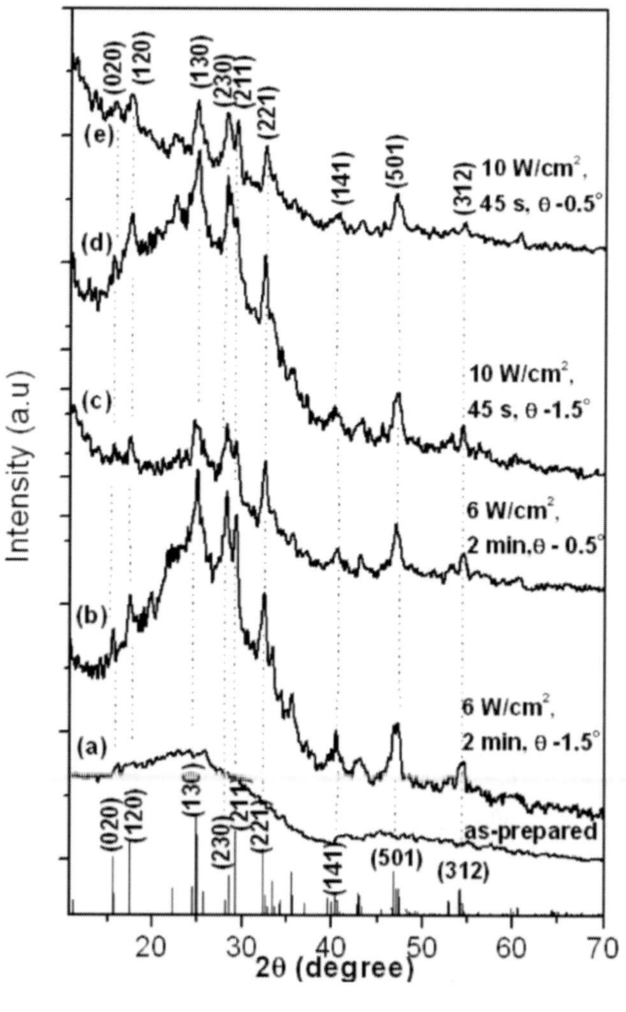

Figure 8. The XRD patterns (grazing incidence mode) of Sb_2S_3 thin film (a) as-prepared (b) and (c) irradiated with laser beam of power density of 6 W/cm^2 for 2 min, (d) and (e) irradiated with laser beam of power density of 10 W/cm^2 for 45 s (both using the expanded beam of 1 cm diameter). In each case, the lower pattern is that recorded at θ=0.5° and the upper one is that at θ=1.5°. The standard lines (PDF# 42-1393) corresponding to the stibnite phase is overlapped with (a) for reference.

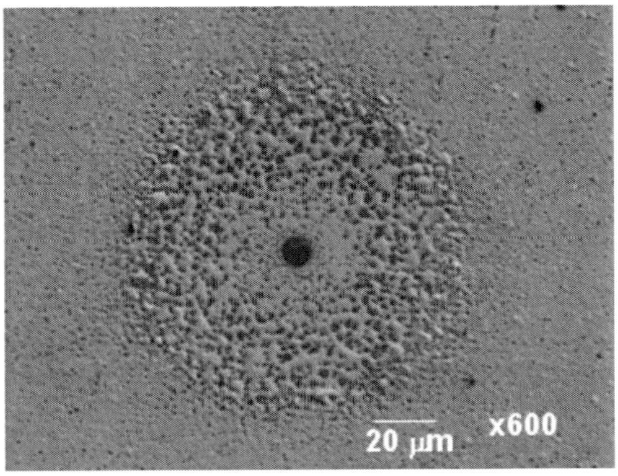

Figure 9. SEM of Sb_2S_3 thin film (550 nm) irradiated with direct beam of moderate power (91 W/cm^2 beam diameter of 2.25 mm, irradiation time 10 s): under low magnification to view the morphology of the irradiated spot completely.

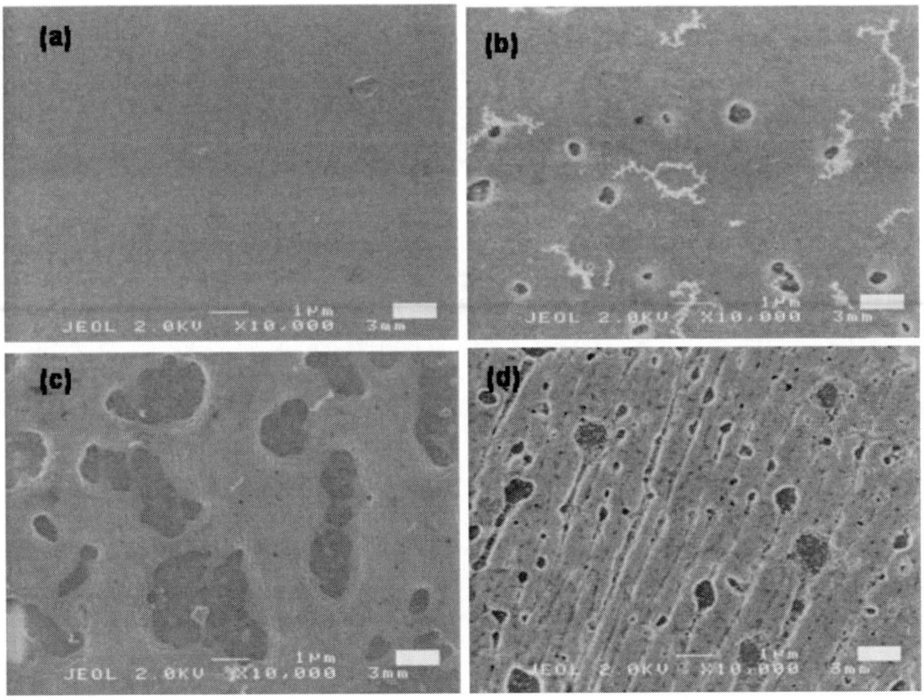

Figure 10. SEM of Sb_2S_3 thin film of 550 nm thickness (a) as prepared (b) irradiated with laser of power density 25 W/cm^2 (c) irradiated with laser of power density 91 W/cm^2 (d) irradiated with laser of power density 126 W/cm^2 (all using direct beam of diameter of 2.25 mm, irradiation time 10 s). All the images are recorded with the same magnification: 2 keV, x10,000.

The morphological features of the thin films after laser irradiation are illustrated by the scanning electron micrographs. Figure 9 shows the entire photo-darkened spot formed by direct beam of power density 90 W/cm^2 (moderate power) and a part of non-irradiated sample area. Laser power leads to the formation of Sb_2S_3 droplets with an average size of a few microns. Figure 10 depicts the micrographs for Sb_2S_3 thin films irradiated for 10 s with direct laser beam of power densities 25, 90 and 126 W/cm^2 in comparison with that for the non-irradiated film. Formation of dendrite type structures, beginning of 'inflating' or melting forming some compact 'fiber like' structures and distinct channel like features are present depending on the power density. This type of morphology of laser induced crystallized Sb_2S_3 thin films is very different from the spherical grains of polycrystalline thin films obtained by thermal treatment in nitrogen as shown in section 1.1.3.

Morphological change due to irradiation using an expanded laser beam (1 cm beam diameter, 6 W/cm^2 for 2 min) of Sb_2S_3 film is given in Figure 11 (a-c). Figures (a) and (b) correspond to the center part of the dark region in figure 1 under different magnifications. Any special structures or inflation due melting are not present in this case in contrary to those in the films irradiated with direct beam. Coalescence due to laser induced annealing and subsequent grain growth are observed in this case. Figure (c) shows the morphology of the dark region near the edge of the dark region and the features are similar as in the case of the center region. These figures illustrate that the morphology of the laser induced area by expanded beam is uniform having nearly similar grains.

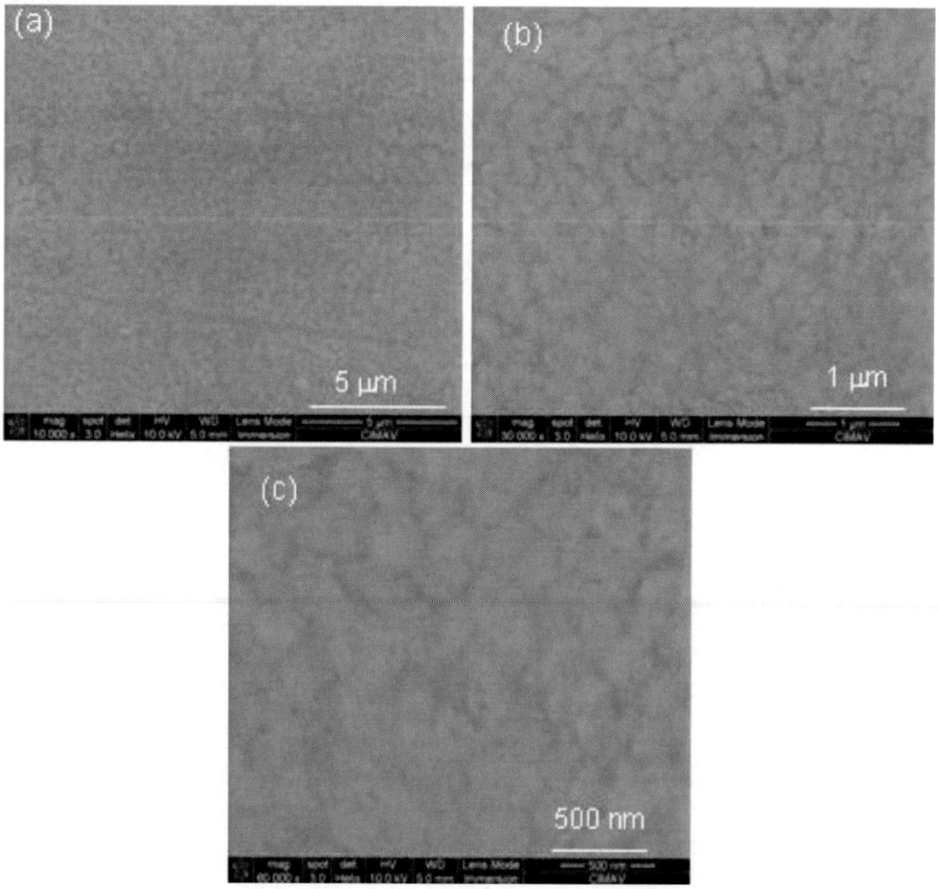

Figure 11. SEM of Sb_2S_3 thin film irradiated with laser of power density 6 W/cm^2 (a) and (b) correspond to the center part of the dark region in figure 1 under different magnifications (c) that of near the edge of the dark region.

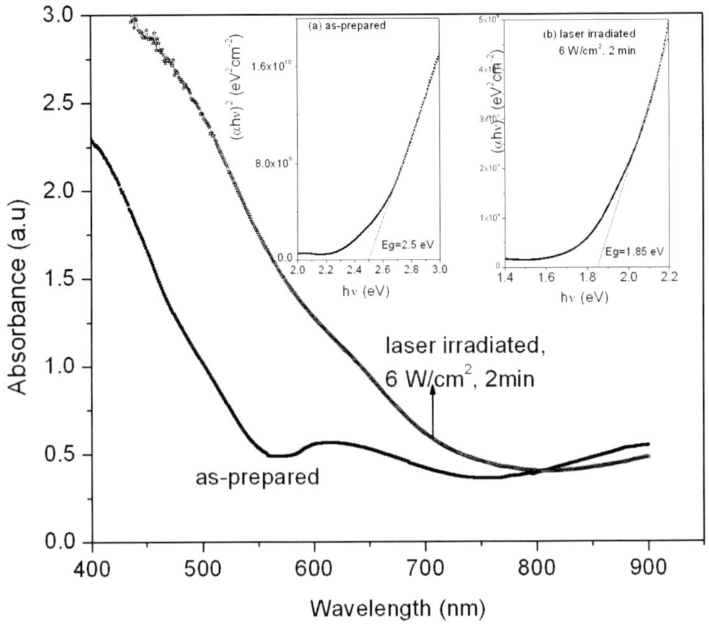

Figure 12. (a) Comparison of optical absorption spectra of Sb_2S_3 thin films before and after laser irradiation. In the inset evaluation of optical band gap.

1.2.2. Modifications in Optical Properties

The changes in optical properties of laser irradiated Sb_2S_3 thin films in comparison with the as-prepared thin films is shown here. Figure 12 depicts the optical absorbance spectra for as-prepared and laser irradiated thin films in the wavelength range of 400-900 nm. Curves represent the absorbance spectra for both as-prepared and irradiated Sb_2S_3 thin films (expanded beam of power density 6 W/cm^2). The optical band gap values were evaluated as explained in section 1.1.4 and the values are 2.50 eV and 1.85 eV for the Sb_2S_3-5h thin films before and after laser irradiation respectively, given in the inset. The red shift for the band gap value E_g for the laser irradiated films can be attributed to enhancement in the particle size during the optically induced crystallization of the thin films.

Thus these studies on laser induced modifications of the physical properties of the chemical bath deposited Sb2S3 thin films imply their use as an optical recording media.

1.2.3. In-situ Device Formation by Laser Exposure

Production of a photovoltaic (**PV**) device by laser irradiation of mutilayers is mentioned here, an *in situ* method. By laser irradiation of carbon coated Sb_2S_3 thin films (Glass/Sb_2S_3/C), carbon doping of Sb_2S_3 was achieved. Interestingly, when a multilayer structure, (Glass/ITO/CdS/Sb_2S_3/C) was irradiated with laser for 2 minutes, an expanded laser beam of diameter approximately 0.5 cm (5 W power, 532 nm DPSS laser) at ambient atmosphere, a p-n junction photovoltaic structure was formed. Cadmium sulfide thin films (~50 nm) was prepared from a containing chemical bath containing $CdCl_2$, sodium citrate,

NH_4OH and thiourea at 70 °C. This p-n junction formed by p-Sb_2S_3:C and n-type CdS showed open circuit voltage, V_{oc}=500 mV and short circuit current density J_{sc}=0.5 mA/cm^2 under illumination by a tungsten halogen lamp, as shown in figure 13.

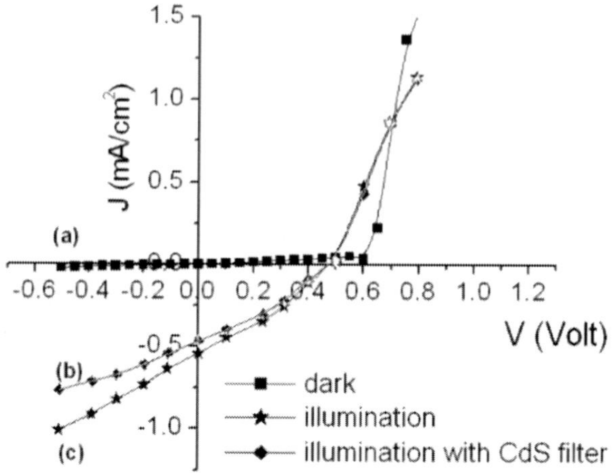

Figure 13. J-V characteristics of C:Sb_2S_3/CdS/ITO/Glass p-n junction formed by laser irradiation of C/Sb_2S_3/CdS/ITO/Glass multilayers (a) in the dark (b) under illumination from ITO side.

In coming sections, we will discuss on ternary Sb based compounds such as $AgSbSe_2$ and $CuSbS_2$ thin films produced using chemical bath deposited Sb2S3 thin films as precursor source of Sb.

2. Silver Antimony Selenide ($AgSbSe_2$)

2.1.1. Introduction

Silver antimony selenide is a I-V-VI_2 compound semiconductor with disordered NaCl cubic structure in which silver and antimony occupy metal sublattice. This compound and its alloys either in single-crystal form or in thin-film form have received considerable interest owing to their optical and electronic properties. They are attractive phase- change (PC) materials used as a switching medium in rewritable optical memories [19]. $AgSbSe_2$ in thin films form possesses an optical absorption coefficient of 10^4 cm^{-1} with a band gap of ~1 eV [20], and hole mobility of 1500 cm^2/Vs [21], suggesting it as a suitable photovoltaic absorber material [22]. In this section, we explain one method to prepare this material in the thin film form, their structural properties and their application in PV technology.

2.1.2. Properties and Applications

Silver antimony selenide ($AgSbSe_2$) thin films were prepared by heating sequentially deposited antimony sulphide (Sb_2S_3), silver selenide (Ag_2Se) and Ag thin films (Glass/Sb_2S_3/Ag_2Se/Ag) in close contact with a selenium thin film (Glass/Se). Sb_2S_3 thin film (500 nm) was prepared on glass substrate from a chemical bath containing $SbCl_3$ and $Na_2S_2O_3$ as mentioned in section 1.1, followed by Ag_2Se layer (150 nm) from a solution containing $AgNO_3$ and Na_2SeSO_3, Also, selenium thin films (300 nm) from an acidified solution of Na_2SeSO_3, at room temperature on glass substrates [20]. Ag thin film (10 nm) was deposited by thermal evaporation. The annealing temperature was 350 °C in vacuum (10^{-3} Torr) for 1 hour. X-ray diffraction analysis showed that the thin films formed were polycrystalline with cubic structure as given in figure 14. These AgSbSe2 thin films showed p-type conductivity and the carrier density was of the order of 10^{12} /cm^3, as evaluated from Hall effect measurements.

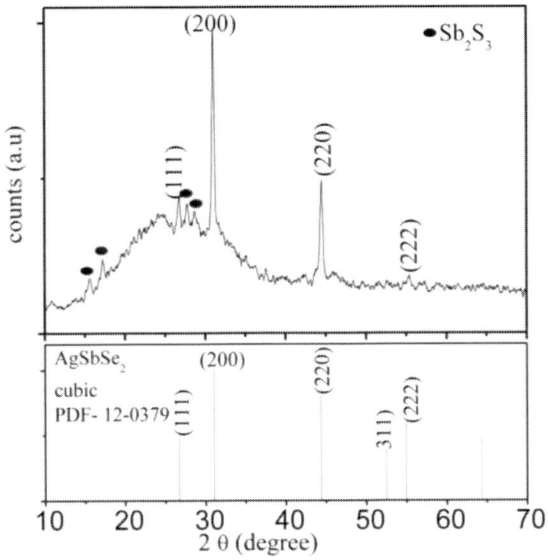

Figure 14. X-ray diffraction analysis of $AgAbSe_2$ thin films which shows that the thin films formed were polycrystalline with cubic structure.

These AgSbSe2 thin film was incorporated in a PV structure: Glass/ITO/CdS/$AgSbSe_2$/Ag. This structure was illuminated from the CdS side using a tungsten halogen lamp. From the J-V curve (Figure 16), V_{oc}= 370 mV and J_{sc}=0.5 mA/cm^2 were obtained [22].

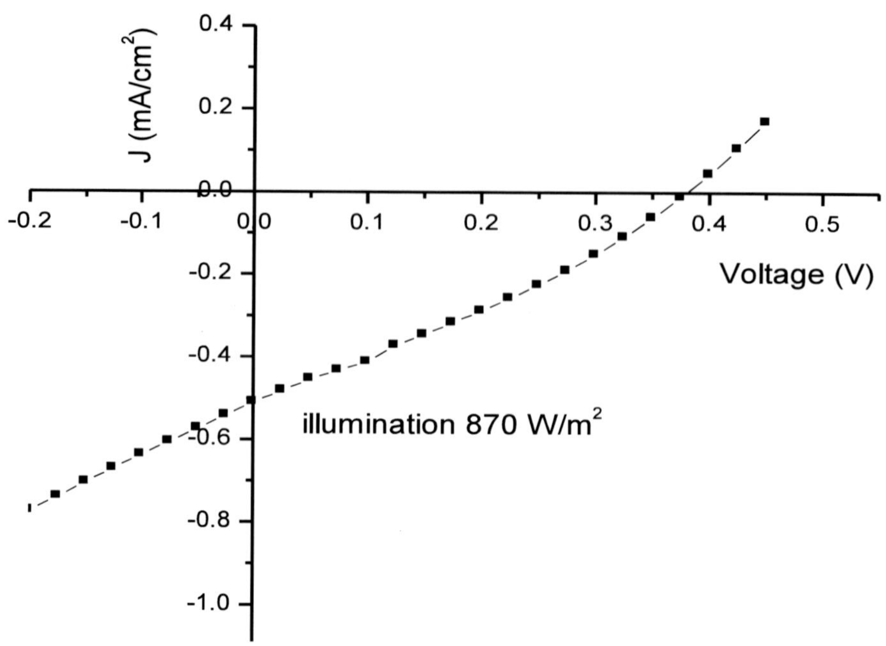

Figure 15. J-V characteristics of PV structure, Glass/ITO/CdS/$AgSbSe_2$/Ag.

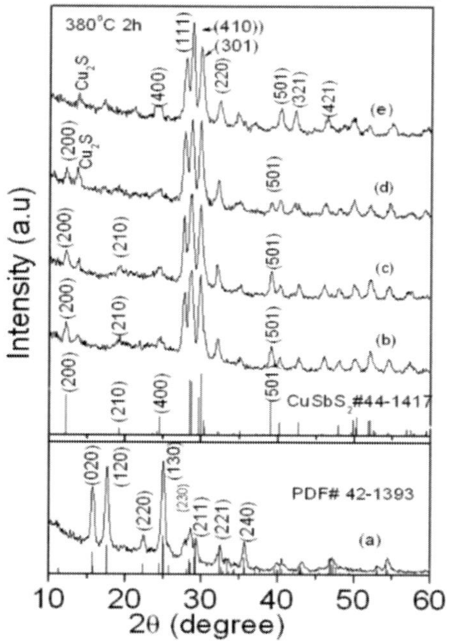

Figure 16. XRD patterns of samples formed by heating Sb_2S_3/Cu thin films at 380 °C in vacuum (10^{-3} Torr) for 2 h, varying Cu layer thickness (a) Sb_2S_3 (b) Sb_2S_3 with Cu 10 nm (c) Sb_2S_3 with Cu 20 nm (d) Sb_2S_3 with Cu 30 nm (e) Sb_2S_3 with Cu 50 nm. These patterns were recorded in the grazing incidence mode at θ= 1°. Standard patterns corresponding to both Sb_2S_3 and $CuSbS_2$ are included.

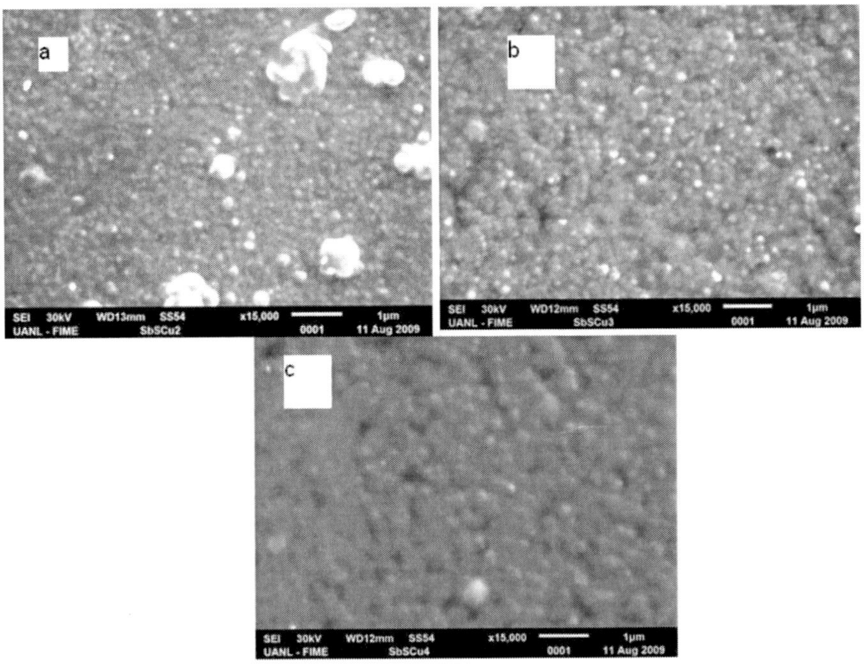

Figure 17. Scanning electron micrographs for Sb_2S_3/Cu thin films with Cu layer thickness 50 nm (a) as prepared (b) heated at 380 °C for 1 h (c) heated at 380 °C for 2 h.

3. $CuSbS_2$ Thin Films [23]

3.1. Introduction

Copper antimony sulfide ($CuSbS_2$) is one of the important semiconductors with narrow band gap showing potential applications in various optoelectronic devices such as infrared detectors and solar cells [24]. $CuSbS_2$ possesses an optical band gap of (E_g) 1.5 eV [25], the optimum value for an absorber material in solar cells. Preparation of this material and some results of their characterization are given here.

3.2. Preparation and Properties

In this section, we explain the preparation of copper antimony sulfide ($CuSbS_2$) thin films by heating Sb_2S_3/Cu multilayer in vacuum. Sb_2S_3 thin film was prepared from a chemical bath containing $SbCl_3$ and $Na_2S_2O_3$ salts as explained in section 1.1. A copper thin film was deposited on Sb_2S_3 film by thermal evaporation and Sb_2S_3/Cu layers were subjected to annealing at different conditions. Copper thin films of different thickness were thermally evaporated onto Sb_2S_3 (300 nm) coated glass substrates at high vacuum (10^{-5} Torr) Structure, morphology, optical and electrical properties of the thin films formed by varying Cu layer thickness and heating conditions are given here.

XRD analysis of Glass/Sb_2S_3/Cu multilayered samples heated at 380 °C for 2 h in low vacuum illustrates the presence of distinct peaks corresponding to $CuSbS_2$ phase arise as marked in figure 16. It is to be noted that as the copper layer thickness increases, additional reflection arises showing the presence of Cu_2S as indicated for the thin films with 30 nm, 40 nm and 50 nm copper layer.

Morphology of Sb_2S_3/Cu thin film, Cu layer thickness 50 nm, before and after heating is illustrated in figure 18. As prepared Sb_2S_3/Cu films shows spherical copper rich particles and irregular shaped clusters (figure 18(a)). Spherical particles are ~ 100 nm in diameter. After 1 h heating at 380 °C, the morphology $CuSbS_2$ thin film formed is given in figure 18(b). Here, the clusters are diffused by heating and the surface is more uniform. After 2 h of heating at 380 °C, the spherical particles have grown to bigger size, ~200 nm and the surface is free of any irregular shaped clusters or growth implying improvement in the surface uniformity of the films formed after 2 h heating compared to the 1 h heated samples.

Optical absorption spectra of the Sb_2S_3/Cu annealed at 380 °C for 2 h are given in figure 19 (a). From the corresponding Tauc plot, the optical band gap of $CuSbS_2$ thin films formed by thermal diffusion of 50 nm Cu into Sb_2S_3 was evaluated and the value is 1.5 eV as seen in figure 19 (b).

The electrical properties of these thin films were analyzed using Hall effect measurements for $CuSbS_2$ films formed at 380 °C by diffusing Cu layer thickness 30 nm and 50 nm into Sb_2S_3 showed p-type conductivity. The conductivity values are 0.50 $(\Omega cm)^{-1}$ and 60 $(\Omega cm)^{-1}$ for the films formed with Cu 30 nm and 50 nm respectively. Photoresponse measurements for $CuSbS_2$ samples formed at different conditions are given figure 20. All the samples showed significant photoresponse. Persistent photoconductivity is observed in $CuSbS_2$ samples formed by diffusing Cu layer thickness of 30 nm. For the sample with Cu 20 nm at 380 °C, the persistent behavior disappeared, may be due to complete diffusion of Cu into Sb_2S_3.

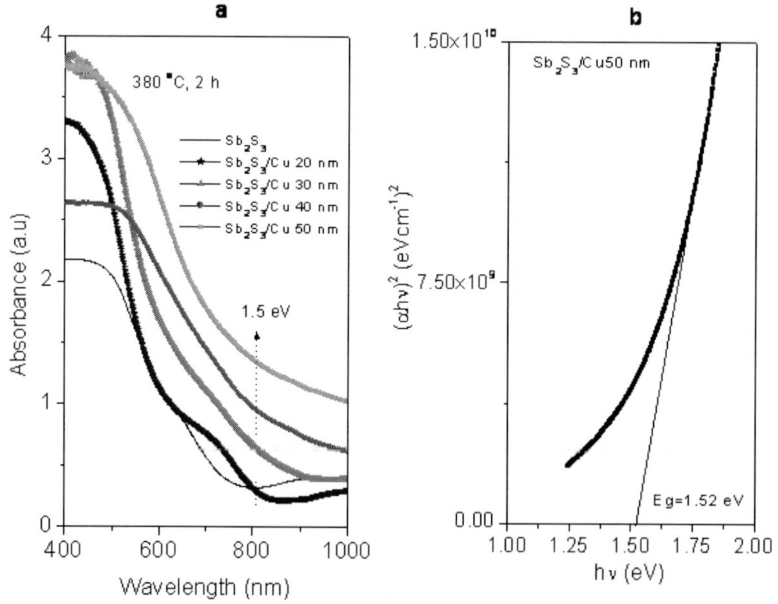

Figure 18. (a) Optical absorption spectra for Sb_2S_3/Cu thin films heated at 380 °C in vacuum (10^{-3} Torr) for 2h, different Cu layer thickness (b) Evaluation of optical band gap (Tauc plot) for $CuSbS_2$ thin films with Cu layer thickness 50 nm.

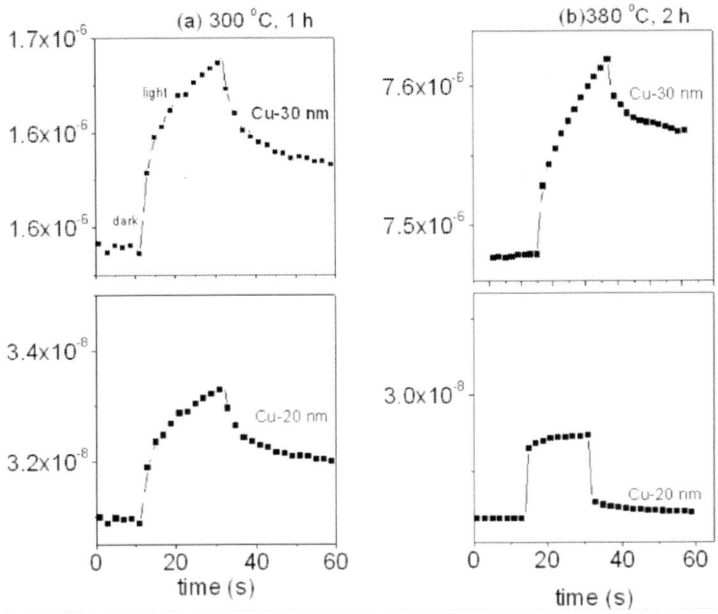

Figure 19. Photocurrent response for Sb_2S_3/Cu thin films heated at 300 °C and 380 °C.

The above mentioned optical and electrical properties suggest the application of $CuSbS_2$ thin films as PV absorber material.

Conclusion

In this chapter we discussed the preparation and characterization of binary and ternary antimony chalcogenide thin films such as Sb_2S_3, $AgSbSe_2$ and $CuSbS_2$, based on results obtained in our laboratory. The optical and electrical properties of these thin films imply their applications in various opto-electronic devices.

References

[1] E. Perales, G. Lifante, F. Agullo-Rueda, C de las Hares. *J. Phys. D: Appl. Phys.* 2007, 40, 2440-2444.

[2] C.D Lokhande, B.R. Sankapal, R.S. Mane, H.M Pathan, M. Muller, M. Giersig, V. Ganesan, *Appl. Surf. Sci.* 2002, 193, 1-10.

[3] O. Savadogo, K.C. Mandal, *Solar Energy Mater. Solar Cells* 1992, 26, 117-136.

[4] B.B. Nayak, H.N. Acharya, T.K Chaudhuri, G.B. Mitra, *Thin Solid Films* 1982, 92, 309-314.

[5] B.B. Nayak, H.N. Acharya, *Thin Solid Films* 1984, 122, 93-103.

[6] R. S. Mane, C. D. Lokhande, *Mater. Chem. Phys.* 2003, 78, 385-392.

[7] O. Savadogo, *Solar Energy Mater. Solar Cells* 1998, 52, 361-368.

[8] S. Messina, M.T.S. Nair, P.K. Nair, *Thin Solid Films* 2007, 515, 5777-5782.

[9] G. Hodes, *Chemical Solution Deposition of Semiconductor Films*, Marcel Dekker Inc., New York, (NY), 2002, 88-134.

[10] M.T.S Nair, Y. Peña, J. Campos, V.M. Garcia, P.K. Nair, *J. Electrochem. Soc.* 1998, 145, 2113- 2120.

[11] K. Bindu, A.M. Arato, E. Cardenas, T.K Das Roy, G. Alan Castillo, *Appl. Surf. Sci.* 2008, 254 3200-3206.

[12] B. D. Cullity, R. S. Stock, *Elements of X-Ray Diffraction*, Third ed., Prentice Hall, New York, (NY), 2001, 96-102.

[13] E. Cardenas, A. Arato, E. Perez-Tijerina, T.K. Das Roy, G. Alan Castillo, B. Krishnan, *Solar Energy Mater. Solar Cells* 2009, 93, 33– 36.

[14] S. Shaji, A. Arato, J. J. O'Brien, J. Liu, G. Alan Castillo, M I Mendivil Palma, T.K. Das Roy, B. Krishnan, *J. Phys. D: Appl. Phys.* 2010, 43, 075404-

[15] T. Sameshima, H. Watakabe, N. Andoh, S. Higashi, *Thin Solid Films* 2005, 487, 63-66.

[16] C. Boutopoulos, P. Terzis, I. Zergioti, A.G. Kontos, K. Zekentes, K. Giannakopoulos, Y.S Raptis, *Appl. Surf. Sci.* 2007, 253, 7912-7916.

[17] R. Saleh, N.H. Nickel, K.V Maydell, *Thin Solid Films* 2005, 487, 89-92.

[18] D. Klinger, J. Auleytner, A. Barez, D. Zymierska, *J. Alloys. Comp.* 2004, 362, 265-268.

[19] H. Soliman, D. Abdel-Hady, E. Ibrahim, *J. Phys.: Condens. Matter* 1998, 10, 847-856.

[20] O Madelung, *Semiconductors other than Group IV Elements and III –V Compounds (Data in Science and Technology)*, Berlin-Springer 1992, 72-73.

[21] K. Bindu, J. Campos, M.T.S. Nair, A. Sanchez, P.K. Nair, *Semicond. Sci. Technol.* 2005, 20, 496-504.

[22] J. Oswaldo, S. Shaji, A. Arato, T.K Das Roy, G.A. Castillo and B. Krishnan, *Mater. Res. Soc. Symp. Proc.* 2009, 1165, 1165-M08-25.

[23] C. Garza, S. Shaji, A. Arato, E. Perez Tijerina, G. Alan Castillo, T.K. Das Roy, B. Krishnan, *Solar Energy Mater. Solar Cells* 2010 (article in press), doi:10.1016/j.solmat.2010.06.011.

[24] J. Zhou, G. Bian, Q. Zhu, Y. Zhang, C. Li, J. Dai, *J. Solid State Chem.* 2009, 182, 259–264.

[25] S. Manolache, A. Duta, L. Isac, M. Nanu, A. Goossens, J. Schoonman, *Thin Solid Films* 2007, 515, 5957–5960.

In: Antimony: Characteristics, Compounds and Applications ISBN 978-1-62100-598-8
Editor: Manijeh Razeghi

Chapter 7

ANTIMONY OXIDE GLASSES AND THEIR NANOCOMPOSITES FOR OPTICAL, PHOTONIC AND NANOPHOTONIC APPLICATIONS

Tirtha Som and Basudeb Karmakar
Glass Science and Technology Section, Glass Division,
Central Glass and Ceramic Research Institute,
Council of Scientific and Industrial Research (CSIR, India),
196, Raja S. C. Mullick Road, Kolkata, India

1. INTRODUCTION

Contemporary glass research focuses on the development of specialized or unusual glasses as a novel media for photonic, plasmonic and nanophotonic technologies by incorporating luminescent rare-earth ions, metal nanoparticles, semiconductor nanoparticles, etc and their hybrids [1-3]. Research in these fast emerging areas require development of new glass compositions and in this respect, unlike the conventional silicate, borate and phosphate glasses, heavy metal oxide glasses like bismuth, tellurite, lead, and/or antimony acting as glass structure network formers is expected to play a dramatic role as they possess some exciting properties like high refractive index, large transmission window, large non-linear optical properties, low phonon energy and high dielectric constant.

There has been considerable investigation on Bi-, Te- and Pb- glasses but Sb-based glasses remain widely unexploited [4]. Antimony(III) oxide (Sb_2O_3) has mainly been used as refining or fining agent for glass melts. Glass forming ability of $Sb_2O_3$was first predicted by Zachariasen [5] in 1932 but it was 1939 when Kordes first succeeded to prepare a few milligrams of antimony glass overcoming the problems of preparation [6]. Since then very few authors have investigated this glass and very few measured properties have been published [7-14]. To the best of knowledge there is no report of preparation of monolithic (bulk) antimony oxide glasses containing more than 50 mol% of Sb_2O3. Literature reports show that they are obtained in pulverized forms. This because Sb3+ cation has weak field strength (0.73), Sb_2O_3 has high volatization rate and antimony glasses have very high tendencies to crystallize [4].

We have developed new monolithic Sb_2O_3 based glasses having compositions and for the first time could incorporate in antimony glasses trivalent RE ions and metallic nanoparticles separately and as hybrids and study their optical and other properties [15-22] for applications in photonics and nanotechnology. This chapter presents some very interesting results of our recent basic research and new aspects of Sb_2O_3 glasses.

2. Experimental

2.1. Preparation of the Glasses and Nanocomposites

The raw materials were antimony(III) oxide, Sb_2O_3 (GR, 99%, Loba Chemie) and potassium metaborate, $KBO_2.xH_2O$ (15.7% H_2O, Johnson Matthey). 20 g of glasses were prepared by melting the well mixed batches of calculated composition in a high purity silica crucible at 900°C for 10 min with intermittent stirring for 30 sec in air in a raising hearth electric furnace. All the molten samples were cast into carbon plates in air and annealed at 260°C for 3h to remove thermal stresses followed by slow cooling (1°C/ min) down to room temperature.

For synthesis of RE^{3+} doped glasses and nano-metal doped glasses, the required quantities of europium(III) oxide, Eu_2O_3 (99.9%, Alfa Aesar), Samarium(III) oxide, Sm_2O_3 (99.99 % Indian Rare Erath), erbium(III) oxide, Er_2O_3 (99.99%, Laborat), and gold(III) chloride, $HAuCl_4.xH_2O$, (49 % Au, Loba Chemie) were added in excess to the mother (base) glass batch and melted similarly.

2.2. Characterization of the Glasses and Nanocomposites and Measurements of Properties

The bulk densities (error ± 0.8 %) of the samples were measured by Archimedes method using toluene (a non-polar hydrophobic solvent) as immersion liquid because water may have some corrosive effect dissolving the antimony glasses to some extent.

The refractive indices and other theoretically predicted respective glass properties were determined using SciGlass (Glass Properties Information System, Version 6.7) software following the Priven-2000 method.

The softening point (T_s) (accuracy $\pm$2 °C) of were measured by a glass softening point system (Harrop/Labino, Model SP-3A).

The coefficient of thermal expansion (CTE, α in the temperature range 50-150°C), glass transition temperature (T_g) and deformation temperature (T_d) were measured using a horizontal vitreous silica dilatometer (DIL 402C, Netzch-Gerätebau GmbH, Bavaria, Germany) with heating rate 5 K/min. These values were reproducible with ±1°C for all samples.

Dielectric constant (accuracy $\pm$ 0.7%) was measured at room temperature using a LCR meter (Hioki 3532-50 LCR HiTESTER, Japan) at 1 MHz frequency by probe method after coating opposite sides of square shaped polished samples (cross-sectional area of 10×10 mm^2 and thickness 2 mm) with a thin layer of conducting silver paint and dried at 140 °C to serve as electrodes.

The infrared transmission spectra (FT-IRTS) were measured by Fourier transform infrared spectrometer (FTIR 1615, Perkin-Elmer Corporation, Norwalk, USA) (resolution ± 2 cm^{-1}) and for infrared reflection spectra measurements a specular reflectance attachment accessory was attached with the FTIR instrument at an incident angle of 15°.

The UV-Vis absorption/ transmission spectra (accuracy ± 0.1 nm) were obtained with a double-beam spectrophotometer (Lambda 20, Perkin-Elmer Corporation, Norwalk, USA).

The fluorescence spectra (accuracy ± 0.2 nm) were measured with a Spex, Fluorolog 2 fluorescence spectrophotometer (New Jersey, United States) with 150 W Xenon lamp as excitation source and a photomultiplier tube as a detector. The excitation slit (1.25 mm) and emission slit (0.5 mm) were kept same for all samples.

The X-ray diffraction patterns of the bulk samples were recorded in an X'pert Pro MPD diffractometer (PANalytical, Almelo, The Netherlands) using Ni-filtered Cukα radiation with step size 0.05° (2θ) step time 0.5 sec, from 10 to 80°.

Field emission scanning electron microscopy (FESEM) images of the fracture surfaces of glasses and nanocomposites were taken with Gemini Zeiss Supra™ 35VP model (Carl Zeiss Microimaging GmbH, Berlin, Germany) after etching the samples in 5 vol% HNO_3 for 1 min.

TEM was done using a Jeol (JEM 2010) operating at an accelerating voltage of 200 kV. Sample for TEM were prepared by ultrasonication (dispersion) of finely powdered nanocomposite samples in acetone, placing a drop of the supernatant dispersion on a carbon coated copper grid and drying under IR lamp at room temperature. The SAED images have been taken by focusing the electron beam on the metal nanoparticles.

3. Results and Discussion

3.1. Low Phonon Monolithic K_2O-B_2O_3-Sb_2O_3 (KBS) Glasses: Structure and Properties

Composition and some physical properties of the samples are indicated in Table 1. The samples having higher Sb_2O_3 content are yellow in color due to host absorption caused by the transition between HOMO (Sb 5s + O 2pπ) and LUMO (Sb 5p) [11].

The X-ray diffraction patterns of the samples in comparison to the raw material Sb_2O_3 used in batch are shown in Fig. 1. Antimony oxides, in general, are known to exist in a wide range of compositions and display interconvertable polymorphism. The two common forms are senarmontite (cubic) and valentinite (orthorhombic) consisting of Sb_4O_6 molecules and chains of SbO_3 trigonal pyramids respectively [23,24]. The polymorphic forms of Sb_2O_4 are the orthorhombic α-phase (cervantite) and there is also high temperature monoclinic β-phase [23,24]. The XRD patterns indicate that when the natural when the natural and conditional glass former (B_2O_3 and Sb_2O_3 respectively) are present in near-equal amounts, crystallization sets in and devitrified monoliths are obtained. This is probably due to the large difference in field strengths of Sb^{3+} (F = 0.73), B^{3+} (F = 1.34) and alkali K^+ (F =0.13) ions [4].

Table 1. Nominal composition and some properties of the KBS smaples

Sample No.	Composition (mol %)			Color	Form (Nature)
	K_2O	B_2O_3	Sb_2O_3		
B1	5	5	90	Deep yellow	Translucent large broken pieces (devitrified)
B2	10	10	80	Yellow	Transparent monolith (glass)
B3	15	15	70	Yellow	Transparent monolith (glass)
B4	20	20	60	Pale -yellow	Transparent vmonolith (devitrified)
B5	25	25	50	Colorless	Transparent monolith (devitrified)
B6	30	30	40	Colorless	Transparent monolith but surface easily attacked by moisture (devitrified)
B7	35	35	30	White	Opaque small broken pieces (devitrified)

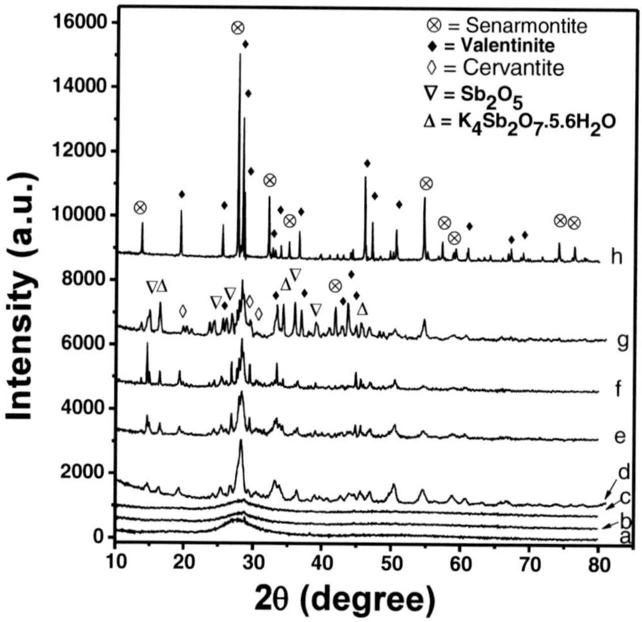

Figure 1. X-ray diffractograms of samples (a) B1, (b) B2, (c) B3, (d) B4, (e) B5, (f) B6, (g) B7 (for composition see Table 1) and (h) Sb_2O_3 used as starting material is shown for comparison.

The XRD results clearly support valentinite as the dominant phase with small amount of senarmontite, cervantite and potassium antimony hydroxide. Orthorhombic Sb_2O_3 is known to crystallize in the space group Pccn (D_{2h}^{10}) and is built up of infinite polymeric Sb-O-Sb chains running along the *c* axis with Sb-O distances of 20 Å. As a result we get some amazing microstructures which are usually not observed in other heavy metal oxide glasses. Their FESEM images of devitrified glasses B4-B6 depict gradual change of microstructures from granule clustered to cell-wall-like to house-of-disks respectively which consists mainly of nanocrystallites of valentinite Sb_2O_3 and have been left after elimination of borate-rich phase by acid.

Figure 2. Representative FESEM images of antimony oxide glasses (a) B4 (scale 200 nm) showing cell-wall like microstructure and (b) B6 (scale 2 μm) showing disk-like microstructures which accumulates to develop rose-flower-like microstructure.

Structural investigation using the Fourier Transform IR transmission (FT-IRTS) spectra (KBr pellet method), particularly in the fingerprint region (400-1500 cm^{-1}) in comparison with those of raw materials (Fig. 3), provide information about structure (short range order) of antimony oxide glasses. The bands located at 692, 592, 546 and 492 cm^{-1} corresponds to symmetric stretching, asymmetric stretching, symmetric bending and asymmetric bending Sb-O-Sb vibrations of valentinite form of SbO_3 trigonal pyramids with C_s or C_2 symmetry, respectively [9,11,24] thereby emphasizing the closeness of the glass structure to valentinite [9,11,24]. The small peak at 954 cm^{-1} is due to Sb-O stretching vibrations of senarmontite form of SbO_3 [9,11]. It may also be due to coincides with Si-O stretching vibrations of SiO_4 tetrahedra which were contaminated from silica crucibles during the melting process [9]. This was also found by energy dispersive X-ray (EDX) spectra of the glasses. The contamination of small amount of silica is expected to aid in glass formation.

The peak at 1231 cm^{-1} arises due to B-O stretching vibrations of $(BO_3)^{3-}$ unit in metaborate chains [25, 26]. The region between 1000-1200 cm^{-1} and 1200-1500 cm^{-1} is generally attributed to asymmetric stretching vibrations of the B-O-B bond of tetragonal $[BO_4]$ and trigonal $[BO_3]$ unit respectively. The FT-IRTS spectra also support intense crystallization with decreasing Sb_2O_3 content in the samples. The peak at 1630 cm^{-1} arises due to bending vibrations of O-H group of H_2O molecules. The broad band around 3392 cm^{-1} corresponds to the stretching vibration of O-H group. While the small peak at 3277 cm^{-1} is due to hydrogen bonded OH ($-O-H^{\delta+}...O^{\delta-}-$) and the small peak at 3554 cm^{-1} corresponds to free O-H group to the glass network [27]. The presence of water in the form of hydroxyl groups in the glasses is thus suggested.

Thus antimony based glass structure resembles valentinite form of Sb_2O_3 having SbO_3 pyramids with the coordination number of Sb as 3. Oxygen atoms are present in three corners (Sb-O bond distance 2.0 Å) and the lone pairs of electrons of Sb^{3+} at the fourth corner [6]. The coordination polyhedra are joined by sharing corners to form double infinite chains with the lone pair pointing out from the chains. Such chains are held together by weak secondary Sb-O bonds with lengths greater than 2.6Å. The third oxygen in each SbO_3 unit takes part in Sb-O-B type bond formation [4, 28]. The presence of common meta-centers and B-O-B vibrational bands in the IR spectra of the glasses supports the statement. Thus the structure essentially consists of long chains of entangled SbO_3 units. The introduction of the glass modifier K_2O probably breaks some linkages (B-O-B bonds, Sb-O-B bonds) and thereby the local symmetry of the glass network. This leads to the formation of non-bridging oxygen ions (NBOs) and introduction of dangling bonds (coordination defects).

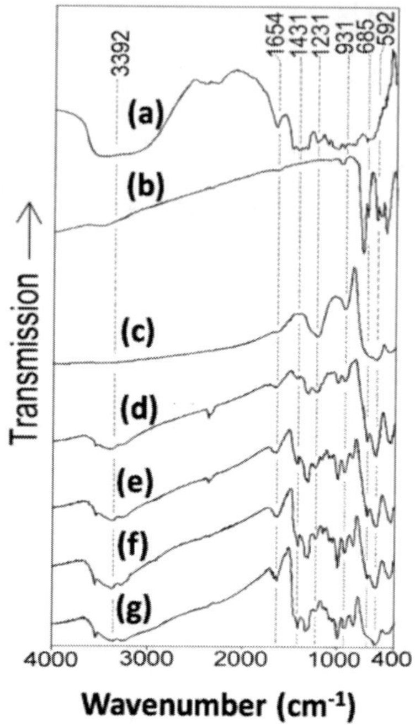

Figure 3. Infrared transmission spectra (FT-IRTS) of (a) $KBO_2.xH_2O$, (b) Sb_2O_3, (c) B2, (d) B3, (e) B4, (f) B5 and (g) B6 (for composition see Table 1). Spectra of raw materials (a) and (b) are shown for comparison.

During melting at high temperature (900ºC) small part of Sb^{3+} (in Sb_2O_3 used as the raw material) is expected to be converted to Sb^{5+} by accepting oxygen from air due to its multivalent characteristics according to the following reaction:

$$Sb_2O_3 + O_2 = Sb_2O_5 \tag{1}$$

These Sb^{5+} ions enter into the glass as singly positive [SbO_4] 4-coordinated units. The presence of positive [SbO_4] would provide charge balance for the negative [BO_4] units [9,11]. The existence of Sb_2O_5 crystals has been confirmed by XRD analysis. Earlier structural investigations of Sb_2O_3-B_2O_3 systems have shown that Sb5+ value reaches almost 15% of glasses containing 70 mol% Sb_2O_3 [13].

The phonon energy (ħω or bond vibration energy of the building units) is an important glass property. The lower the phonon energy, the higher is the efficiency of unconversion of the doped RE^{3+} ions. The infrared reflection spectrum (IRRS) can to determine the phonon energy of glasses as alternation to Raman or phonon side band spctra. In IRRS, the main and the highest intensity band is assigned as the phonon energy which is established to be 602 cm^{-1} for Sb_2O_3 glasses (Fig. 4) and this value is very close to the theoretically predicted Sb-O-Sb stretching vibration of 605 cm^{-1} [24]. It is observed that the intensity of the 602 cm^{-1} band decreases with decrease of Sb_2O_3 content and the intensity of the 1207 cm^{-1} band increases indicating an increase in borate content. As a consequence, the IRRS curves show an isosbestic point at 777 cm^{-1} manifesting an existence of equilibrium between the two principle species, antimony and boron.

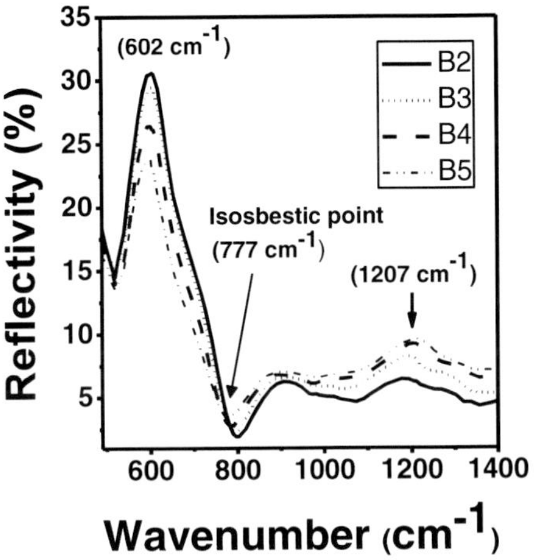

Figure 4. Infrared reflection spectra (FT-IRRS) of monolithic antimony oxide glasses and devitrified glasses: B2, B3, B4 and B5 (for composition see Table 1).

Table 2. Comparison of phonon energies of various glasses

Glass	Phonon energies (cm^{-1})	References
Chalcogenide	200-300	[29]
Fluoride	500-600	[29]
Antimony oxide	600	Present study
Tellurite	600-850	[29]
Germanate	800-975	[29]
Silicate	1060-1150	[29]
Phosphate	1200-1350	[29]
Borate	1480-1340	[29]

Table 2 compares the relative phonon energies of various glasses with that of antimony glass. It is thus established that Sb_2O_3 glasses have low phonon energy analogous to that of fluoride glasses and are impending candidates for upconversion when doped with rare-earth (RE^{3+})-ions. We shall discuss the lasing properties of some RE^{3+} ions in antimony glass in the next sub-section and that they may be used as solid state laser materials in place of fluoride glass wherever suitable as they can be synthesized in ambient atmosphere, where as the fluoride glasses require sophisticated atmosphere controlled furnace.

Sb_2O_3 glasses have wide transmission window from ~393 nm (UV cut-off) to ~6700 nm (IR cut-off). The later is due to heavier atomic mass of Sb atom which contributes to the reduced strength or force constant of Sb-O bond by weak columbic interactions.

Interestingly, the UV-Vis-NIR transmission spectra of KBS antimony glasses also yield isosbestic point at 393 nm (Fig. 5). This is because the position of the cutoff wavelength or fundamental absorption edge red-shifts towards higher wavelength by the increasing influence of non-bridging oxygen ions (NBOs) with increasing K_2O content [4]. While there is a decrease in transmittance of the glasses containing higher concentration of Sb_2O_3 due to high polarisability of Sb^{3+} cation with a lone pair of electrons and high polarization the O^{2-} anion which in turn increases the total electronic polarizability. A direct consequence is the molar refraction and refractive indices increases causing a greater loss of transmittance due to scattering.

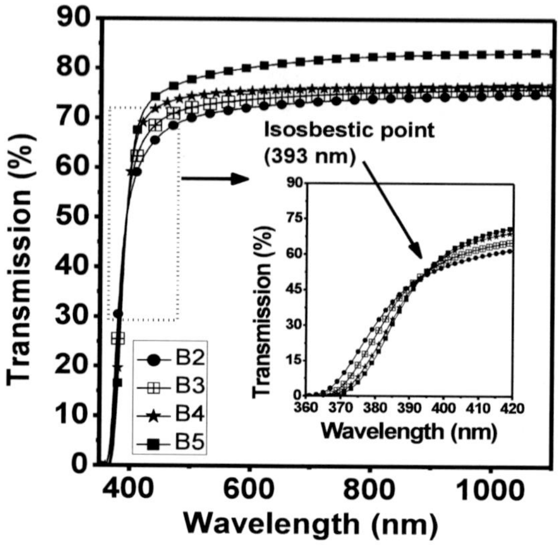

Figure 5. UV-visible transmission spectra of the KBS samples (thickness: 1.0 mm, for composition see Table 1). Inset shows isosbestic point at 393 nm.

The band gap, E_{opt} (eV), of KBS glasses can be evaluated from the transmission spectra by plotting $(\alpha h\upsilon)^2$ vs. $h\upsilon$ [where absorption coefficient is $\alpha(\upsilon)$ in Tauc's plot], considering allowed direct transitions, and extrapolating the linear portion of the curve to intersect the photon energy ($h\upsilon$) axis at zero absorption [30]. The decrease in value of E_{opt} from 3.22 eV down to 3.15 eV with decreasing concentration of Sb_2O_3 understood in terms of structural changes in the glass system. In the present ternary system both Sb_2O_3 and B_2O_3 acts as network formers while K_2O as network modifier. The incorporation of K_2O increases the quantity of NBOs. Since NBOs are more easily excited than bridging oxygen, E_{opt} decreases with addition of K_2O and removal of Sb_2O_3. Thus these glasses are wide gap materials.

The KBS antimony glasses are found to possess high refractive index and dispersion (Table 3). They possess a very unique position in the legendary Abbe diagram of optical glasses (Fig. 6) and exhibit the potential of replacing the lead oxide (PbO) containing glasses with respect to environmental friendliness and low softening point.

The experimental softening point (T_s), glass transition temperature (T_g), coefficient of thermal expansion (CTE), and dielectric constant (ε) of KBS glasses have been found to vary in the ranges 331-392°C (± 2 °C), 234-264°C (± 1 °C), 201-222 (± 1) × 10^{-7} K^{-1} and 12.4-14.5 (± 0.1) respectively (Table 3). The T_s and T_g are found to obey the empirical relation:

$$T_s = 3\,T_g / 2 \qquad (2)$$

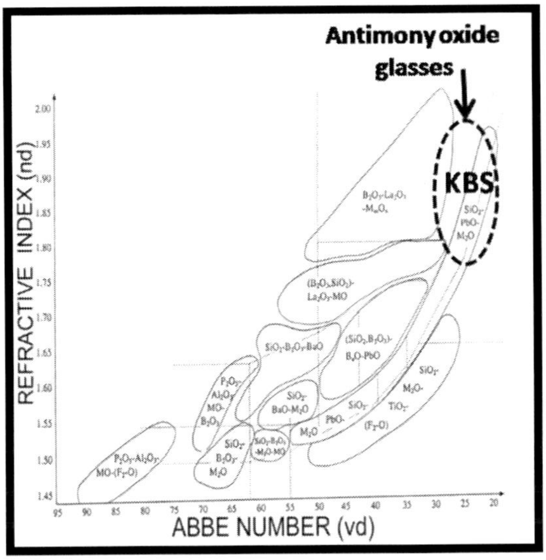

Figure 6. Position of KBS antimony glasses in the legendary abbe diagram.

Similar results have been reported for other Sb_2O_3 containing glasses [13]. These properties are found to be directly controlled by their fundamental property like covalent character and optical basicity of the glasses which are found to increase with increase in Sb_2O_3 content (Table 3).

The extent of covalent bonding character may be calculated as [31]:

$$\text{Covalent character (\%)} = \exp\left[-0.25\,(\Delta\chi)^2\right] \times 100 \tag{3}$$

where $\Delta\chi$ is the electronegativity of the composite, that is, the electronegativity difference ($\chi_A - \chi_C$) of the anions and the cations following the Pauling's scale [32]. It is found to vary in the range 48.55-40.55 % and decreased with decrease in Sb_2O_3 content.

The optical basicity, as proposed by Duffy and Ingram [33,34], is used as parameter to determine the acid-base properties of the glass in terms of the electron density carried by oxygen. It represents the average electron donation capacity of the oxide (II) species in the medium after the polarization of their electron charge clouds by constituent cations. The theoretical (ideal) optical basicity (Λ_{th}), is calculated as [33-35]:

$$\Lambda_{th} = X(Sb_2O_3)\,\Lambda(Sb_2O_3) + X(B_2O_3)\,\Lambda(B_2O_3) + X(K_2O)\,\Lambda(K_2O) \tag{4}$$

where, $X(Sb_2O_3)$, $X(B_2O_3)$ and $X(K_2O)$ are the equivalent fractions based on the proportion of oxygen, each oxide contributes to the overall glass stoichiometry and $\Lambda(Sb_2O_3) = 1.18$, $\Lambda(B_2O_3) = 0.42$ and $\Lambda(K_2O) = 1.4$ are the basicities assigned to individual oxides [35].

Both Sb^{3+} and K^+ have considerably less field strength (0.73 and 0.13 respectively) than B^{3+}(1.34) [4]. Consequently they play a major role of decreasing the T_s and T_g. Besides covalent bonds are less strong than ionic bonds and generally, the compounds with a more covalent bonds or covalent character posses lower softening points (T_s) and glass transition temperatures (T_g). However, one should also consider the influence of structural changes on these properties as well.

Table 3. Some of experimental and calculated properties of the KBS antimony glasses and devitrified glasses

Property	Sample identity[a]/ Property value				
	B2	B3	B4	B5	B6
Measured Properties					
OH content (ppm), (± 0.4)	125	133	141	150	
Density ($g.cm^{-3}$), (± 0.04)	4.7650	4.5560	4.3467	4.0985	3.8658
Softening point (T_s, ± 2 ºC)	331	341	356	378	392
Glass transition temperature ($T_{g,}$, ± 1) ºC)	234	246	255	259	264
Thermal expansion coefficient (CTE (± 1), $\alpha \times 10^{-7}$)	222	219	206	204	201
Dielectric constant (± 0.1)	14.5	14	13.3	13.2	12.4
Calculated Properties					
Covalent character, % (± 0.03)	48.55	46.50	44.49	42.50	40.55
Theoretical optical basicity, Λ_{th} (± 0.002)	1.072	1.035	0.994	0.950	0.901
Band gap (from UV-vis spectra)	3.22	3.19	3.17	3.15	-
Refractive index	2.001	1.948	1.892	1.835	1.775
Abbe number	23.12	23.86	24.80	26.01	27.65
Total molar polarizability, α_m (Å^3) (±0.007)	10.40	9.60	8.77	7.97	7.10
Metallization criterion (M) (±0 .001)	0.4998	0.5178	0.5376	0.5590	0.5827

[a] For composition see Table 1.

The CTE values are seen to increases with increase of Sb_2O_3 concentration. Generally the CTE increases as bond strength reduces. This manifests that Sb-O bonds are weaker than B-O bonds. In order to form a strong covalent bond, the participating orbitals of both the bond forming atoms should be comparable in size and energy and must have same symmetry with respect to the bond axis for the electron clouds to undergo a considerable overlap. Generally network bond strengths and glass forming-regions of heavy metal oxide glasses are small compared to conventional network formers. This is justified by the fact that heavy metal cation-oxygen bond strengths are relatively weaker, give rise to low fundamental vibration frequencies and thereby large IR transmission spectra. Another evidence is that in vitreous state B_2O_3 has interionic distance of 1.47 Å while in Sb_2O_3 it is about 2.0 Å [4,36]. Also in Sb_2O_3 glasses, the SbO_3 polyhedral chains are interlinked by weak secondary Sb-O bonds having lengths greater than 2.6Å [36]. In fact, the bond strengths of Sb-O and B-O are 89 ± 20 and 192.7 ± 1.2 kcal. mol^{-1} [36].

The dielectric constant values are also found to increase with increase of Sb_2O_3 content. The dielectric properties of materials are the inherent effect associated to the mechanism of polarization of the permanent and induced electrically charges by an external applied electric field. Here the KBS antimony glasses exhibits a relatively higher value of dielectric constant (ε) than the vitreous silica ($\varepsilon = 3.8$), soda-lime silicate ($\varepsilon = 7.2$) and borosilicate glasses ($\varepsilon =$ 4.1–4.9) [37]. This is essentially due to the spontaneous high ionic polarization of Sb^{3+} ions along with its lone pair of $5s^2$ electrons ($\alpha_{Sb3+} = 1.111$ Å^3) [35] under an applied electric field.

Due to non-centrosymmetric structure and high polarizability, antimony glasses are expected to possess large optical non-linearity [38]. The theoretical calculation of the metallization criterion (M) value (table 3) gives an indication of large NLO property of antimony glasses. It is well established that the third-order non-linear susceptibility (χ^3) value increases and M-value decreases [39]. The small metallization criterion means that the width of both valence and conduction bands become large, resulting in narrow band gap and increased tendency for metallization of the glass.

3.2. Photoluminescent Properties of RE-doped Antimony Oxide Glasses

Low phonon glasses doped with rare-earth (RE) ions have emerged as a significant category of solid-state luminescent materials with promising and ever-increasing applications as compact visible and NIR lasers, broad band amplifiers, light emitting devices, color display panels, optical data storage, sensors, optical communications etc [31,41,42]. Typically, the multiphonon relaxation rate is three orders of magnitude lower in low phonon fluoride glasses than in silicates. High phonon glasses can provide nonradiative decay pathways to suppress luminescence. On the contrary, low phonon glasses lower the non-radiative de-excitation processes that compete for upper state laser population. As an effect, many transitions which are non-observable in silicate glasses become active in fluoride glasses. This eventually led to obtain upconversion in low phonon glasses and to acquire short wavelength visible lasers upon pumping with NIR laser by involving two photons simultaneously [31].

Among the RE^{3+} ions capable of exhibiting photoluminescence upconversion, Sm^{3+} and Er^{3+} are the most suitable ones due to their favorable energy level structures [40,41] while Eu^{3+} is an important activator for down-shifted red light [42].

The antimony glass having composition (mol%) $15K_2O$-$15B_2O_3$-$70Sb_2O_3$ were doped with 0.3, 0.7 and 1.0 wt % Sm_2O_3 and Er_2O_3 (in excess) and studied for photonic applications.

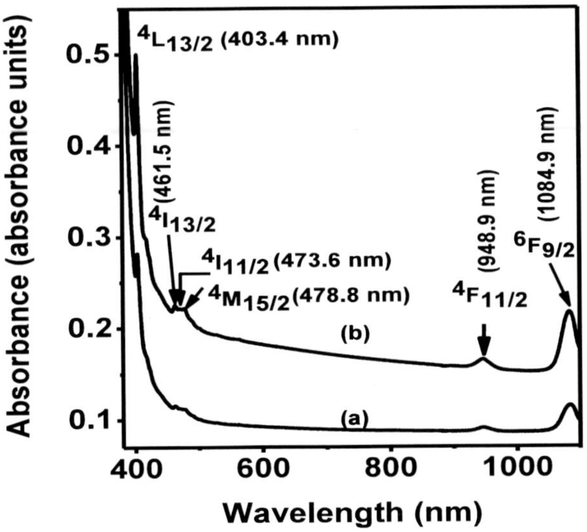

Figure 7. Absorption spectra of (a) 0.3 and (b) 0.7 wt % Sm_2O_3 doped $15K_2O$-$15B_2O_3$-$70Sb_2O_3$ (mol %) glass in the range 380-1100 nm showing the various transitions arising from the ground state term $^6H_{5/2}$ (sample thickness: 2 mm).

The ultraviolet-visible-near infrared (UV-Vis-NIR) absorption spectra of Sm^{3+}-doped glasses show six absorption bands from the ground state $^6H_{5/2}$ of Sm^{3+} ion (Fig. 7). The absorption bands give an approximation of the various energy levels of Sm^{3+} ion in antimony glass.

The upconversion emission spectra of Sm_2O_3 doped KBS antimony glasses (Fig. 8) show three strong upconverted bands are observed to be centered at 566 (green, weak), 602 (orange, weak) and 636 (red, very strong) nm originated from the $^4G_{5/2} \rightarrow {}^6H_{5/2}$ ($\Delta J = 0$, zero-zero band, forbidden transition), $^4G_{5/2} \rightarrow {}^6H_{7/2}$ ($\Delta J = 1$, magnetic dipole transition) and $^4G_{5/2} \rightarrow {}^6H_{9/2}$ $\Delta J = 2$, electric dipole transition) anti-Stokes transitions respectively and a very weak red emission band at 649 nm due to $^4G_{5/2} \rightarrow {}^6H_{11/2}$ ($\Delta J = 3$, forbidden transition) upon excitation at 949 nm [17]. The ratio of intensities between the $^4G_{5/2} \rightarrow {}^6H_{9/2}$ and $^4G_{5/2} \rightarrow {}^6H_{7/2}$ transitions is used to determine the intensity parameter, η_{Sm} [17].

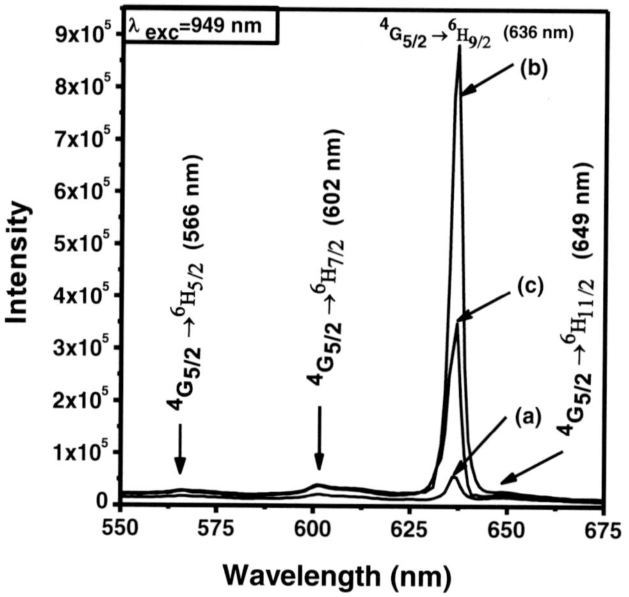

Figure 8. Upconversion fluorescence spectrum of: (a) 0.3, (b) 0.7 and (c) 1.0 wt % Sm_2O_3-doped $15K_2O$-$15B_2O_3$-$70Sb_2O_3$ (mol %) glasses upon excitation at λ_{ex} = 949 nm radiation.

Similar emission bands at 566, 602 and 652 nm due to stokes transitions are also observed for normal (down conversion) fluorescence process upon excitation at 403 nm. For both cases, η_{Sm} reaches to a maximum at 0.7 wt% Sm_2O_3 concentration (12.02×10^{19} Sm^{3+} ions/cm^3). The upconversion can be explained considering the flowing processes operating between different energy levels thereby populating the metastable radiative storage level, $^4G_{5/2}$ [17]:

$^6H_{5/2} \rightarrow {}^6F_{11/2}$ [Ground state absorption, GSA]

$^6F_{11/2} \rightarrow {}^4I_{11/2}$ [Excited state absorption, ESA]

$({}^6F_{11/2}, {}^6F_{11/2}) \rightarrow ({}^4I_{11/2}, {}^6H_{5/2})$ [Energy transfer, ET]

$({}^6F_{9/2}, {}^6F_{9/2}) \rightarrow ({}^4G_{5/2}, {}^6H_{5/2})$ [Energy transfer, ET]

$({}^4I_{9/2} \rightarrow {}^4G_{5/2})$ and $({}^6F_{7/2} \rightarrow {}^6F_{11/2})$ [Resonant cross-relaxation, RCR]

In the case of upconversion, the intensity of ${}^4G_{5/2} \rightarrow {}^6H_{9/2}$ electric dipole transition is found to be about 23 times larger than that of ${}^4G_{5/2} \rightarrow {}^6H_{7/2}$ magnetic dipole transition. The full width at half maximum (FWHM) of the ${}^4G_{5/2} \rightarrow {}^6H_{9/2}$ transition at 0.7 wt % Sm_2O_3 is 4.2 nm. These results are quite contrary to the spectra obtained for conventional glasses where the ${}^4G_{5/2} \rightarrow {}^6H_{7/2}$ (orange) is usually the most intense one. These two facts illustrate the hypersensitive behavior of the ${}^4G_{5/2} \rightarrow {}^6H_{9/2}$ (red) transition and indicate that Sm^{3+} ions are in a highly polarizable chemical environment possessing a low symmetry geometric configuration. The calculated molar polarizability and refractive index of KBS antimony glass are found to be 9.598 Å^3 and 1.947 respectively. These values are much higher than that of silica glass (2.965 Å^3 and 1.46 respectively) [27]. So the strong upconverted red emission at 636 nm is ascribed to the forced electric dipole transition $({}^4G_{5/2} \rightarrow {}^6H_{9/2})$ indicating the asymmetric geometrical environment of Sm^{3+} ion. The narrow FWHM means that Sm^{3+} environment distribution is small. Another reason for very strong red fluorescence may be attributed to the large non-linear optical (NLO) properties of antimony glasses [9]. Because additional electric field induced by NLO glass host may also make its contribution to the allowed electric dipole ${}^4G_{5/2} \rightarrow {}^6H_{9/2}$ transition which results in red fluorescence enhancement [17].

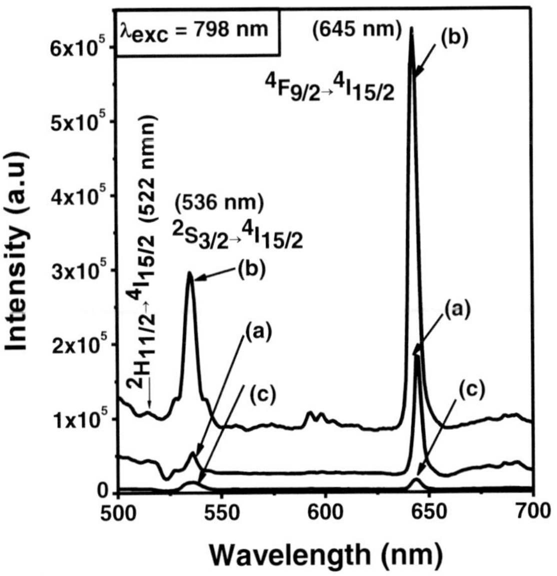

Figure 9. Upconversion fluorescence spectra of: (a) 0.3, (b) 0.7, and (c) 1.0 wt % Er_2O_3-doped antimony glasses upon excitation at λex = 798 nm radiation.

The other feature of the upconverted spectra is that the ${}^4G_{5/2} \rightarrow {}^6H_{9/2}$ transition bands at 636 nm have encountered a blue shift of 18 nm compared to the down conversion band at 652 nm for the same transition. This suggests that the upconverted Sm^{3+} ions are located in different unusual or distorted environments in KBS antimony glasses, i.e., they have different symmetry sites and orientation compared to those of Sm^{3+} ions participated in downconversion process [17]. The difference in the intensity of the yellow-green, orange and red fluorescence bands has happened due to the difference in the radiative transition rates from $^4G_{5/2}$ to the $^6H_{5/2}$, $^6H_{7/2}$ and $^6H_{9/2}$ energy levels respectively.

The Er_2O_3 doped antimony glasses show a total of eleven absorption bands from the ground state $^4I_{15/2}$ term. Upon excitation with 798 nm, The Er_2O_3 glass also exhibit two upconverted bands at 536 (green, medium), 645 (red, strong) nm due to $^4S_{3/2} \rightarrow {}^4I_{15/2}$ and $^4F_{9/2} \rightarrow {}^4I_{15/2}$ anti-Stokes transitions respectively and a weak band at 522 nm (green) due to $^2H_{11/2} \rightarrow {}^4I_{15/2}$ transition (Fig 9). Similar emission bands are also observed in normal (down conversion) fluorescence spectra on excitation at 407 and 377 nm.

The Er^{3+} ion in antimony glass show intense red upconversion on contrary to strong green upconversion usually observed in conventional glasses. This can be explained by considering the mechanism of upconversion.

The strikingly greater red emission is possibly due to predominant cooperative energy transfer (CET) process operating between two neighbouring Er^{3+} ions from the very beginning even at low Er^{3+} ion concentration as [18]:

$$({}^4F_{7/2}, {}^4I_{11/2}) \rightarrow (2\ {}^4F_{9/2})$$

Secondly, Two excited Er^{3+} ions at $^4F_{7/2}$ and $^4I_{11/2}$ levels may also interact by resonant cross relaxation (RCR) process: ($^4F_{7/2} \rightarrow {}^4F_{9/2}$) and ($^4F_{9/2} \leftarrow {}^4I_{11/2}$) and thereby enhance the population of the $^4F_{9/2}$ state.

Thirdly, co-operative energy transfer (CET) phenomenon is accountable for the enhanced red emission as one Er^{3+} ion decays from the $^4S_{3/2}$ state to the $^4F_{9/2}$ state and a second ion is excited to $^4F_{9/2}$ state from $^4I_{9/2}$ state [18]:

$$({}^4S_{3/2}, {}^4I_{9/2}) \rightarrow (2\ {}^4F_{9/2})$$

The energy gaps of E_A ($^4S_{3/2} \rightarrow {}^4F_{9/2}$) and E_B ($^4F_{9/2} \leftarrow {}^4I_{9/2}$) are matched with a small red shift of 242 cm^{-1}. Based on the Frank-Condon principle and lattice relaxation theory, a small red shift between E_A and E_B is favorable for such type of energy transfer. The CET process leads to improved population of the $^4F_{9/2}$ level and consequently superior red radiative emission. This fact indicates non-random distributed of E^{r3+} within the glass network even at a low concentration. The structure of the KBS antimony glass obviously plays a major role in suitable close distribution of the Er^{3+} ions within the interstitial spaces of the glass framework which facilitates the CET process irrespective of concentration. Another possible cross-relaxation energy transfer scheme includes [18]:

$$({}^2H_{11/2}, {}^4I_{11/2}) \rightarrow (2{}^4F_{9/2})$$

The above CET mechanism leads to partial quenching of the green and improvement of the red emission. Thus the red emission is enhanced at the cost of the green emissions and hence the green emissions never exceed the red emission in antimony glasses. These processes are depicted in the energy level diagram of Er^{3+} (Fig. 10).

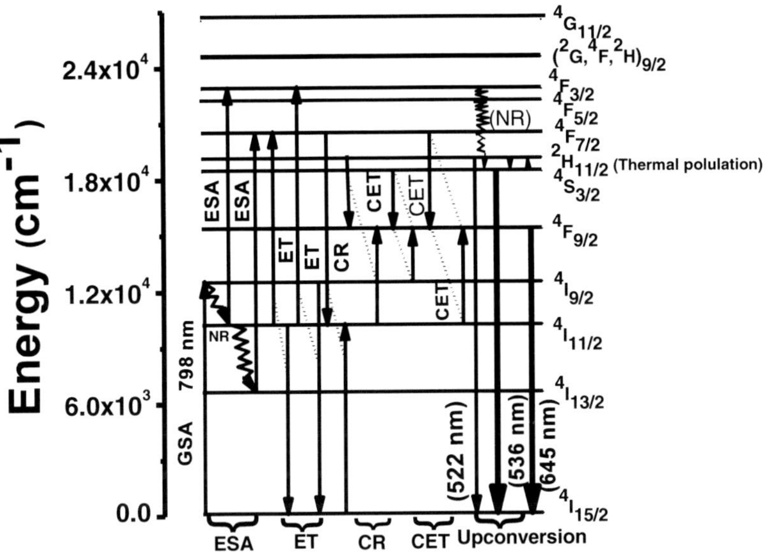

Figure 10. Partial energy level diagram of Er^{3+} ion in $15K_2O$-$15B_2O_3$-$70Sb_2O_3$ (mol %) glass showing transitions in ground state absorption (GSA), excited state absorption (ESA), energy transfer (ET), cross relaxation (CR) and co-cperative energy transfer (CET) processes and upconversion fluorescence emissions at 522, 536 and 645 nm under excitation λ_{ex} = 798 nm radiation (R and NR represent the radiative and nonradiative transitions respectively).

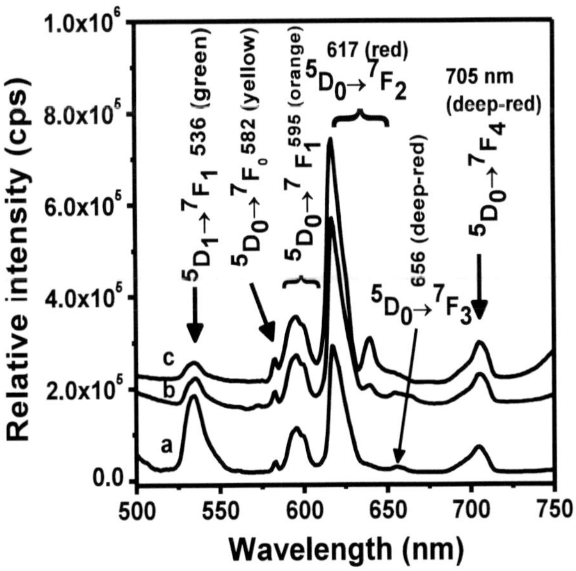

Figure 11. Emission spectra of Eu_2O_3-doped KBS antimony glasses upon excitation at λex = 393 nm radiation: (a) E-1, (b) E-2, and (c) E-3.

The presence of OH groups may also assist in intensified red emission and dminished green emission. A similar observation has been recorded by De et al. [43]. The OH groups generate high energy vibrational modes (3259 cm^{-1}). The $^4I_{11/2}$ and $^4I_{13/2}$ are intermediate energy levels responsible for the green and red emissions respectively (see Fig. 10). The energy gap between these two energy levels is ~ 3655 cm^{-1}. The high energy OH stretching vibration facilitates the $^4I_{11/2} \rightarrow ^4I_{13/2}$ multiphonon relaxation process more efficiently compared to the intrinsic phonons (600 cm^{-1}) in bulk KBS antimony glass. Since the upconversion intensity depends on the population of the intermediate states, the presences of OH groups in the KBS antimony glass, therefore, results in an enhanced red emission and diminished green emission.

To study the downconversion (fluorescence) property of Eu^{3+} in antimony glass (Fig. 11), each of the KBS base glasses having composition (mol %) xK_2O-xB_2O_3-(100-2x) Sb_2O_3 glasses, where x = 15, 20 and 25 were doped with 0.7 wt% (in excess) with Eu_2O_3 (numbered as E-1, E-2 and E-3 respectively) and excited with 393 nm wavelength.

In glassy materials, due the absence of a center of symmetry and long-range periodic arrangement of atoms, amalgamation of the 4f orbitals with an opposite parity orbitals take place. This rise to the ED transitions. The $^5D_0 \rightarrow ^7F_2$ transition (electric dipole allowed) are forced by the crystal field environment in the vicinity of the Eu^{3+} ions. It is hypersensitive (ΔJ=2) transition and its intensity is very sensitive to the local environment. It shows two distinct main peaks (617 and 639 nm) due to the Stark splitting of the $^7F_2$ state, meaning that the Eu^{3+} ions are located in drastically disordered lower symmetry sites. The $^5D_0 \rightarrow ^7F_1$ emission transition (magnetic dipole, (ΔJ=1)) of Eu^{3+} ions is forbidden under selection rules, has intensity independent of the host environment and can be used as a reference. It also exhibit small Stark splittings (595 and 600 nm) depending upon the ligand fields being experienced by them. The ratio IR = $I(^5D_0 \rightarrow ^7F_2)/I(^5D_0 \rightarrow ^7F_1)$ of the two integrated fluorescence intensities is commonly used as a measure of the asymmetry around the rare earth sites. Higher is its value, more would be the distortion from the inversion symmetry [16,44]. The value of IR is found to be 2.9, 4.0 and 4.3 for E-1, E-2 and E-3 respectively. This happens due to devitrification of the glasses (Fig. 1) with decreasing Sb_2O_3 content despite the increase in covalency and optical basicity. The ratio provides valuable information about the red color enrichment compared to the orange emission in developing strongly red luminescent optical systems.

The fluorescence spectra (Fig. 11) explicitly show the effect of lowering of phonon energy on the luminescence of Eu^{3+}. The $^5D_1 \rightarrow ^7F_1$ green transition is observed only in low phonon glasses [42] and its intensity decreases with decreasing Sb_2O_3 content. Under 393 nm excitation, the Eu^{3+} ions are excited to $^5L_6$ level by ground state absorption (GSA) from which multiphonon relaxation takes place mainly resulting in the population of the $^5D_0$ and $^5D_1$ levels due to the low phonon energy of the glass. Multi-phonon relaxation from the $^5D_0$ level to the next lower level $^7F_6$ does not take place because of large energy difference (ΔE ~12290 cm^{-1}) between the two levels. Due to the low phonon energy of antimony glasses, minimum of three and twenty photons are required for bridging the $^5D_1 \rightarrow ^5D_0$ and $^5D_0 \rightarrow ^7F_6$ multiphonon relaxation process. Due to the high phonon orders (phonons $\geq$ 3) required, possibility of non-radiative multiphonon bridging decreases and results in $^5D_1 \rightarrow ^7F_1$ (green) emission.

3.3. Plasmonic Properties of Nano Metal (Au)-doped Antimony Oxide Glasses

Metal-glass nanocomposites witness a historic background. The 4th Century A. D. Roman glass makers are originally credited to exploit the optical properties of metal nanoparticles for designing colourful medieval cathedral windows. In 1689, the German alchemist Johannes

Kunckel first documented the history of the gold-ruby glass [45]. But modern works on preparation of metal-glass nanocomposites and scientific evaluation of their optical (plasmonic) properties were pioneered by Doremus around 1964 [46]. Thereafter the metal-glass nanocomposites kept receiving significant exposures and are currently occupy a significant area of material science and nanotechnology [47-53].

Preparation of nano metal doped conventional glass systems are not simple and demand multi-step techniques like sol-gel process, metal-dielectric co-sputtering deposition, direct metal-ion implantation, radiofrequency magneton sputtering, pulsed laser deposition, ion-exchange of thin plates followed by long time heat treatment at high temperatures in reducing (hydrogen) atmosphere or UV-light/X-ray/^{60}Co γ-radiation or high energy laser/synchrotron irradiation [45-53]. These methodologies face several limitations like tendency of sample damage due to exposure to high intensity radiation, formation of this layers of near-spherical nanoparticles restricted near the surface, applicable mainly to high melting and high softening glasses.

Dichroic glasses (reflect and transmit different light wavelengths) are generally produced by deformation of embedded spherical NPs into ellipsoidal NPs by intense irradiation with ultrashort laser pulses [54,55] or stretching metal-doped glasses in their softening range [56]. Formation of elliptical nanoparticles is an essential requirement to develop dichroic glasses [54,55] and but now it is a well acknowledged fact that prolate ellipsoid Ag NPs having an aspect ratio around 1.2 exhibits the phenomenon of dichroism [57].

Glasses containing nanosized metal NPs are increasingly being appreciated for their potential optoelectronic and photonic applications as optical data recording disks and memory devices, optical waveguides, optical switches based on their nonlinear optical properties, photochromatic and ophthalmic lenses, color glass recycling industry, three-dimensional multicolored industrial art objects, display devices, dichroic polarizers, etc [45-56]. In all these applications the size, shape, number density, and distribution of the nanoparticles critically determine performance and properties of the nanocomposites.

Glass is also often acknowledged as the ideal optically transparent matrix for encapsulating mono-disperse metal nanoparticles for practical applications. Metal-glass nanocomposites possess some inherent advantages like low cost of fabrication, the ease of processing in desirable shapes and sizes, high stability, thermal resistance, mechanical strength, ability to withstand high intensity radiation, preventing air oxidation of metal nanoparticles, etc. The possibility of of tailoring the behaviour of the encapsulating mother glass may creates possibility of unprecedented applications in nanotechnology. So a significant part of the applied research on glass is therefore currently focused on development of new cost-effective methodologies for large-scale production and controlling the behaviour of these glass-based nanomaterials.

Sb_2O_3 is a well known mild (selective) reducing agent (Sb^{5+}/Sb^{3+}, $E^0 = 0.649$ V). Here we exemplify, for the first time, how this property can be judiciously used for single-step synthesis of nano metal-glass nanocomposites (Au nanoparticles embedded glass) and RE-nano metal-glass hybrid nanocomposites (such as Eu^{3+} ions and Au nanoparticles co-embedded in antimony oxide glass in sub-section 3.4). Trivalent Au (introduced in the batch as $HAuCl_4.xH_2O$) is easily reduced to Au^0 (Au^{3+}/Au^0, $E° = 1.498V$) during the melting by trivalent Sb which itself is oxidized to pentavalent Sb. Thus a straightforward, low-cost strategy for the fabrication of bulk metal-glass nanocomposites in a single-step is exemplified without incorporation of any other external reducing agent. The overall reaction $3Sb^{3+}$ +

$2Au^{3+} \rightarrow 3Sb^{5+} + 2Au^{0}$ is spontaneous reduction having E° = 1.05 V and a free energy (ΔG value) around -608 KJ. In principle, the E° of polyvalent elements in glass melts may be different from those in aqueous solutions and are temperature-dependent. But the E° values for antimony glasses and other required species at high temperature are not available in the literature, therefore for simplicity of understanding and considering the relative difference of E^0 values may be unaltered and that the trend or movement of alteration of the E^0 values of different redox pairs at high temperature of the glass melt would not be much different than in aqueous solution at room temperature, here we have used the room temperature E° for simple systems at equilibrium with air.

Noble metal nanoparticles of sizes below the electron mean free path incorporated in dielectrics exhibit the phenomenon of "*surface plasmon resonance*" (SPR) [3]. In presence of incident light, the free electrons in the conduction bands collectively oscillates to and fro with respect to the positive core and when the incident electromagnetic frequency is resonant with the frequency of oscillation, intense absorption bands in the visible-near-UV region results. In case of the small spherical particles this collective oscillation of the electrons is called the dipole plasmon resonance to distinguish from plasmon excitation that can occur in bulk metal or metal surfaces. The oscillation frequency is determined by four factors: the density of electrons, the effective electron mass, and the shape and size of the charge distribution. In addition, the surrounding dielectric medium also exerts considerable influence [3,58]. So the position of the SPR band in these nanocomposites may be tuned by monitoring the refractive indices of the glass matrices, size, concentration and interparticle distance of the nanoparticles [3,58].

To study the effect of Au concentration on SPR band in monolithic K_2O-B_2O_3-Sb_2O_3 (KBS) glasses, the following compositions were selected. Some of their properties are also listed in Table 4.

All the Au-doped antimony glass nanocomposites are dichroic, that is, they transmit blue light and reflect the brown light (Fig. 12). The intensity of dichroism increases with increase in Au concentration.

Table 4. Composition and some properties of Au^0-antimony glass dichroic nanocomposites

Sample identity no.	Concen-tration of Au (wt %)[a]	Color of transmitted light	Color of reflected light	SPR band position, λ_{max} (±1, nm)
B	-	Yellow	Yellow	-
G1	0.001	Blue	Brown	610
G2	0.003	Blue	Brown	610
G3	0.1	Blue	Brown	680
G4	0.3	Blue	Reddish- brown	681

[a] Base glass (B) composition (mol %): $15K_2O$-$15B_2O_3$-$70Sb_2O_3$.

The dichroism possibly originates due to elliptical Au nanoparticles in these nanocomposites, as seen from the TEM image (Fig. 13a). The representative TEM image (Fig. 13a) (of nanocomposites G-2) shows of the Au^0 NPs nanoparticles majority of which have elliptical. The major axis (diameter) of the NPs ranges from 3 to 13 nm (± 3 %) while the average aspect ratio of the elliptical Au^0 NPs is found to be about 1.2. Here the mechanism of deformation or the formation of elliptical nanoparticles by the simple single-step melt-quench process is probably due to the influence of the high viscosity of the molten Sb_2O_3 based matrix on the Au^0 colloids. The SAED image shows the presence of (111) and (200) crystallographic planes of Au^0 and consents well with XRD patterns.

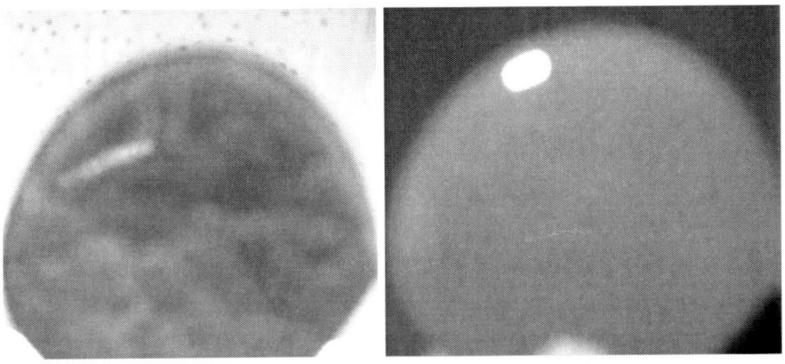

(a) (b)

Figure 12. Photograph showing dichroic behavior of Au-antimony glass nanocomposites: (a) blue in transmitted light and (b) brown in reflected light.

Prolate ellipsoid metal NPs are known to exhibit the phenomenon of dichroism due to the difference in polarizations along the major (longitudinal) and minor (transverse) axes of a polarizable ellipsoidal nanoparticle in presence of electromagnetic wave [54-57]. It has also been observed that for such elliptical (elongated) Ag NPs the SPR peak obtained from the interaction of transversely polarized light is different from the SPR peak position acquired in presence of longitudinally polarized light corresponding to transverse and longitudinal electron oscillations (with respect to the particle principal axis) respectively [54-57]. So it is possible to selectively excite each of them by using light with oscillating electric field parallel and perpendicular to the major axis of the ellipsoids (polarization dependent optical response), consequently such dichroic nanocomposites are therefore possible candidates for displays or encoding security systems and for making dichroic glass polarizers [3,54,54].

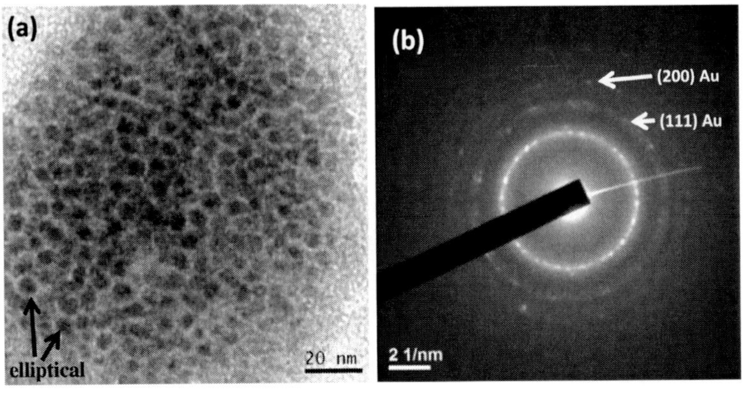

Figure 13. (a) TEM image of nanocomposite G-2 and (b) the SAED pattern of a Au nanoparticles (for composition see Table 4).

The XRD spectra provide another evidence of the presence of Au nanocrystals in the nanocomposites (Fig. 14). In the XRD band (curve-a), particularly the hump between 25° to 35° indicates its amorphocity due to presence of uncrystallized Sb_2O_3 or other constituents in the base glass (B). The sharp peak at 2θ = 38.4021° (d = 2.34216 Å) and 44.1634° (d = 2.04906 Å) (curves b-e) is assigned as (111) and (200) diffractions of fcc Au^0 nanoparticles (JCPDS, card file no. 4-0784) embedded within the antimony borate glass matrix.

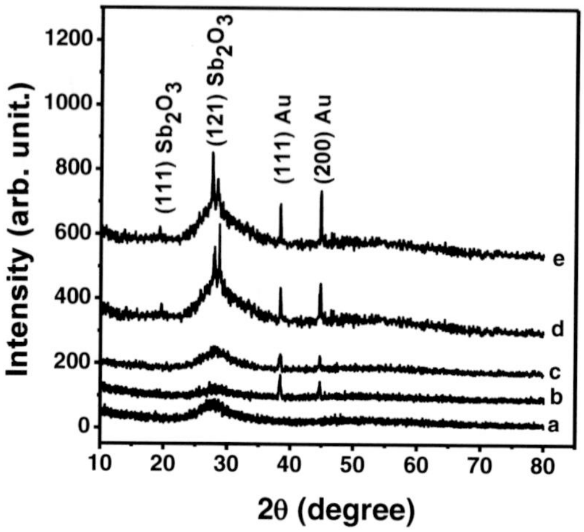

Figure 14. XRD pattern of (a) base glass B having composition (mol %): $15K_2O$-$15B_2O3$-$70Sb_2O_3$, (b) G1, (c) G2, (d) G3, and (e) G4 showing nano Au^0 and crystallized valentinite type phases of Sb_2O_3 (for composition see Table 1).

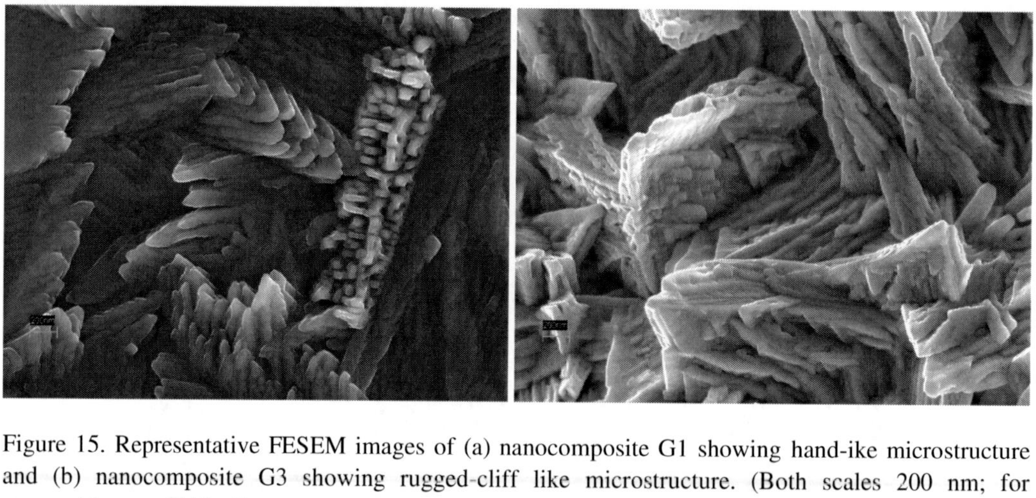

Figure 15. Representative FESEM images of (a) nanocomposite G1 showing hand-ike microstructure and (b) nanocomposite G3 showing rugged-cliff like microstructure. (Both scales 200 nm; for composition see Table 4).

In the XRD spectra (Fig. 14), the sharp peaks at 2θ = 19.5366° (d = 4.54013 Å), 25.5436° (d = 3.48443 Å) and 28.5159° (d = 3.12764 Å) corresponds to (110), (111) and (121) diffractions of valentinite form of Sb_2O_3 crystals (JCPDS, card file no. 11-689) created within the antimony borate glass matrix due to crystallization of some Sb_2O_3 of the matrix by very high concentrations of metallic Au precipitated throughout the body of the glass and acting as heterogeneous nuclei and catalyzing the crystallization of Sb_2O_3 in K_2O-B_2O_3-Sb_2O_3 system [60]. The representative FESEM images (Fig. 15 a and b) of nanocomposites G1 and G3 shows distinct microstructure tuning with increasing Au concentration. Such crystallization results because the (111) plane of Au crystal resembles the (012) and (200) planes of valentinite with a disregistry of approximately 14% and 4% respectively. This falls well within the maximum disregistry of 10 to 15 % suggested for good nucleation catalysis [61].

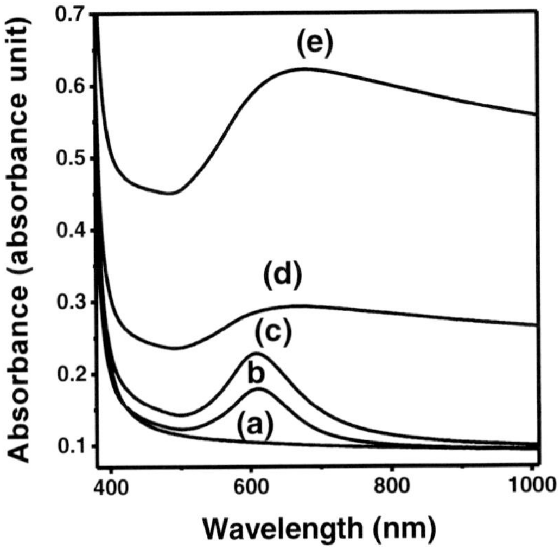

Figure 16. UV-Vis absorption spectra of (a) based glass B and nanocomposites (b) G1, (c) G2, (d) G3 and (e) G4 (for composition see Table 4).

The UV-Visible absorption spectrum (Fig. 16) is one of the most sensitive tools to detect the formation of plasmonic metal NPs. The UV-Vis-NIR absorption spectrum of the undoped glass (curve-a) shows absence of any features but the absorption spectra of the Au doped NCGs (curves b-e) display well-defined broad plasmon (SPR) absorption bands (610-681 nm) characteristic of nano sized Au^0.

Typically the SPR for nano Au in sodaline silicate glass (refractive index ~ 1.5) appear around 520 nm for near-spherical nanoparticles. But here in KBS antimony glass the maxima of the plasmon peaks (λ_{max}), listed in Table 4, experience a distinctive red-shift towards higher wavelength (from 610 nm to 681 nm) with increase in Au^0 concentration (from 0.001 to 0.3 wt %). At higher concentration the SPR band gradually broadened and their tails have extended up to 1100 nm. The curves are very similar to that obtained by Pardiñas-Blanco (Fig. 3.11) [62]. This type of complex system where a high refractive index dielectric matrix encompasses closely spaced elliptical nanoparticles having electromagnetic coupling, the need arises for theory that can describe the electrodynamics of such nanoparticles [63].

According to the Mie theory, for a collection of small spherical non-interacting metallic spheres (low particle density) in the quasi-static regime of radius (R) having a dielectric function (complex) that depend on the angular frequency of light ω, $\varepsilon_{Au}(\omega) = \varepsilon_1(\omega) + i\varepsilon_2(\omega)$, embedded in a medium of dielectric constant ε_m (real), and sizes much smaller than the wavelength of light ($D = 2R << \lambda$) (about 25 nm for gold particles), only the dipole oscillations contribute significantly and the extinction (= absorption + scattering) coefficient C_{ext} is given by [63]:

$$C_{ext} = [24\pi^2 R^3 \varepsilon_m^{3/2} / \lambda] \times [\varepsilon_2 / \{(\varepsilon_1 + 2\varepsilon_m)^2 + \varepsilon_2^2\}] \quad (5)$$

where N is the number of particles, V_{np} the volume of each particle in the quasi-static regime ($V_{np} = 4\pi R^3/3$). The real (ε_1) and imaginary (ε_2) parts of the dielectric function $\varepsilon_{Au}(\omega)$ may be expressed according to the Drude model as [63,64]:

$$\varepsilon_1(\omega) = \varepsilon_\infty - \omega_p^2/(\omega_p^2 + \omega_d^2) = \varepsilon_\infty - \lambda^2/(\omega_p^2 + \omega_d^2) \quad (6)$$

$$\varepsilon_2(\omega) = \omega_p^2 \omega_d/\omega(\omega^2 + \omega_d^2) = \lambda(\lambda^2 + \lambda_d^2)/\lambda_p^2 \lambda_d \quad (7)$$

where ε_∞ is the high-frequency value of the dielectric function or optical dielectric function of the metal and $\omega_p = 2\pi c/\lambda_p = (n_{eff} e^2/m_{eff}\, \varepsilon_o)^{1/2}$ is the metal's bulk plasma frequency while λ_p is the metal's bulk plasma wavelength, $\omega_d = V_F/R_{bulk}$ is the damping or relaxation frequency, n_{eff} is the free electron concentration in the metal, m_e is effective mass of conduction electrons, V_F is the Fermi velocity of the metallic electrons, R_{bulk} is the mean free path of the oscillating conduction band electrons and ε_o is vacuum dielectric constant or free space permeability. Such particles will retaliate as dielectric media in an optical field. If ε_i is very small or weakly dependent on ω then, from Eq. (5), the absorption maximum corresponding to the resonance condition is produced when $\varepsilon_1 = -2\varepsilon_m$, leading to a disappearing denominator. Hence, a SPR absorption is produced at optical frequency ω at which the resonance condition $\varepsilon_1 = -2\varepsilon_m$ is fulfilled [63,64].

When $\omega_p >> \gamma$ (damping), the SPR absorption maxima λ_{max}, is susceptible to the changes of refractive index, n_m of the surrounding medium as [63,64]:

$$\lambda_{max} = \lambda_p^2 (\varepsilon_\infty + 2\varepsilon_m) = 2\pi c\, m_{eff}\, \varepsilon_o\, [(\varepsilon_\infty + 2n_m^2)]^{1/2} / n_{eff} e^2 \quad (8)$$

where $n_m = (\varepsilon_m)^{1/2}$ Thus, the SPR peak exhibits a red-shift with increase in medium refractive index [63,64].

To study the effect of medium refractive index on plasmon band of Au in monolithic K_2O-B_2O_3-Sb_2O_3 (KBS) glasses, the following compositions were selected (Table 5).

Table 5. Composition and some properties of Au^0-devitrified antimony glass dichroic nanocomposites

Nano-composite no.	Glass composition (mol %)			Concentration of Au (wt%)[a]	Refractive index	SPR maxima, λ_{max} (nm)	Transmitted color	Reflected color
	K_2O	B_2O_3	Sb_2O_3					
G5	10	10	80	0.003	2.0006	611	Green	Reddish-brown
G6	15	15	70	0.003	1.9477	610	Bluish green	Brown
G7	20	20	60	0.003	1.8925	599	Blue	Brown
G8	25	25	50	0.003	1.8349	587	Bluish-violet	Brown

Since the Au concentration remains constant, the tuning of the SPR wavelength is essentially accomplished by varying the content of high refractive index Sb_2O_3. With increase of refractive index of the matrix from 1.8349 to 2.0006, there is a red shift of 24 nm of SPR band of Au. This results in different colored nanocomposites (Fig. 17).

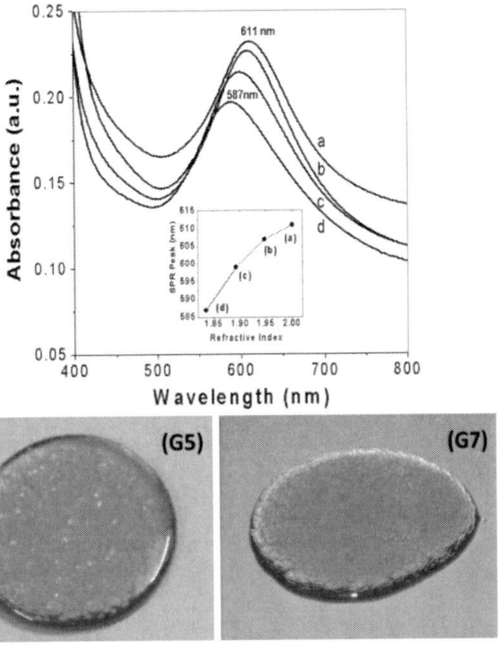

Figure 17. UV-Vis absorption spectra of nanocomposites (a) G5, (b) G6, (c) G7, and (d) G8. Inset plot shows the red shift of Au^0-SPR band with increase in refractive index of the embedding matrix. Photographs show the transmitted color of as-prepared G5 (green) and G7 (blue) (for composition see Table 5).

The effects of the medium refractive index on the optical properties of noble metal nanostructures have predominantly captured rising fundamental and technological interest in recent times. Consequently, tuning of the SPR bands of Au and Ag nanoparticles have been experimentally in a variety of surroundings including solutions, polymers and sol-gel derived hybrid matrices [3, 63, 65,66]. Most of metal-glass composites to date have been comprised of spherical NPs inserted within silicate or phosphate matrices where to make a wide variation of refractive index is very difficult. Consequently, refractive index controlled plasmon tuning has hardly been realized within glass matrices. But antimony glasses provide the unique opportunity of plasmon tuning by controlling the fraction of Sb_2O_3 (high refractive index material) in glasses.

Within the dipole approximation, the Gans theory, which is an extension of Mie theory, is applicable for small ellipsoidal (prolate and oblate) particles. Here the plasmon resonance may split into two distinct modes depending on the aspect ratio. Such splitting is the result of surface curvature which decides the restoring force or depolatization effect that acts on the resonating conduction band electrons. Consequently, the extinction cross-section, C_{ext}, of ellipsoidal particle is directly proportional to its imaginary part of polarizability, α_{ED} as [63,64]:

$$C_{ext} = 2\pi n_h I(\alpha_{ED})/ \lambda \quad (9)$$

where

$$\alpha_{ED} = V\varepsilon_m \frac{\varepsilon_{Au} - \varepsilon_m}{\varepsilon_m + L_j(\varepsilon_{Au} - \varepsilon_m)} \quad (10)$$

j = x, y, z denotes the principal axes of the ellipsoid, the volume of the ellipsoidal nanoparticle, $V_{ED}=(4R_xR_yR_z)/3$ and Lj are depolarization factors given by [3,63,64]:

$$L_x = [(1-e^2)/e^2][-1 + (1/2e)\ln(1+e)/(1-e)] \quad (11)$$

$$L_{y,z} = (1-L_x)/2 \quad (12)$$

where $e = [1-(b/a)^2]^{1/2}$ is the eccentricity of the ellipsoid. For the degenerate case of a sphere e = 0, or radius $R_x = R_y = R_z$, $Lj = 1/3$. under these conditions, the birefringence disappears. Thus oriented ellipsoids exhibit strong polarization-dependent optical spectra. Simply, nanocomposites exhibits dichroic behavior, that is, one color in the transmitted light and a different color in the reflected light [63]. Consequently, the new SPR position is:

$$\varepsilon_{Au} = (-1/Lj + 1)\,\varepsilon_m, \quad (13)$$

It is thus established SPR is intimately controlled by the nanoparticle size, shape, refractive index of the dielectric host (here glass) and other proximal NPs.

3.4. Nano Metal: Rare Earth Ion- Antimony Glass Hybrid Nanocomposites for Enhanced Photoluminescence Applications

Nano metal: RE ion-glass hybrid nanocomposites have very recently received startling importance as potential candidates of nanophotonic applications. These intelligent solid state materials exhibiting enhanced photoluminescence are promising for applications in some current thrust areas of materials science like advanced displays, light emitting diodes, solar cells (energy harvest) and communications [3].

SPR by metallic nanoparticles not only result in strong absorption bands but also tremendous enhancement of the local electromagnetic field and intense light scattering [3]. The local electromagnetic field enhancement may be so strong that luminescent species in the vicinity of the plasmonic noble metal nanoparticle (~ 10 nm apart) may feel up to a 100 times more intense field as opposed to the direct excitation in a plasmon-free environment [67]. These enhanced near-fields around metallic nanostructures have already fostered a variety of intriguing applications such as non-linear optics, surface-enhanced Raman scattering (SERS), single-molecule detection (spectroscopy), nano-metal enhanced fluorescence (NMEF) and phosphorescence (NMEP), etc [68].

Originally inspired by the discovery of SERS in 1974, Malta et al. [69] first demonstrated the luminescence enhancement of the Eu^{3+} ions by silver-nanoparticles in fluoro-borate glasses prepared by conventional melting and annealing technique which incorporated a small quantity of reducing agent (1.5-3 wt% Sb_2O_3). The phenomenon was explained as silver plasmonic enhancement effect and long range electromagnetic interactions associated with SPR. Their results were further established by Hayakawa et al [70] who observed luminescence enhancement of Ag NPs–Eu^{3+} ions co-doped sol-gel derived silica glasses (synthesized by using partially hydrolyzed tetrakis(hydroxymethyl) phosphonium chloride as the reducing agent and annealing in hydrogen atmosphere) and found that irrespective of Ag concentration, the decay from the $^5D_0$ level of Eu^{3+} almost have the same life-time.

Thereafter, several researchers worldwide took up NMEL studies of Er^{3+}, Sm^{3+}, Tb^{3+}, Eu^{3+}, Dy^{3+}, etc ions by monometallic Au or Ag NPs in glasses and other dielectric offering the possibilities of plamon induced nanphotonics engineering [67,71-77]. However, the simultaneous studies also led to several controversies related to metal–RE interactions in a dielectric host and the mechanism of photoluminescence enhancement.

Strohhöfer and Polman [71] reported the photoluminescence enhancement of Er^{3+} ions implanted in boro-silicate glasses that had been subjected to $Na^+ \leftrightarrow Ag^+$ ion exchange via absorption at a defect related to a pair of silver ions/atoms and energy transfer towards Er^{3+}. Matarelli et al. [72] described electric-dipole induced energy transfer process from ionic silver aggregates, particularly dimmers, accounting the enhancement of 1.52 μm emission of Er^{3+} in silver-doped silicate glasses fabricated by ion-exchange technique. These contrasting conclusions are probably because when metal ions are introduced in glasses by conventional ion-exchange or ion-implantation they are very mobile and with heat treatment have a strong tendency to aggregate. This results not only in formation of nanoparticles, but single ions, cluster of few atoms (dimers, trimers, etc.), and other luminescent optical centers, examples are Ag^+, $(Ag_2)^+$, $(Ag^+)_2$, $(Ag_3)^{2+}$ for silver. Moreover low temperature small annealing schedules results in large fraction of the metal to form sub-nm aggregates that are two small to have a continuous density of states necessary for evolution of SPR typical of larges free-electron metal nanoparticles.

Recent studies of nanometal enhanced rare-earth luminescence in glasses have revealed plasmonic induced local field enhancement as the primary cause particularly when the pumping excitation or emission is resonant or lies near the SPR wavelength [73-77]. Energy transfer from metal nanoparticles has also been regarded as a contributing effect when excitation or emission is non-resonant with SPR provided the chosen metal concentration is below the threshold concentration beyond which quenching may occur [73-77].

Although, the luminescence of Eu^{3+} ion by gold and silver nanoparticles have been studied in bismuth and tellurite glasses [75,76], but there is no report of nanometal enhanced Eu^{3+} luminescence in antimony glasses. The selective reduction capability of Sb_2O_3 enables the synthesis of nano Au:Eu^{3+}-KBS glass nanocomposites in a single-step. Although Sb_2O_3 reduces Au(III) to Au^0, but its reducing potential is not strong enough to reduce Eu^{3+} to its lower valent states $E^o_{Eu3+/Eu0} = -1.991$ V, $E^o_{Eu3+/Eu2+} = -0.36$ V and $E^o_{Eu2+/Eu0} = -2.812$ V] [78].

The effect of Au nanoparticles of various concentrations on the luminescence (downconversion) of Eu^{3+} ion in KBS antimony glass, having base glass composition (mol%) $15K_2O$-$15B_2O_3$-$70Sb_2O_3$ is shown in Fig. 18.

The $^5D_0 \rightarrow {}^7F_1$ orange transitions are magnetic-dipole (MD) allowed and are not much sensitive to changes in the crystal field environment of the host matrix. But $^5D_0 \rightarrow {}^7F_2$ red emission transitions are allowed by electric-dipole (ED) and are influenced by the crystal field environment in the vicinity of the Eu^{3+} ions. Their amplitudes are susceptible to changes in the polarizability of the ligand and reduction of the local symmetry around the Eu^{3+} ions.

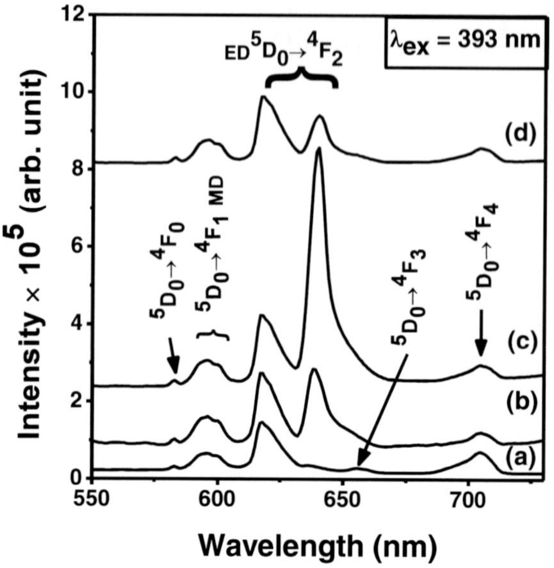

Figure 18. Fluorescence spectra of (a) 0.3 wt% Eu^{3+}-doped, (b) 0.003 wt% Au and 0.3 wt% Eu^{3+}-codoped, (c) 0.03 wt% Au and 0.3 wt% Eu^{3+}-codoped and (d) 0.3 wt% Au and 0.3 wt% Eu^{3+}-codoped KBS glass, under excitation wavelength at λex = 393 nm. The bases of the emission curves b, c and d have been uplifted for clear visibility.

The formation of Au nanocrystallites introduces local crystalline environment and perturbs the local site symmetry resulting in well resolved Stark's splitting of the hypersensitive (ΔJ=2) $^5D_0 \rightarrow ^7F_2$ red emission transition (617 and 640 nm) due to crystal field effect. The principal result of the current study is that the 640 nm (deep-red) fluorescence of Eu^{3+} initially drastically enhances in presence of Au^0 NPs and then diminishes at very high Au concentration. The enhancement of ED transition is as high as 8 folds while the MD transition is much less affected. This indicate that Eu^{3+} ions are located at least into two different crystalline environments, namely, a clustered crystalline environment (due to coalescence of some Au nanoparticles) and dispersed one. The emission spectra are indicative of the rare-earth sites. Earlier Kassab et al [75,76] have showed that the dominant hypersensitive $^5D_0 \rightarrow {}^7F_2$ electric dipole (red) transition is increased by 10 fold by Au nanoparticles while the less sensitive $^5D_0 \rightarrow {}^7F_1$ magnetic dipole (orange) transition is enhanced only by 5 folds.

We believe that the enhanced luminescence is primarily due to local field enhancement (LFE) around the Eu^{3+} ions induced by SPR of Au NPs. The SPR results in generation of electromagnetic waves at the interface of the dielectric medium and metal. Concentration of these waves results in enhanced local fields around the metal nanoparticles. The enhancement is dramatic at the interparticle region called "hot spots". Any luminescent species at the vicinity of metal nanoparticles or located at the "hot spots" experience enormous increase in their excitation rate compared to the incident field. Consequently, their emission also drastically increases. At very high Au concentration there is quenching due to reverse energy transfer from Eu^{3+} to Au and plasmonic re-absorption.

According to the classical electromagnetic theory, local field enhancement factors for ellipsoidal particles are given as [79]:

$$\eta_x(\omega) = \frac{\varepsilon_m L_{LR}}{\varepsilon_{Au} - \varepsilon_m + L_{LR}\left(\varepsilon_m + i\frac{4\pi^2 V}{\lambda^3}(\varepsilon_m - \varepsilon_{Au})\right)} \quad (14)$$

$$\eta_{y,z}(\omega) = \frac{\varepsilon_m \frac{2L_{LR}}{L_{LR} - 1}}{\varepsilon_{Au} - \varepsilon_m + \frac{2L_{LR}}{L_{LR} - 1}\left(\varepsilon_m + i\frac{4\pi^2 V}{\lambda^3}(\varepsilon_m - \varepsilon_{Au})\right)} \quad (15)$$

where η_x, $\eta_{z,y}$ are the enhancement factors for the electric field vectors directed along the axis of revolution of ellipsoid (x) and perpendicular to this axis, respectively. L_{LR} is lightning rod factor due to enhanced filed contribution from sharp edges (or rough surfaces) of anisotropic nanostructures. The field enhancement factor, η is defined as the ratio of the amplified local field and the incident field, $|\eta(\omega)| = |E_{loc}|/|E_i|$.

We may eliminate the possibility of energy transfer from Au nanoparticles to Eu^{3+} ions as the excitation wavelength does not overlay the SPR absporption. Moreover, our luminescence experiments indicated the absence of any other optical centers. Apart from the Eu^{3+} emission bands, any other emission band was not obtained under excitation at various wavelengths which could be attributed to any ionic species of Au. This is because we used exceedingly large quantity of reducing agent (89.25 wt% Sb_2O_3) compared to $HAuCl_4.H_2O$ (0.003-0.3 wt% Au).

The Au:Sm^{3+}-KBS and Au:Er^{3+}-KBS glasses synthesized by similar single step technique also exhibit intensified upconversion emissions due to local field effect of Au nanoparticles [21,22].

CONCLUSIONS

A novel series of antimony trioxide based monolithic glasses and monolithic devitrified glasses in the systems K_2O-B_2O_3-Sb_2O_3 (KBS) having low melting, low softening point, high thermal expansion and high dielectric constant are developed. Their properties are explained with evaluation of fundamental properties like covalent character and optical basicity. The glasses are impending candidates for electrical packaging industry. It is also established that they possess low phonon energy (600 cm^{-1}) comparable to fluoride glasses. On doping with Sm_2O_3 they emit extraordinary red upconverted light, with Er_2O_3 both intense green and red upconversion emissions are observed with NIR excitations. On doping with Eu_2O_3, green, orange and red down-shifting emissions are exhibited upon UV excitation the intensity of which is dependent on the concentration of Sb_2O_3 in the glasses. RE-doped antimony glass are found to be promising candidates for visible lasers and display applications and possible substitutes for fluoride glass. The mild (selective) reducing property of Sb_2O_3 enables the single-step synthesis of nano metal-glass nanocomposites and RE-nano metal-glass hybrid nanocomposites for the first time. Dichroic Au-doped antimony glass nanocomposites exhibiting shape, size, inter-particle distance and refractive index dependent plasmon absorption bands may be used for various plasmonic applications. Namo metal- rare earth ion hybrid antimony glass nanocomposites exhibiting intense photoluminescence plasmon induced due to local field enhancement of metal nanoparticles, may be used for various nanophotonic applications promising new materials for various nanophotonic applications, like in solar cells, light emitting diodes (LEDs), advanced displays, etc. These pioneering findings so-far explored in rare-earth doped Sb_2O_3 glasses and Sb_2O_3 glass based nanometal-glass nanocomposites are expected to set a new prototype field of unusual glasses.

ACKNOWLEDGEMENTS

TS expresses her sincere gratitude to Council of Scientific and Industrial Research (CSIR-India) for award for NET-SRF research grant under sanction number 31/015(0060)/2007-EMR-1. The authors thank the technical supports provided by the infrastructural facility of CGCRI, Kolkata and TEM facility at IACS, Kolkata.

REFERENCES

[1] Yamane, M. and Asahara, Y. (2000). *Glasses for Photonics*. Cambridge University Press.

[2] Gonella, F. and Mazzoldi, P. (2000). "Metal Nanocluster Composite Glasses." In: *Handbook of Nanostructured Materials and Nanotechnology*. Nalwa H. S., Ed., vol 4, San Diego, Academic Pres.

[3] Prasad, P. N. (2004). *Nanophotonics*. New Jersey, Wiley, pp. 129-151.

[4] Vogel, W. (1992). *Glass Chemistry*. Berlin, Springer-verlag.

[5] Zachariasen, W. H. (1932). "The atomic arrangements in glass." *Journal of the American Chemical Society* 54, 3841-3851.

[6] Kordes, E. (1939). "Physikalische-chemische untersuchungen über feinbau von gläsern, III (Physicochemical researches on fine structure of glasses: III Binary and pseudobinary glasses without appreciable packing effect)." *Zeitschrift für Physikalische Chemie B* 43, 173-190.

[7] Hedden W. A. and King, B. W. (1956). "Antimony oxide glasses." *Journal of the American Ceramic Society* 39, 218-222.

[8] Bednarik J. F. and Neely, J. A. (1982). "Thermal properties of single component antimony oxide glass." *Physics and Chemistry of Glasses* 23, 204-205.

[9] Terashima, K., Hashimoto, T., Uchnio, T., Kim, S. and Yoko, T. (1996). "Structure and nonlinear optical properties of Sb_2O_3-B_2O_3 binary glasses." *Journal of the Ceramic Society of Japan* 104, 1008-1014.

[10] Nalin, N., Poulain, M. J., Poulain, M. A., Ribeiro, S. J. L. and Messaddeq, Y. (2001). "Antimony oxide based glasses." *Journal of Non-Crystalline Solids* 284, 110-116.

[11] Nalin, M., Messaddeq, Y., Ribeiro, S. J. L., Poulain, M., Briois, V., Brunlkaus, G., Rosenhahn, C., Mosel, B. D. and Eckert, H. (2004) "Structural organization and thermal properties of the Sb2O3–SbPO4 glass system." *Journal of Materials Chemistry* 14, 3398-3405.

[12] Holland, D., Hannon, A. C., Smith, M. E., Johnson, C. E., Thomas, M. F. and Beesley, A. M. (2004) "The role of Sb^{5+} in the structure of Sb_2O_3-B_2O_3 binary glasses-an NMR and Mössbauer spectroscopy study." *Solid State Nuclear Magnetic Resonance* 26, 172-179.
[13] Chatlani S. and Shelby J. E. (2006). "Properties of arsenic and antimony borate glasses." *Physics and Chemistry of Glasses: European Journal of Glass Science and Technology B* 47, 288-293.
[14] Zhang, B., Chen, Q., Song, L., Li, H., Hou, F. and Zhang, J. (2008). "Fabrication and properties of novel low-melting glasses in the ternary system ZnO-Sb_2O_3-P_2O_5." *Journal of Non-Crystalline Solids* 354, 1948-1954.
[15] Som, T. and Karmakar, B. (2010). "Structure and properties of low-phonon antimony glasses and nano glass-ceramics in K_2O-B_2O_3-Sb_2O_3 system." *Journal of Non-Crystalline Solids* 356, 987-999.
[16] Som, T. and Karmakar, B. (2010). "Optical properties of Eu^{3+}-doped antimony-oxide-based low phonon disordered matrices." *Journal of Physics: Condensed Matter* 22, 035603 (11 pages).
[17] Som, T. and Karmakar, B. (2008). "Infrared-to-red upconversion luminescence in samarium-doped antimony glasses." *Journal of Luminescence* 128, 1989–1996.
[18] Som, T. and Karmakar, B. (2009). "Efficient green and red fluorescence upconversion in erbium doped new low phonon antimony glasses." *Optical Materials* 31, 609–618.
[19] Som, T. and Karmakar, B. (2010). "Surface plasmon resonance and enhanced fluorescence application of single-step synthesized elliptical nano gold embedded antimony glass dichroic nanocomposites." *Plasmonics* 5, 149-159.
[20] Som, T. and Karmakar, B. (2009). "Plasmon tuning of nano-Au in dichroic devitrified antimony glass nanocomposites by refractive index control." *Chemical Physics Letters* 479, 100–104.
[21] Som, T. and Karmakar, B. (2010). "Enhanced frequency upconversion of Sm^{3+} ions by elliptical Au nanoparticles in dichroic Sm^{3+}: Au-antimony glass nanocomposites." *Spectrochimica Acta Part A* 75, 640–646.
[22] Som, T. and Karmakar, B. (2009). "Enhancement of Er^{3+} upconverted luminescence in Er^{3+}: Au-antimony glass dichroic nanocomposites containing hexagonal Au nanoparticles" *Journal of the Optical Society of America B: Optical Physics* 26, B21-B27.
[23] Orman R. G. and Holland, D. (2007). "Thermal phase transitions in antimony (III) oxides." *Journal of Solid State Chemistry* 180, 2587-2596.
[24] Miller P. J. and Cody, C. A. (1982). "Infrared and raman investigation of vitreous antimony trioxide." *Spectrochimica Acta Part A: Molecular and Biomolecular Spectroscopy* 38, 555-559.
[25] Kamitsos, E. I., Patsis, A. P. and Chryssikos, G. D. (1993). "Infrared reflectance investigation of alkali borate glasses." *Journal of Non-Crystalline Solids* 152, 246-257.
[26] Motke, S. G., Yawale, S. P. and Yawale, S. S. (2005). "Infrared spectra of zinc doped lead borate glasses." *Bulletin of Materials Science* 25. 75-78.
[27] Scolze, H. (1991). *Glass Nature, Structure and Properties.* New York, Springer-Verlag, p. 145.
[28] Reddy, M. S., Raju, G. N., Nagarjuna, G. and Veeraiah, N. (2006). "Structural influence of aluminium, gallium and imdium metal oxides by means of dielectric and spectroscopic properties of CaO-Sb_2O_3-B_2O_3 glass system." *Journal of Alloys and Compounds* 464, 472-482.

[29] Reisfeld, R. (2005). *Nanostructured and Advanced Materials*. Vaseashta, et al. Eds., Springer, p. 77.

[30] Mott N. F. and Davis E. A. (1979). *Electronic processes in Non-Crystalline Materials.* 2nd Edition, Oxford, Clarendon Press.

[31] Karmakar, B. (2005). "IRRS, UV-Vis-NIR absorption and photoluminescence upconversion in Ho^{3+}-doped oxyfluorophosphate glasses." *Journal of Solid State Chemistry* 178, 2663-2672.

[32] Pauling, L. (1960). *The Nature of the Chemical Bond.* 3rd Edn., NewYork, Cornell University Press.

[33] Duffy, J. A. and Ingram, M. D. (1991). *Optical Properties of Glass.* Kreidl, N et al. Eds., Westerville, OH, American Ceramic Society.

[34] Duffy, J. A. and Ingram, M. D. (1992). "Comments on the application of optical basicity to glass". *Journal of Non-Crystalline Solids* 144, 76-80.

[35] Dimitrov, V. and Komatsu, T. (2002). "Classification of simple oxides: A polarizability approach" *Journal of Solid State Chemistry* 163, 100-112.

[36] Shackelford, J. F., Alexander, W. and Park J. S. (Eds.). (1994) *CRC Materials Science and Engineering Handbook.* Boca Raton, CRC Press, pp. S-14-21.

[37] Blech, I. A. (1986). *Properties of Materials in Electronics Engineering Handbook.* 2nd edition. Fink, D. G. and Christiansen D. Ed., New York, McGraw-Hill, pp. 6–30.

[38] de Araujo, R. E., de Araujo, C. B., Poirier, G., Poulain, M. and Messaddeq, Y. (2002). "Nonlinear optical absorption of antimony and lead oxyhalide glasses." *Applied Physics Letters* 81, 4694 – 4696.

[39] Dimitrov V. and Komatshu, T. (1999). "Electronic ion polarizabillity, optical basicity and metal (or nonmetal) binding energy of simple oxides." *Journal of the Ceramic Society of Japan* 107, 879-886.

[40] Biju, L P. R., Jose, G., Jyothy, P. V., Thomas, V. and Unnikrishnan, N. V. (2005). "Upconversion fluorescence in Sm^{3+} -doped zinc phosphate glassy matrix." *Journal of Modern Optics* 52, 2687-2694.

[41] Tripathi, G., Rai, V. K. and Rai, S. B. (2007). "Upconversion and temperature sensing behavior of Er^{3+} doped Bi_2O_3-Li_2O-BaO-PbO tertiary glass." *Optical Materials* 30, 201-206.

[42] Lin, H., Tanabe, S., Lin, L., Yang, D. L., Liu, K., Wong, W. H., Yu, J. Y. and Pun, E. Y. B. (2006). "Infrequent blue and green emission transitions from Eu^{3+} in heavy metal tellurite glasses with low phonon energy." *Physics Letter A* 358, 474-477.

[43] Xu, S., Fang, D., Zhang, Z. and Jiang, Z. (2005). "Effect of OH^- on the upconversion luminescence of Er^{3+}-doped oxyhalide tellurite glasses." *Journal of Solid State Chemistry* 178, 2159-2162

[44] Pisarski, W. A., Pisarska, J., Mączka, M. and Ryba-Romanowski, W. (2006). "Europium-doped lead fluoroborate glasses: Structural, thermal and optical investigations." *Journal of Molecular Structure* 792–793, 207-211.

[45] W.A. Weyl, Coloured Glasses, Society of Glass Technology, Sheffield, 1951.

[46] Doremus, R. H. (1964). "Optical properties of small gold particles." *The Journal of Chemical Physics* 40, 2389-2396.

[47] Eichelbaum, M., Rademann, K., Weigel, W., Löchel, B., Radtke, M. and Müller, R. (2007). "Gold-ruby glass in a new light: On the microstructuring of optical glasses with synchrotron radiation." *Gold Bulletin* 40(4), 278-282.

[48] Gonella F. (2007). "Metal nanocluster composite silicate glasses." *Rev.Adv.Mater.Sci.* 14, 134-143.
[49] de Lamaestre, R. E., Béa, H., Bernas, H., Belloni, J. and Marignier, J. L. (2007). "Irradiation-induced Ag nanocluster nucleation in silicate glasses: Analogy with photography." *Physical Review B* 76, 205431 [18 pages].
[50] Shin, J., Jang, K., Lim, K. –S., Sohn, I. –B., Noh, Y. –C. and Lee, J. (2008). "Formation and control of Au and Ag nanoparticles inside borate glasses using femtosecond laser and heat treatment" *Applied Physics A* 93, 923–927.
[51] Dubiel, M., Hofmeister, H. and Wendler, E. (2008). "Formation of nanoparticles in soda-lime glasses by single and double ion implantation." *Journal of Non-Crystalline Solids* 354, 607–611.
[52] Penninkhof, J. J., Polman, A., Sweatlock, L. A., Maier, S. A., Atwater, H. A., Vredenberg, A. M. and Kooi, B. J. (2003). "Mega-electron-volt ion beam induced anisotropic plasmon resonance of silver nanocrystals in glass." *Applied Physics Letters* 83, 4137-4139.
[53] Zhang, J., Dong, W., Sheng, J., Zheng, J., Li, J., Qiao, L. and Jiang, L. (2008). "Silver nanoclusters formation in ion-exchanges glasses by thermal annealing, UV-laser and X-ray irradiation." *Journal of Crystal Growth* 310, 234-239.
[54] Hofmeister, H., Drost, W. –G. and Berger, A. (1999). "Oriented prolate silver particles in glass-characteristics of novel dichroic polarizers." *Nanostructured Materials* 12 207-210.
[55] Podlipensky, A., Abdolvand, A., Seifert, G. and Graener, H. (2005). "Femtosecond laser assisted production of 3D structures in composite glass containing Ag nanoparticles." *Applied Physics A* 80, 1647-1652.
[56] Land, E. H. (1943). "Light polarizer and process of manufacture." *US Patent* 2, 319, 816.
[57] Fort, E., Ricolleau, C. and J. Sau-pueyo, (2003) "Dichroic thin films of silver nanoparticle chain arrays on facetted alumina templates." *Nano Letters* 3, 65-67.
[58] Kelly, K. L., Coronado, E., Zhao, L. L. and Schatz, G. C. (2003). "The optical properties of metal nanoparticles: The influence of size, shape, and dielectric environment." *The Journal of Physical Chemistry B* 107 (2003) 668-677.
[59] Dirix, Y., Bastiaansen, C., Caseri, W. and Smith, P. (1999). "Oriented pearl-necklace arrays of metallic nanoparticles in polymers: A new route toward polarization-dependent color filters." *Advanced Materials* 11, 223-227.
[60] Beall G. H. and Duke, D. A. (1983). *"Glass Ceramic Technology."* In: *Glass: Science and Technology.* Uhlman D. R. and Kreidl N. J. Eds., vol 1, San Diego, Academic Press.
[61] Mcmillan, P.W. (1979). *Glass–ceramics.*, London, Academic Press.
[62] Pardiñas-Blanco, I, Hoppe, C. E., López-Quintela, M.A. and Rivas, J. (2007). "Control on the dispersion of gold nanoparticles in an epoxy network." *Journal of Non-Crystalline Solids* 353, 826–828.
[63] Pérez-Juste, J., Pastoriza-Santos, I., M. Liz-Marzán, L., and Mulvaney, P. (2005). "Gold nanorods: Synthesis, characterization and applications." *Coordination Chemistry Reviews* 249, 1870–1901.
[64] Link S. and El-Sayed, M. A. (1999). "Spectral properties and relaxation dynamics of surface plasmon electronic oscillations in gold and silver nanodots and nanorods." *The Journal of Physical Chemistry B* 103, 8410-8426.

[65] De S. and De, G. (2008). In situ generation of Au nanoparticles in UV-curable refractive index controlled SiO_2-TiO_2-PEO hybrid films." *The Journal of Physical Chemistry C* 112, 10378–10384.

[66] Mulvaney, P. (1996). "Surface plasmon spectroscopy of nanosized metal particles." *Langmuir* 12, 788-800.

[67] Eichelbaum, M. and Rademann, K. (2009). "Plasmonic enhancement or energy transfer? On the luminescence of gold-, silver-, and lanthanide-doped silicate glasses and its potential for light-emitting devices, *Adv. Funct. Mater.* 19, 1–8.

[68] Zhang, J. Z. (2009). "*Optical properties and spectroscopy of nanomaterials*". Singapore, World Scientific Publishing Co. Pte. Ltd.

[69] Malta, O. L., Santa-Cruz, P. A., De Sá, G. F. and Auzel, F. (1985). "Fluorescence enhancement induced by the presence of small silver particles in Eu^{3+} doped materials." *Journal of Luminescence* 33, 261-272.

[70] Hayakawa, T., Selvan, S. T. and Nogami, M. (1999). "Field enhancement effect of small Ag particles on the fluorescence from Eu^{3+}-doped SiO_2 glass." *Applied Physics Letters* 74, 1513 (3 pp).

[71] Strohhöfer, C. and Polman, A. (2002). "Silver as a sensitizer for erbium." *Applied Physics Letters* 81, 1414 (3 pp).

[72] Mattarelli, M., Montagna, M., Vishnubhatla, K., Chiasera, A., Ferrari, M. and Righini, G. C. (2007). "Mechanisms of Ag to Er energy transfer in silicate glasses: A photoluminescence study." *Physical Review B* 75, 125102.

[73] Fukushima, M., Managaki, N., Fujii, M., Yanagi, H. and Hayashi, S. (2005). "Enhancement of 1.54-μm emission from Er-doped sol-gel SiO_2 films by Au-nanoparticles doping." *Journal of Applied Physics* 98, 24316 (4 pp.).

[74] Pan, Z., Ueda, A., Aga Jr., R., Burger, A., Mu, R. and Morgan, S.H. (2010). "Spectroscopic studies of Er^{3+} doped Ge-Ga-S glass containing silver nanoparticles." *Journal of Non-Crystalline Solids* 356, 1097-1101.

[75] de Almeida, R., da Silva, D. M., Kassab, L. R.P. and de Araújo, C. B. (2008). "Eu^{3+} luminescence in tellurite glasses with gold nanostructures." *Optics Communications* 281, 108–112.

[76] Kassab, L. R. P., da Silva, D. S., de Almeida, R. and de Araújo, C. B. (2009). "Photoluminescence enhancement by gold nanoparticles in Eu^{3+} doped GeO_2–Bi_2O_3 glasses." *Applied Physics Letters* 94, 101912.

[77] Jiménez, J. A., Lysenko, S., Liu, H., Fachini, E. and Cabrera, C. R. (2010). "Investigation of the influence of silver and tin on the luminescence of trivalent europium ions in glass." *Journal of Luminescence* 130, 163–167.

[78] Lide, D. R. Ed. (1975). *CRC Handbook of Chemistry and Physics.* 75th edn, Boca Raton, CRC Press, pp.8-21-34.

[79] Varnavski, O. P., Goodson, T., Mohamed, M. B. and El-Sayed, M. A. (2005). "Femtosecond excitation dynamics in gold nanospheres and nanorods." *Physical Review B* 72, 235405.

In: Antimony: Characteristics, Compounds and Applications
Editor: Manijeh Razeghi
ISBN 978-1-62100-598-8

Chapter 8

HIGH TEMPERATURE INTERACTIONS OF ANTIMONY WITH NICKEL

***Olga A. Marina*[1] *and Larry R. Pederson*[2]**
[1]Pacific Northwest National Laboratory, Energy Efficiency Division
PO Box 999, MS: K2-44, Richland, WA, US
[2]North Dakota State University, Center for Nanoscale Science and Engineering
Dept. 4310, PO Box 6050, Fargo, ND, US

ABSTRACT

In this chapter, the surface and bulk interactions of antimony with the Ni-based anodes in solid oxide fuel cells (SOFC) are discussed. High fuel flexibility is a significant advantage of SOFCs, allowing the direct use of fossil and bio fuels without a hydrogen separation unit. Synthesis gas derived from coal and biomass consists of a mixture of hydrogen, carbon monoxide, carbon dioxide, and steam, but finite amounts of tars and trace impurities such as S, Se, P, As, Sb, Cd, Pb, and Cl are also often present. While synthesis gas is commonly treated with a series of chemical processes and scrubbers to remove the impurities, complete purification is not economical. Antimony is widely distributed in coals. During coal gasification, antimony is volatilized, such that contact with the SOFC anodes and other SOFC parts, e.g., interconnect, current collecting wires, fuel gas supplying tubing, is likely. This chapter addresses the following topics: high temperature Ni – Sb interactions; alteration phase, Ni_3Sb, Ni_5Sb_2, NiSb, formation; thermochemical modeling; impact of Sb on the electrocatalytic activity of Ni toward the fuel oxidation and the presence of other impurities (arsenic, in particular); converted anode structural instability during long-term SOFC operation; and comparison with nickel heterogeneous catalysts.

LIST OF SYMBOLS AND ABBREVIATIONS

ASR	Area specific resistance
EBSD	Electron back-scatter diffraction
EDS	Energy dispersive spectrometry

LEED	Low-energy electron diffraction
SEM	Scanning electron microscopy
SOFC	Solid oxide fuel cell
XPS	X-ray photoelectron spectroscopy
YSZ	Yttria-stabilized zirconia

1. Introduction

Coal and biomass may be converted to synthesis gas, a mixture consisting principally of hydrogen, carbon monoxide, carbon dioxide, steam, and methane, and then used in the production of electricity. In integrated gasification combined cycle (IGCC) technology, impurities are reduced in concentration prior to combustion, with substantial environmental and efficiency advantages [1]. Synthesis gas may also be utilized in fuel-flexible solid oxide fuel cells (SOFCs), where electricity production efficiencies greater than 50 percent (higher heating value) have been projected, even with carbon sequestration [2-4]. However, minor and trace impurities including As, Cd, Hg, P, Pb, Sb, and Se tend to be volatilized extensively during gasification [5-9]. For IGCC technology, these impurities are primarily of environmental and regulatory concern, whereas poisoning of the nickel-based anode is an additional issue for SOFC systems [6]. Other SOFC components, tubing, heat exchangers, and interconnect plates may also be affected by interactions with impurities in fuel gases [10].

Here we address issues specific to the interaction of trace concentrations of antimony in coal and biomass-derived synthesis gas with nickel-based electrocatalysts. Antimony is a common trace impurity in most of the world's coals [11] and in the earth's upper crust [12]. Even in sub-ppm concentrations in synthesis gas, Sb has been reported to react with Ni to form $NiSb_x$ reaction products, eventually leading to diminished SOFC performance [13, 14]. Further, current warm gas cleanup methods are projected to be of limited effectiveness in removal of this impurity [6, 8, 10]. The following addresses the composition and form of Sb in various coals, speciation of Sb in synthesis gas, the thermochemistry of Sb interactions with Ni, and the results of SOFC performance studies in synthesis gas containing Sb.

2. Antimony Sources, Speciation and Reactivity

2.1. Antimony in Coal

Antimony is widely distributed in coals, typically in concentrations of 0.05 to 10 ppmw world-wide [11]. In comparison, the average concentration in the upper crust is 0.4 ppmw [12]. In U.S. coals, the average Sb concentration has been estimated at 1.2 ppmw [15], which corresponds to an average of ~0.07 ppmv in synthesis gas derived from coal [8]. Substantially higher concentrations have been reported for specific coal deposits: nearly 350 ppmw for a Late Permian coal from Inner Mongolia [16], nearly 2000 ppmw for brown coals from the Pavlovka deposits in Russia [17], more than 2300 ppmw for Middle Miocene bituminous coals from Turkey [18], and more than 3800 ppmw for a Late Permian coal from southwest China [19, 20]. Significant differences in mean Sb concentrations have been reported depending on coal age, with coals formed in the Late Permian containing the highest Sb concentrations and those from the Carboniferous period containing the least [16].

Antimony is primarily found associated with inorganic constituents in coals, especially with sulfur [11, 15, 21-23]. Like As, Hg, and Cd, Sb is commonly found in pyrites [11, 15, 21, 22]. The relatively rare mineral ullmanite (NiSbS) was identified in a coal bed in Great Britain [23], which has possible implications for fuel cell performance. Antimony is also often closely associated with heavy metals such as Pb and Bi [16]. Though Sb is also often found in organic matter [24], it is not clear whether this is the result of an organic affinity or is due to the dispersal of fine inorganic particles in organic material [11, 22, 25].

2.2. Antimony Speciation in Synthesis Gas

Three principal types of gasifiers have been developed to produce synthesis gas from coal: the moving bed, fluidized bed, and entrained flow [2, 26-28]. Entrained flow gasification, with short residence times and high operating temperatures, is a likely technology to be combined with solid oxide fuel cell systems [2, 26]. The resulting synthesis gas composition will depend on gasifier type, operating conditions, and of course the source of the coal used. Table 1 provides a range of synthesis gas compositions obtained using dry or slurry-fed, entrained flow gasifiers with outlet temperatures in the range of 1300-1700K [2].

Table 1. Synthesis gas composition for dry or slurry-fed entrained flow gasifier [2]

Gas Component	Composition, Volume %
H_2	26-30
CO	29-63
CO_2	1.5-14.5
H_2O	2.0-25.8
CH_4	0.0-0.6
N_2	0.5-4.3
Ar	0.7-1.1

Trace impurities in coal have been classified into three groups [6, 29]: *Class 1* elements are least volatile and likely to remain in the ash; *Class 2* elements will partition between solid and gas phases; and *Class 3* elements are highly volatile and are unlikely to condense [5]. Class 2 elements have been identified to be of most concern to fuel cell performance [6, 8], and include antimony, arsenic, cadmium, lead, mercury, phosphorus, selenium, and vanadium. Some of these components can form alteration phases with nickel under the SOFC operating conditions.

Antimony speciation in the gas phase is strongly affected by temperature and the presence of other impurities in synthesis gas, particularly As and S. Figure 1a compares fugacities of Sb-containing gas species versus temperature in synthesis gas with no other impurities present. Calculations were performed using HSC Chemistry version 6.1, which employs free energy minimization methods to estimate equilibrium concentrations [30]. The species SbO_2H_2(g) added following Heames et al. [31], and parameters for SbO(g) were taken from the Landolt-Börnstein database [32]. Although introduced as 1 ppmv stibine (SbH_3(g)), antimony is expected to be predominantly in the form Sb_4(g) and SbO_2H_2(g) below ~650°C, as Sb_2(g) from ~650 to 1000°C, and as Sb(g) and SbH(g) above ~1000°C.

With the introduction of 1 ppmv arsine (AsH_3(g)) and 1 ppmv hydrogen sulfide (H_2S(g)), both of which are well-know impurities in synthesis gas derived from coal [5-8], various Sb_xAs_y species are expected to predominate over a broad temperature range, as shown in Figure 1b. Also formed was Sb_2S_3(g), in negligibly low concentrations. Figure 1c shows the results of calculations performed with 10 ppmv H_2S(g) and 1 ppmv SbH_3(g) in synthesis gas (AsH_3(g) absent), where Sb_2S_3(g) was predominant at the lowest temperatures. Formation of gaseous species containing Sb and other impurities such as As and S helps to ensure that Sb is fully volatilized in synthesis gas.

Table 2. *Class* 2 Trace Components of Gasified Coal and Estimated Concentrations Following Warm Gas Clean-Up [6]

Component	Concentration, ppmv
Antimony	0.07
Arsenic	0.6
Cadmium	0.011
Lead	0.26
Mercury	0.025
Phosphorus	1.91
Selenium	0.15

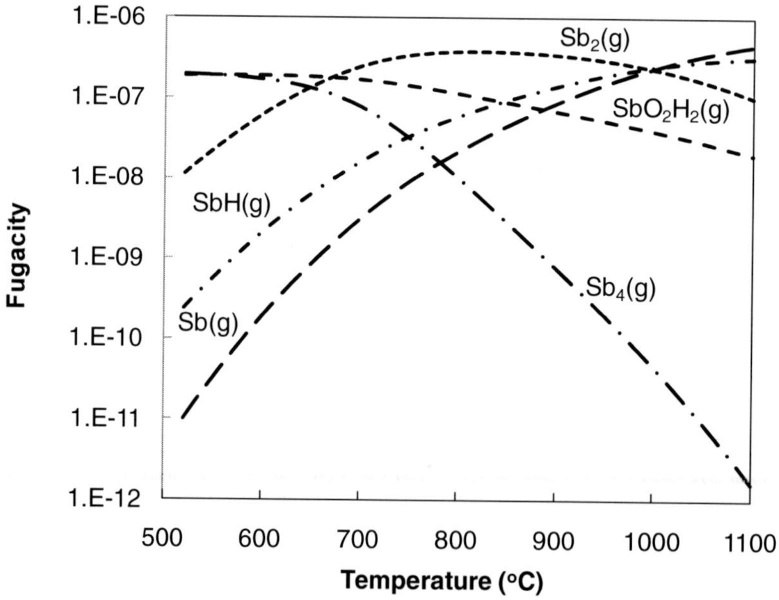

Figure 1a. Calculated gas phase species for 1 ppmv SbH_3 in synthesis gas versus temperature.

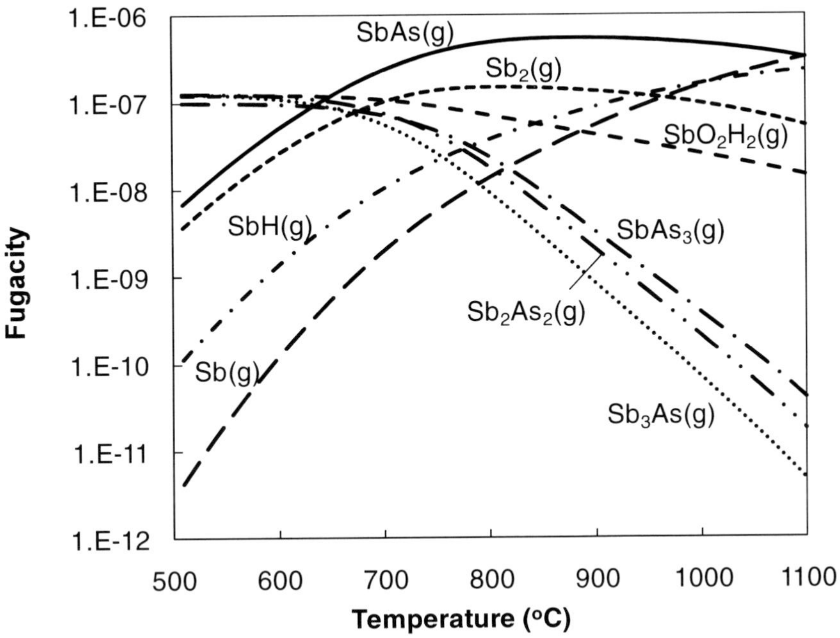

Figure 1b. Calculated gas phase species for 1 ppmv SbH_3, 1 ppmv H_2S, and 1 ppmv AsH_3 in synthesis gas versus temperature.

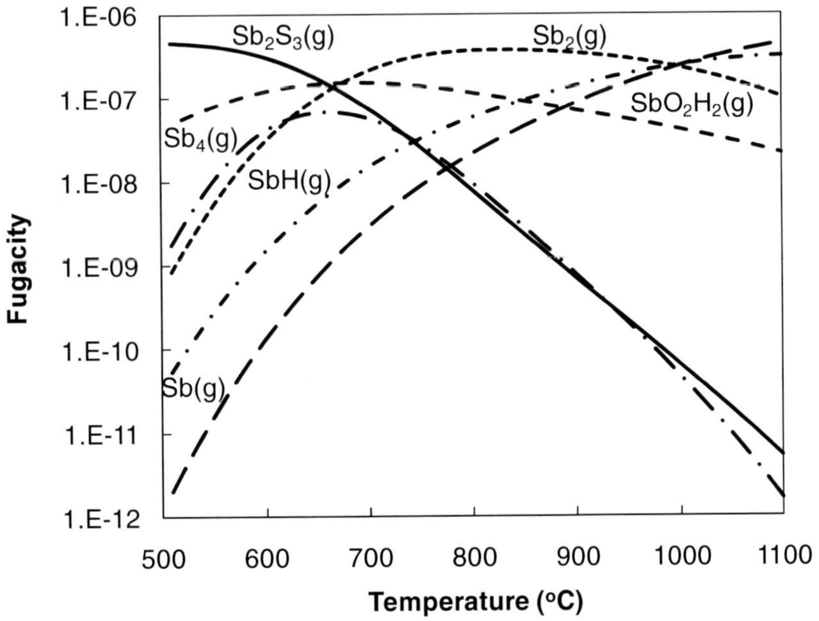

Figure 1c. Calculated gas phase species for 1 ppmv SbH_3 and 10 ppmv H_2S in synthesis gas versus temperature.

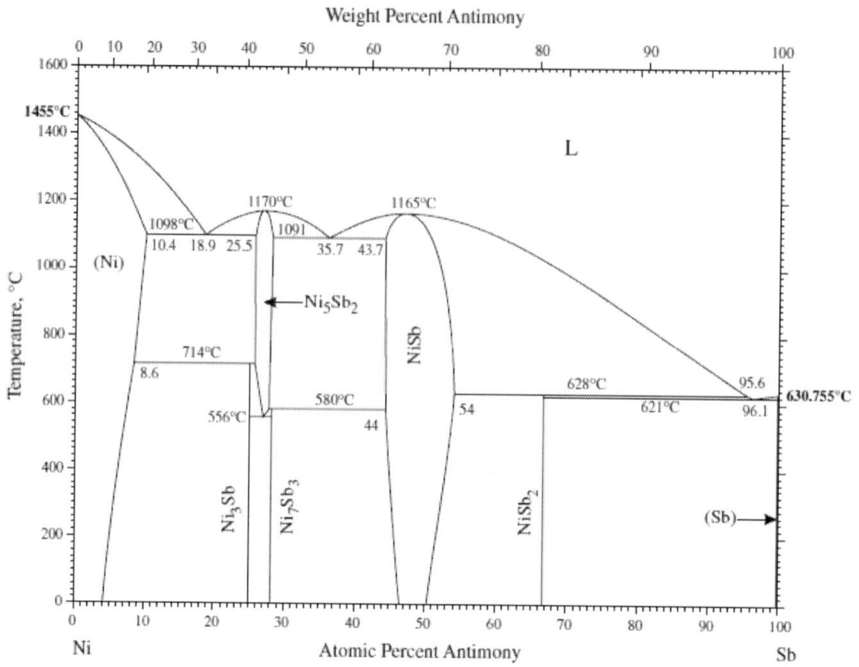

Figure 2. Phase diagram for the Ni-Sb system (from [33], with permission).

2.3. Antimony - Nickel Phase Relations

The phase diagram for the nickel-antimony system is quite complex, with multiple intermediate phases possible. Okamoto [33] recently updated the phase diagram for the nickel-antimony system and this latest version is shown in Figure 2. Phases of potential importance to SOFC anode interactions include Ni-Sb solid solution (Ni(ss)), Ni_3Sb, Ni_5Sb_2, and NiSb. Sources are not in complete agreement for the Ni–Sb system: while the work of Okamoto et al. show Ni_5Sb_2 solid solution persisting to high temperature [33], Cao et al. [34] and Zhang et al. al. [35] show greater stability for Ni_3Sb at high temperature.

At high temperatures relevant to SOFC operation, the phase NiSb exhibits the lowest enthalpy of formation of all intermediate phases in the Ni–Sb binary system [35], and thus is most likely to be formed. Figure 3 shows the formation of NiSb crystals after 75 hour exposure to 1000 ppm Sb in moist hydrogen at 800°C. NiSb has the nickel arsenide structure and crystallizes in the hexagonal space group $P6_3/mmc$. The structure can be described as an *hcp* array of antimony in which the metal fills all of the octahedral holes. This system exhibits quite high solid solubility of antimony in nickel at 9.0 atomic percent at 800°C, which leads to changed catalytic properties of nickel [36] and should certainly lead to changes in the electrocatalytic activity of the fuel cell anode. The phase Ni_7Sb_3 is expected to appear only at temperatures less than 582°C, which is well below the typical operating temperature of an SOFC and thus is unlikely to be formed.

Sub-ppm concentrations of antimony in coal gas are expected to react with nickel to form new solid phases over a broad temperature range. Minimum antimony fugacities required to form the solid phase NiSb as a function of temperature are shown in Figure 4. Phase boundaries for solid phases Ni(Sb)ss, Ni_3Sb, and/or Ni_5Sb_2, all of which are solid solutions, were not calculated because of a lack of thermodynamic data. The presence of additional impurities in synthesis gas, especially As-containing gas species, raises the minimum concentration of Sb needed to form NiSb, also included in Figure 4. This change is attributed to the predicted predominance of $Sb_xAs_y(g)$ species (see Figures 1a and 1b), which increases the stability of Sb in the gas phase.

Figure 3. Nickel wire converted to the accumulation of Ni–Sb crystals after 75 hour exposure to 1000 ppm Sb in moist hydrogen at 800°C.

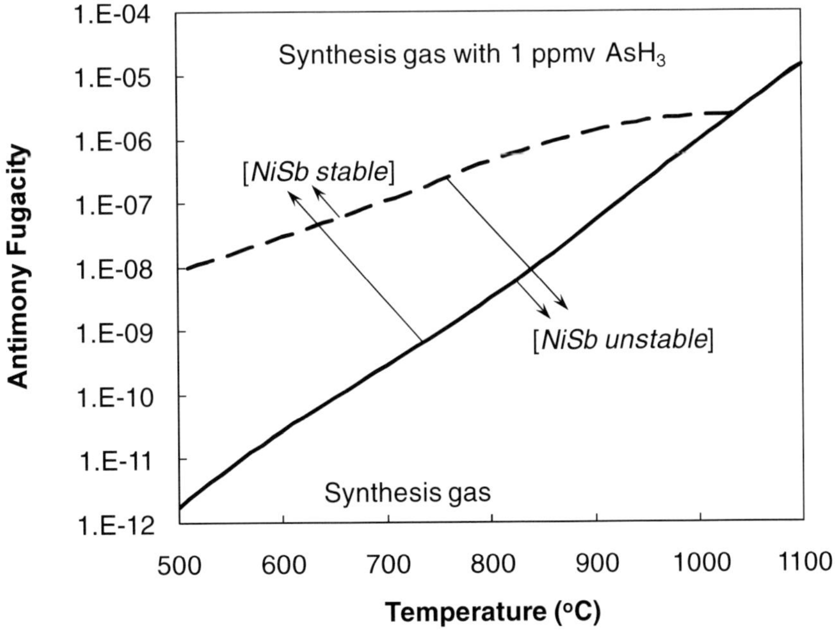

Figure 4. Phase boundary versus temperature for NiSb in synthesis gas with and without 1 ppmv AsH_3.

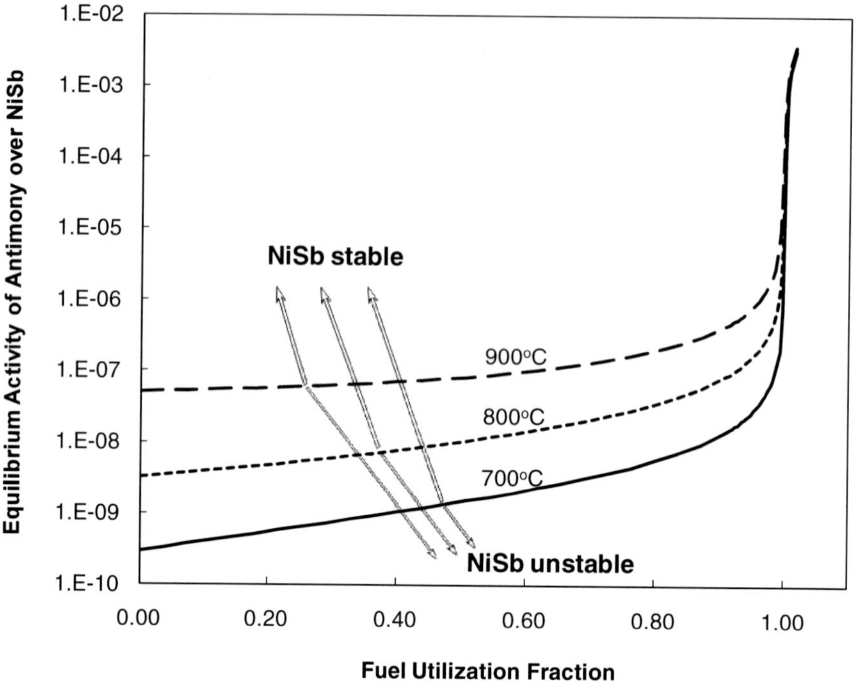

Figure 5. Minimum Sb concentrations required to form NiSb in synthesis gas versus fuel utilization (reproduced with permission from [13]).

That the stability of NiSb is expected to decrease with increased fuel utilization, as shown in Figure 5, is an important finding and quite different from behavior that we have reported due to P, As, and especially Se exposure [37] [38] [39]. For Se exposure, we have found that high anodic polarization in an SOFC can result in the formation of nickel selenide phases at the active interface, at H_2Se concentrations well below those needed to form alteration phases elsewhere in the anode support [39]. Locally high fuel utilization at the anode/electrolyte interface can result from high anodic polarization. Similarly, Kishimoto et al. predicted that high anodic polarization could lead to nickel sulfide formation at the active interface but nowhere else within the anode [40]. For Sb exposure, high current densities render NiSb and perhaps other reaction products less likely to form.

2.4. Nickel Surface Interactions with Antimony

Surface interactions of antimony with nickel has been studied in detail, mainly because of the extensive commercial use of Sb additives as passivating agents in the catalytic crackers in refineries in the fluidized catalytic cracking process (FCC) [41-43] [44, 45]. Nickel impurities, if not captured, catalyze undesirable hydrogen generation and coking reactions decreasing gasoline yield [43-46].

When deposited onto Ni, antimony forms ordered structures. In particular, at temperatures above 500 K, $p(\sqrt{3} \times \sqrt{3})R30^{\circ}$ –Sb/Ni(111) and $p(\sqrt{16/7} \times \sqrt{16/7})R19^{\circ}$ – Sb/Ni(111) structures were observed by the low-energy electron diffraction (LEED) [47]. In dilute alloys, surface enrichment of nickel in antimony driven by a low surface free energy and large differences in ionic radii has been observed [36].

When the Sb is not in solid solution, surface enrichment of antimony on bulk nickel metal has been observed after hydrogen reduction [36]. The low surface free energy and large difference in ionic radii provide a strong force for the surface segregation of antimony to the surface of nickel. Alloying in general is known to dramatically alter catalytic and chemisorption properties of metal catalysts because of the associated geometric to electronic effects [48, 49]. This has been observed for even slightly exothermic or endothermic alloys, i.e., systems in which the mutual perturbation of the alloy components was weak and no substantial changes in the electronic structure was expected. Indeed, the catalytic properties of nickel were altered by antimony additions. Hydrogen and CO chemisorption on Ni catalysts was effectively suppressed by the presence of antimony [50, 51]. Goldwasser et al. [51] explained these observations in terms of a Ni–Sb alloy formation with surface enrichment in Sb. Parks et al. [50] attributed antimony poisoning to: (a) geometric blocking of nickel sites by the antimony, (b) modification of the electronic properties of Ni surface atoms by Sb, and (c) the equilibrium between Sb interacting with the support and with nickel determined the amount of Sb available to passivate the nickel.

3. Antimony Effects on SOFC Performance

3.1. Sb Interactions with Metallic Components

Oxidation-resistant alloys find use as interconnect materials, heat exchangers, and gas supply tubing in SOFC systems, especially when operated at temperatures below ~800°C. When fueled with synthesis gas derived from coal or biomass, these metallic components are also exposed to impurities contained in those fuel sources and can provide a sink for Sb impurities and thus complicate experimental studies of impurity interactions with the electrode or catalyst.

Interaction of antimony with an alumina-forming high nickel alloy Inconel 600 and alumel have been studied at SOFC operating temperatures by Marina et al. [10]. Thermally grown oxide scale layers on Ni-Cr alloys are found to be ineffective in suppressing Ni–Sb interactions, probably because of the open porosities and the presence of Ni or NiO in the oxide scales. In comparison, chromia scale layers on ferritic stainless steels Crofer 22 APU and SS 441, however, suppressed Fe – Sb interactions [10]. Non-chromia forming scales, including alumina, silica and manganese oxides, grown on alumel were also ineffective in suppressing Ni-Sb interactions. As illustrated in Figure 6, alumel wires exposed to synthesis gas with 1 ppm Sb formed extensive Ni – Sb secondary phases agglomerated into large crystallites, while no Si, Mn, or Al compounds with antimony were detected. A passivation layer rich in Ni, Al, Mn and Si, likely present in the form of oxides, was ineffective in protecting the alloy from the interactions with Sb.

Thus, when performing laboratory tests with strongly interacting impurities, the following should be taking into account:

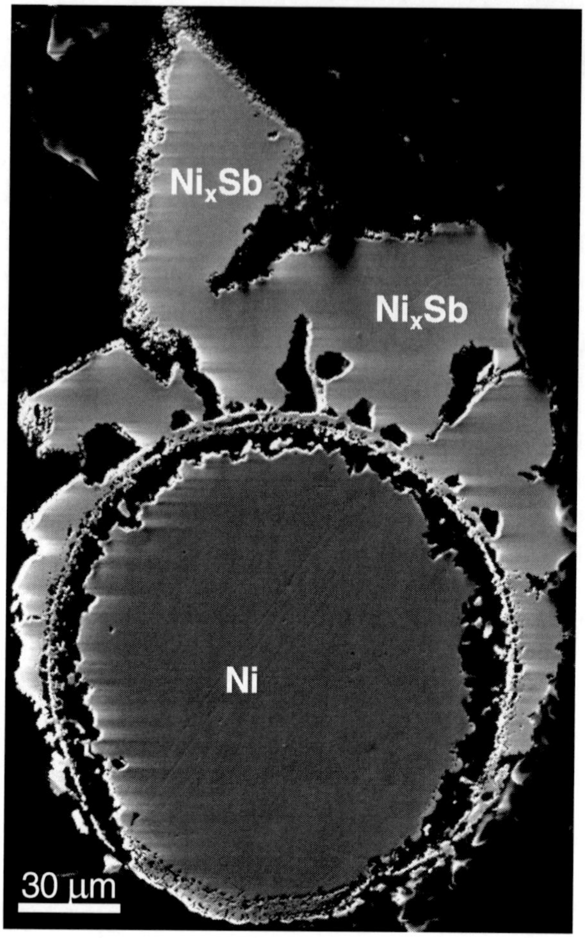

Figure 6. SEM image of the alumel wire after 800 hour test at 700°C in synthesis gas with 1 ppm Sb (reproduced with permission from Elsevier, [10]).

(1) All gas tubing must be protected to avoid the impurity intake. Alternatively, non-metal tubing can be used.
(2) No stainless steel chambers or stainless steel fittings should be used in the warm/hot zone as they can provide a sink for Sb impurities, which may lead to the underestimated effects of the impurities on the anode. If re-used, they can release the impurities during the control tests with clean anode and contaminate them.
(3) Cell tests are preferred to be repeated at least several times to insure to reproducibility.
(4) The use of different fuel cell configurations (anode-supported and electrolyte-supported cells, different thicknesses) are preferred to either accelerate the degradation.
(5) Flow-through and flow-by coupons could be useful to determine the extent of the reaction.
(6) For post-tests analyses, the sample should be cooled in the reducing gas without the impurities to avoid further interactions with the nickel in the anode or speciation changes at different gas environments and surface oxidation.

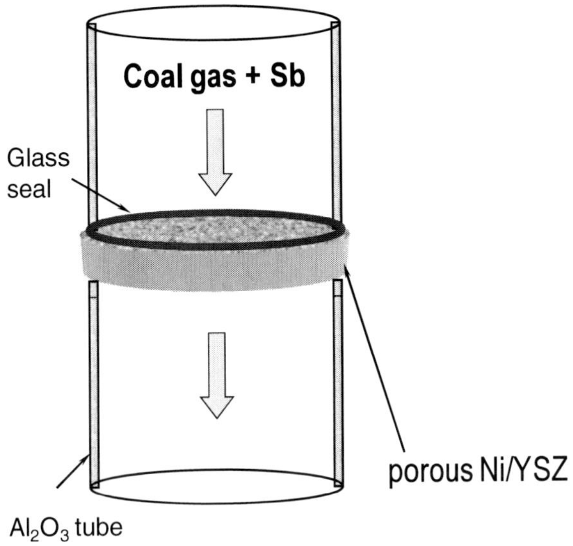

Figure 7. Flow-through coupon test configuration.

3.2. Sb Interactions with Nickel/YSZ Cermet without Electrical Load

Porous Ni/YSZ discs without a dense electrolyte layer were exposed to synthesis gas containing 10 or 100 ppm Sb in a flow-through configuration (Figure 7) and the extent of the Sb reaction with Ni and the depth of Sb penetration was assessed from polished cross-sections using SEM/EDS [13]. Antimony reacted strongly with nickel forming large Ni–Sb crystals on the coupon surface, Figure 8. As determined by SEM, Ni from the Ni/YSZ migrated from to the surface in the form of Ni–Sb and an almost bare YSZ layer was left below the crystalline products. If exposed to 1 or 10 ppm Sb for 2 hours only, no solid Sb reaction products were found by SEM, but Sb on the Ni/YSZ surface was detected by XPS in the amounts close to the antimony solubility limit in nickel. Additionally, porous Ni/YSZ coupons were tested in flow-by tests, in a configuration similar to that of a cell within a planar SOFC stack. Synthetic coal gas containing 0.1 or 1 ppm Sb flowed lengthwise across the coupon at flow rates similar to those utilized in a stack. As shown in Figure 9, strong interactions between Sb and nickel resulted in effective capture of Sb at the cell inlet and secondary phase formation. The uptake was non-uniform along the cell length and limited to a distance of few millimeters from the gas inlet.

3.3. Sb Interactions with Operating SOFCs

Bulk and surface interactions of the Ni/YSZ anodes with antimony have been studied both in anode-supported (500-900 μm thick cermet) and electrolyte-supported (30-35 μm tick cermet) button cells at 700-850°C [13, 14, 52]. Because of the relatively thick anode support, an anode-supported cell design allowed an assessment of structural rearrangements that may have occurred during short-term and long-term exposures. Electrolyte-supported cells, due to a smaller nickel inventory, provide a more rapid response to the impurities.

Figure 8. Cross section (left) and surface (right) SEM images of the Ni/YSZ flow-through coupon after exposure to hydrogen with 100 ppm of Sb at 700°C for 75 hours. Dark gray area in the upper part of the cross-sectional image represent porous YSZ without any nickel. Unaltered Ni/YSZ is light grey.

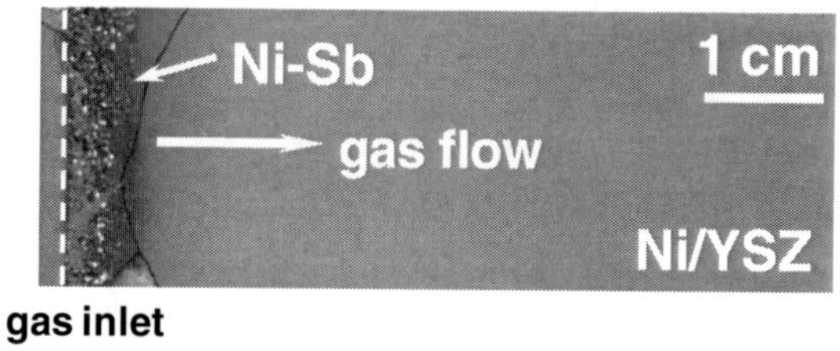

Figure 9. Digital image of the part of the Ni/YSZ flow-by coupon after 560 hour exposure to synthesis gas with 0.1 ppm Sb at 800°C. Coupon cracked during dissembling and not during testing. (with permission from Elsevier, from [13]).

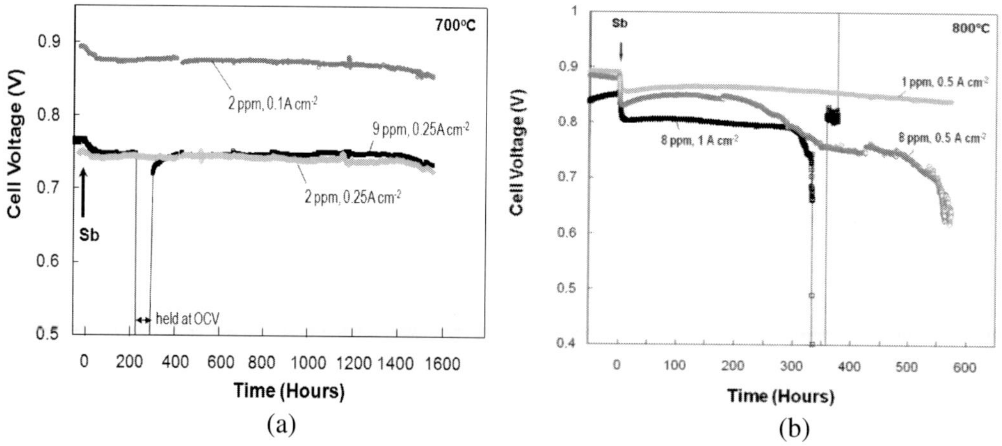

Figure 10. Performances of anode-supported cells in synthetic coal gas with 1-9 ppm Sb at (*a*) 700°C and (*b*) 800°C at given current densities. Time of Sb addition is moved to the origin (with permission from Elsevier, from).

3.3.1. Tests with Anode-supported Cells

Bao et al. [52] investigated Ni/YSZ anodes in a gas mixture of 30.6% H_2, 30.0% CO, 27.6% H_2O, and 11.8% CO_2 with 8 ppm Sb (SbO) at 750, 800 and 850°C and a constant current load of 0.3 A/cm^2. No degradation was observed within a 100 hour exposure at 700°C, and the current – voltage characteristics showed a very slight decline at very high current levels. At 800 and 850°C, only slight degradation was observed with a constant rate of 0.03 mW/cm^2 h (within 160 h) and 0.04 mW/cm^2 h (within 120 h), respectively. Bao et al. [52] concluded that Sb present in the coal-derived gas at the typical level would not cause significant degradation to the cell performance.

Marina et al. [13, 14] conducted Ni/YSZ anode tests at 700 and 800°C in hydrogen or synthetic coal gas 33% H_2, 22% CO, 23% CO_2, and 22% H_2O containing 0.5-9 ppm Sb at a constant current ranging from 0.1 to 1 A cm^2 and low fuel utilizations, <10%. A rapid but minor performance decrease to a new steady-state was always observed on the Sb introduction. Figures 10 a,b illustrate cell potentials recorded at 700 and 800°C at a constant current for a set of anode-supported cells from the moment when different amounts of Sb were added to the fuel gas. At each temperature, an almost immediate ca. 20-60 mV cell potential drop was observed as compared to control cells operated on clean synthesis gas. The area specific resistance (ASR) of cells was increased by ~10-15% at 700°C and 15-25% at 800°C that indicated a minor anode poisoning by Sb. Changes in ASR were essentially independent of Sb concentration and were reversible. When the Sb source was removed after several hours of testing, the cell resistance was slowly decreased to approximately the initial value, suggesting that Sb diffusion into the Ni bulk or Sb desorption may have occurred [14]. While prolonged operation with 1-9 ppm Sb at 700°C did not lead to immediate performance losses, SOFC operation at 800°C with 8 ppm Sb caused more degradation. As concluded from the extensive impedance analysis, the ASR increase was due to an anode polarization increase. A significant increase in the ohmic resistance was also observed at higher Sb concentrations and longer exposure times [13].

Cell components such as current-collecting wires can affect the extent of reactions that occur between the Ni/YSZ cermet anode and synthesis gas containing Sb. An image of an anode-supported cell after the 550 hour test at 800°C, given in Figure 11, shows that the anode surface was heavily converted to Ni – Sb phases. In these particular tests, current collecting wires were coated with glass to avoid any loss of electrical contact due Ni–Sb interactions [10]. As given in Figure 11 b, the current collecting wire served as a shield to the anode and protected the Ni/YSZ underneath the wire. This indicates that the Ni–Sb interactions are so strong, that the presence of any obstacles between the solid Ni and the gaseous Sb caused a "shielding effect". The Ni below the "shield" was not altered, while the unprotected Ni was converted to the Ni – Sb phases. Without a glass coating, the current collecting wire could also serve a sink for Sb impurities [10], which may lead to an underestimate of the effects of anode poisoning.

As determined by post-test SEM/EDS and EBSD analyses, Ni – Sb interactions led to the extensive Ni–Sb formation accompanied by severe agglomeration and coalescence with the Ni–Sb crystallite sizes ranging from 3-7 μm inside the anode support to several hundred microns on the anode surface. For comparison, an average Ni particle size in the Ni/YSZ is approximately 1 μm. The Ni–Sb crystallite size generally increased with Sb concentration and time of exposure. Because of the strong interactions, a rather sharp boundary between the altered and unaltered Ni/YSZ inside the anode support was clearly observed, Figure 12. Minor penetration of Sb below the region of visible transformation was limited to a few microns. While the main altered phase was NiSb, because of the Sb concentration decline toward the boundary between converted and unconverted regions, other Ni – Sb phases, Ni_5Sb_2, Ni_3Sb and solid solution were also identified by EBSD analysis. The surface crystallites were also comprised of at least two Ni–Sb phases. The phase in the upper part of the crystal was recognized as NiSb, and present as a single crystal. The phase underneath NiSb was Ni_5Sb_2 or a mixture of Ni_5Sb_2 and NiSb. Similar morphologies were observed on Ni/YSZ anodes tested at 800°C. Although the test duration at 800°C was a factor of 3 shorter than that at 700°C, approximately similar amounts of the NiSb phase were revealed by the SEM analysis, indicating stronger interactions at higher temperatures [13].

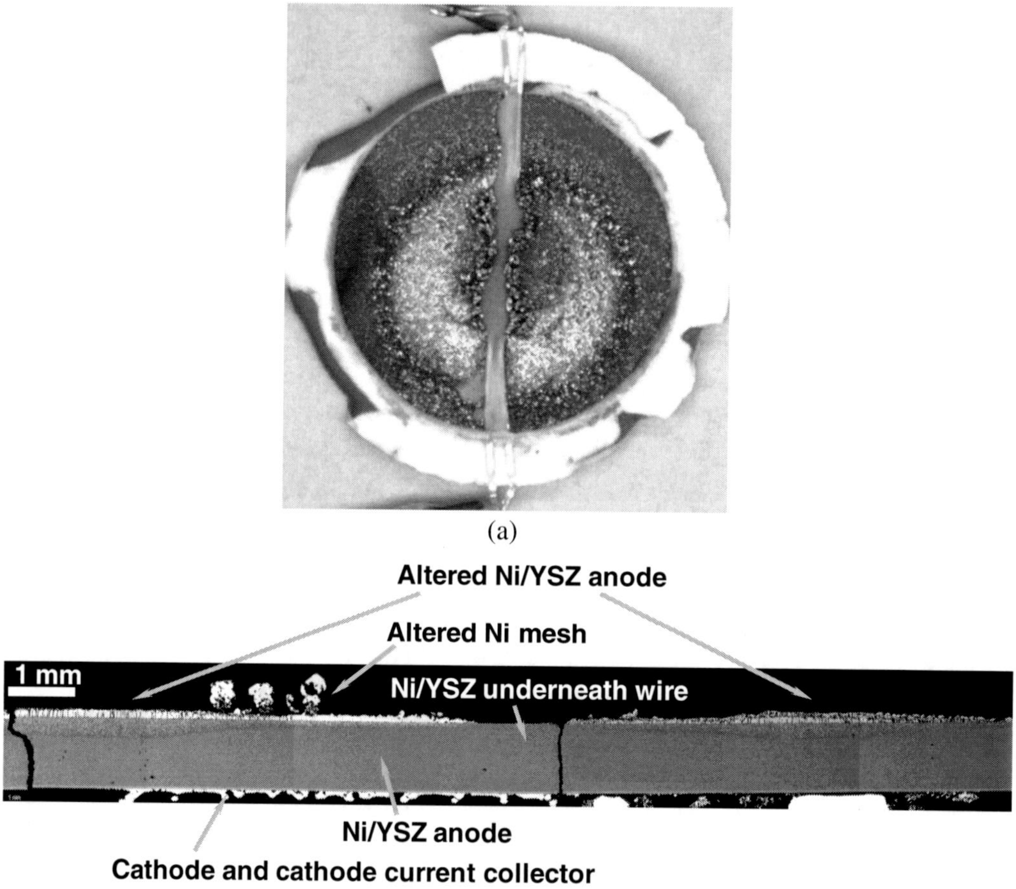

(b)

Figure 11. *(a)* Ni/YSZ anode-supported button cell after the 580 hour test with 1 ppm of Sb at 800°C at 0.5 A/cm^2. Large shiny crystallites are Ni – Sb. *(b)* Cross sectional SEM image of the part of the anode-supported cell illustrating the "shielding effect". The Ni/YSZ area below the current collecting wire, initially present in the center and removed prior to the SEM analyses, is not altered. The unprotected Ni/YSZ area around is affected by Sb to form large NiSb crystallites. Cracks occurred after the test during cell dismounting (adopted from [13], with permission).

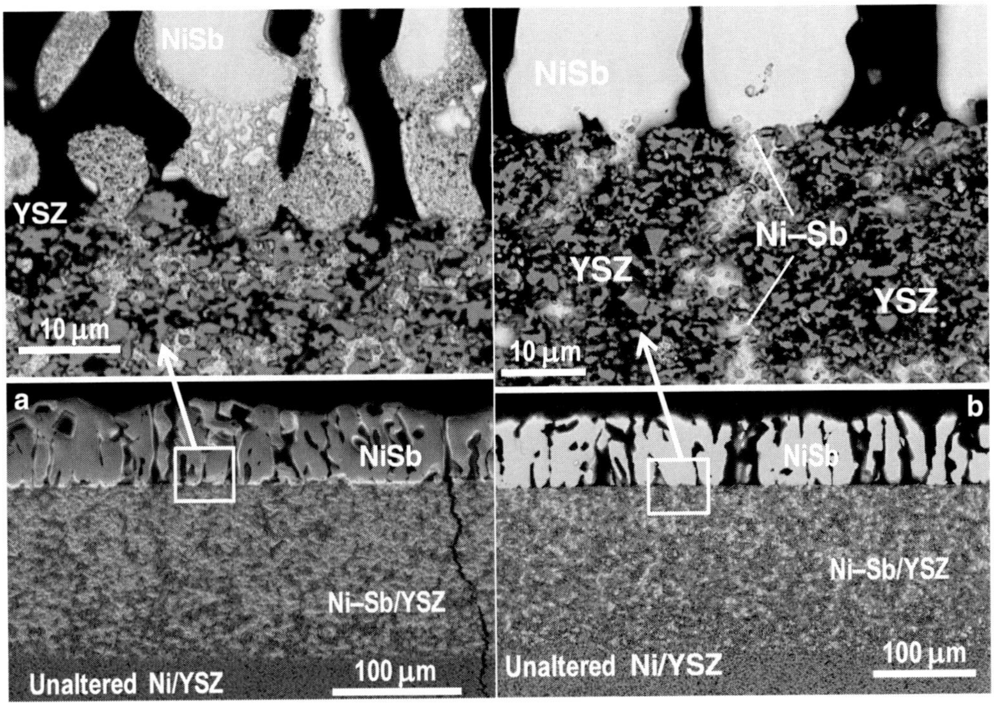

Figure 12. Cross sectional SEM images and an X-ray elemental color map of the upper part of anode-supported cells after exposure to (*a*) 1 ppm and (*b*) 9 ppm Sb for 1560 hours at 700°C. Images are positioned such, that the fuel gas entered from the top perpendicular to the surface. The Ni from Ni/YSZ has reacted with Sb and migrated to the surface to form large NiSb crystals shown in light gray (or purple in the color mix map). The Ni–Sb phase inside the anode support is a mixture of NiSb, Ni_5Sb_2, and Ni_3Sb phases. The Ni–Sb/Ni boundary is not as sharp as pictured due to minor Sb penetration below the regions of visible transformation. The YSZ is dark gray or green, Ni is gray or red, and the pores are black. (Adopted from [13] with permission)

No morphological changes at the active anode/electrolyte interface were observed for the anode-supported cells for the test duration, even for long exposures above 1000 hours in the presence of Sb concentrations as high as 9 ppm.

3.3.2. Tests with Electrolyte-supported Cells

Electrolyte-supported cell performance in the presence of 0.5-6 ppm Sb in the fuel gas was reported by Marina et al. [13]. Approximately 35 μm thick Ni/YSZ anodes on the YSZ electrolytes were tested at 700 and 800°C and a constant current of 0.1 A/cm^2 or constant voltage of 0.8 V and showed a response to Sb similar to that of anode-supported cells, though degradation occurred more quickly. At 700°C, an almost immediate but small, ~20 mV, drop in cell voltages was observed in the presence of Sb, as shown in Figure 13. Poisoning was reversible and cell voltage recovered, if the Sb was removed from the fuel gas after short (few hours) exposures. Only little further change were observed during the following 650 hours of continuous exposure to 0.8 and 2.5 ppm Sb, and substantially more degradation was observed with 5 ppm Sb for exposures greater than 200 hours. The ohmic component of overall cell resistance was reported to be most affected. At high Sb exposures, the electrodic resistance was affected adversely as well.

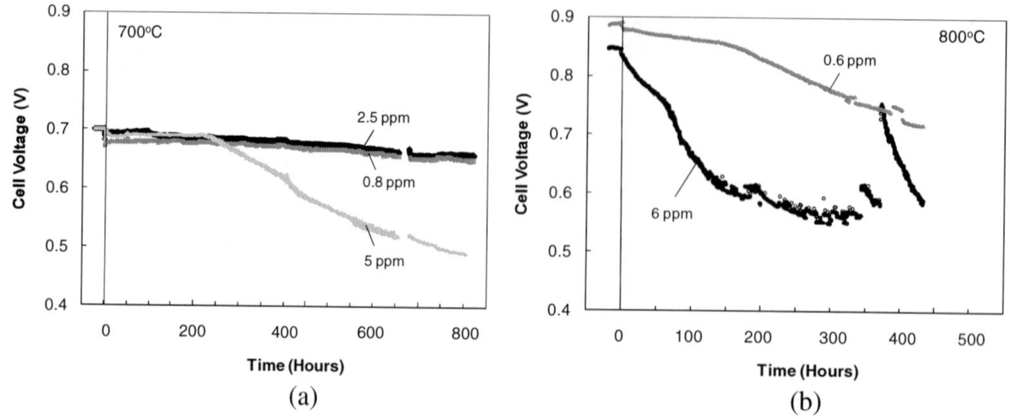

Figure 13. Cell voltages at a constant current of 0.1 A/cm^2 of the electrolyte-supported cells at 700°C (*a*) and 800°C (*a*) synthetic coal gas with given concentrations of Sb. Time when Sb was added was moved to the origin. (adopted from [13] with permission).

Electrolyte-supported cells exposed to Sb at 800°C showed more rapid degradation than those at 700°C under otherwise identical conditions [13], Figure 13. Rates of performance loss slowed with exposure time, reaching an apparent steady-state (where the cell ASR was doubled) in about 200 hours. After approximately 350 hours, sudden sporadic voltage increases and decreases were observed. As determined by post-test SEM/EDS analysis, the anode surface was converted into Ni–Sb phases that coalesced into large grains on the surface of the anode and within the anode support. Furthermore, the anode layer just below those crystals was altered progressively depending on the Sb concentration, temperature and test time and the Sb distribution inside the anode was not uniform, Figure 14.

A summary of EDS analyses of anodes tested at 700°C is given in Figure 15. NiSb was found throughout the anode (together with YSZ) for cells exposed to 5 ppm Sb, whereas the Ni transformation in cells exposed to 0.8 and 2.5 ppm Sb was not as complete, and a mixture of NiSb, Ni_5Sb_2 and Ni_3Sb with some amount of a Ni–Sb solid solution near the YSZ electrolyte was identified. The presence of three compounds, Ni_3Sb, Ni_5Sb_2 and NiSb, was further confirmed by the EBSD analysis. The anode morphology was significantly affected, especially during long-term tests with higher Sb concentrations, and the electrical network continuity was compromised.

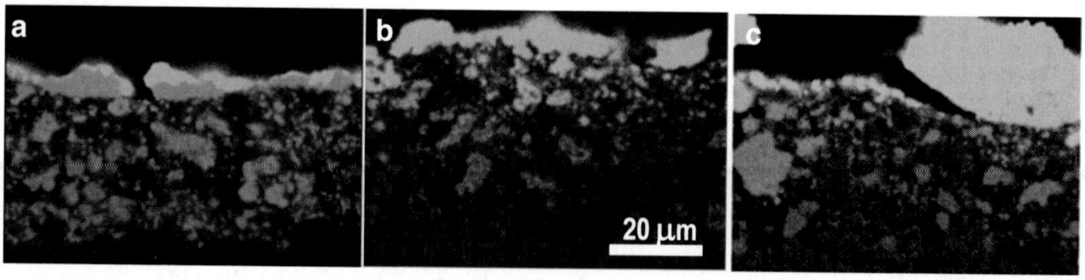

Figure 14. Sb penetration into the 35 µm thick Ni/YSZ anode in electrolyte-supported cells after 440 hour tests in synthetic coal gas with (*a*) 0.6 ppm, (*b*) 2 ppm, (*c*) 6 ppm Sb at 800°C (Adopted from [13], with permission).

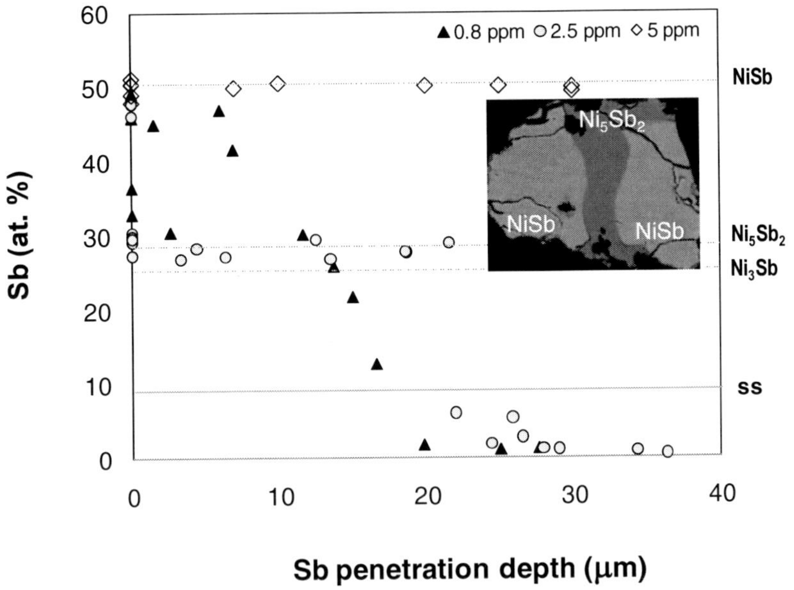

Figure 15. Sb penetration into the 35 μm thick Ni/YSZ anode in electrolyte-supported cells after 820 hour test at 700°C with 0.8, 2.5 and 5 ppm Sb in coal gas. The amount of Sb in solid solution (ss) and possible Ni–Sb phases are suggested by gray lines. Insert illustrates the presence of two Ni–Sb compounds on the anode surface, one with 49±1 at.% Sb and another with 29±1 at.% Sb. Twinning and cracks inside the crystal indicate on stresses during the crystal growth (Adopted from [13], with permission).

It has been predicted from the thermochemical calculations that a minimum gas phase concentration of antimony of ~0.5 ppb and 5 ppb is needed to form NiSb in coal gas at 700°C and 800°C, respectively, for 0% fuel utilization [13]. The stability of NiSb is expected to decrease with increased fuel utilization [39].

3.4. Anode Degradation Mechanism

Two principal processes contribute to the Ni anode deactivation and degradation in the presence of Sb: (1) adsorption of Sb onto Ni surfaces, which lowers the electrocatalytic activity of the anode, and (2) reaction of Sb and Ni to form binary alloys, which leads to coalescence of Ni within the anode support and possible loss of electrical connectivity. These processes are summarized schematically in Figure 16. Both anode- and electrolyte-supported cells rapidly responded to Sb present in the synthesis gas and a new, less active steady state was established. This rapid electrochemical response suggests fast Sb transport to the triple phase boundary and adsorption on the Ni sites that are active in the electrochemical fuel oxidation. Such responses were similar to those due to exposure to H_2S or H_2Se, but the absolute voltage drop at a given current density was much smaller in the present study. Adsorption was reversible: when the Sb source was removed from the fuel gas in a couple of hours, the cell resistance slowly returned to the initial value. Although no Ni–Sb secondary bulk phases at the active anode/electrolyte interface for the test duration were observed in post-test analyses, XPS analyses confirmed that adsorbed Sb species were in metallic form rather than oxidized. Further, XPS analyses showed a surface concentration of Sb on Ni well in excess of the solid solubility limit, consistent with Sb surface segregation as observed previously. Thus, initial reaction products are limited to surface adsorption layers and solid solution formation.

Initial changes in polarization resistance were principally electrodic in nature, were independent of the Sb concentration, and were affected by current density. Independence with respect to the Sb concentration during initial exposure relates to the formation of a monolayer, identified in previous studies as $p(\sqrt{3} \times \sqrt{3})R30^{\circ}$ –Sb/Ni(111) and $p(\sqrt{16/7} \times \sqrt{16/7})R19^{\circ}$ –Sb/Ni(111) structures using LEED [47]. With continued exposure, multilayer adsorption and the formation of new solid phases becomes possible, depending on the activity of Sb in coal gas.

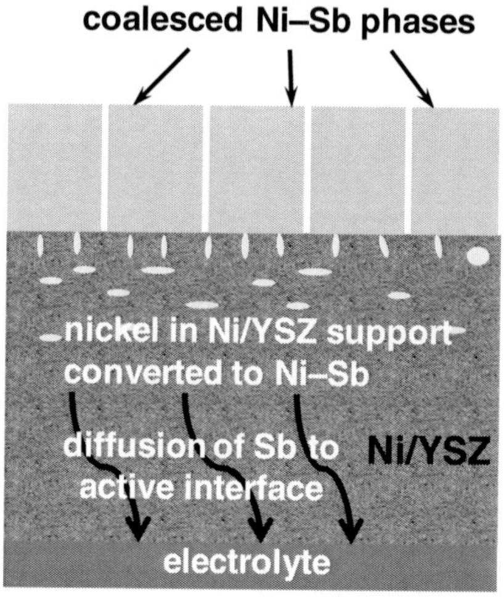

Figure 16. Schematic diagram of the reaction of the Ni/YSZ anode with antimony in coal gas. Ni–Sb phases form in outer portions of the anode support. Surface diffusion of antimony to the active interface occurs as well, which is responsible for initial performance losses. (Adopted from [13], with permission).

As the time of exposure progressed, a second stage of degradation became apparent where performance losses increased substantially, sometimes to the point of complete cell failure. Extensive formation of solid reaction products, particularly NiSb, was observed on the surface of the anode and within the anode support. From impedance spectroscopy results, it is clear that increases in ohmic losses were dominant. Since the electrical conductivity of NiSb is high [53], the most probable explanation involves the loss of the electrical conduction path between the particles (percolation loss), similar to that observed in tests with phosphorus and arsenic. Abrupt performance losses and gains (Figures 10 and 13) also support the electrical percolation loss/gain mechanism. Lack of continuity between Ni–Sb particles could be seen in Figures 12.

An intrinsically lower activity of Ni–Sb phases toward the electrochemical fuel oxidation can be eliminated as a possible explanation for these observations, especially given that NiSb was never formed within the active anode of an anode-supported cell. Further, while solid reaction products were formed at the anode/electrolyte interface of electrolyte-supported cells, the NiSb phases retained moderate activity for the fuel oxidation. Despite most of the anode layer being converted to NiSb, only small performance losses ensued. The electrocatalytic activity of the Ni and Ni–Sb anode cannot be compared directly: in the electrolyte-supported cells where Sb did reach the triple phase boundary, substantial morphological changes inside the active anode also became apparent and likely were responsible for the irreversible degradation.

As determined by the SEM/EDS and EBSD analyses, Ni_3Sb, Ni_5Sb_2, and NiSb, were present in the anode support of the anode-supported cells and coalesced into large grains on the surface or much smaller crystals inside the anode support. In such cells, the Ni–Sb agglomeration was different from that observed for Ni–As and Ni–P, where large agglomerates inside the anode support caused numerous stresses and cracks and caused cell failure. Considering the Ni and NiSb densities of 8.902 g cm^{-3} and 8.74 g cm^{-3} at 25°C, respectively, the volume change on Ni conversion to NiSb is substantial, about 45%. The large difference should contribute to large volume expansion and enormous stress created within the anode support. Indeed, twin structures, sheer effect, twining, and deformation, and splitting one metal particle into several, all indicative of stress, were observed, especially with the electrolyte supported cells. As illustrated in the insert in Figure 15, some Ni–Sb particles had numerous cracks, which is indicative of stress, and some consisted of two phases.

Without cleanup, the average concentration of Sb in coals world-wide is sufficiently high and interactions with the Ni anode of an SOFC are expected. Antimony interacts with the Ni anode in two principal ways: formation of a surface adsorption layer, which results in small and reversible performance losses, and formation and coalescence of binary Ni–Sb solid phases, which can lead to cell failure through loss of electrical connectivity in the anode support and is of greatest importance. At 800°C and low fuel utilization typical of the fuel inlet, the minimum Sb activity in coal gas needed to react to form binary Ni–Sb solid phases is approximately an order of magnitude smaller than the expected average concentration in coal gas. That minimum Sb concentration decreases with increased temperature and with increased fuel utilization.

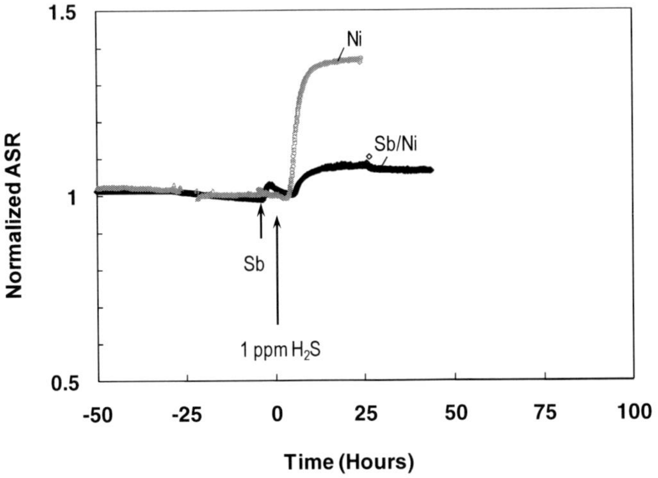

Figure 17. Change in area specific resistances (ASR) of cells with a standard Ni/YSZ anode (*Ni*) and that pre-treated with 1 ppm Sb (*Sb/Ni*) and 10 ppm Sn (*Sn/Ni*) in the presence of 1 ppm H_2S in moist hydrogen. All tests were performed at a constant voltage of 0.7 V at 800°C. The time when H_2S was added was moved to the origin (Adapted from [14], with permission of The Electrochemical Society).

The time required to form an adsorption layer of Sb on Ni is quite short. Assuming an average nickel particle size of 1 microns and an anode thickness of 500 microns, the surface area of nickel is ~3000 cm^2 per cm^2 of the anode cross sectional area. To completely cover the surface with an adsorption layer (√3x√3) rotated 30° (1/3 monolayer), assuming 10^{15} surface sites cm^{-2}, approximately $7 \cdot 10^{-7}$ moles Sb cm^{-2} would be required. With a fuel gas flow rate of ~1 cm^3 gas cm^{-2} cell area s^{-1} used in present tests, corresponding to a typical fuel utilization of 10%, and an Sb concentration in coal gas of 1 ppm, available surface sites should be saturated within ~two hours, which is in good agreement with experimental observations. For a large SOFC stack operating at a fuel utilization of 90% and a Sb concentration of 0.07 ppm consistent with the expected average, available surface sites are expected to be saturated within 100-200 hours. These losses in performance due to adsorption layer formation are very small, and probably can be ignored.

Later stage degradation due to the formation of binary Ni–Sb phases imposes more strict requirements. These reaction products can coalesce to form large grains on the surface of the anode and within the anode support, are the most visible of degradation processes. It is straightforward to estimate the time for conversion of Ni in the anode to NiSb, the product observed most extensively. Assuming an Sb concentration in coal gas of 0.07 ppm (expected average for U.S. coals, but recognizing that some coal deposites contain considerably higher Sb concentrations than the average), an anode thickness of 500 μm, a Ni/YSZ volume ratio of 40/60, a porosity fraction of 0.3, and 90% fuel utilization at a power output of 0.5 W cm^{-2}, approximately 1 million hours would be required for full anode conversion. For anode-supported cells, failure often occurred well before antimony-containing bulk phase formation reached the active anode/electrolyte interface. Failure of these cells occurred abruptly, whereas there was little evidence of cell degradation prior to complete failure. The mode of failure clearly is related to loss of electrical connectivity within the anode support. Failure with as little as 10% of the anode converted to reaction products was observed in laboratory tests, which corresponds to an estimated lifetime of 100,000 hours. Further, as was shown in Figure 9, Sb maybe captured preferentially at the fuel inlet of the stack, so damage could be localized within a small portion of the stack.

3.5. Antimony Exposure as a Means of Mitigating Sulfur Poisoning of Ni/YSZ

The formation of Ni – Sb bimetallics has been shown to be useful for controlling or preventing sulfur poisoning of Ni anodes [14]. Ni is very sensitive to the presence of sulfur even in low amounts of H_2S, 0.1-1 ppm [54-60]. Sulfur poisoning was effectively inhibited by alloying the nickel with Sb or by forming at least one monolayer of Sb on the nickel surface at the triple-phase boundary [14]. The response to 1 ppm H_2S was larger than that to 1 ppm Sb, but still by a factor of 3-4 lesser than the performance loss of the standard Ni/YSZ anode, Figure 17. Electrochemical impedance spectra revealed only small electrodic resistance increase on the Ni–Sb/YSZ anode, Figure 18. Neither Sb, nor S were detected on the anode surface by post test SEM/EDS analyses, indicating minimal diffusion of either into the bulk Ni and no formation of higher order binary solid phases under the mild exposures [14]. However, Sb was detected on the Ni surface using XPS, Figure 19. Surface concentrations of Sb was well above that expected in solid solutions with Ni [33] and was consistent with previously observed tendency of Sb to segregate to both surfaces and grain boundaries [61,

62]. The XPS results established lower relative concentrations of sulfur on both alloys compared to those on unaltered nickel: 3.4 atomic percent on Ni–Sb/YSZ compared to 14.9 atomic percent on Ni/YSZ. Ni surface enrichment with Sb suppressed reversible sulfur adsorption by an approximately factor of four. Sulfur remained concentrated on the surface rather than having diffused into the Ni grains: in all cases, the concentration of sulfur was reduced by half following sputtering to a depth of ~ 20 nm.

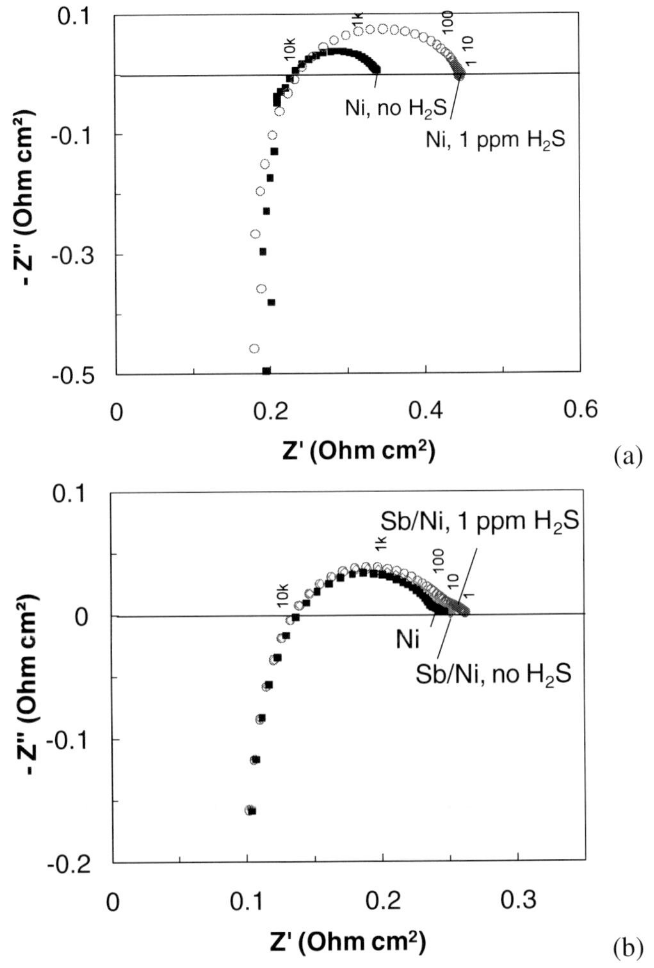

Figure 18. Complex impedance spectra of (*a*) the Ni/YSZ anode-supported cell and (*b*) cell with the Ni/YSZ anode pre-treated for 1 hour with 1 ppm Sb. Spectra were obtained in moist hydrogen at 800°C at a bias voltage of 0.7 Volt (Adapted from [14], with permission of The Electrochemical Society).

Because the addition of Sb weakens CO chemisorption, the use of bimetallic anodes may be beneficial in suppressing of coking when operating on the hydrocarbon fuel. Carbon deposition occurs mainly on Ni ensembles consisting of several neighboring Ni atoms. To prevent the formation of a Ni ensemble, Ni may be alloyed with another metal and Ni–Sb alloys have been shown to suppress the nickel activity for coke formation [43, 44]. The stability of the Ni–Sb/YSZ electrodes for coking in the presence of methane was demonstrated [14].

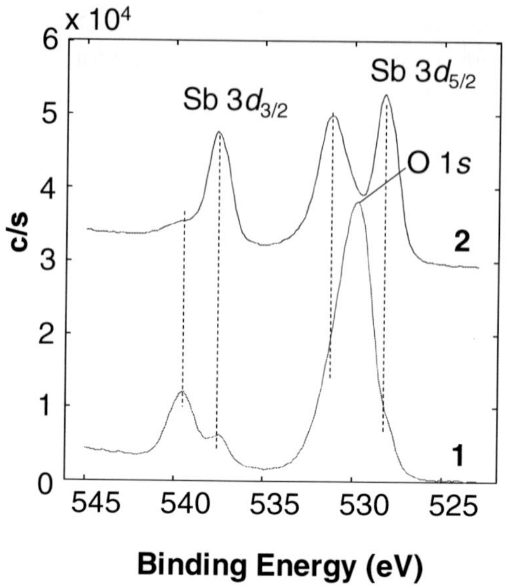

Figure 19. High resolution photoemission spectra of the Sb 3*d* and O 1*s* region (*1*) before and (*2*) after 5 min 2 kV Ar^+ sputter (sputter ~20 nm) obtained from the SbNi/YSZ surface. (Adapted from [14], with permission of The Electrochemical Society).

CONCLUSION

Antimony is broadly distributed in coals and biomass. When coal or biomass is gasified, antimony will be primarily converted to the gas phase, and is expected to be difficult to capture by warm gas cleanup methods. The performance of SOFCs operated on gasified coal or biomass containing even sub-ppm concentrations of Sb is expected to be degraded, due to reactions of the Ni/YSZ anode to form a series of nickel antimonide phases Ni_3Sb, Ni_5Sb_2, NiSb. Immediate though small initial changes in fuel cell performance are attributed to the formation of a surface sorption layer at the active interface, similar to that of sulfur. Irreversible, long-term anode degradation is attributed to the formation and aggolomeration of Ni–Sb reaction products. Ni interactions with Sb are sufficiently strong that most Sb is captured near the fuel inlet. Short-term exposure of the Ni/YSZ anode to low concentrations of Sb is effective in inhibiting sulfur poisoning of the fuel cell.

ACKNOWLEDGEMENT

The authors would like to acknowledge Dr. D.J. Edwards for performing SEM analyses and helpful discussions.

REFERENCES

[1] *Fuel Cell Handbook*. EG&G Technical Services, Inc., 2004, U.S. Department of Energy-Office of Fossil Energy. Morgantown, WV, USA.

[2] Kivisaari, T., Bjornbom, P., Sylwan, C., Jacquinot, B., Jansen, D., de Groot, A. The feasibility of a coal gasifier combined with a high-temperature fuel cell. *Chem. Eng. J.* 2004, 100(1-3), 167-180.

[3] Williams, M. C., Strakey J. P., Surdoval W. A. The US Department of Energy, Office of Fossil Energy stationary fuel cell program. *J. Power Sources*, 2005, 143(1-2), 191-196.

[4] Williams, M. C., Strakey, J. P., Surdoval, W. A., Wilson, L. C., Solid oxide fuel cell technology development in the US. *Solid State Ionics,* 2006, 177(19-25), 2039-2044.

[5] Helble, J.J., Mojtahedi, W., Lyyranen, J., Jokiniemi, J, Kauppinen, E. Trace element partitioning during coal gasification. *Fuel*, 1996, 75(8), 931-939.

[6] Trembly, J. P., Gemmen, R. S., Bayless, D. J. The effect of IGFC warm gas cleanup system conditions on the gas–solid partitioning and form of trace species in coal syngas and their interactions with SOFC anodes. *J. Power Sources*, 2007, 163, 986-996.

[7] Gibbs, B.M., Thompson, D., Argent, B. B. Mobilisation of trace elements from as-supplied and additionally cleaned coal: Predictions for Ba, Be, CdCo, MoNb, Sb, V and W. *Fuel*, 2008, 87(7), 1217-1229.

[8] Martinez, A., Gerdes, K., Gemmen, R., Postona, J. Thermodynamic analysis of interactions between Ni-based solid oxide fuel cells (SOFC) anodes and trace species in a survey of coal syngas. *J. Power Sources*, 2010, 195(16), 5206-5212.

[9] Font, O., Querol, X., Izquierdo, M., Alvarez, E., Moreno, N., Diez, S., Alvarez-Rodriguez, R., Clemente-Jul, C., Coca, P., Garcia-Pena, F. Partitioning of elements in a entrained flow IGCC plant: Influence of selected operational conditions. *Fuel*, 2010, 89(11), 3250-3261.

[10] Marina, O. A., Pederson, L. R., Coyle, C. A., Edwards, D. J., Chou, Y.-S., Cramer, C. N. Interaction of coal-derived synthesis gas impurities with solid oxide fuel cell metallic components. *J. Power Sources*, 2011, 196, 636-643.

[11] Swaine, D.J., *Trace elements in coal*. 1990, London, Butterworths.

[12] Rudnick, R.L., Gao, S. In *Composition of the Continental Crust*, Holland, H.D. Turkian, K.K., Eds., Elsevier, Holland, 2006, p. 1-64.

[13] Marina, O. A., Pederson, L. R., Coyle, C. A., Thomsen, E. C., Nachimuthu, P., Edwards, D.J. Electrochemical, structural and surface characterization of nickel/zirconia solid oxide fuel cell anodes in coal gas containing antimony. *J. Power Sources*, 2011, 196, 4911-4922.

[14] Marina, O. A., Coyle, C. A., Engelhard, M. H., Pederson, L.R. Mitigation of sulfur poisoning of Ni/zirconia SOFC anodes by antimony and tin. *J. Electrochem. Soc.* 2011, 158(4), B424-429.

[15] Finkelman, R. B., In *Organic Geochemistry*, M.H. Engel, S.A. Macko, Eds., Plenum Press, New York, 1993.

[16] Qi, C. C., Liu, G. J., Chou, C. L., Zheng, L. Q. Environmental geochemistry of antimony in Chinese coals. *Sci. Total Environ.* 2008, 389(2-3), 225-234.

[17] Seredin, V. V., Anomalous trace elements contents in the Spetsugli germanium deposit (Pavlovka brown coal deposit) southern Primorye: Communication 1. Antimony. *Lithology and Mineral Resources*, 2003, 38(2), 154-161.

[18] Karayigit, A. I., Spears, D. A., Booth, C. A. Antimony and arsenic anomalies in the coal seams from the Gokler coalfield, Gediz, Turkey. *Int. J. Coal Geol.* 2000, 44(1), 1-17.

[19] Dai, S. F., Zeng, R. S., Sun, Y. Z. Enrichment of arsenic, antimony, mercury, and thallium in a Late Permian anthracite from Xingren, Guizhou, Southwest China. *Int. J. Coal Geol.* 2006, 66 (3), 217-226.

[20] Sun, R., Liu, G., Zheng, L., Chou, C.-L. Geochemistry of trace elements in coals from the Zhuji Mine, Huainan Coalfield, Anhui, China. *Int. J. Coal Geol.* 2010, 81(2), 81-96.

[21] Dvorniko, A. G., Tikhonen, E. G. Peculiarities of admixture element distribution in iron sulfides from coals of different structures of Central Donetsk Basin. *Geokhimiya*, 1973, 9, 1416-1419.

[22] Goodarzi, F., Swaine, D. J. Chalcophile elements in Western Canadian coals. *Int. J. Coal Geol.*, 1993, 24(1-4), 281-292.

[23] Spencer, L. J., On the occurrence of alstonite and ullmanite in a barytes-witherite vein at the New Brancepeth colliery near Durham. *Mineral Magazine*, 1910, 15, 302-311.

[24] Eskenazy, G. M., Geochemistry of arsenic and antimony in Bulgarian coals. *Chem. Geology*, 1995, 119(1-4), 239-254.

[25] Swaine, D. J., The organic-association of elements in coals. *Organic Geochemistry*, 1992. 18(3): p. 259-261.

[26] Gemmen, R. S., Trembly, J. On the mechanisms and behavior of coal syngas transport and reaction within the anode of a solid oxide fuel cell. *J. Power Sources*, 2006, 161(2), 1084-1095.

[27] Kristiansen, A., *Understanding Coal Gasification* Vol. IEACR/86, London, 1996, No 1-3.

[28] Takematsu, T., Maude, C. *Coal gasification for IGCC power generation.* Vol. IEACR/37. London, 1991.

[29] Higman, C., van der Bergt, M. *Gasification*. Elsevier, Amsterdam, 2003.

[30] Roine, A., HSC *Chemistry*, 6.12 ed., Outokumpu Research Oy, Pori, Finland.

[31] Heames, T. J., Williams, D. A., Bixier, N. E., Grimley, A. J., Wheatley, C. J., Johns, N. A., Domagala, P., Dickson, L. W., Alexander, C. A., Osborne-Lee, L., Zawadzki, S., Rest, J., Mason, A., Lee, R. Y. Victoria: *A Mechanistic Model of Radionuclide Behaviour in the Reactor Coolant System Under Severe Accident Conditions*, 1992, NUREG/CR-5545 SADN90-0756 Rev.1, Sandia National Laboratories.

[32] *Thermodynamic Properties of Compounds, SbO to* SO_2. Vol. Volume IV/19A: Pure Substances, SpringerMaterials - The Landolt-Börnstein Database.

[33] Okamoto, H., Ni-Sb (Nickel-Antimony). *J. Phase Equilib. Diffus*. 2009, 30(3), 301-302.

[34] Cao, Z. M., Takaku, Y., Ohnuma, I., Kainuma, R., Zhu, H. M., Ishida, K. Thermodynamic assessment of the Ni-Sb binary system. *Rare Metals*, 2008, 27(4), 384-392.

[35] Zhang, Y., Li, C., Du, Z., Guo, C, A thermodynamic assessment of Ni-Sb system. *Calphad*, 2008. 32(2), 378-388.

[36] Dreiling, M. J., Schaffer, A. M. Interaction of antimony with reduced supported nickel catalysts. *J. Catal.* 1979, 56(1), 130-133.

[37] Coyle, C. A., Marina, O. A., Thomsen, E. C., Edwards, D. J., Cramer, C. N., Coffey, G. W., Pederson, L. R. Interactions of nickel/zirconia solid oxide fuel cell anodes with coal gas containing arsenic. *J. Power Sources*, 2009, 193, 730-738.

[38] Marina, O. A., Coyle, C. A., Thomsen, E. C., Edwards, D. J., Coffey, G. W., Pederson, L. R. Degradation mechanisms of SOFC anodes in coal gas containing phosphorus. *Solid State Ionics*, 2010, 181(8-10), 430-440.

[39] Marina, O. A., Pederson, L. R., Coyle, C. A., Thomsen, E. C., Edwards, D. J. Polarization-Induced Interfacial Reactions between Nickel and Selenium in Ni/Zirconia SOFC Anodes and Comparison with Sulfur Poisoning. *J. Electrochem. Soc.*, 2011, 158(1), B36-B43.

[40] Kishimoto, H., Horita, T., Yamaji, K, Brito, M. E., Xiong, Y. P., Yokokawa, H. Sulfur poisoning on SOFC Ni anodes: thermodynamic analyses within local equilibrium anode reaction model. *J. Electrochem. Soc*. 2010, 157(6), B802-B813.

[41] Occelli, M. L., Metal-resistant fluid cracking catalysts - 30 years of research. *ACS Symposium Series*, 1991, 452, 343-362.

[42] Occelli, M. L., Naraghi, S. M., Krishnan, V., Suib, S. L. The effect of tin and antimony addition on the performance of dual function cracking catalyst (dfcc) mixtures. *J. Catal.* 1992, 135(1), 325-331.

[43] Corma, A., Grande, M. S., Iglesias, M., Delpino, C., Rojas, R.M. Nickel passivation on fluidized cracking catalysts with different antimony complexes. *Appl. Catal. A: Gen.* 1992, 85(1), 61-71.

[44] Nielsen, R. H., Doolin, P. K. Metals passivation. *Stud. Surf. Sci. Catal.*, 1993, 76, 339-384.

[45] Karakhanov, E. A., Kovaliova, N. F., Lysenko, S.V. Comparison of the Sb, Sn, and Bi compound efficiency in the Ni passivation on the cracking catalysts. Vestnik Moskovskogo Universiteta, Seriya 2, *Khimiya*, 1998, 39(6), 418-421.

[46] Stakheev, A. Y., Grigoryan, A. A., Shpiro, E. S., Lysenko, S. V., Korolkov, N. S., Karakhanov, E. A., Minachev, K. M. Study of Processes of Deactivation of Cracking Catalysts by Heavy-Metals and the Mechanism of Their Passivation .2. Xpes Study of the State of Nickel and Antimony and Their Distribution in Cracking Catalysts. Bulletin of the Academy of Sciences of the USSR, *Division of Chemical Science*, 1988, 37(12), 2414-2418.

[47] Krupski, A., Mroz, S. Properties of ultrathin Sb layers on the Ni(111) face. *Surf. Rev. Lett.*, 2003, 10(1), 65-72.

[48] Verbeek, H., Sachtler, W. M. H. The study of the alloys of platinum and tin by chemisorption. *J. Catal.*, 1976, 42(2), 257-267

[49] Ponec, V., Surface composition and catalysis on alloys. *Surf. Sci.,* 1979, 80(1), 352-366.

[50] Parks, G. D., Surface Segregation in $Ni_{0.96}Sb_{0.04}$. Appl. Surf. Sci., 1980, 5(1), 92-97.

[51] Goldwasser, M. R., Rojas, D., Goldwasser, J. A study of the Ni---Sb interaction in a rare-earth Y-zeolite. *J. Catal.*, 1992. 135(2), 596-608.

[52] Bao, J. E., Krishnana, G. N., Jayaweeraa, P., Perez-Marianoa, J., Sanjurjoa, A. Effect of various coal contaminants on the performance of solid oxide fuel cells: Part I. Accelerated testing. *J. Power Sources*, 2009, 193(2), 607-616

[53] Katsuyama, S., Watanabe, M., Kuroki, M., Maehata, T., Ito, M. Effect of NiSb on the thermoelectric properties of skutterudite CoSb3. *J. Appl. Phys.*, 2003, 93(5), 2758-2764.

[54] Matsuzaki, Y., Yasuda, I. The poisoning effect of sulfur-containing impurity gas on a SOFC anode: Part I. Dependence on temperature, time, and impurity concentration. *Solind State Ionics*, 2000, 132(3-4), 261-269.

[55] Sasaki, K., Susuki, K., Iyoshi, A., Uchimura, M., Imamura, N., Kusaba, H., Teraoka, Y., Fuchino, H., Tsujimoto, K., Uchida, Y., Jingo, N. H_2S poisoning of solid oxide fuel cells. *J. Electrochem. Soc.*, 2006, 153(11), A2023-A2029.

[56] Christiansen, N., Hansen, J. B., Holm-Larsen, H., Linderoth, S., Larsen, P. H., Hendriksen, P. V., Hagen, A. Solid oxide fuel cell development at Topsoe Fuel Cell A/S and Riso National Laboratory. *ECS Trans.*, 2007, 7(1), 31-38.

[57] Zha, S., Cheng, Z., Liu, M. Sulfur poisoning and regeneration of Ni-based anodes in solid oxide fuel cells. *J. Electrochem. Soc.*, 2007, 154(2), B201-B206.

[58] Mukerjee, S., Haltiner, K., Kerr, R., Chick, L., Sprenkle, V., Meinhardt, K., Lu, C., Kim, J. Y., Weil, K. S. Solid oxide fuel cell development: latest results. *ECS Trans.*, 2007, 7(1), 59-65.

[59] Hansen, J. B., Correlating sulfur poisoning of SOFC nickel anodes by a Temkin isotherm. *Electrochem. Solid State Lett.*, 2008, 11(10), B178-180.

[60] Galea, N. M., Lo, J. M. H., Ziegler, T. A DFT study on the removal of adsorbed sulfur from a nickel(111) surface: Reducing anode poisoning. *J. Catal.*, 2009, 263(2), 380-389.

[61] White, C. L., Padgett, R. A. Effect of antimony additives on the fracture of Ni at 600°C. *Acta Metall.*, 1983, 31(7), 1005-1012.

[62] Padgett, R. A., White, C. L. Retardation of Grain Boundary Self-Diffusion of Nickel Doped with Antimony and Tin. *Scripta Metallurgica*, 1984, 18, 459-462.

In: Antimony: Characteristics, Compounds and Applications
Editor: Manijeh Razeghi
ISBN 978-1-62100-598-8

Chapter 9

ADSORPTION OF ANTIMONY AT METAL ELECTRODES AND THEIR APPLICATION IN ELECTROCATALYSIS

Qi-Hui Wu[1] and Shi-Gang Sun[2]
[1]Department of Physics, La Trobe University, Bundoora 3083, VIC, Australia
[2]State Laboratory of Physical Chemistry of Solid Surfaces, Xiamen University, Xiamen, China

ABSTRACT

Antimony (Sb) is a semimetallic chemical element which can exist in two forms: the metallic form and the non-metallic form. The Sb has been widely used in the fields of polycondensation catalysis and electrocatalysis. In this chapter the application of Sb in the electrocatalysis has been reviewed. The adsorption of Sb on metal electrodes (e.g. Au and Pt) will change their surface atomic structures, electronic properties, and consequently influence the adsorption and oxidation of carbon monoxide (CO) or other small organic molecules (e.g. formic acid) on these Sb modified metal electrodes. Firstly, the adsorption properties of Sb on the metal electrodes via both under potential deposition and irreversible adsorption and the influences on their surface atomic and electronic structures are introduced. Secondly, the adsorption and electrochemical oxidation of CO on the Sb decorated Pt electrodes is discussed. This is because the fact that CO is the stable poison species in the oxidation of most small organic molecules. It is well known that the adsorption of foreign adatoms will greatly inhibit the dissociative reactions of small organic molecules on metal electrode surfaces. Finally the electrocatalytic process of formic acid on the Sb modified Pt electrodes is presented.

1. INTRODUCTION

Electrocatalysis is defined as the study of how electrochemical reactions can be accelerated at electrodes. This often requires the surfaces of electrodes to be modified in some ways in order to change their surface atomic structures, electronic properties, and surface chemical compositions to consequently improve their electrocatalytic properties. The enhancement in electrocatalytic properties of metal electrodes modified by monolayer (ML) or sub-ML foreign adatoms through underpotential deposition (UPD) or irreversible adsorption (IRA) has been recognized few decades ago. By using these two modification methods, it is possible to implement surface modifications of the electrocatalysts in an elegant and well-controlled way. Moreover, these bi-element surfaces can serve as models for the design of new catalysts for industrial purposes. UPD is an electrodeposition of a metal cation to an electrode surface at a potential less negative than the equilibrium potential for the reduction of this metal. IRA is realized by immersing the electrode into the cation containing solution; it is ascribed to chemical interaction of cations with electrode surface -OH species [1, 2]. The deposition/adsorption amount of the foreign cations can be easily controlled by the reaction time and the cation density of solution with both methods.

The electrochemical oxidation of small organic molecules that are potential fuels is one of the central catalytic issues in the development of low temperature direct organic molecular fuel cells [3-6], but the significant overpotential for their oxidation severely limits the fuel cells' efficiencies. The most important paradigm in research on the mechanism of the electrocatalytic oxidation of small organic molecules is the dual pathway mechanism introduced by Capon and Parsons [7, 8]: one via reactive intermediate, and the other via blocking intermediate. CO is the most prominent intermediate in the oxidation of the organic molecules. Recent studies of these foreign element modified metal electrode surfaces and their electrocatalytic properties using new in situ surface structure sensitive techniques (e.g. electrochemical scanning tunneling microscopy (ECSTM) and Fourier transform infrared (FTIR) spectroscopy) have greatly improved the understanding of these systems. The mechanism of the catalytic action of foreign element adlayers in the oxidation of organic molecules has been interpreted by using several models including the "third-body" effect, the bifunctional catalyst mechanism, and the electronic effect. Third-body effect is equivalent to the "ensemble" effect in heterogeneous catalysis, which is based on the role of foreign adatoms in blocking the surface sites for the site reactions that generate poisoning species, or in blocking the adsorption of the inhibiting species, which requires more than one surface site for adsorption [9]. The bifunctional mechanism has been considered by Cathro [10] and Motto et al. [11]. In this mechanism the electrode surface is considered to have two types of sites that play distinct roles in the reaction. The substrate breaks the bonds in organic molecules upon adsorption, while the adatom can adsorb oxygen-containing species that can oxidize efficiently adsorbed intermediates such as CO. Oxygen adsorption on the adatoms is expected to occur at potentials that are negative than the potential of oxygen adsorption on the metal electrode. Electronic effects can result from the change of work function of the metal surface altered by the adsorption of foreign atoms. This is well established at the metal/vacuum interface. Pb adlayers on Au cause a pronounced decrease in the work function with increasing Pb coverage [12]. A similar change can be expected at the electrode surface. The potential of zero charge (PZC) of the electrode surface, which is directly related to the work function, has been determined for the Pb adlayer on Au(111) [13]. This will greatly change the chemical reaction energy of organic molecules on these modified metal electrodes.

In this chapter, we summarize the recent progresses in adsorption of Sb on the Au and Pt electrode surfaces, and the oxidation of CO and formic acid at the Sb decorated Pt (Sb/Pt) electrodes through the combination of in situ spectroscopy, kinetic modeling, and theoretical calculations, because Sb is the most intensively studied foreign atoms adsorbed/deposited on these two metal electrodes owing to the significant enhancement in electrocatalytic activity of the Sb decorated Pt and Au electrodes. Sb adatom coverage (θ_{Sb}) is defined in two ways throughout literature. One definition consists in that the coverage is defined as the ratio between the number of the Sb inhibited and the total number of substrate surface atoms; another definition is that the coverage is the ratio of the number of Sb adatoms and total number of substrate surface atoms. The latter definition is becoming used more often. For adatoms of large atomic radius, the difference in the coverage values measured from these two definitions can be very large. In order to simplify the comparison of results, and to preserve the link to original work when the coverage is given using the first definition, approximate coverage calibrated by using the second approach is given. All the voltages mentioned below are referenced to a saturated calomel electrode (SCE) except special indication.

2. Adsorption of Sb on Au and Pt Electrodes

Surface structural and electronic properties of metal electrodes decorated by foreign adatoms are expected to be considerably different from either those of the substrate or the adsorbate bulk materials. The structure of the top surface layer will be that of the adlayer, determined by the substrate surface structure, adatom size and coverage, substrate-adsorbate interaction and in some systems, by interaction with coadsorbed anions in the solution. Major advances in the understanding of foreign atom adlayer structural behavior have been achieved by recent applications of in situ scanning probes. In general, for UPD, the adatoms are neutral species, while for IRA, the surface structures are often affected by the coadsorbed anions and commensurate adlayer structures will form.

2.1. Sb Decorated Au (Sb/Au) Electrodes

2.1.1. Sb/Au(100) Electrode

Itaya and co-workers [14] have found that on Au(100) the IRA oxygenated Sb (SbO^+) layer is a uniform terrace with a quasi (2 × 2) structure. Figure 1 shows the schematic illustration of the SbO^+ adlayer. In this model, about 60% of the SbO^+ species form clusters near 4-fold sites, as indicated by solid squares. Some of the rest 40% of the SbO^+ species are placed separately at 4-fold sites, and the others are found to interact with neighboring SbO^+ so that they are moved away from 4-fold sites (Figure 1). The distance between those interacting SbO^+ species varies from 0.48 to 0.55 nm. From the number of SbO^+ species counted in the STM images, the coverage was derived as being equal to 0.25. This coverage agrees with that obtained from the cyclic voltammetry (CV) using the oxidation and reduction charge of the adsorbed Sb species in the potential range between 0.06 and 0.26 V. Their results also implied that when SbO^+ was electrochemically reduced to Sb in the cathodic scan to 0.06 V, the positions of individual Sb species remained unchanged; thus, the SbO^+ species were reduced at their original sites without diffusion. When the potential was scanned back in the anodic direction to 0.26 V, no change in the adlayer structure was observed.

When the UPD Sb was carried out onto the Au(100) at 0.0 V, the deposited model of Sb atoms is presented in Figure 2a. Solid lines outline the unit cell, the shorter sides are in the <001> direction, whereas the longer sides are titled by 72° with respect to the shorter sides.

The average lengths of the sides of the unit cell are √2 and √5 times of the Au-Au distance, respectively, which indicates that the Sb adlayer on Au(100) has a (√2 × √5) structure. Upon UPD at 0.06 V, the model of the deposited Sb atoms is shown in Figure 2b. A unit cell is also outlined by solid lines. In contrast to that UPD at 0.0 V, the longer sides of the unit cell are in the <001> direction and the shorter sides are titled by 72° against the longer sides. The average lengths of the sides of the unit cell are 2√2 and √5 times of the Au-Au distance, indicating a (2√2 × √5) pattern of the deposited Sb atoms.

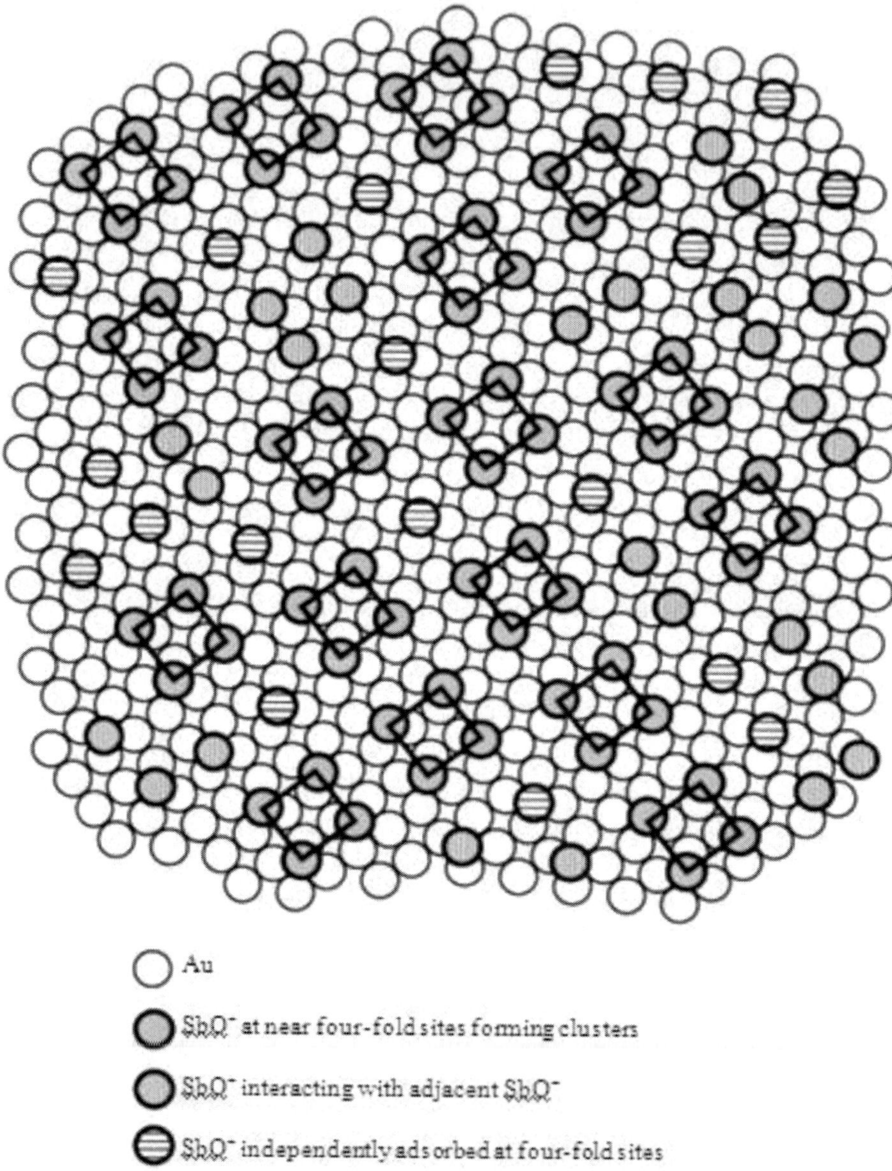

Figure 1. Schematic representation of SbO^+-modified Au(100). From Ref. [14]. © 2004, American Chemical Society.

Due to different modification methods, different adsorbed models for the Sb species are expected. The voltammetric studies on the depositing behaviour of Sb on Au(100) electrodes via IRA or UPD have also been carried out by Jung and Rhee [15]. The coverage of the Sb layer was 0.33 using IRA and 0.43 using UPD at 0.15 V. They proposed that the sites for UPD process are the oxygen vacancies generated during the reduction of the irreversible exogenous adsorbates, which suggested that the adsorption models for UPD and IRA may be different. This can be also proved by the Sb adsorption models presented in Figures 1 and 2.

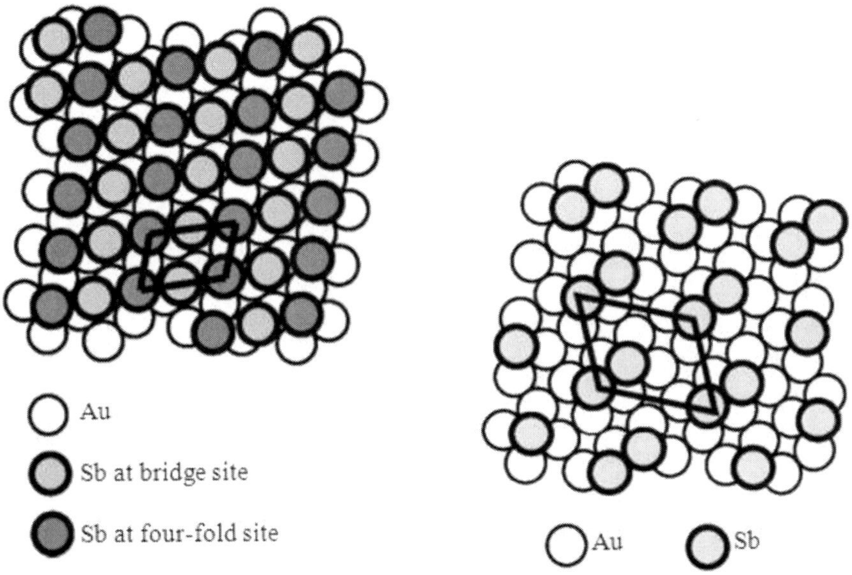

Figure 2. Structure model for an Sb adlayer adsorbed on Au(100) forming a ($\sqrt{2} \times \sqrt{5}$) structure at an electrode potential 0 V and a ($2\sqrt{2} \times \sqrt{5}$) structure at an electrode potential 0.06 V. From Ref. [14]. © 2004, American Chemical Society.

2.1.2. Sb/Au(111) Electrode

Mao and co-workers have observed that the adlayer of Sb on Au(111) showed different morphologies depending on the deposition methods using ECSTM [16, 17]. Specifically, the IRA Sb layer was a uniform layer of SbO^+ covered with sulfate ions, while the UPD Sb layer was an island of an Au-Sb alloy, when the deposition of Sb was carried out in H_2SO_4 solutions. When IRA and UPD took place sequentially, a wormlike network structure was formed. Morphology of UPD Sb has been investigated in detail, which induced a relaxation of Au(111) ($\sqrt{3} \times 22$) reconstruction to corrugation lines of trigonal networks. The triangles typically appear pairwise with an average length of the baselines of about 20.2 ± 1.0 nm and a height of the triangles of about 10.6 ± 1.4 nm at a corrugation height of about 0.05 nm. The baselines of the trigonal network rotate by 30° with respect to the corresponding double rows of the ($\sqrt{3} \times 22$) reconstruction and align along one of the three <110> directions. An atomic resolution image revealed an apparent ($\sqrt{3} \times \sqrt{3}$) structure in the regions with hcp stacking sites enclosed by the triangles, which is attributed to the atomically dispersed surface alloy structure of Sb with the topmost Au atoms [18]. Similar relaxation structures were observed with decreasing coverage of IRA Sb by anodic stripping, although the size of the relaxation is alleviated. The surface alloying is confirmed by the creation of holes at the Au electrode, which are confined at the connections between the two triangles. The results are rationalized as a consequence of compromise between two opposing effects: directional bonding via electron back donation, which tends to lift the reconstruction, and dislocation-mediated alloying, which favors the formation of relaxed reconstruction. [17].

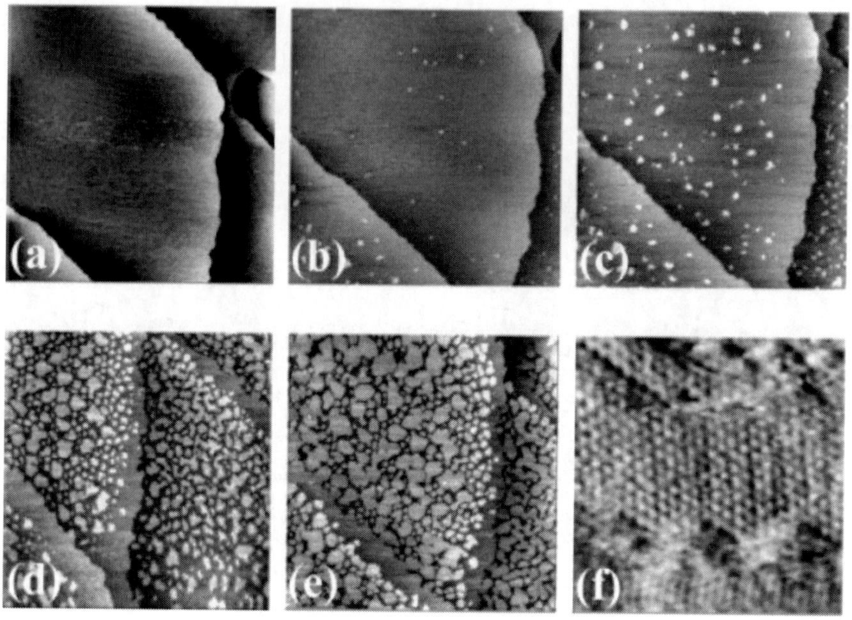

Figure 3. Sequential ECSTM images observed during the irreversible adsorption of Sb at open circuit potential in the Sb3+ containing 0.05 M H_2SO_4 solution. Image (a) was obtained before the addition of an aliquot of the Sb containing H_2SO_4 solution. The irreversible adsorption times for image (b), (c), (d), and (e) are 0, 4, 10, and 26 minutes, respectively. Image (f) is an atomically resolved image. Scan size: (a(e) 300 (300 nm2 and (f) 10 (10 nm2. Tip bias: (a~e) -7.65 mV and (f) -0.27 mV. Tunneling current: (a(e) 0.88 nA and (f) 0.81 nA. From Ref. [20]. © 2005, American Chemical Society.

Further, the adsorption process of IRA Sb on Au(111) at open circuit potential (approximately -0.05 V) was investigated using in situ ECSTM by Jung and Rhee [19, 20]. The adlayer of oxygenated Sb nucleates and grows on terraces and at step edges. The oxygenated Sb domains on terraces are round-shaped islands, whose size ranges from 3 to 6 nm in diameter. The adlayers originating from step edges, on the other hand, grow perpendicularly to the steps to show terrace-like uniform domains. There is no island of Sb on such terrace-like domains. In the long term, the island-shaped domains and the terrace-like domains are inter-connected to produce a networked monolayer of oxygenated Sb. The ultimate morphology at the open circuit potential was a network adlayer with a $(\sqrt{3} \times \sqrt{3})$ R30° atomic arrangement. Figure 3 is a series of ECSTM images observed during Sb IRA on Au(111) at open circuit potential. Figure 3a is an image of a clean and well ordered Au(111) surface, showing the characteristic double rows of the $(\sqrt{3} \times 22)$ reconstruction. Upon addition of an aliquot of the Sb_2O_3-saturated 0.05 M H_2SO_4 solution (the final concentration of Sb_2O_3 was roughly 10^{-6} M), small islands and terrace-like features of oxygenated Sb, as shown in Figure 3b, formed within 2-4 min time scale on the terrace and step edges, respectively. The formation of the oxygenated Sb domains is accompanied by the disappearance of the double rows of reconstructed Au(111). In the following image (Figure 3c), the growth of the islands and the generation of new islands are clearly seen. Addition of another aliquot of the Sb_2O_3-saturated 0.05 M H_2SO_4 solution, as shown in Figure 3d, accelerated the growth of the existing islands and facilitated further nucleation of new islands. Ultimately, the islands were interconnected to show a network structure (Figure 3e). Figure 1f is a typical atomic image of an oxygenated Sb island, which does not experience any electrochemical treatment. The repeating period of the spots was 0.50 ± 0.03 nm in a hexagonal symmetry, revealing that the unit cell of the oxygenated Sb island is $(\sqrt{3} \times \sqrt{3})R30°$. This particular local surface structure of 0.33 ML Sb coverage indicates that one of the constituent elements of the oxygenated Sb(III) [19] or SbO^+ [14] was imaged under the employed experimental condition. When the oxygenated layer is reduced, the adsorption features, such as the island, shrunk or disappeared depending on their sizes. This process was interpreted in terms of an alloy formation of Sb and Au. But not all of the Sb atoms were involved in the alloy formation. Interestingly, the alloyed and unalloyed Sb domains both showed $(\sqrt{3} \times \sqrt{3})R30°$ atomic structures with a spot-spot distance of 0.50 ± 0.05 nm, but they have different brightness in ECSTM images. During oxidation of the reduced Sb layers at 0.2 V, the alloyed and unalloyed domains of Sb behave in different ways: the alloyed Sb was stripped to a soluble species to leave pits, while the unalloyed Sb became an oxygenated adspecies, which desorbed very slowly. A long oxidation leads to an Au(111) covered with pits and islands of (1 × 1) without any adsorbed Sb.

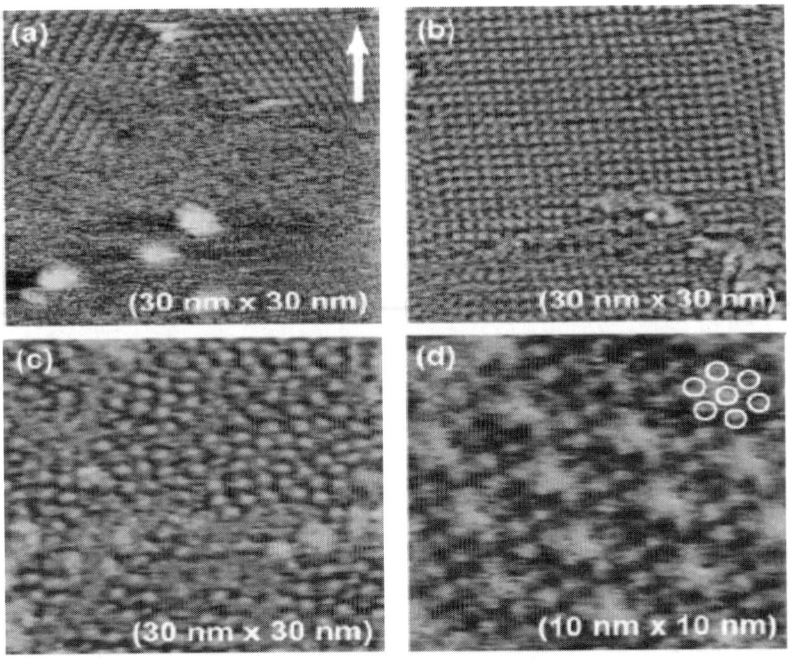

Figure 4. STM images of the $SbCl_2^+$ adsorption on Au(111) in MBIC-ALCl$_3$ +5 mM $SbCl_3$. (a) E is scanned from 1.2 to 1.10 V (vs Al/Al(III)) and (b) E = 1.10 V (vs Al/Al(III)). (c) Large-scale and (d) high-resolution STM images acquired at E = 1.05 V (vs Al/Al(III)). From Ref. [21]. © 2010, American Chemical Society.

The electrodeposition of Sb on Au(111) in an acidic chloroaluminate ionic liquid, 1-methyl-3- butylimidazolium chloride-aluminum chloride (MBIC-$AlCl_3$, 42:58 mol %) containing 5 mM $SbCl_3$, has been investigated by cyclic voltammetry and STM [21]. Upon decreasing the potential from 1.3 V (vs Al/Al(III)), desorption of $AlCl_4^-$ starts [22, 23], the STM image becomes fuzzy, and a new domain appears gradually (as shown in Figure 4a). Figure 4b shows a typical STM image acquired at 1.10 V (vs Al/Al(III)). A well-ordered structure extends all over the Au(111) surface. The lattice constants extracted from STM image are $a = 1.0 \pm 0.1$ nm, $b = 1.0 \pm 0.1$ nm, and R = 82 ± 2°, indicating that this structure is different from previous reported results [24, 25]. Although the physical origin of this ordered superstructure is still unclear at the moment, a possible explanation is attributed to a less closely packed Sb adlayer or the adsorption of $SbCl_2^+$ on Au(111). The latter seems more reasonable since the $SbCl_2^+$ cation is the main Sb(III) species in the acidic ionic liquid [26]. This is reasonable also because the strong electrostatic repulsion would prevent the clustering of such highly charged species. Upon the potential going more negative, structural transition occurs. As shown in Figure 4c, a Moiré-like pattern is finally formed at 1.05 V (vs Al/Al(III)). The image shows a larger repeat spacing of ~2.5 nm, which is attributed to the lattice mismatch between the adsorbate and substrate. Figure 4d depicts a higher resolution STM image, revealing the structural details. Obviously, each cluster consists of a set of small bright pots, which are indicated by white circles and arranged in hexagonal lattice. To determine the content of the unit cell is not straightforward because the structural details are missing from the STM images. However, careful inspection has revealed a large separation of ~0.7 nm between two neighboring spots. This value is much bigger than those of iodine and PF_6^- on Au(111) [27], indicating that the bright spots of the clusters cannot be formed with elemental Sb, instead of $SbCl_2^+$. It is noted that the ($\sqrt{3} \times \sqrt{3}$) and ($\sqrt{7} \times \sqrt{7}$) structures reported previously could not be observed in this study [28]. Since the ion species for these experiments were not identical, but differed in chemical composition and acidity, this can have a strong impact on the reduction potentials and on the Sb IRA and as well as on the UPD process.

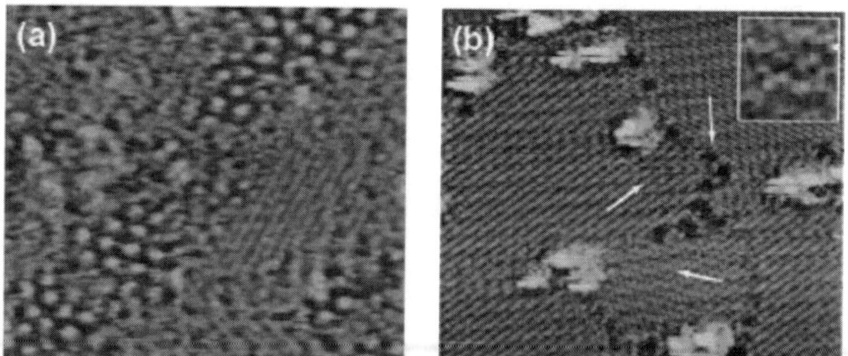

Figure 5. STM images of Sb UPD on Au(111) in MBIC-$ALCl_3$ +5 mM $SbCl_3$: (a) from 1.05 to 1.0 V and (b) E = 1.0 V (vs Al/Al(III)). The inset is a close-up STM image. From Ref. [21]. © 2010, American Chemical Society.

When the potential is continuously decreased to 1.0 V (vs. Al/Al(III), in the UPD range of Sb), a sudden smoothing of surface has been observed (see in Figure 5b). In each domain, the atomic structure is characterized by periodic rows (indicated by arrows) with a repeat distance of 0.9 ± 0.1 nm. The rows are composed of two pairs of aligned short bars with a periodicity of 0.5 ± 0.1 nm and aligned along the <110> directions of the Au(111) surface. A (4 × 9) structure can be concluded, regardless of the constituents of the unit cell. Reducing the potential to 0.87 V (vs. Al/Al(III), see in Figure 6a), the deposition of Sb becomes obvious and a few nuclei are formed both on the terraces and along the step edges of Au(111). As the deposition process is continued, the nuclei develop into anisotropic nanostripes. Interestingly, the nanostripes are confined to a width of about 2.0 nm and to a height of about 1.0 nm, and lie randomly on the Au(111) surface. This observation is comparable to the Sb deposition on Cu(100) [29]. However, the width and orientations of nanostripes are different in the two systems. This is possibly attributed to different lattice constant for the two substrates, Au(111) and Cu(100). As the deposition process is continued at 0.8 V (vs. Al/Al(III), see in Figure 5b), the number of deposited nanostripes increases rapidly, which spread all over the Au(111) surface, then 3D clusters begin to grow on top of the nanostripes.

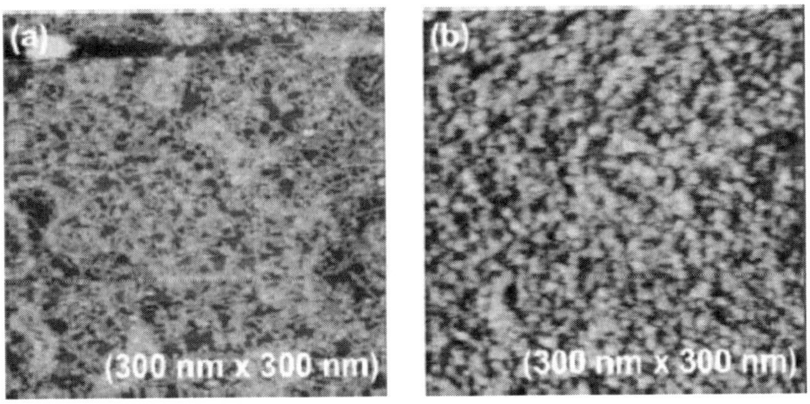

Figure 6. STM images of Sb UPD on Au(111) in MBIC-ALCl$_3$ +5 mM SbCl$_3$: (a) E = 0.87 V and (b) E = 0.8 V (vs Al/Al(III)). From Ref. [21]. © 2010, American Chemical Society.

2.1.3. Sb/Polycrystalline Au Electrode

The electrochemical processes of Sb(III) species on the polycrystal Au electrode were studied by CV, electrochemical impedance spectroscopy (EIS), and electrochemical quartz crystal microbalance (EQCM) [30, 31]. The CV data revealed that IRA Sb on Au electrode yielded oxidation and reduction features at about 0.15 V, EQCM data indicated that Sb species were stable on Au electrode in the potential region from -0.25 to 0.18 V; the adsorption of Sb further inhibited the adsorption of water and anion on Au electrode at low electrode potentials. Sb_2O_3 species were suggested to form on the Au electrode at 0.18 V. At a potential, higher than 0.20 V the Sb_2O_3 species could be further oxidized to Sb(V) oxidation state and desorbed gradually from the Au electrode. The electrochemical reduction of SbO^+ on the Au electrode in the Sb(III)-contained solution includes three parts: the irreversible reduction of adsorbed SbO^+, the reversible UPD of dissociative SbO^+, and the irreversible overpotential deposition of dissociative SbO^+. While the reduction of Sb(III) species in the presence of the complex reagent (tartaric acid $C_4H_4O_6^{2-}$) can be conjectured as comprised of four processes: the reduction of adsorbed SbO^+ to Sb, the UPD of dissociative SbO^+, the reduction of absorbed $[Sb_2(C_4H_4O_6)]^{2+}$ to Sb, and the reduction of dissociative $[Sb_2(C_4H_4O_6)]^{2+}$ to Sb. Further EIS studies implied that the three electrons are not gained simultaneously during the formation of Sb in these four processes.

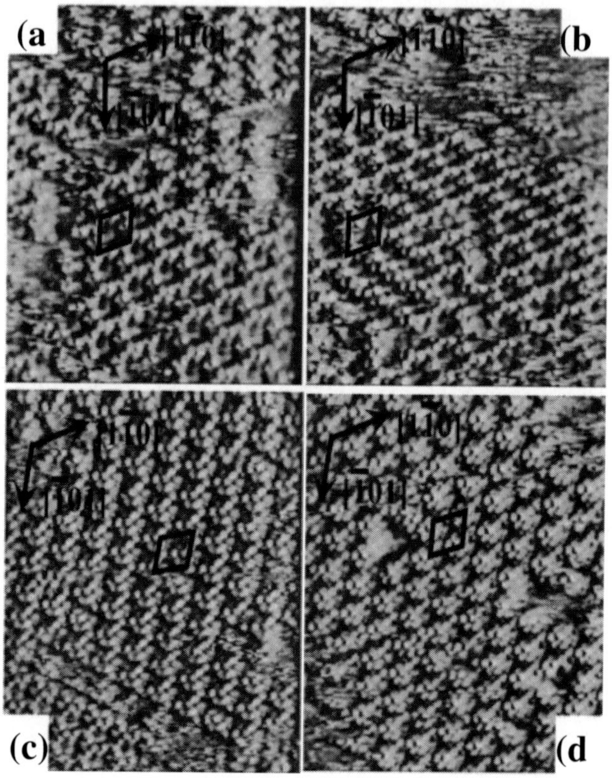

Figure 7. Micrographs of the pristine Sb layer on Pt(111) in (4 × 3) arrangements. The micrographs were obtained without potential control. Scan size: 10 nm × 10 nm. Tip bias: 0.878 mV; set point: 1.7 nA. From Ref. [35]. © 2006, American Chemical Society.

2.2. Sb/Pt Electrodes

2.2.1. Sb/Pt(111) Electrode

It is well established that Pt surfaces adsorb irreversibly a wide number of foreign atoms, such as Bi, As, and Sb due to their unique electronic properties. The influence of IRA Sb on the hydrogen adsorption and evolution on Pt(111) surface has been detected [32]. The data indicated that: Sb atoms remain stable on the surface and there is no proof of their hypothetical mobility; the inhibition of the adsorption of H could be explained by using only geometric effects. It seems that there are no long-rang electronic interactions, and in all cases the adlayer structure is ordered without random distributions or bidimensional island formation. This fact should be a consequence of the existence of lateral repulsive forces between adatoms. These repulsive forces are favored because of the weakness of Sb-Sb bonds, which has a relatively small enthalpy of vaporization. At low and intermediate coverage, IRA Sb on Pt(111) electrodes has a similar behavior to that of Bi adsorbed Pt(111).

The voltammetric profile exhibits a peak at 0.45 V that corresponds to the surface oxidation of Sb to Sb(III) in form of an oxygenated species. The Sb adlayer can be desorbed at potentials above 0.75 V without disruption of the long range order of the Pt(111) surface [33]. At high coverage, a different behavior for the Sb modified electrodes has been found. Adsorbed Sb forms likely a surface alloy, probably by place exchange with the surface Pt atoms. The desorption of the Sb leads to a disruption of the long range order of the Pt(111) surface, in a similar way (although at lower extent) to that obtained by successive cycles in the oxygen adsorption region. When the alloy is formed, a new oxidation wave for formic acid is observed at 0.16 V [34] and indicates a high increase of the catalytic activity of the surface toward formic acid oxidation.

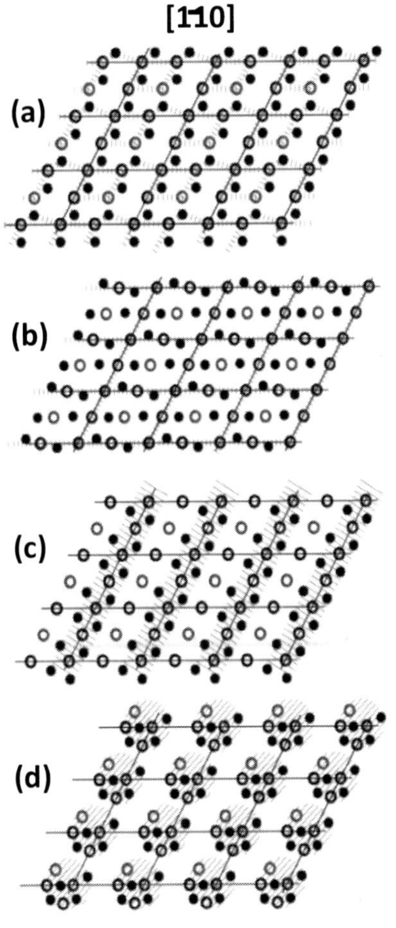

Figure 8. Schematics of the (4 × 3) structures of the pristine Sb layer: (a) (4 × 3)-I, (b) (4 × 3)-II, (c) (4 × 3)-III, and (d) (4 × 3)-IV. The open circles represent Sb(III) ions, and the filled circles stand for O^{2-} ions. The gray open circles of Sb(III) in (b), (c) and (d) denote the unobservable Sb(III) ions, while the ones in (a) designate the dim spots at the corners of the parallelograms. From Ref. [35]. © 2006, American Chemical Society.

ECSTM has been used to study IRA Sb on Pt(111) at various potentials [35]. At an open circuit potential (0.41 V), well-ordered structures of SbO^+ were found: four (4 × 3) structures and one (2√3 × 2√3)R30° structure. In Figure 7, there are four atomic images of Sb irreversibly adsorbed on Pt(111) in 0.05 M H_2SO_4 solution without potential control. The lattice parameters of the unit cells in the images were measured to be 1.11 ± 0.03 nm in the $[1\bar{1}0]$ direction and 0.84 ± 0.01 nm in the $[\bar{1}01]$ direction, respectively, which correspond to a (4 × 3) structure taking one Pt-Pt distance (a = 0.277 nm) into account. Although the unit cells of the four images were identical, the details in the arrangement of the observed spots within each unit cell were different from each other. Specifically, the (4 × 3) structure of Figure 7a (designated as (4 × 3)-I) was an array of small parallelograms (0.56 ± 0.01 nm in the $[1\bar{1}0]$ direction and 0.37 ± 0.02 nm in the $[\bar{1}01]$ direction), while that of Figure 7b (denoted as (4 × 3)-II) showed eight bright spots along the sides of the unit cell. The (4 × 3) structure of Figure 7c (designated as (4 × 3)-III), on the other hand, consisted of the wide rows of spots in the $[\bar{1}01]$ direction and the isolated bright spots between them. In Figure 7d (denoted as (4 × 3)-IV), the aggregates of seven spots were arranged in the (4 × 3) pattern. A close examination of the ECSTM images disclosed that within the (4 × 3) unit cells, there were seven atomic spots: three bright spots and four dim spots. One exception was the (4 × 3)-I structure, whose unit cell consisted of three bright spots and five dim spots. Probably, the dim spots at one of the corners of the parallelograms in Figure 7a would not be observable in the images in parts b-d of Figure 7 because of an extremely low tunneling efficiency of those particular atoms. Such an absence of STM spots despite existence would be exemplified with the atomic structures of I on Pt (111) [36, 37]. Then, the number of the spots, including the unobservable spot, in one (4 × 3) unit cell was eight, which indicates that the spot coverage, regardless of their brightness, was 0.66. Because the electrochemically measured coverage of Sb was 0.33, half of the eight spots must be Sb atoms and the other half must be O atoms. In other words, the coverages of Sb and O were both 0.33; thus, the chemical identity of the irreversibly adsorbed Sb was a two-dimensional oxide of SbO^+, as proposed on Au(111) and Au(100) mentioned previously. The positive charge of the oxide monolayer might be compensated by the adsorption of HSO_4^- ion as proposed for the monolayer of IRA TeO^{2+} on Pt(111) [38]. The schematic models in Figure 8 are proposed for the (4 × 3) structures observed in Figure 7. Here, the bright and dim spots were arbitrarily assigned to Sb and O atoms, respectively. In the models for the (4 × 3)-I, (4 × 3)-II, and (4 × 3)-III structures, the Sb atoms (open circles) are arranged in a (4 × 3) pattern in Figures 8a-8c. Phenomenologically, however, the Sb atoms at the center of the (4 × 3) unit cell (gray open circles) were not observable with STM, except in the (4 × 3)-I structure. It should be noted that the Sb atoms at the center of the unit cell in the (4 × 3)-IV structure are not observable with STM, and are presumably located close to the corners of the (4 × 3) unit cell to account for the local stoichiometry of SbO^+. On the other hand, the respective locations of the O atoms (filled circles) were determined based on the dim spots in Figure 7. Figure 9 illustrates another atomic image of Sb irreversibly adsorbed on Pt(111) and a proposed model. The respective lattice parameters of the unit cell in Figure 9a were 0.96 ± 0.03 nm in the $[1\bar{2}1]$ direction and 0.95 ± 0.05 nm in the $[\bar{2}11]$ direction, indicating that the unit cell in Figure 9 was (2√3 × 2√3)R30°. In the unit cell, there were three bright spots and four dim spots, so that according to the same argument as in the (4 × 3) structures, the coverages of Sb and O were 0.33, respectively. In Figure 9b, the Sb atoms are arrayed in a (2√3 × 2√3)R30° pattern, equivalent to a (√3 × √3)R30° pattern, with the central Sb atom of low electron tunneling efficiency, and the O atoms are located at appropriate positions according to the dim spots in Figure 9a. Furthermore, a few discernible but unidentifiable structures were observed and proposed to be transient structures to the well-ordered ones. Upon electrochemical reduction, the O atoms in the oxide layer were stripped to leave a layer of Sb, a part of which was rearranged to compact electrochemically inactive clusters below 0.25 V. When an additional amount of Sb was UPD onto the Pt(111) covered with irreversibly adsorbed Sb, a full monolayer of the incommensurate √2*a* rows, presumably incommensurate (√2 × √2), was formed, so that the whole surface remained electrochemically inactive. The origin for the presence of electrochemically inactive Sb was concluded to be the narrow rooms between the Sb atoms in the √2*a* row not to accommodate the incoming O atoms during oxidation.

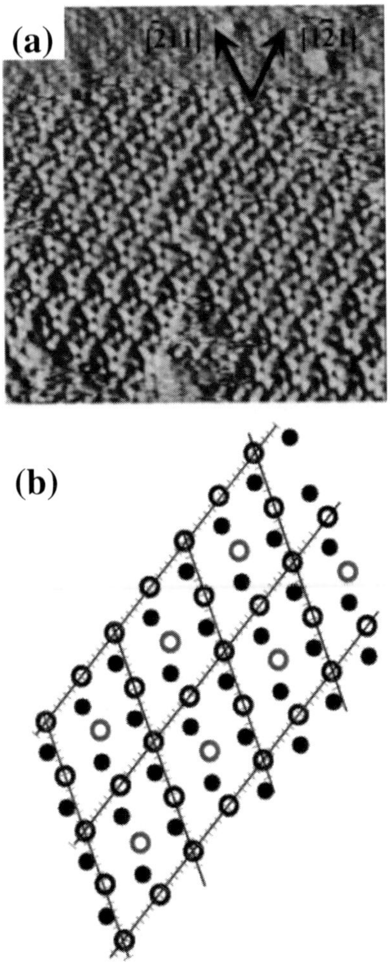

Figure 9. (a) A micrograph of the pristine Sb layer on Pt(111) in a (2√3 × 2√3)R30° arrangement and (b) its corresponding schematic. The open circles represent Sb(III) ions, and the filled circles stand for O^{2-} ions. Scan size: 10 nm × 10 nm. Tip bias: 0.878 mV; and set point: 1.7 nA. From Ref. [35]. © 2006, American Chemical Society.

2.2.2. Sb/Pt(100), (110), (320) and (331) Electrodes

Rhee et al. reported contrasting electrochemical behavior of IRA Sb on Pt(100) and Pt(111) single crystal electrodes [39]. The saturation charge of the IRA Sb on Pt(100) was 206±3.1 μC cm^{-2} (θ_{Sb} = 0.33) in 0.05 M H_2SO_4 solution. In the Sb_2O_3-saturated solution, the charge of Sb on Pt(100) was increased to be 244±8.8 μC cm^{-2} (θ_{Sb} = 0.39), which means that additional Sb is underpotentially deposited during the reduction of the IRA oxygenous Sb. On Pt(111), on the other hand, the saturation charge of the IRA Sb was observed to be 275±8.7 μC cm^{-2} (θ_{Sb} = 0.38) and the charges associated with the redox feature of the irreversible Sb adsorbate decreased during cyclic voltammetric scans, depending on the Sb concentration of the contacting solution. This particular behavior of Sb on Pt(111) is interpreted in terms of the formation of an electrochemically inactive Sb species as discussed in last paragraph. Behaviors of IRA Sb adatoms on Pt(100), Pt(110), Pt(320) and Pt(331) single crystal were investigated by Sun et al. as well [40]. It was determined that Sb adatoms are stable at potentials below 0.45 V on Pt(100) and Pt(110), below 0.40 V on Pt(320), and below 0.35 V on Pt(331). Different Sb coverage was obtained conveniently by partially stripping Sb from saturation coverage. They have demonstrated that the redox behaviors of Sb adatoms and the coadsorption properties of Sb with H depend strongly on the orientations of the Pt single crystal electrodes. The behaviors of surface redox process of adsorbed Sb atoms varied with the surface atomic arrangement of Pt single crystal electrodes. The oxidation characteristics of adsorbed Sb were measured at 0.37 V on surface sites of (100) symmetry and at 0.20 V on surface sites of (111) symmetry. These results indicate that the coverage of IRA Sb is affected by the atomic structures of the Pt substrates, which is confirmed the above conclusion reported by Rhee et al.

2.2.3. Sb/polycrystalline Pt Electrode

Combined studies of EQCM and CV by Wu et al. demonstrated that Sb adatoms are stable on polycrystalline Pt electrode in the potential range between -0.25 and 0.50 V (vs SCE), and the saturation coverage of Sb is 0.34 [41]. Owning to the advantage of EQCM, the $Sb(OH)_3$ species were determined as the adsorbed products of Sb oxidation up to 0.50 V. At a potential higher than 0.50 V the Sb(III) can be oxidized further and will desorb progressively from Pt electrode surface. In order to measure quantitatively the number of electrons transfer in surface Sb oxidation, Wu et al. have designed an programmed CV as shown in Figure 10: (1) the first cycle of potential cycling was limited between -0.25 and 0.50 V to confirm the adsorption of Sb; (2) the second cycle was conducted from -0.25 to 1.10 V to partially remove the adsorbed Sb; (3) the third and forth cycles were carried out again between -0.25 and 0.50 V to estimate the remain quantity of Sb. The oxidation current of adsorbed Sb in the double layer region was decreased in the third cycle, then it was increased slightly and stabilized in the fourth cycle. The slightly increase of Sb oxidation current in the fourth cycle indicates that part of Sb species desorbed at potentials above 0.50 V in the second cycle can readsorb on Pt surface. By integrating the j–E curves in the double layer region the decrease of electric charge (ΔQ_{Sb}) due to the partially desorbed Sb between the first and the fourth cycle of potential cycling has been measured to be 75.6 mC cm^{-2}. The df curve was shifted up from the first to the third cycle, manifesting a decrease in surface mass due to partial desorption of Sb in the second cycle. The df curve moved down from the third to the fourth cycle in confirming the readsorption of part of desorbed Sb. The difference of df (Δ(df)) between the first and the fourth cycle can be used to calculate the quantity of Sb desorbed from Pt surface, which is 5.7 Hz at 0.1 V. The Δ(d*f*) was measured at 0.1 V because the hydrogen adsorption–desorption process is ended at this potential, as a consequence the interference of replacement of H by water to the determination of Δ(d*f*) can be avoided. With the value of ΔQ_{Sb} and Δ(d*f*), the authors calculated the number of electron transfer (*n*) in Sb oxidation for a potential scan up to 0.50 according to the Sauerbrey equation [42],

$$n = \left(\frac{C_f M_{Sb} \Delta Q_{Sb}}{F|\Delta(df)|} \right)$$

where the M_{Sb} is the molar mass of Sb, and *F* is the Faraday constant. By introducing the values of ΔQ_{Sb} and Δ(d*f*), a average value of 3.02 ± 0.06 has been evaluated for the charge transfer from several repeat experiments, which indicates that the oxidation of Sb on the Pt electrode is a three-electron transfer process.

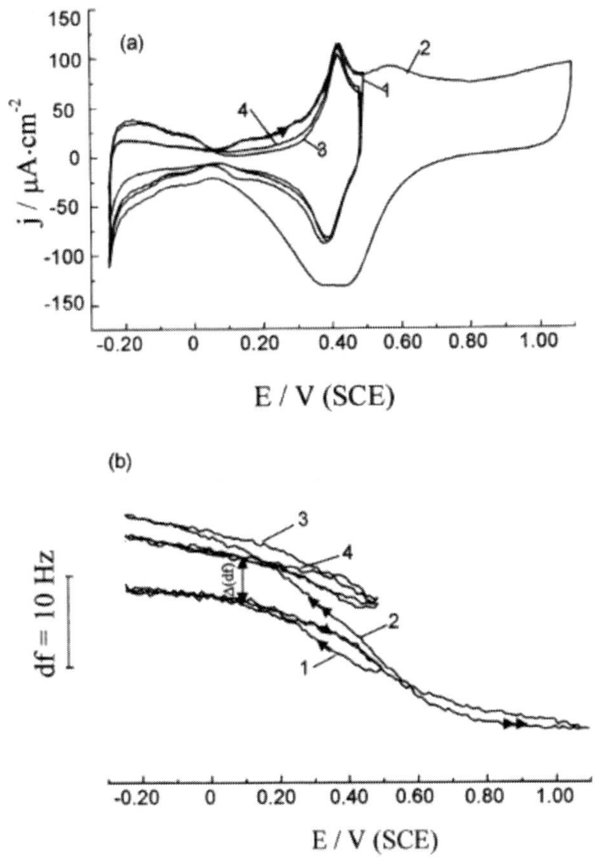

Figure 10. (a) Voltammograms and (b) df variations of Sb/Pt electrode recorded for determining the number of electron transfer in Sb oxidation in 0.1 M $HClO_4$ solution. Sweep rate: 50 mV s^{-1}. From Ref. [41]. © 2000, Elsevier.

3. Electrocatalytic Oxidation of CO on Sb/Pt Electrodes

CO is a model adsorbate for studying molecular adsorption in catalysis and electrocatalysis. CO is a strongly bonded intermediate in methanol, ethanol, and particularly formic acid oxidation. Poisoning of Pt fuel cell catalysts by CO is undoubtedly one of the most severe problems in fuel cell anode catalysis. In addition, the oxidation of CO is also of great significance for hydrogen-oxygen fuel cells in which the hydrogen stream is contaminated by small trace of CO, which is the additional motivation for studying its oxidation. Not surprisingly, CO electrooxidation is one of the most intensively studied electrocatalytic reactions, and there is a continued search for CO-tolerant electrocatalyst materials that are able to either bind CO weakly but still oxidize small organic molecules or that oxidize CO at significantly reduced overpotential. Foreign element adlayers on the Pt electrodes affect the adsorption behavior of CO and can induce enhancement of CO oxidation. There are essentially two ways to study electrochemical CO oxidation. The first one, called stripping, involves pre-adsorbing a monolayer of CO and then oxidizing it from a solution that does not contain CO, in either a voltammetric scan or a potential-step experiment. The second way to study electrochemical CO oxidation is continuous or bulk CO oxidation, that is, with a fixed concentration of CO in solution. We will discuss mainly CO stripping on Pt single crystal electrodes, because this makes it easier to study the oxidation mechanism of CO on the electrodes.

The positive effect has been observed on CO stripping due to the presence of IRA Sb on Pt(100) electrodes. Sb adsorption as low as a coverage of 0.03 can shift the CO stripping peak potential 60 mV towards lower potential values as compared with bare Pt(100) under the same conditions [34, 43]. The positive effect of the Sb adatoms may be interpreted through a bifunctional effect: the oxidized form of the adatom may supply the oxygen required for CO stripping. This behaviour is similar to that observed on Bi or As modified Pt(111) electrodes in which the redox process of the adatom takes place at potential lower than those of CO oxidation [44, 45]. An analogous explanation has been given for CO oxidation on glassy $Pt_{66}Sb_{34}$ alloys [46]. The onset of CO oxidation is reduced by about 200 mV in comparison with pure polyoriented Pt. It has been demonstrated that the oxidation of adsorbed and solution CO species on Pt and Sb/Pt electrodes undergoes different surface processes. The electrocatalytic effect of adsorbed Sb on the oxidation of CO consists mainly of the negatively shift of oxidation potential. This is because the activation energy of CO oxidation is decreased on a Sb/Pt electrode in comparison to the same reaction carrying out on a Pt electrode [41]. Kizhakevariam and Weaver [34] also found that Sb inhibits the adsorption of CO on Pt(100) and Pt(111) by decreasing the two-fold binding geometries deduced from the relative CO vibration band intensities. The optimum catalysis for Pt(100) was observed at $\theta_{sb} = 0.35$. At higher Sb coverages there is a shift of the Sb oxidation step towards lower potentials. Similar shifts are observed on Pt(111) electrodes with high coverates of Bi or As even in absence of adsorbed CO, and have been interpreted by the existence of more compressed adatom layers. If this assumption is valid it means that CO adsorption locally affects the compactness of the adatom adlayer. It can be realized that at Sb coverage near saturation only a small part of the Sb redox process is shifted, because the surface is only able to adsorb a small amount of CO. The part of the surface, in which CO adsorption is not possible because it is covered by the compact Sb adlayer, shows the same redox behaviours as in absence of CO. The modification of the redox behaviour of the adatom agrees with the existence of mixed adlayer Sb-CO on Pt(100) electrodes as suggested previously from FTIR results [34]. More insight about the complex behaviour of Sb-CO adlayers can be obtained from partially covered surfaces. When the surface amount of adsorbed CO is small the Sb redox process remains practically unaffected, suggesting that CO preferentially adsorbs on Pt atoms away from those bonded to Sb. When the amount of adsorbed CO increases, the oxidation of the adatom is clearly shifted towards lower potential values, indicating that the Sb adlayer becomes affected by the presence of the adsorbed CO.

It is interesting to point out that poisoning by CO, dissociating from the small organic molecules, does not block totally the Pt adsorption sites for hydrogen. It thus suggests that its formation requires the existence of ensembles of a number of neighboring Pt sites, in which the dissociative adsorption step could occur. It has been shown that IRA adatoms can inhibit CO formation from formic acid dissociative adsorption. The extent of this effect strongly depends on the crystallographic orientation of the substrate, and the nature of the foreign adatom. This effect can be rationalized as a simple third body effect: the adsorbed adatoms block the surface sites and the resulting adlayer structure has not enough neighboring Pt sites in which formic acid dissociative adsorption could take place. However, the basic hydrogen evolution reaction mechanism does not change with adatom adsorption, while its rate is lowered as a consequence of poisoning [32]. Sb adsorption causes significant up-shifting of the CO band formed from formic acid oxidation, which suggests that dissociative chemisorption is triggered at adjacent sites, electronically modified by Sb. The enhancement has been interpreted in line with third body and electronic effects [34]. However, good agreement was found between the simulated effects and experimental data for the Sb/Pt(100) system where Sb was assumed to obey the third body effect [47].

4. Electrocatalytic Oxidation of Formic Acid on Sb/Pt Electrodes

Electrochemical oxidation of formic acid on Pt single crystal electrodes can be considered as a model reaction in electrocatalysis [48, 49]. Direct formic acid fuel cells (DFAFCs) are very promising green power sources for portable electronic devices [50, 51]. Nowadays intensive investigations have been conducted to understand the mechanism of formic acid electrooxidation at the Pt-based surfaces [52-60], and to develop high efficient Pt-based catalysts in light of the mechanism. A "dual pathway" mechanism has been proposed [57, 59]. One of them leads to direct formation of CO_2, and the other involves the formation of dissociated CO, which poisons the metal surface. Both paths are structure sensitive [61]. Currently, the challenge for developing new Pt-based catalysts is to redirect the reaction from intermediate pathway to direct pathway through alloying or surface decoration with a foreign adatoms [40, 47, 62]. Among the elements of group V, Sb is the most investigated one for surface modification on Pt electrodes, and has proved to be highly effective for facilitating the direct pathway due to the so-called "third-body" or "electronic effect". Satisfactory modification is achieved so far via either UPD or IRA followed by potential controlled desorption of partial adatoms, in addition to the use of Pt-based alloys and intermetallics. In fact, the Sb-modified Pt/C catalyst has been used in practical DFAFC, yielding significant improvement in overall cell performances [62, 63].

Regarding the catalytic activity of the low Sb coverages on Pt(111) electrodes, these surfaces are able to inhibit poison formation from formic acid through a long range effect. For the direct oxidation of formic acid, Sb adsorption increases the catalytic current for formic acid oxidation at 0.26 V (the same potential value as on Pt(111)). At high coverages, adsorbed Sb forms likely a surface alloy, probably by place exchange with the surface Pt atoms. When the alloy is formed, a new oxidation wave for formic acid is observed at -0.06 V, indicating a high increase of the catalytic activity of the surface toward formic acid oxidation [33]. The electrochemical oxidation of formic acid via reactive intermediates on Sb/Pt(110), Sb/Pt(320), and Sb/Pt(331) electrodes were investigated quantitatively by employing the potential step technique and in situ FTIR spectroscopy [40, 64]. It has been revealed that the dissociative adsorption of formic acid can be inhibited by the presence of Sb on the Pt single crystal surfaces. Significant catalytic effects towards formic acid oxidation were observed on Pt single crystal electrodes modified with Sb adatoms, which consist of (1) the inhibition of dissociative adsorption of formic acid, (2) the enhancement of oxidation current, and (3) the negative shift of oxidation potential that was measured about 220 mV on Sb/Pt(110), 110 mV on Sb/Pt(320), and 100 mV on Sb/Pt(331) electrodes. However, surprisingly neither enhancement of oxidation current nor negative shift of oxidation potential can be observed on Sb/Pt(100) electrode. This result suggests that electronic effect is the main effect presented on Pt(110), Pt(320) and Pt(331) surface upon Sb modification, while geometric effect is considered to the major effect on Pt(100) electrode. It has been attributed to the electronic effect that the presence of a small amount of Sb on Pt(110), Pt(320) and Pt(331) can inhibit largely hydrogen adsorption and enhance significantly electrocatalytic activity of these electrodes for formic acid oxidation. Specially, the kinetics of formic acid oxidation on the Sb/Pt(110) electrode has been studied [64]. The in situ FTIRS studies provided new information at a molecular level. At potentials below 0.0 V the IR bands of CO_2 near 2345 cm^{-1} and the carbonyl band around 1724 cm^{-1} are the main bands observed from spectra recorded on Sb/Pt(110) electrodes, which are indications for the direct oxidation of formic acid via the reactive intermediates. However, under the same conditions on a bare Pt(110) electrode, a bipolar IR band around 2058 cm^{-1} appears in the spectra. This bipolar band is assigned to IR absorption of linear CO species derived from the dissociative adsorption of formic acid. The kinetics of formic acid oxidation via the reactive intermediates on Sb/Pt(110) electrode can be described by the rate of CO_2 production (k_{CO2}), which has been evaluated from in situ time-resolved FTIRS data. The values of k_{CO2} vary with both Sb coverage and the oxidation potential. The activity of Sb/Pt(110) electrode towards formic acid oxidation has been arranged according to k_{CO2} values; an order of Sb/Pt(110) ($\theta_{Sb} = 0.126$) > Sb/Pt(110) ($\theta_{Sb} = 0.372$) > Pt(110) was found at the same oxidation potential. This result indicates that the fully Sb-modified Pt electrode would not have the highest electrocatalytic properties, which suggests that the oxidation of formic acid still needs the surface Pt atoms. It has determined that at its saturation coverage the Sb adatom can inhibit three surface Pt sites on Pt(100), Pt(110) and Pt(331) and four surface Pt sites on Pt(320). The results revealed that the dissociative adsorption of formic acid can be suppressed by the presence of Sb on platinum single crystal surfaces via a geometric effect. The peak potential of formic acid oxidation on Sb/Pt(hkl) were all converged in a narrow potential range between 0.25 and 0.30

V, which implied that the adsorption of Sb on the Pt(hkl) planes may alter the reaction activation energy.

Further Sun and Yang have investigated the effects of Sb adsorption on the kinetics of formic acid oxidation on the Pt(100), Pt(110), Pt (111), Pt (320), and Pt(331) surfaces using CV, potential step technique, and in-situ FTIR spectroscopy [65]. The kinetics of formic acid is faster on bare surfaces of Pt(100), Pt(110), and Pt(320) than on their Sb-modified surfaces, but it is slower on clean surfaces of Pt(111) and Pt(331) compared with their corresponding Sb-modified surfaces. The apparent activation energies of formic acid oxidation on these Sb-modified Pt single crystal electrodes have been determined, which are 12.80±0.32, 18.74±0.39, 16.68±0.35, -7.88±0.29, and -11.69±0.24 for Sb/Pt(100), Sb/Pt(110), Sb/Pt(320), Sb/Pt(111) and Sb/Pt(331), respectively. The positive values for Sb/Pt(100), Sb/Pt(110), and Sb/Pt(320) electrodes give rise to an increase in the apparent activation energy of formic acid when Pt(100), Pt(110), and Pt(320) surfaces are decorated with Sb adatoms. The negative values for Sb/Pt(111) and Sb/Pt(331) electrodes result in a decrease of the apparent activation energy of formic acid oxidation, indicating that the adsorption of Sb adatoms on Pt(111) and Pt(331) surfaces will enhance the electrocatalytic activity. These results throw a new light on revealing the origins of electrocatalytic effects of Sb adatoms, and upon understanding the surface structure effects of Sb/Pt single crystal electrodes on the kinetics of formic acid oxidation.

Poison formation reaction from formic acid has been studied on different platinum stepped surfaces (Pt(755) and Pt(533)) modified by IRA Sb [66]. For the high-index Pt electrodes, two different sites can be distinguished on these surfaces: step and terrace sites. These two different sites will have different properties for poison formation, and the selective deposition of the Sb adatoms on these sites will allow establishing the role of the sites in the poison formation reaction. Surfaces with narrower terraces exhibit a linear decrease in the total poison accumulation with the Sb coverage, whereas, for the surfaces with wide terraces, the total amount of poison diminishes sharply after all the step sites have been covered by the adatoms. The results have been interpreted in terms of the electronic properties of the adatom and the surface. The main conclusion is that in surfaces with wide terraces, poison formation from formic acid only takes place on the step sites. The presence of a step causes the appearance of net dipole on the surface, with the positive part of the dipole on the upper part of the step and negative part of the dipole on the lower part of the step [67]. It is obvious that the electronic modification of the surface not only affects the step but also reaches some of the atoms of the terrace [68].

For the adatoms that decorate the step, the initial stages of adsorption do not lead to a significant diminution of the total poison accumulation. As long as some step sites remain uncovered by the adatoms, the poison formation reaction will take place on them. However, these CO molecules can diffuse to the terrace sites. Meanwhile, the step site is once more free from CO, and formic acid dissociation can take place again on this site, generating a new CO molecule adsorbed on the step sites. Since the number of sites covered by the adatom is small (for the Pt(755) surface, the number of terrace sites is six times number of step sites), the total accumulation of poison does not diminish significantly. The situation changes drastically when all the step sites have been covered by Sb adatom. In this moment, poison formation cannot take place, and the total poison accumulation should be zero. In the real situations, diminution of poison formation after all the step sites is steep but not as sharp as predicted by the model and that observed for Pt(111) electrodes modified by the same adatoms [69]. This deviation could be related to the difficulty of covering the last step sites in a real situation, or because the existence of some marginal poison formation on the (111) terraces. For the adatoms that do not decorate the step, the situation is much simpler. The deposition of the adatoms on the terrace will lead to a diminution of the available sites for poison accumulation but leaving unaltered the possibility of poison formation on the step sites, since the step sites are covered in the last stages of the adatom deposition. Therefore, the total poison accumulation exhibits a linear dependence on Sb coverage. On these cases, the total inhibition of poison formation is reached when the surface has completely covered by the adatoms. This result probably reflects the existence of an ensemble effect in the poison accumulation reaction. If the distribution of adatoms on the terrace is such that isolated platinum sites (platinum sites totally surrounded by adatoms) are found, these sites will be unable to be covered by the poison, since the CO cannot diffuse through the adatom boundary, and the total poison accumulation will reach a zero value. However, this surface will be able to adsorb hydrogen and, thus, show the presence of free platinum sites.

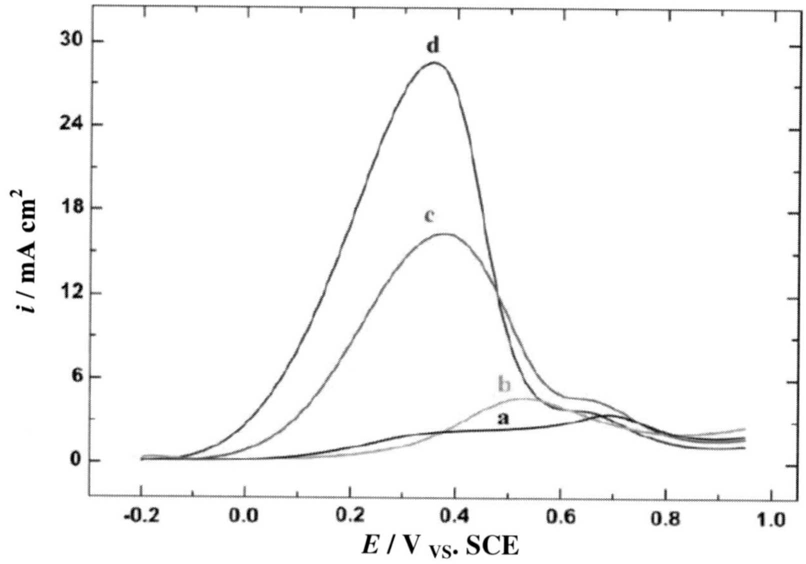

Figure 11. Anodic linear sweep voltammograms of (a) unmodified Pt/C, (b) Pt-Ru/C, (c) Sb-Pt/C (IRA), and (d) Sb-Pt/C (MUPD) coated glass carbon electrodes in 0.5 M H_2SO_4 + 0.5 M HCOOH at 50 mV s^{-1}. From Ref. [63]. © 2009, Elsevier.

In situ electrochemical surface-enhanced infrared absorption spectroscopy (EC-SEIRAS) together with a periodic density functional theory (DFT) calculation has been used to investigate the mechanism of formic acid electrooxidation on Sb/Pt electrodes, which were electrochemically deposited on the Au/Si substrates [70]. EC-SEIRAS measurement revealed that the main formic acid oxidation current on Sb/Pt electrode is ca. 10-fold enhanced as compared to that on clean Pt electrode, mirrored by nearly synchronous decrease of the CO and formate surface species, suggesting a "non-formate" oxidation as the main pathway on the Sb/Pt electrode. On the basis of the calculations from periodic DFT, the catalytic role of Sb adatoms can be rationalized as a promoter for the adsorption of the CH-down configuration but an inhibitor for the adsorption of the O-down configuration of formic acid, kinetically facilitating the complete oxidation of formic acid into CO_2. From Bader charge analysis, the adsorption of Sb on Pt causes a surface dipole of $[Sb]^{\delta+}$- $[Pt]^{\delta-}$, which contributes to the enhanced adsorption of CH-down configuration via electrostatic interaction. Besides, DFT calculation also confirms that CO adsorption is weakened on Sb/Pt surfaces, indicating that the adsorption of Sb on Pt can help to relieve the CO poisoning on Pt electrodes. The authors pointed out that the combined EC-SEIRAS and first-principles study may be generally extended to the investigation of electrocatalytic oxidation of formic acid on other efficient adatom modified Pt surfaces.

A new facile approach towards developing superior Pt-based catalysts for formic acid electrooxidation has been proposed, which was realized with a mimetic UPD (MUPD) of Sb on 60 wt% Pt/C powders to attain a favorable coverage [63]. The catalytic activity towards formic acid electrooxidation on the above Sb-modified Pt/C catalyst far exceeds that on commercial Pt–Ru/C or Sb-modified Pt/C through traditional irreversible adsorption. Figure 11 shows the anodic linear sweep voltammograms for different catalyst-coated glass carbon electrodes in 0.5 M H_2SO_4 + 0.5 M formic acid. It can be seen that for the unmodified Pt/C catalyst the oxidation current is very weak at low potentials (curve a) due to CO poisoning through the intermediate pathway. The use of the Pt–Ru/C catalyst has a limited improvement over the electrocatalysis of HCOOH (curve b). However, on Pt–Sb/C catalysts the electrooxidation currents are significantly enhanced at lower potentials because of the so-called ‘‘third-body" or ‘‘electronic effect", with the direct pathway being predominant. The Pt–Sb/C (MUPD) exhibits nearly doubled peak current as compared to the Pt–Sb/C (IRA). The long-term electrocatalytic activities of these catalysts were further explored by polarizing different catalyst-coated glass carbon electrodes at 0.25 V in 0.5 M H_2SO_4 + 0.5 M formic acid for 1200 s. It should be stressed that the Pt–Sb/C (MUPD) catalyst exhibits the highest current density, demonstrating the feasibility and the extendibility of the MUPD. This electroless approach is generally applicable to all types of Pt surfaces, in particular suited for upgrading Pt/C for practical anode catalysts of direct formic acid fuel cells.

CONCLUSIONS

The adsorption/deposition of Sb on the Au and Pt electrodes and their application on the CO and formic acid electrocatalytic oxidation were reviewed. Also the reduction and oxidation processes of the Sb species on the Au and Pt electrodes have been discussed. The literature results indicated that the adsorption model of Sb species on the metal electrodes strongly depends on the electrode surface structure, the modification method (IRA or UPD), the Sb coverage, the electrode potential, and the anion species in the Sb containing solution or electrolyte. The modification of Sb on the metal electrodes will change their surface atomic structure, electronic and electrocatalytic properties. The electrochemical oxidation of CO on the Sb/Pt electrodes is greatly enhanced by negatively shift of the oxidation potential due to the decrease of activation energy of CO oxidation on the Sb/Pt electrode. The electrocatalysis of formic acid on the Sb/Pt is realized by: (1) the inhibition of dissociative adsorption of formic acid into CO; (2) the enhancement of oxidation current; and (3) the negative shift of oxidation potential for formic acid. The results also showed that different Pt single crystal substrates have different behavior on the Sb adsorption, and consequently on the formic acid oxidation. The mechanism of formic acid on the Sb/Pt electrode has been proposed based on the FTIR data and theoretical calculations. Very high electrocatalytic property was found on the Sb-modified Pt/C electrode.

As demonstrated in this review, the adsorption of Sb on metal surfaces is a model system to investigate the electrocatalytic property of a metal electrode upon its surface modification. The coverage of Sb adatoms is conveniently controllable through the IRA process, and the effects of surface structure (chemical, electronic and atomic arrangement) can be probed using different methods such as electrochemistry, scanning probe spectroscopy, X-ray photoelectron spectroscopy, and FTIR spectroscopy. It may expect that the knowledge gained in the fundamental studies of Sb adsorption and its electrocatalytic effects could be contributed in designing and fabricating electrocatalysts of high performances.

ACKNOWLEDGEMENT

Shi-Gang Sun thanks the Natural Science Foundation of China for financial support (grants: 20833005, 21021002).

REFERENCES

[1] Clavilier, J.; Feliu, J.M.; Aldaz, A. *J. Electroanal. Chem.* 1988, 243, 419-433.
[2] Kozawa, A. *J. Inorg. Chem.* 1961, 21, 315-324.
[3] Wu, Q.H.; Yang, Y.Y.; Zhou, Z.Y.; Xiao, X.Y.; Chen, S.P.; Sun, S.G.; Chen, G.L. *Spectrosc. Spect. Anal.* 2000, 20, 612-613.
[4] Yang, Y.Y.; Wu, Q.H.; Zhou, Z.Y.; Zheng, M.S.; Chen, S.P.; Sun, S.G. *Spectrosc. Spect. Anal.* 2000, 20, 765-767.
[5] Yang, Y.Y.; Zhou, Z.Y.; Wu, Q.H.; Zheng, M.S.; Gu, Y.J.; Chen, S.P.; Sun, S.G. *Chem. J. Chin. Univ.* 2001, 22, 1201-1204.
[6] Wu, Q.H.; Li, N.H.; Sun, S.G. *J. Phys. Chem. B* 2006, 110, 11383-11390.
[7] Capon, A.; Parsons, R. *J. Electroanal. Chem.* 1973, 44, 1-7.
[8] Capon, A.; Parsons, R. *J. Electroanal. Chem.* 1973, 44, 239-254.
[9] Angerste, H.; MacDougall, B.; Conway, B.E. *J. Electrochem. Soc.* 1973, 120, 756-766.
[10] Cathro, K.J. *J. Electrochem. Soc.* 1969, 116, 1608-1614.
[11] Shibata, M.; Motoo, S. *J. Electroanal. Chem.* 1986, 202, 137-146.
[12] Koel, B.E.; Sellidj, A.; Paffett, M.T. *Phys. Rev. B* 1992, 46, 7846-7856.
[13] El-Aziz, A.M.; Kibler, L.A.; Kolb, D.M. *Electrochem. Commun.* 2002, 4, 535-539.
[14] Hara, M.; Inukai, J.; Yoshimoto, S.; Itaya, K. *J. Phys. Chem. B* 2004, 108, 17441-17447.
[15] Jung, G.; Rhee, C.K. *J. Electroanal. Chem.* 1997, 436, 277-280.
[16] Wu, Q.; Shang, W.-H.; Yan, J.-W.; Mao, B.-W. *J. Mol. Catal. A: Chem.* 2003, 199, 49-56.
[17] Wu, Q.; Shang, W.-H.; Yan, J.-W.; Xie, Z.X.; Mao, B.-W. *J. Phys. Chem. B* 2003, 107, 4065-4069.

[18] Yan, J.-W.; Wu, Q.; Shang, W.-H.; Mao, B.-W. *Electrochem. Commun.* 2004, 6, 843-848.
[19] Jung, C.; Rhee, C. K. *J. Electroananl. Chem.* 2004, 566, 1-5.
[20] Jung, C.; Rhee, C.K. *J. Phys. Chem. B* 2005, 109, 8961-8966.
[21] Wang, F.-X.; Pan, G.-B.; Liu, Y.-D.; Xiao, Y. *J. Phys. Chem. C* 2010, 114, 4576-4579.
[22] Pan, G.B.; Freyland, W. *Electrochim. Acta* 2007, 52, 7254-7261.
[23] Borissov, D.; Aravinda, C.L.; Freyland, W. *J. Phys. Chem. B* 2005, 109, 11606-11615.
[24] Mann, O.; Aravinda, C.L.; Freyland, W. *J. Phys. Chem. B* 2006, 110, 21521-21527.
[25] Fu, Y.C.; Yan, J.W.; Wang, Y.; Tian, J.H.; Zhang, H.M.; Xie, Z.X.; Mao, B.W. *J. Phys. Chem. C* 2007, 111, 10467-10477.
[26] Habboush, D.A.; Osteryoung, R.A. *Inorg. Chem.* 1984, 23, 1726-1734.
[27] Pan, G.B.; Freyland, W. *Chem. Phys. Lett.* 2006, 427, 96-100.
[28] Aravinda, C.L.; Freyland, W. *Chem. Commun.* 2006, 16, 1703-1705.
[29] Magnussen, O.M. *Chem. Rev.* 2002, 102, 679-725.
[30] Li, F.-H.; Wang, W.; Gao, J.-P.; Wang, S.-Y. *J. Electrochem. Soc.* 2009, 156, D84-D91.
[31] Wu, Q.-H.; Zhen, C.-H.; Zhou, Z.-Y.; Sun, S.-G. *Acta Phys.-Chim. Sin.* 2008, 24, 201-204.
[32] Gomez, R.; Fernandez-Vega, A.; Feliu, J.M.; Aldaz, A. *J. Phys. Chem.* 1993, 97, 4769-4776.
[33] Climent, V.; Herrero, E.; Feliu, J.M. *Electrochim. Acta* 1998, 44, 1403-1414.
[34] Kizhakevariam, N.; Weaver, M.J. *Surf. Sci.* 1994, 310, 183-197.
[35] Zhao, J.; Jung, C.; Rhee, C.K. *J. Phys. Chem. B* 2006, 110, 10814-10821.
[36] Schardt, B.C.; Yau, S.-L.; Rinaldi, F. *Science* 1989, 243, 1050-1053.
[37] Yau, S.-L.; Vitus, C.M.; Schardt, B.C. *J. Am. Chem. Soc.* 1990, 112, 3677-3679.
[38] Rhee, C.K.; Kim, D. *J. Eletroanal. Chem.* 2001, 506, 149-154.
[39] Jung, G.; Park, H.; Rhee, C.K. *J. Electroanal. Chem.* 1998, 453, 243-247.
[40] Yang, Y.-Y.; Sun, S.-G.; Gu, Y.-J.; Zhou, Z.-Y.; Zhen, C.-H. *Electrochim. Acta* 2001, 46, 4339-4348.
[41] Wu, Q.-H.; Sun, S.-G.; Xiao, X.-Y.; Yang, Y.-Y.; Zhou, Z.Y. *Electrochim. Acta* 2000, 45, 3683-3690.
[42] Sauerbrey, G. *Z. Phys.* 1964, 178, 457-471.
[43] Fernandez-Vega, A.; Feliu, J.M.; Aldaz, A.; Clavilier, J. *J. Electroanal. Chem.* 1989, 258, 101-113.
[44] Hayden, B.E. *Catal. Tod.* 1997, 38, 473-481.
[45] Herrero, E.; Rodes, A.; Perez, J.M.; Feliu, J.M.; Aldaz, A. *J. Electroanal. Chem.* 1995, 393, 87-96.
[46] Sun, S.G.; Noel, J.; Lipkowski, J.; Altounian, Z. *J. Electroananl. Chem.* 1990, 278, 205-216.
[47] Leiva, E.; Iwasita, T.; Herrero, E.; Feliu, J.M. *Langmuir* 1997, 13, 6287-6293.
[48] Jarvi, T.D.; Stuve, E.M. In Electrocatalysis; Editors, Lipkowski, J.; Ross P.N. Wiley-VCH, New York, 1998, Chapter 3, p.75-153.
[49] Haan, J.L.; Stafford, K.M.; Masel, R.I. *J. Phys. Chem. C* 2010, 114, 11665-11672.
[50] Yu, X.W.; Pickup, P.G. *J. Power Sources* 2008, 182, 124-132.
[51] Zhu, Y.; Kang, Y.Y.; Zou, Z.Q.; Zhou, Q.; Zheng, J.W.; Xia, B.J.; Yang, H. *Electrochem. Commun.* 2008, 10, 802-805.

[52] Markovic, N.M.; Gasteiger, H.A.; Ross Jr., P.N.; Jiang, X.D.; Villegas, I.; Weaver, M.J. *Electrochim. Acta* 1995, 40, 91-98.
[53] Lu, G.Q.; Crown, A.; Wiekoski, A. *J. Phys. Chem. B* 1999, 103, 9700-9711.
[54] Okamoto, H.; Kon, W.; Mukouyama, Y. *J. Phys. Chem. B* 2004, 108, 4432-4438.
[55] Chen, Y.X.; Ye, S.; Heinen, M.; Jusys, Z.; Osawa, M.; Behm, R.J. *J. Phys. Chem. B* 2006, 110, 9534-9544.
[56] Samjeske, G.; Miki, A.; Ye, S.; Osawa, M. *J. Phys. Chem. B* 2006, 110, 16559-16566.
[57] Samjeske, G.; Osawa, M. *Angew. Chem., Int. Ed.* 2005, 44, 5694-5698.
[58] Chem, Y.X.; Heinen, M.; Jusys, Z.; Behm, R.J. *Angew. Chem., Int. Ed.* 2006, 45, 981-985.
[59] Chen, Y.X.; Heinen, M.; Jusys, Z.; Rehm, R.J. *Langmuir* 2006, 22, 10399-10408.
[60] Uhm, S.; Lee, H.J.; Lee, J. *Phys. Chem. Chem. Phys.* 2009, 11, 9326-9336.
[61] Lamy, C.; Leger, J.M.; Clavilier, C.; Parsons, R. *J. Electroanal. Chem.* 1983, 150, 71-77.
[62] Lee, J.K.; Jeon, H.; Uhm, S.; Lee, J. *Electrochim. Acta* 2008, 53, 6089-6092.
[63] Peng, B.; Wang, J.Y.; Zhang, H.X.; Lin, Y.H.; Cai, W.B. *Electrochem. Commun.* 2009, 11, 831-833.
[64] Yang, Y.-Y.; Zhou, Z.-Y.; Sun, S.-G. *J. Electrochem. Chem.* 2001, 500, 233-240.
[65] Yang, Y.-Y.; Sun, S.-G. *J. Phys. Chem. B* 2002, 106, 12499-12507.
[66] Macia, M.D.; Herrero, E.; Feliu, J.M. *Electochim. Acta* 2002, 47, 3653-3661.
[67] Smoluchowski, R. *Phys. Rev.* 1941, 60, 661-674.
[68] Gomez, R.; Feliu, J.M.; Aldaz, A.; Weaver, M.J. *Surf. Sci.* 1998, 410, 48-61.
[69] Herrero, E.; Fernandez-Vega, A.; Feliu, J.M.; Aldaz, A. *J. Electroanal. Chem.* 1993, 350, 73-88.
[70] Peng, B.; Wang, H.-F.; Liu, Z.-P.; Cai, W.-B. *J. Phys. Chem. C* 2010, 114, 3102-3107.

In: Antimony: Characteristics, Compounds and Applications ISBN 978-1-62100-598-8
Editor: Manijeh Razeghi

Chapter 10

IMPROVED P-TYPE ANTIMONY TELLURIDE ALLOYS (SB_2TE_3) FOR THERMAL CONVERTERS

J. P. Carmo, M. F. Silva, J. F. Ribeiro, L. M. Gonçalves, and J. H. Correia
University of Minho, Department of Industrial Electronics
Campus Azurém, Guimarães, Portugal

ABSTRACT

This focus of this chapter is the presentation of antimony-based materials for thermoelectric applications. Examples of thermoelectric applications are (1) the electrical energy from temperature gradients and (2) cooling devices, but both using solid-state converters. These converters use pairs of thermoelectric p- and n-type materials. In this context, this chapter presents all the fabrication and characterization details of the p-type material. The antimony-based material selected to serve as p-type structures was the Antimony telluride (Sb_2Te_3). The Sb_2Te_3 is a chalcogenide and its thin-film deposition requires a precise controlled process to achieve the highest possible thermoelectric figure-of-merit, *ZT*. This goal is achieved with the co-evaporation of antimony (Sb) and telluride (Te) technique. The optimizations of the Sb_2Te_3 thin-films deposition on a flexible substrate are presented. It is also analyzed the influence of the parameters involved in the depositions (e.g., the temperature which the substrate is subjected, and the evaporation rates of Sb and Te), and their impacts in the final composition of Sb_2Te_3 thin-films.

Some well known techniques of analysis were employed during the measurements for characterizing the quality of Sb_2Te_3 thin-films: X-ray diffraction (XRD), energy-dispersive X-ray spectroscopy (EDX), X-ray photoelectron spectroscopy (XPS) and Raman spectroscopy. The measurements done at the room temperature of 300 K for the best deposited samples showed Sb_2Te_3 thin-films with Seebeck coefficients up to 200 $\mu V.K^{-1}$, an in-plane electrical resistivity in the range [8-15] $\mu\Omega.m$, a carrier concentration in the range $[1\text{-}7]\times10^{19}$ cm^{-3}) and an Hall mobility in the range [120-180] $cm^2.V^{-1}.s^{-1}$.

INTRODUCTION

The thermoelectricity is the conversion of temperature differences in electric energy, and vice-versa. Three effects are included in the thermoelectricity, e.g., the Seebeck, Peltier and Thomson effects. Thomas Seebeck (1770-1831) discovered that a temperature difference between the junction of two different materials makes an electric voltage to arise, and an electric current will start to flow when an electric circuit is closed (Seebeck, 1822). This effect is quantified by the Seebeck coefficient, α [μVK^{-1}]:

$$\alpha = \frac{\Delta V}{\Delta T} \tag{1}$$

Essentially, the Seebeck is used for two type of applications: temperature sensors and thermoelectric (TE) generators (Seebeck, 1822) – see the Figure 1.

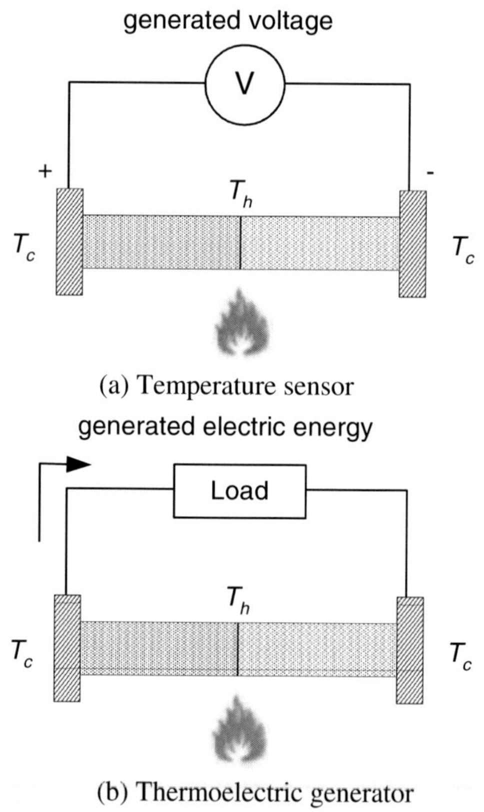

Figure 1. Seebeck effect, used as temperature sensor or energy generator (b).

Jean Peltier (1785-1845) verified in 1834 that the flow of an electric current through the junction connecting two different materials, makes an amount of thermal energy to be released or absorbed by this junction – see the Figure 2 -, and the temperature in this same junction rises or decreases, respectively. Also, he observed that the released or the absorbed energy was proportional to the electric current. This effect is quantified by the Peltier coefficient, π. Some years later, William Thomson (later known as Lord Kelvin) established the relation between the Peltier and Seebeck coefficients:

$$\pi=\alpha.T \tag{2}$$

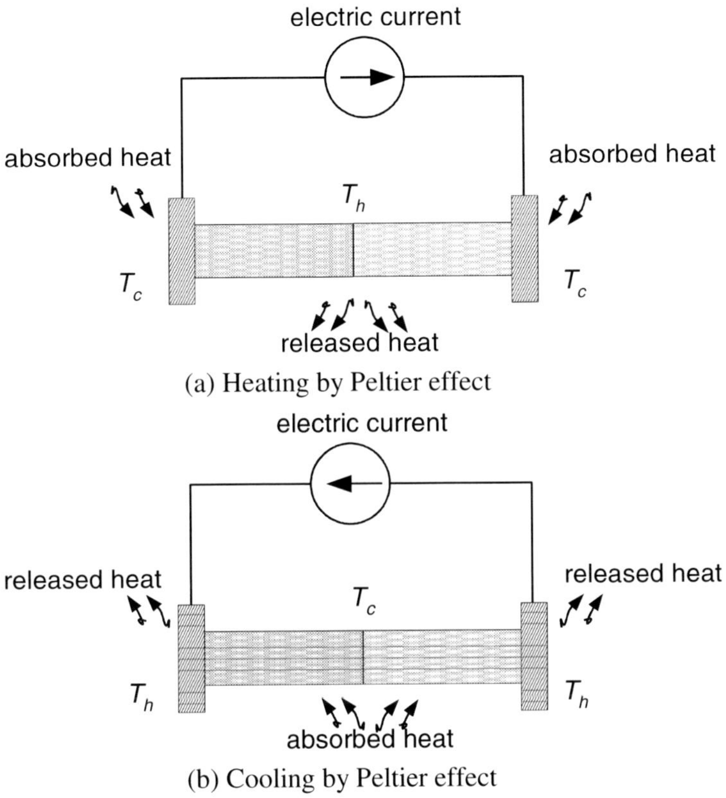

Figure 2. Using the Peltier effect for heating or cooling.

Motivation

The thermoelectricity renewed interest with the rise of new materials, both in academic research and in industry. The interest of these areas include cooling applications (for use in laser modules, biomedical applications, chip spot-cooling) and energy generation (in energy harvesting systems) for use in autonomous microsystems, where the use of batteries is not adequate. However and up to now, the research is being essentially focused in the way to get new materials with higher performance (using nanostructures), or to improve the performance of the conventional thermoelectric converters (whose final target is the refrigeration in domestic applications and energy generation in cars by recovering heat lost in the exhaust gases from engine). The integration of these converters with electronic circuits (using the same techniques used to fabricate microsystems) is still in a start-up phase. This chapter presents the fabrication technology of thermoelectric materials made of p-type thin-films of antimony tellurides for use in cooling, infrared radiation detection, and in thermoelectric energy scavenging systems.

MODELLING THERMOELECTRIC MICROSYSTEMS

A planar Peltier microcooler is usually composed of many pairs of PN junctions, like one represented on Figure 3.

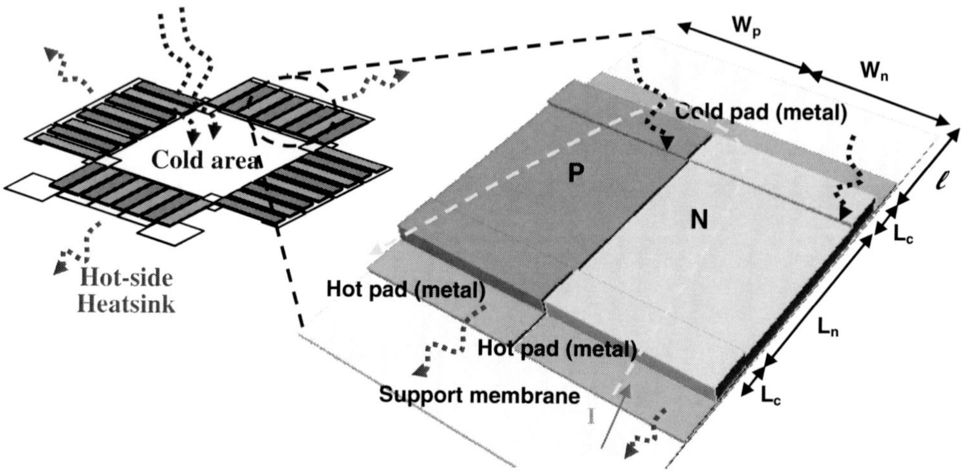

Figure 3. A single cell Peltier planar cooler made of n=16 thermoelectric p-/n-type pairs. Two thermoelectric elements (p- and n-type) connected in series by metal pads stand on top of a support membrane.

The performance of a thermoelectric converter is determined by the figure-of-merit of the materials used (Peltier, 1834):

$$ZT = \frac{\alpha^2 T}{\rho\kappa} \tag{3}$$

where α [VK^{-1}] is the Seebeck coefficient, ρ [Ωm] is the electrical resistivity, κ [Wm^{-1}K^{-1}] is the thermal conductivity and T the temperature. The performance parameter ZT is very useful in Seebeck sensing devices, such as infrared thermal detectors. Another factor is more appropriate for specifying the performance of a TE microconverter, which is the power factor, PF [WK^{-2}m^{-1}]. The PF is defined as the electric power per unit of area through which the heat flows, per unit of temperature gradient between the hot and the cold sides, and is expressed as:

$$PF = \frac{\alpha^2}{\rho} \tag{4}$$

The device illustrated in the Figure 3 is a planar thermoelectric converter that was fabricated on top of a kapton foil with a thickness of 25 μm. The maximum power provided by a thermoelectric device with n thermoelectric pairs when subjected to a thermal gradient ΔT [K] is given by:

$$P_{MAX} = \frac{[n^2(\alpha_p - \alpha_n)^2 - \Delta T]^2}{4nR} \tag{5}$$

The Figure 4 shows the maximum power and the maximum voltage that can be provided by a four pair thermoelectric device made of p- and n-type materials with figures-of-merit, ZT, at room temperatures of 0.5 and 0.84, and power factors of 2.81×10^{-3} $WK^{-1}m^{-2}$ and 4.87×10^{-3} $WK^{-1}m^{-2}$, respectively. The p-type materials are made of antimony tellurides and are the focus of this chapter.

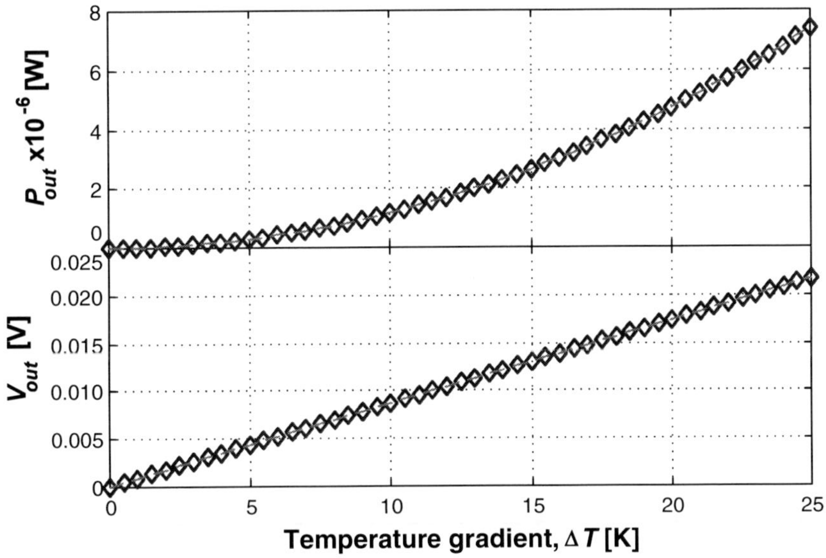

Figure 4. For a thermoelectric converter with four pairs: the maximum power and the maximum voltage at its output terminals.

Antimony Telluride Depositions

This section focuses on the deposition and characterization of p-type Sb_2Te_3 thin films. It was demonstrated before that the direct evaporation of the bulk materials results in a compositional gradient along the film thickness due to the large differences in vapour pressure of antimony (Sb) and tellurium (Te) (da Silva *et al* 2002, da Silva *et al* 2005). Other deposition processes have been reported in the literature for the deposition of Sb_2Te_3 thin films. Co-sputtering, electrochemical deposition, metal-organic chemical vapour deposition or flash evaporation are some examples. Zou *et al* (Zou *et al*, 2002) fabricated p-type Sb_2Te_3 films, and reported the influence of substrate temperature and evaporation rate of the materials during the film growth, on the electronic properties. In the present work, the influence of growing parameters (evaporation rates of Sb and Te and substrate temperature) is studied in detail. Glass, silicon and polyimide were used as substrates, with good film-to-substrate adhesion. However, for thermoelectric applications, 25 μm-thick Kapton film was chosen as substrate because of the low thermal conductivity (as low as 0.12 $W.m^{-1}.K^{-1}$) and appropriate value of thermal expansion coefficient (12×10^{-6} K^{-1}) which closely matches the thermal expansion coefficient of the telluride films, thus reducing residual stress and increasing adhesion.

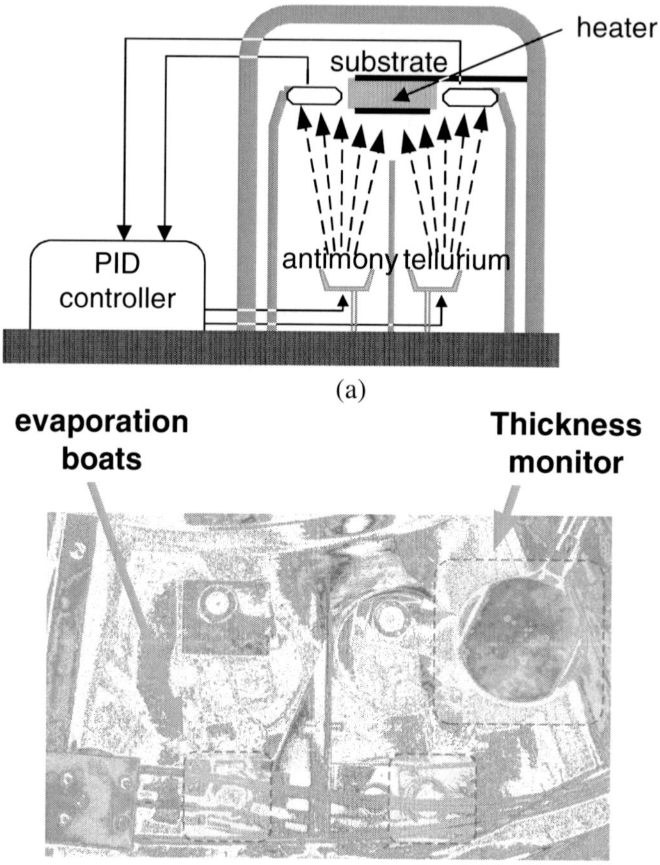

(b)

Figure 5. (a) The co-evaporation system; (b) the boats and the mass sensors placed inside the co-deposition chamber. These figures were reproduced with previous authorization of the publisher (Elsevier Science Direct).

Experimental

The antimony telluride thin-films were deposited by co-evaporation using resistive heating in a high-vacuum chamber - see the Figure 5(a). The power applied to each evaporation boat (made of molybdenum) was controlled independently by way of two Proportional Integral Derivative (PID) (Zou *et al*, 2002) controllers to maintain a constant deposition rate at a certain setting during deposition. The Sb/Bi evaporation flow rate, Fr_{Sb}/Fr_{Bi} [nm.s^{-1}], and Te evaporation flow rate, Fr_{Te} [nm.s^{-1}], allow tuning of the ratio of the evaporation flow-rates, R, which is given by: $R=Fr_{Te}/Fr_{Sb}$ or $R=Fr_{Te}/Fr_{Bi}$, which is basically the amount of Te divided by the amount of Sb that is actually being deposited at the substrate surface during co-evaporation.

Each PID controller reads the deposition rate from a thickness monitor (acting as a mass sensor) and was designed to compute the power necessary to be applied to the corresponding evaporation boat in real time in order to maintain the desired evaporation rate. The signals at the output of each mass sensor (see the Figure 6 that shows the schematic of this read-out circuit) are periodic with an oscillation frequency, f [Hz]. The amount of material accumulated in the surface of each mass sensor is proportional to the thickness, h [m], of deposited Te (or Sb). Such a relation is given by $h=(N_Q d_Q)/(\pi d_F Z f)\times\text{arctg}\ [Z.\tan(\pi(f_Q-f)/f_Q)]$, with $Z=[(d_Q S_Q)/(d_F S_F)]^{1/2}$, where N_Q=1668 Hz.m is the frequency constant of a crystal made of quartz, d_Q=2688 Kg.m^{-3} is the density of the quartz material, d_F [Kg.m^{-3}] is the density of deposited material (Te or Sb), Z is a material constant, f_Q [Hz] is the output frequency before the deposition, S_Q=300 Kg.m^{-1}s^{-2} is the shear modulus of the quartz material, and S_F is the shear modulus of deposited material. The values of d_F and Z-factor for Te and Sb are: 6250 Kg.m^{-3} at Z=0.90 and 6620 Kg.m^{-3} at Z=0.77. A metal sheet was placed between the two boats to partially separate the flows from the two evaporated species to prevent mixing of the materials at the quartz crystals - see the Figure 5(b). Large boats (baffled boxes with a volume of 4 cm^3) were used to keep the evaporation rates stable. The substrates were heated to temperature setpoints, T_{sub} [°C], between 150 °C and 220 °C.

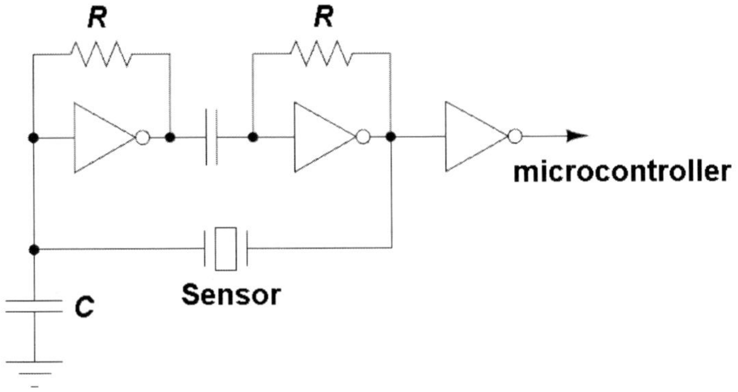

Figure 6. The schematic of the readout circuit of the mass sensors.

Each evaporation boat is heated by an AC electrical signal with voltage in the range 2-6 V and current in the range 50-60 A, respectively. The PID controllers were implemented, using the P89C51RD2 microcontroller from Philips Semiconductors. Basically, the PID controllers adjust the power by switching of two independent sine waves at a particular phase angle. The photograph of the control circuit (e.g., the two PID controllers) is in the Figure 7(a). The Figure 7(b) shows the schematic of the electronic circuit responsible to do the phase detection of the power line voltage. The electronic circuit illustrated in the Figure 7(c) is the schematic of the triac switching circuit, whose switching angle is controlled by the microcontroller with the help of an optocoupler. This circuit is responsible for controlling the voltage applied to the primary winding of a power transformer. The secondary winding of the transformer is directly connected to the evaporation boat. Finally, the Figure 7(d) shows the controlled 50 Hz voltage supplied to one of the evaporation boats.

(a)

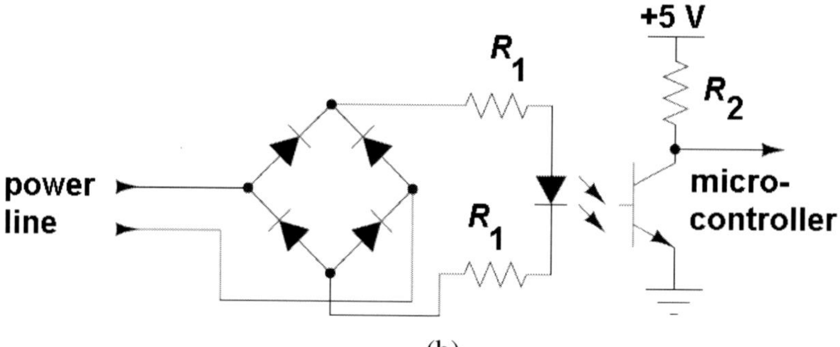

(b)

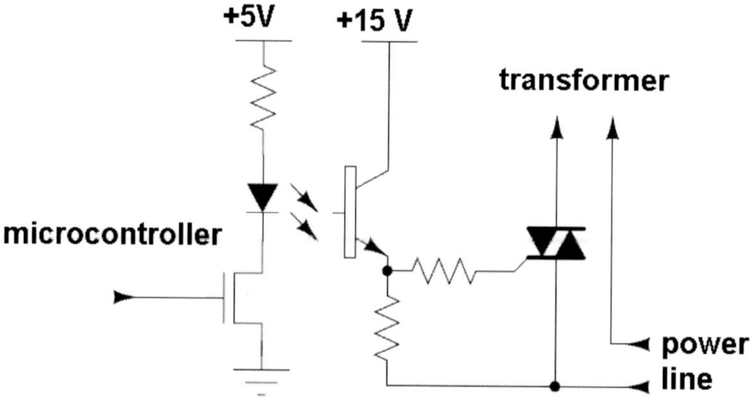

(c)

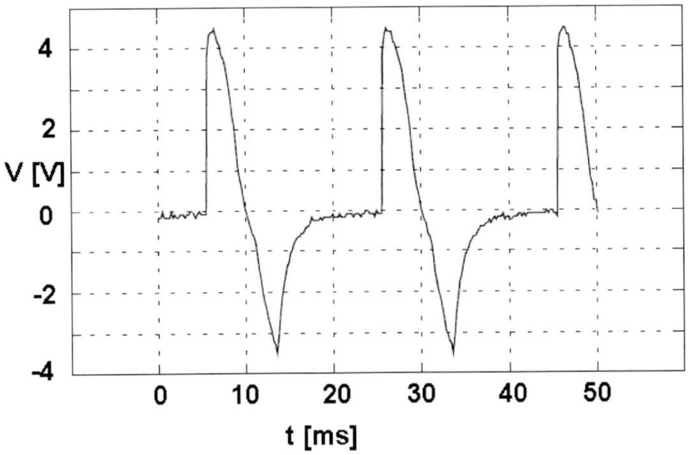

(d)

Figure 7. (a) A photography showing the control circuit. (b) The schematic of the circuit responsible for doing the phase detection of the power line voltage. (c) The schematic of the power switching circuit. (d) The voltage acquired from one of the evaporation boats, for a conduction angle of 100º.

Results

Several films of Sb_2Te_3 were deposited with different flow rate ratios, R, and substrate temperatures, T_{sub}, in order to study the effect of these parameters on the thermoelectric properties of the films. R was set by fixing the Sb evaporation flow rate, Fr_{Sb}, at 2 $Ås^{-1}$ and varying the Te evaporation flow rate, Fr_{Te}, in the range 3-9 $Ås^{-1}$. Three series of films were fabricated, at T_{sub}=150, 180 and 220 °C. All films were approximately 1 μm thick (accepting a tolerance of ±30%). The Table 1 summarizes the thermoelectric properties of selected Sb_2Te_3 films.

Table 1. Properties of selected Sb_2Te_3 films. This table was reproduced with previous authorization of the publisher (Elsevier Science Direct)

Film	T_{sub} [°C]	$R=F_{Te}/F_{Sb}$	%Te by EDX	α [μVK^{-1}]	ρ [μΩm]	n [10^{19} cm^{-3}]	μ [cm^2 V^{-1} s^{-1}]	$PF\times10^{-3}$ [WK^{-2}m^{-1}]
#195C	150	1.47	54.5%	91	7.6	12.5	65.5	1.09
#305A	150	1.67	61.4%	140	14.0	2.4	185.8	1.40
#305D	150	2.13	64.7%	191	31.2	2.0	174.3	1.17
#197D	150	2.35	64.4%	137	43.6	4.6	31.1	0.43
#198A	180	1.85	60.4%	134	24.4	9.0	28.4	0.74
#198B	180	2.02	59.1%	158	30.3	1.44	143.2	0.82
#198D	180	2.35	62.4%	156	39.1	2.0	78.8	0.62
#304A	220	1.67	64.1%	81	10.5	25.3	23.5	0.63
#304D	220	2.13	61.2%	133	10.0	3.6	174.3	1.78
#306A	220	2.50	67.3%	156	9.2	4.1	166.4	2.66
#306D	220	3.18	73.5%	188	12.6	4.3	115.2	2.81

X-ray diffraction (XRD) reveals the polycrystalline structure of Sb_2Te_3, as shown in the Figure 8. The peaks agree with the powder diffraction spectra (JCPDS, 1983) for polycrystalline Sb_2Te_3 (15-874) and similar results from XRD analysis can also be found in literature (da Silva, 2005). The difference on X-ray diffraction patterns on the two films analyzed could be due to differences on composition and substrate temperature. Film #305 was deposited at T_{sub}=150 ºC and films #304 and #306 were deposited at T_{sub}=220 ºC. The tellurium content in the composition was 63% of films #304 and #305 and 70% on film #306.

The planes with higher intensity obtained in films deposited at T_{sub}=220 ºC (#304 and #306) are not the same from films deposited at T_{sub}=150 ºC (#305) or from bulk Sb_2Te_3. The substrate temperature during deposition influences the structure and orientation of Sb_2Te_3 thin-films. Tellurium phase is also detected in films with higher Te content (#306). The SEM cross-sectional and surface images of the material also reveal the polycrystalline structure of the films and the effect of substrate temperature on polycrystalline morphology (see the Figure 9).

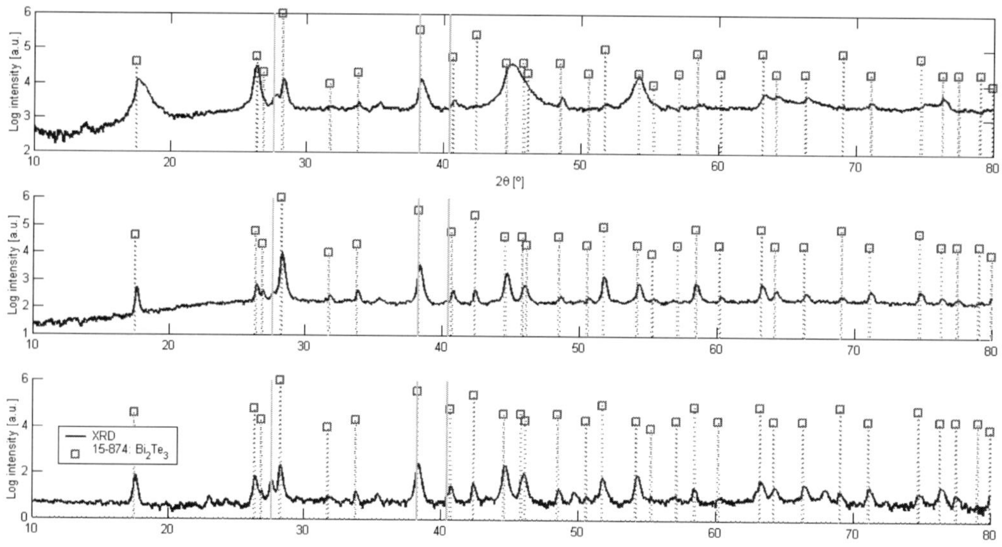

Figure 8. XRD planes of powdered crystalline bulk Sb_2Te_3 detected on films #304, #305 and #306. The gray-shaded vertical lines represent the positions of tellurium (Te) diffractograms on the XRD.

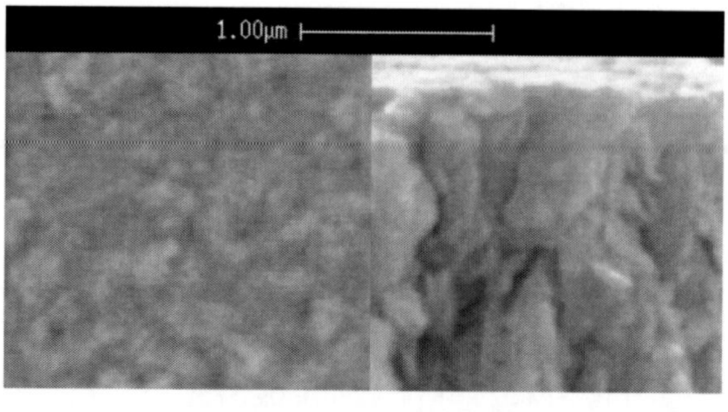

Figure 9. Surface and cross-sectional SEM images of $Sb2_2Te_3$ thin-films deposited on glass.

The XPS analysis was performed at on a Sb_2Te_3 thin-film with 1 μm thickness (Gonçalves *et al*, 2011). Ion etching during analysis was performed at low energy, to avoid heating the films and lose original composition, due to different vapour pressures of Sb and Te. Composition was measured as function of film deep and results present the contamination (carbon) and oxidation of film surface, up to 50 nm deep. Composition measured deeper thickness corresponds to that obtained by EDX. Composition measurements were also confirmed by RBS/PIXIE technique.

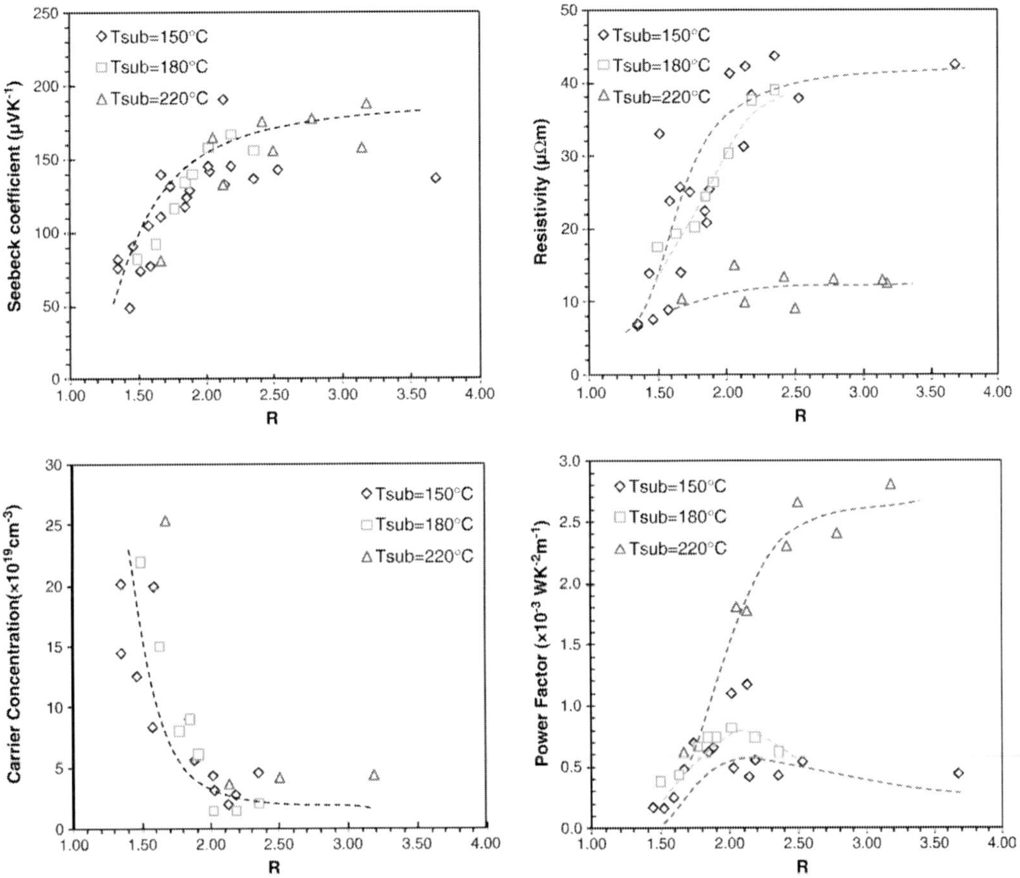

Figure 10. (a) The Seebeck coefficient of Sb_2Te_3 thin films as a function of Te/Sb evaporation flow ratio, R. (b) The electrical resistivity of Sb_2Te_3 as a function of Te/Sb evaporation flow rate ratio, R. (c) The carrier concentration and Hall mobility of Sb_2Te_3 as a function of Te/Sb evaporation flow rate ratio, R. (d) The power factor of Sb_2Te_3 as a function of Te/Sb evaporation flow rate ratio, R. (Gonçalves et al, 2011). This figure was reproduced with previous authorization of the publisher (Elsevier Science Direct).

The film composition is always poorer in Te than the corresponding fraction of evaporated Te, *R*. This effect is more noticeable as T_{sub} increases. At 220 ºC, the vapour pressure of Te is $10^3\times$ higher than the vapour pressure of Sb. Therefore, it is possible to conclude that re-evaporation of Te from the substrate proceeds at a higher rate than Sb. This explains why it is necessary to use a higher value of *R* than the desired atomic ratio of Te/Sb in the final film composition. Since the Te percentage on the films is always small than the corresponding percentage of evaporated Te and it also depends on substrate temperature, it's difficult to obtain the desired stoichiometry. The evaporation rate of tellurium must be adjusted to overcome these factors, in order to obtain the desired final stoichiometry.

The Figure 10(a) shows the dependence of the Seebeck coefficient on the flow rate ratio (*R*), for films deposited at different substrate temperatures (150 ºC, 180 ºC and 220 ºC). It was not noted relevant dependence of the substrate temperature (T_{sub}) on the value of Seebeck coefficient but the value of this coefficient increases as *R* is increased.

The Figures 10(b) and 10(c) show the electrical resistivity, the carrier concentration and the Hall mobility of Sb_2Te_3 films as a function of *R* for different T_{sub}. There is a strong dependence of electrical resistivity with T_{sub}. Lower electrical resistivity is obtained with substrate temperature above 220 ºC. In order to achieve high power factors, *PF*, a low carrier concentration with high mobility is desirable. Films with composition rich in tellurium have lower carrier concentration (less than 5×10^{19} cm^{-3}) and higher Seebeck coefficients. High mobility (higher than 100 cm^{2}V^{-1}s^{-1}) is obtained with *R* between 2 and 3. At each T_{sub} the maximum value of power factor is also obtained at a value of *R* that depends on T_{sub} - see the Figure 10(d).

The maximum *PF* is obtained at *R*=2 for T_{sub}=150 ºC or T_{sub}=180 ºC. For T_{sub}=220 ºC, the maximum *PF* is obtained with *R*=3. On thin films, resistivity and Seebeck coefficient are higher than on bulk crystal, respectively $\rho_{//}\approx1.9\mu\Omega$m $\alpha_{//}$=83 μVK^{-1} (Scherrer *et al*, 2005). On the other way, carrier concentration and Hall mobility was lower than those values on bulk Sb_2Te_3, respectively n=10.6×10^{19} cm^{-3} and μ=313 cm^{2}V^{-1}s^{-1} (Scherrer *et al*, 2005). The influence of substrate temperature on power factor was also analysed (Gonçalves *et al*, 2011) and comparable results obtained by da Silva (da Silva *et al*, 2005) and Zou (Zou *et al*, 2002) on the same materials are also presented. The highest value of *PF*=2.8×10^{-3} WK^{-2}m^{-1} was obtained in films deposited at T_{sub}=220 ºC with Sb evaporation rate of 2 Ås^{-1} and Te evaporation rate of 6.4 Ås^{-1} (*R*=3.2). These evaporation parameters resulted in tellurium rich composition (measured by EDX) of the compound with 25-35% of Sb and 65-75% of Te. The power factor versus the composition, measured by EDX, for different substrate temperature was also analysed (Gonçalves *et al*, 2011). The power factor of Sb_2Te_3 bulk crystal and thermal evaporated thin-film obtained by other authors (da Silva *et al*, 2005) are also plotted. The value of *PF* measured on films deposited by co-evaporation at low substrate temperature is lower than this obtained in bulk materials. But with T_{sub}=220 ºC, this value is near those obtained in bulk Sb_2Te_3. Higher *PF* value was measured with tellurium rich films (70% Te on composition).

The thermal conductivity, κ, was measured in one Sb_2Te_3 film, using the method described by Völklein (Völklein *et al*, 1997). The value of 1.8 Wm^{-1}K^{-1} was obtained. Thermal conductivity is due to electron and phonon contribution effects, respectively κ_e and κ_p (κ=κ_e+κ_p). Electron contribution could be estimated through Wiedemann-Franz law, k_e=LT/ρ_e, where *L* is the Lorenz number, *T* the temperature and ρ_e is the electrical resistivity. Considering L=2.1×10^{-8} V^{2}K^{-2} (Dyck *et al*, 2002) and ρ_e=15 μΩm, at 300 K, κ_e=0.42 Wm^{-1}K^{-1} and κ_p=1.38 Wm^{-1}K^{-1}. These values are comparable to those on bulk Sb_2Te_3 ($\kappa\approx$4 Wm^{-1}K^{-1}, parallel to crystal planes and e $\kappa\approx$1 Wm^{-1}K^{-1}, perpendicular to crystal planes). With the measured value of Seebeck coefficient on this film (α=160 μVK^{-1}) the figure-of-merit value *ZT*=0.3 was obtained.

Discussion

P-type Sb_2Te_3 thermoelectric thin films with high figure-of-merit were fabricated by thermal co-evaporation. The best films have a Seebeck coefficient of 188 $\mu V.K^{-1}$ and a resistivity of 12.6 $\mu\Omega m$ of which represents a power factor of 2.8×10^{-3} $WK^{-2}m^{-1}$. The films were deposited onto glass substrates and flexible Kapton substrates, with good adhesion. The influence of evaporation rates and substrate temperature on composition, Seebeck coefficient, electrical resistivity, Hall mobility and carrier concentration was reported. EDX results show that films with high power factor are obtained when the composition of film is tellurium rich (30% Sb, 70% Te for Sb_xTe_{1-x}). The highest thermoelectric power factor for Sb_2Te_3 thin films was obtained at substrate temperature around 220ºC, Sb evaporation rate of 2 $Ås^{-1}$ and Te evaporation rate of 6.4 $Ås^{-1}$. A thermal conductivity of 1.8 $Wm^{-1}K^{-1}$ as measured and a figure-of-merit at room temperature of ZT=0.3 calculated. These results are comparable with the best published results for the same material (Goncalves *et al*, 2008a).

DEVICES AND APPLICATIONS

The Figure 11 shows a thermoelectric device made with four pairs of p-type (Sb_2Te_3) and n-type (Bi_2Te_3) materials, whose maximal power and voltage behaviors under optimal conditions were presented in the Figure 3. Two applications can be implemented with these thermoelectric converters. The first one is an on-chip array of thermoelectric Peltier microcoolers array (Goncalves *et al*, 2008b). The second application consists on a thermoelectric energy scavenging microsystem to convert the small thermal power available in human-body. A thin-film battery is integrated in the same device as well as ultra low-power electronics to charge the battery and to perform DC-DC conversion (Carmo *et al*, 2009).

Figure 11. A photograph showing a thermoelectric converter made with four pairs of planar thermoelectric p- and n-type materials. The p-type materials are the Sb2Te3 thin-films presented in this chapter.

On-chip array of thermoelectric Peltier microcoolers

As show in the Figure 12, the array of microcoolers can accommodate up to 64 pixels organised in a 8×8 structure (Goncalves *et al*, 2008a). Each pixel can be independently controlled to heat or cool. The Figure 13 illustrates a single pixel cross-section. When a current flows from the n-type thermoelectric element (TE) to the metal cold pad and from this to the p-type TE, by Peltier effect, heat is absorbed in the metal-TE element junctions. The reverse applies to contact pads on electronics, where heating is generated by Peltier effect.

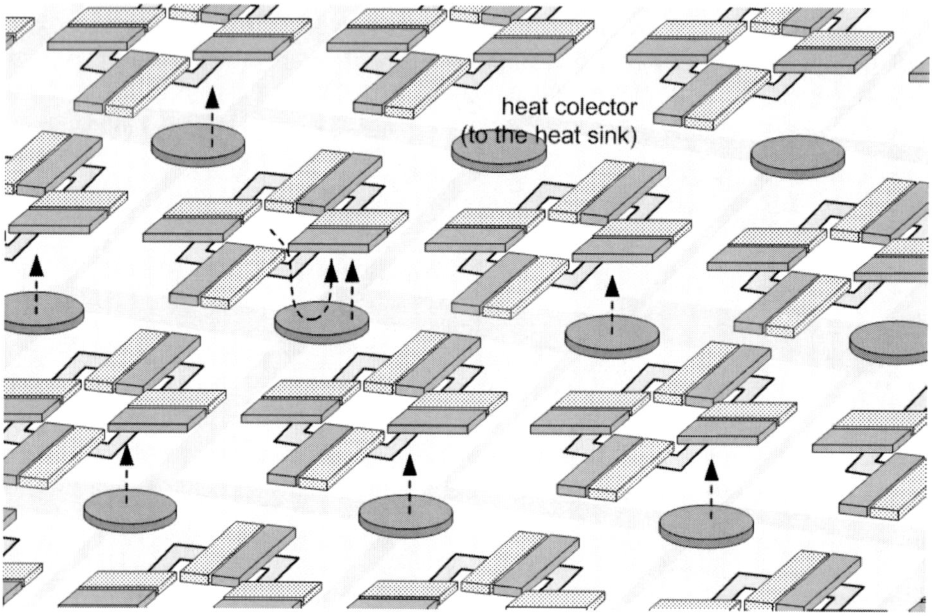

Figure 12. An artist impression showing a part of the microcoolers array. In this artwork, the heat is collected by each pixel (constituted by 4 thermoelectric n-/p-type material pairs) on bottom of the supporting membrane and directed to the heat collectors (on top of the supporting membrane), to be dissipated in a heat sink (the heat flow from the bottom to the top is represented with dashed lines and arrows).

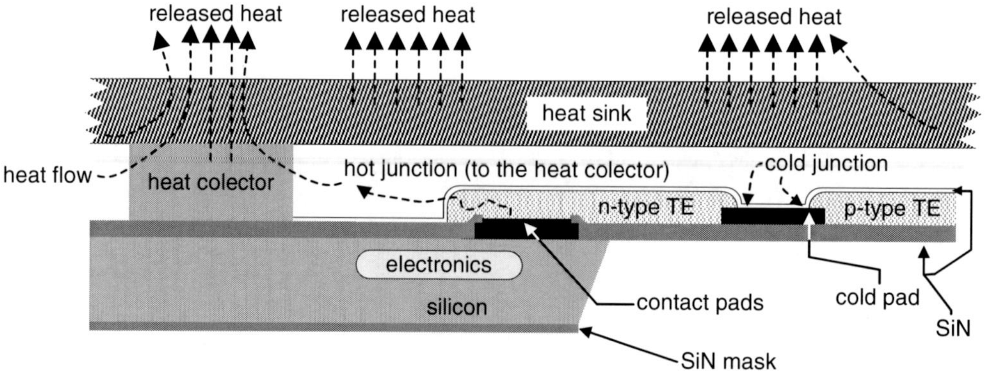

Figure 13. Drawing of a pixel of the microcooler array (it isn't on scale).

A Finite Element Method (FEM) simulation was used to calculate the expected temperature drop on each pixel. A temperature drop of 15 ºC, bellow room temperature was obtained (see the Figure 14). To obtain this cooling capacity, a membrane (200-nm thick) of silicon nitride supports four pairs of thermoelectric elements (40 μm×100 μm×10 μm), powered with 14 mA current. Contact resistivity (between thermoelectric elements and metal pads) of 10^{-10} Ωm^{-2} was assumed on simulations (da Silva *et al* 2004, Birkholz *et al* 1987). Radiation and convection was considered on the cooled surface (10 $Wm^{-2}K^{-1}$). Thermoelectric properties of n-type and p-type elements were considered as achieved on previous experimental results (Goncalves *et al* 2006a, Goncalves *et al* 2006b). The results obtained from the FEM simulation on a single pixel microcooler agree with the theoretical calculations.

A CMOS microchip was designed, with the electronics to address and control each pixel of the array, memorising the state of microcooler. However, the explanation of these electronic circuits is outside of the scope of this chapter.

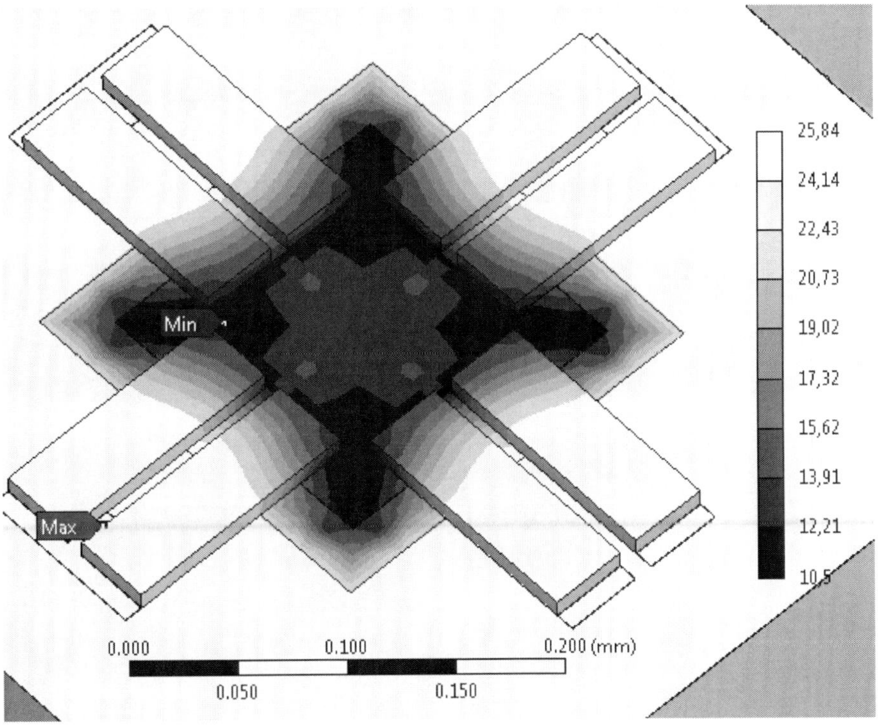

Figure 14. A single pixel microcooler simulation obtained by FEM that shows the possibility to obtain 15 ºC of cooling at the centre of the pixel.

Thermoelectric Converters with Solid-State Batteries

The present-day battery technologies are being outpaced by the ever-increasing power demands from new applications, as well as being inherently safe, batteries of the future will have to integrate the concept of environmental sustainability (Armand *et al*, 2008), thus thin-film solid-state batteries are ideal candidates for such purposes. This type of battery shows a very high life cycle and is intrinsically safe. Also, this is of special concern in the design of thermoelectric energy scavenging microsystem, especially when the thought applications are in the biomedical field, or for long cycle operations without requiring human activity. This meets the target related of the proposed thin-film rechargeable battery, which is to use it together with a single-chip regulated thermoelectric power source, which must operate from low temperature gradients (a minimum temperature difference of 3 °C between ambient and target thermo-source must provide an IC-compatible voltage). In fact, the target objective is to have thermoelectric sources to provide power densities of 10-50 μWcm^{-2}, with temperature gradients of 3-10 °C.

The Figure 15 shows an artwork of a thermoelectric microconverter with a solid-state battery placed on the top. Practical thermoelectric generators connect large number of junctions in series to increase the operating voltage.

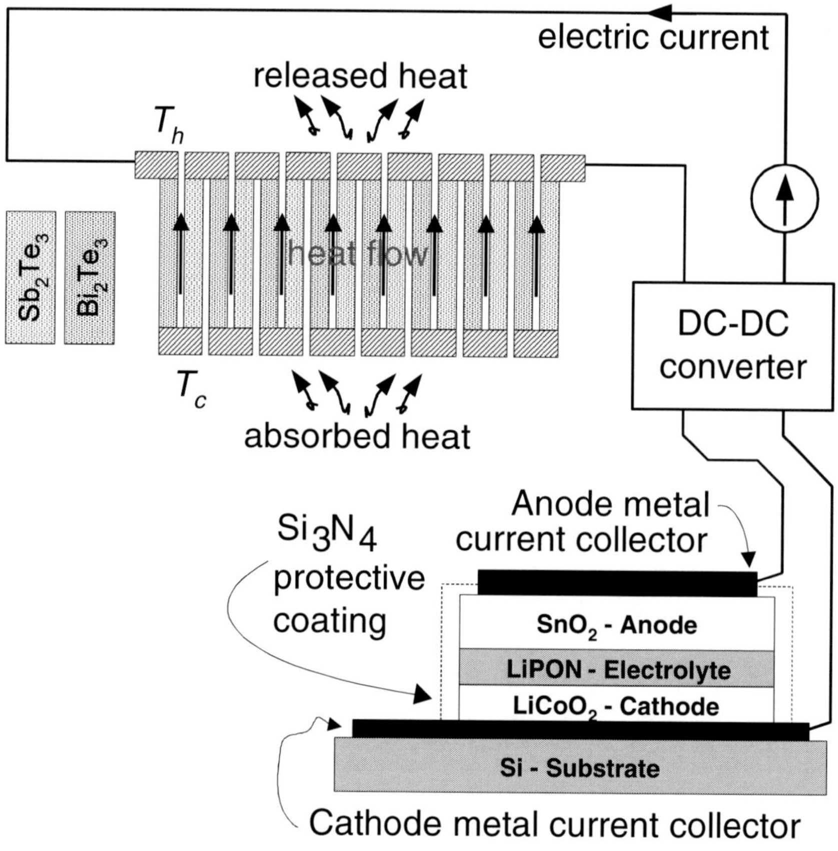

Figure 15. An artist impression showing the integrated power concept, composed by a thermoelectric converter, electronics of power conversion and a solid-state thin-films battery.

A reactive sputtering technique was used in the thin-films depositions. The Figure 15 also shows a cross-section of the proposed thin-film battery. The battery is fabricated onto an insulating substrate, by successive film depositions of the metal current collectors, cathode, electrolyte and anode. The depositions of these layers were made by sputtering (Direct Current, DC and Radio-frequency, RF). A protective coating of silicon nitride (Si_3N_4) is required to prevent the reaction of lithium when the battery is exposed to the air (Dudney *et al*, 1999). A silicon substrate will be used for integration with microelectronics, while at same time providing good thermal contact with heat source and sink.

Lithium cobaltate ($LiCoO_2$) was the compound chosen as cathode, due to its excellent electrochemical cycling stability, which is a result of the structural stability of the material, in which the layered cation ordering is extremely well preserved even after a repeated process of insertion and extraction of lithium ions (Park *et al*, 2008).

For the electrolyte deposition, it was used the lithium phosphorus oxynitride (Li_3PO_4N or Lipon). The material used for the electrolyte deposition is the lithium phosphate, whose chemical formula is Li_3PO_4. After the sputtering deposition with a N_2/Ar plasma a Lipon is obtained. This electrolyte has an acceptable ionic conductivity, which is about 2×10^{-6} Scm^{-1}. In addiction, Lipon films presents an electronic resistivity greater than 10^{14} Ωcm, which helps to minimise the short circuit self-discharge and increasing the battery's life (Dudney, 2005).

Tin dioxide (SnO_2) was chosen for the anode material due to their high-lithium storage capacity and low potential of lithium ion intercalation. A tin dioxide anode can give a maximum theoretical 781 $mAhg^{-1}$ charge-storage capacity (Wang *et al*, 2005). Also, the average resistivity of tin dioxide films are in the range 10-17 mΩ.cm for films with thickness less than 100 nm. However, for values of the thickness greater than 100 nm, the values fall in the range 5-9 mΩcm.

Target applications for energy scavenging systems with thin-films batteries and thermoelectric converters are in wireless nodes of wireless sensors networks. Energy harvested wireless sensors must be powered in a peak basis because a temperature gradient could not always be present, thus the energy must be stored in a rechargeable microbattery integrated in the system (Alahmad *et al*, 2008). Ultra-low power electronics performs DC-DC rectification with a variable conversion factor. The Figure 16 shows a simple step-up converter. The step-up conversion is made with the help of the capacitor C_{up} and the inductor L_{up}. The current at the output of the thermoelectric microdevice charges this capacitor, then the switch SW is systematically closed and open, with a fast frequency. However, it remains closed during a very short time in order to reduce the losses. In order to meet this requirement, the command signal must have a very low duty-cycle to avoid the over-discharge of the capacitor C_{up}. When SW opens, the stored energy in the inductor L_{up} forces the capacitor C_{up} to discharge through the diode D, e.g., a DC rectification is present. Then, the current charges the high-charge-capacity capacitor, C_{store}, which further connects to a DC regulator.

A more interesting solution was implemented, using components off-the-shelf. The Figure 17(a) shows the electronic schematic of a circuit prototype to do DC-DC step-up conversion, which uses the S-882Z and S-8353 circuits from the Seiko Manufacturer (Mateu *et al*, 2007). The photographs in the Figure 17(b) are the bottom and top planes of the final prototype mounted with the several components. This circuit was tested to evaluate its electrical behaviour. The Table II lists the whole set of tests (for several input voltages and currents) and it can be observed (except for the situation where V_{out}=4.50 V) that in spite of the maximum 3.70 V of a lithium battery not be present, the output voltage approaches such a value. The efficiency of the circuit is always above 85%, thus its loss of power inside is acceptable.

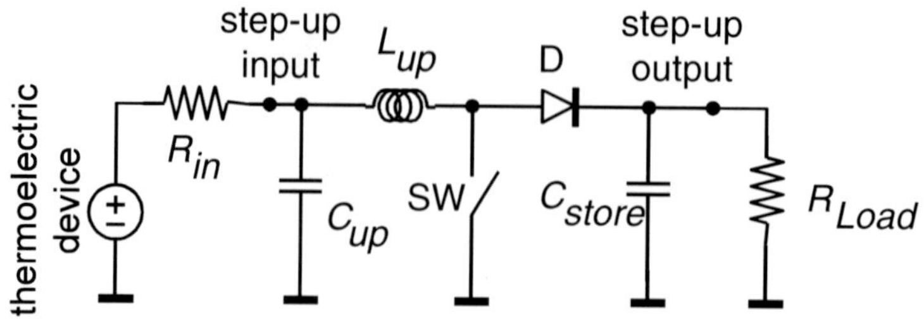

Figure 16. Schematic of a simple step-up circuit.

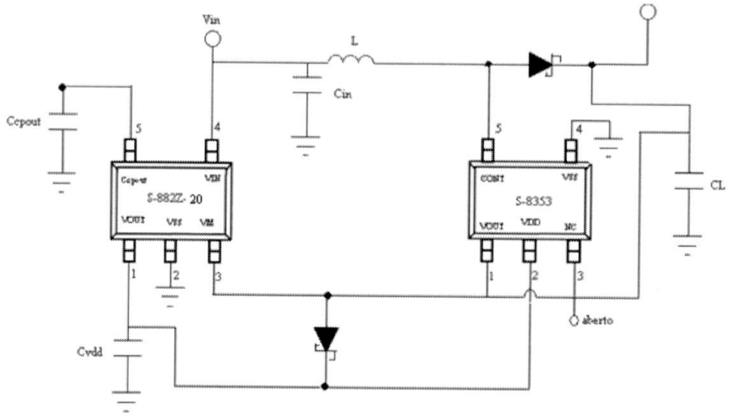

(a)

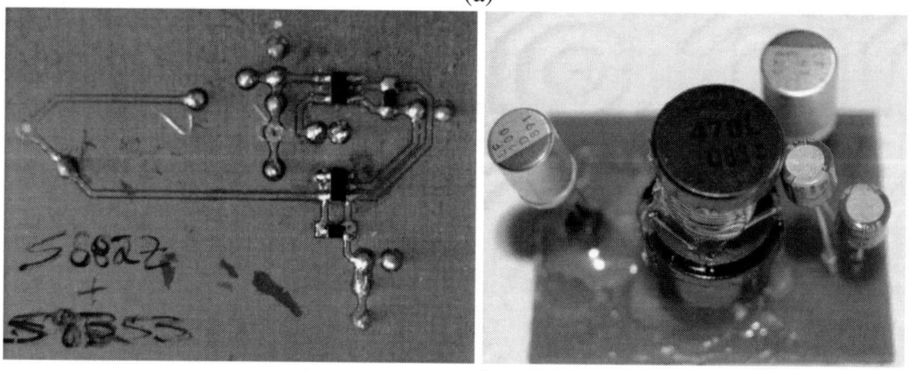

(b)

Figure 17. A circuit prototype of a DC-DC step-up conversion, using the S-882Z and the S-8353 circuits from Seiko: (a) schematic and (b) the bottom and top planes of the fabricated prototype.

Table II. The set of tests done to the DC-DC converter circuit

V_{in} [V]	I_{in} [ma]	V_{out} [V]	I_{out} [ma]	P_{in} [mW]	P_{out} [mW]	η [%]
0.31	40.11	3.20	3.32	12.43	10.62	0.85
0.39	45.30	3.40	5.05	17.57	17.17	0.98
0.45	50.50	3.60	5.91	22.70	20.68	0.91
0.50	70.10	4.50	7.55	35.05	33.97	0.97

Thin-film solid-state batteries show a very high-life cycle and are intrinsically safe. This is of special concern in the design of thermoelectric energy scavenging microsystems, mainly when the thought applications are in the biomedical field or for long cycle operations without requiring human activity.

CONCLUSION AND FUTURE DEVELOPMENTS

In today's best commercial thermoelectric modules (specifically for the compounds made of antimony and tellurium presented in this chapter) the figure-of-merit, *ZT*, is about one, despite many approaches to find compounds with higher performance. In conventional 3D crystalline systems it is difficult to control each of the following are interrelated factors to improve *ZT*.This means that an increase of the Seebeck voltage per unit of temperature, usually results in a decrease of the electrical conductivity. Moreover, a decrease of the electrical conductivity leads to a decrease of the electronic contribution to the thermal conductivity, following the Wiedemann-Franz law. However, if the dimensionality of the material is decreased, the new variable of length scale becomes available for the control of materials properties, due to differences in the density of electronic states. Recent work with antimony/bismuth/tellurium superlattices demonstrated an enhancement in the *ZT* to about 2.4 (Venkatasubramanian *et al*, 2001). Thermoelectric micro devices with high figures of merit, based on superlattices are the key to generate power from low temperature gradients, and to produce high temperature gradients with a low-power supply. This will opens doors to automotive and domestic applications, where high heat flows are present as well as the operation at higher temperatures.

REFERENCES

Alahmad, M., et al. (2008). Evaluation and analysis of a new solid-state rechargeable microscale lithium battery. *IEEE Transactions on Industrial Electronics*, 55, 3391-3401.

Armand, M., and Tarascon, M. (2008). Building better batteries. *Nature,* 451, 652-657.

Birkholz, U., *et al.* (1987). Fast semiconductor thermoelectric devices. *Journal Sensors and Actuators A,* 12, 179-184.

Carmo, J. P., *et al.* (2009). Integrated thin-film rechargeable battery in a thermoelectric scavenging microsystem. in *Proc. Powereng* 2009, Lisbon, Portugal, 359-363.

da Silva, L., and Kaviany, M. (2002). Miniaturized thermoelectric cooler. in *Proc. IMECE 2002 - 2002 ASME International Mechanical Engineering Congress & Exposition.* New Orleans, Louisiana, USA. Paper No. 2-8-1-6.

da Silva, L., and Kaviany, M. (2004). Micro-thermoelectric cooler: interfacial effects on thermal and electrical transport. *Internacional Journal Heat Mass Transfer,* 47, 2417–2435.

da Silva, L., *et al.* (2005). Thermoelectric performance of films in the bismuth-tellurium and antimony-tellurium systems. *Journal of Applied Physics*, 97. 114903.

Dudney, N. J. (2005). Solid-state thin-films rechargeable batteries. *Materials Science & Engineering B,* 116, 245-249.

Dudney, N. J., and Neudecker, B. J. (1999). Solid state thin-films lithium battery systems. *Solid State & Materials Science,* 4, 479-482.

Dyck, J. S., *et al.* (2002). Heat transport in $Sb_{2-x}V_xTe_3$ single crystals. *Physical Review B*, 66, 125206.

Goncalves, L. M., *et al.* (2006a). Optimization of thermoelectric thin-films deposited by coevaporation on plastic substrates. in *Proc. ECT'06*, Cardiff, UK, 2006.

Goncalves, L. M., *et al.* (2006b). Flexible thin-film planar Peltier microcooler. in *Proc. of ICT'06*, Vienna, Austria, 2006.

Goncalves, L. M., *et al.* (2008a). Thermoelectric Micro Converters for Cooling and Energy Scavenging Systems. *Journal Micromechanics and Microengineering,* 18.

Goncalves, L. M., *et al.* (2008b). On-chip array of thermoelectric Peltier microcoolers, *Journal Sensors and Actuators A,* 145-146, 75-80.

Goncalves, L. M., *et al.* (2011). Thermal co-evaporation of Sb2Te3 thin-films optimized for thermoelectric applications, *Thin Solid Films,* Accepted for publication on 2011, 1-6.

JCPDS. (1983). Joint Committee on Powder Diffraction Standards (JCPDS), *Powder Diffraction File*, Pennsylvania.

L. Mateu, et al. (2007). Human body energy harvesting thermogenerator for sensing applications. in *Proc. 2007 International Conference na Sensor Technologies and Applications*, Valencia, Spain.

Park, M., *et al.* (2008). Performance evaluation of printed LiCoO2 cathodes with PVDF-HFP gel electrolyte for lithium ion microbatteries. *Electrochimica Acta,* 53, 5523-5527.

Peltier, J. C. (1834). Nouvelles experiences sur la caloricité des courans électriques. *Annales de Chimie et the Physique,* LVI 56, 371-386.

Scherrer, H., and Scherrer, S.. (1995). Bismuth Telluride, Antimony Telluride and Their Solid Solution. *Handbook of thermoelectrics*, edited by D.M. Rowe, CRC Press, 211-237.

Seebeck, T. I. (1822). Magnetische polarisation der metalle und erze durch temperatur-differenz. *Abhandlungen der Deutschen Akademie der Wissenschaften zu Berlin*, 265-373.

Venkatasubramanian, R., *et al.* (2001). Thin-film thermoelectric devices with high room-temperature figures of merit. *Nature,* 413.

Völklein, F., and Stärz, T. (1997). Thermal Conductivity of Thin Films - Experimental Methods and Theoretical Interpretation. in *Proc. 16th International conference on Thermoelectrics*, 711-718.

Wang, Z., *et al.* (2005). Coating of multi-walled carbon nanotube with SnO2 films of controlled thickness and its application for Li-ion battery. *Journal of Power Sources.*

Zou, H., *et al.* (2002). Peltier effect in a co-evaporated Sb_2Te_3(P)-Bi_2Te_3(N) thin film thermocouple. *Thin Solid Films*, 408, 270-274.

In: Antimony: Characteristics, Compounds and Applications ISBN 978-1-62100-598-8
Editor: Manijeh Razeghi

Chapter 11

COBALT ANTIMONIDE - SYNTHESIS, PROPERTIES AND APPLICATIONS

Elzbieta Godlewska, Kinga Zawadzka and Krzysztof Mars
Faculty of Materials Science and Ceramics, AGH University of Science and Technology

ABSTRACT

Cobalt triantimonide is known for the attractive thermoelectric properties, i.e. good electrical conductivity and thermopower combined with relatively low thermal conductivity. Its applications in thermoelectric devices, such as power generators or coolers, is however limited due to the low thermal and chemical stability. Both components of the compound, cobalt and antimony, exhibit high affinity for oxygen. Several binary and ternary oxide phases can be formed at elevated temperatures and some of the oxidation products are volatile, which implies significant material losses. Numerous efforts have been undertaken to prevent degradation of cobalt triantimonide in service conditions. The best protection developed so far is provided by an aerogel coating, effective in the conditions of dynamic vacuum at temperatures not exceeding 450°C. This chapter will be focused on the synthesis and selected properties of cobalt triantimonide and its derivatives. Oxidation kinetics and mechanism of $CoSb_3$ will be discussed with reference to thermodynamics of the Co-Sb-O system at elevated temperatures. A range of preventive measures will be described to eliminate or reduce the loss of antimony by sublimation or oxidation, comprising porous polymeric-ceramic hybrid coatings, non-porous amorphous coatings and thin compact metallic/ceramic layers deposited by magnetron sputtering. Finally some applications will be highlighted for different types of antimonides.

INTRODUCTION

The Sb-Co binary phase diagram [1] demonstrates a number of phases with different stoichiometry and crystal structure: αCo and εCo solid solutions, Sb, intermetallic phase CoSb(β) with a NiAs structure, $CoSb_2(\gamma)$ with a FeS_2 structure and intermetallic compound $CoSb_3(\delta)$ with a skutterudite structure.

At 695°C, there is an allotropic transformation $\alpha Co \longleftrightarrow \varepsilon Co$. At the temperature of 1113°C and Sb concentration of 25.5 at% there is a eutectic transformation $L \longleftrightarrow \alpha Co + CoSb(\beta)$, while at 623°C: $L \longleftrightarrow CoSb_3(\delta) + Sb$. $CoSb_2(\gamma)$ is formed in a peritectic transformation $L + CoSb(\beta) \longleftrightarrow CoSb_2(\gamma)$ at 931°C, whereas $CoSb_3(\delta)$ in a transformation at 876°C: $L + CoSb_2(\gamma) \longleftrightarrow CoSb_3(\delta)$. This chapter is devoted to synthesis, properties and applications of $CoSb_3$ (δ).

STRUCTURE

$CoSb_3$ belongs to the family of compounds referred to as skutterudites. The name comes from a cobalt arsenide mineral with a general formula: $(Co,Ni,Fe)As_3$, which was discovered in Skutterud in Norway in the nineteenth century. Skutterudites are compounds with a general composition of MX_3, where M is one of the group 9 transition metals - Co, Rh, or Ir, and X is one of the group 15 elements (pnicogens) - P, As, or Sb. The compounds have the same bcc structure and the space group *Im3*. A crystallographic unit cell of binary skutterudite MX_3 consists of eight units of MX_3 (Fig.1). Eight M atoms occupy the *c*-sites, forming a simple cubic sublattice. Twenty-four X atoms are situated at the *g*-sites of the structure. Four X atoms are located at the corners of a square ring (X_4). Out of six planar X_4 rings, two are oriented in the xy-plane, two in the yz-plane, and two in the zx-plane (x, y, and z denote crystallographic axes). The structure is characterized by the lattice constant and two positional parameters *y* and *z* that denote the *g* sites. Both M-X and X-X distances are close to the sum of the covalent radii, indicating strong covalent bonding. The M-M distance is much larger, and M has no nearest neighbor metal atoms, which implies weak bonding between M atoms. The interstitial voids are located at the 2*a* positions (12-coordinated) in the crystal lattice. The crystallographic unit cell can be represented by a formula $\square_2 M_8 X_{24}$, where '' stands for a void in the structure. Filled skutterudites (discussed further in this chapter) are formed by inserting guest atoms such as rare-earth, alkali or alkaline-earth into these voids. Corner-sharing MX_6 octahedra form dodecahedral cages which can host foreign atoms.

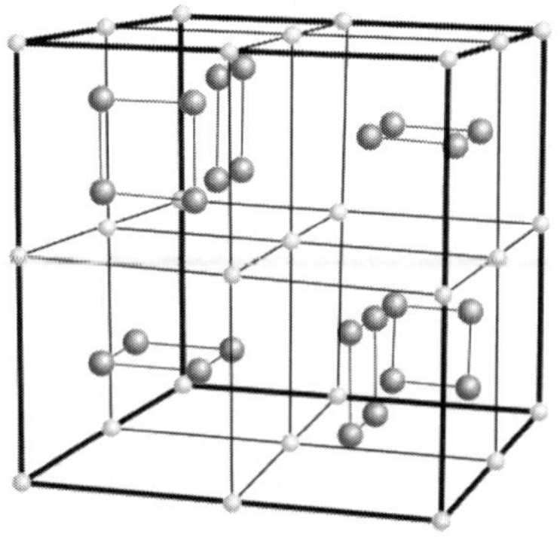

Figure 1. Schematic representation of a crystallographic unit cell of a skutterudite: small light balls being M atoms, dark bigger balls being X atoms.

SYNTHESIS

As follows from the Co-Sb phase diagram [1], $CoSb_3$ undergoes peritectic decomposition at 876 °C and both melting points and vapor pressures of cobalt and antimony differ significantly. An adequate choice of synthesis parameters becomes crucial, especially when account is taken of the fact that all $CoSb_3$–based materials of practical importance are chemically modified (doped/filled skutteridites). Historically, first syntheses relied on melting pure reactants. Later on, a range of other processes was used, such as mechanical alloying (MA), spark plasma sintering (SPS) or thin film technologies.

Melting

Melting of pure reactants in stoichiometric proportions is the most straightforward synthesis route for the skutterudites. Frequently, the melting stage is performed at temperatures lower than peritectic decomposition of $CoSb_3$ and higher than melting point of antimony, i.e. within the range 700 °C - 800 °C. In this regime, the synthesis in the evacuated quartz ampoules extends over several days. Modifications of this method, described in the literature, differ in temperature, duration and number of heating and cooling cycles as well as sintering conditions [2-12].

Melting experiments at temperature exceeding the peritectic decomposition of cobalt triantimonide were conducted to obtain $(Ce/La)_yFe_xCo_{(4-x)}Sb_{12}$ [8,9]. A homogenized mixture of elemental powders was sealed in the evacuated quartz ampoules and annealed first under 600 °C for 3h then at 1050 °C for 20h, and finally quenched into a water bath. The second annealing was performed at 700 °C for 30h to obtain single-phase $CoSb_3$. The final product, was received after hot pressing under 3.5 GPa at 700 °C.

$(Ce/La)_yFe_xCo_{(4-x)}Sb_{12}$ was also successfully synthesized by arc-melting [10-12]. First stage involved melting of Ce and (Fe, Co) to obtain master alloys, CeFe and/or CeCo. Next, the master alloys were loaded into boron nitride crucibles with Sb for induction melting under argon at 1600 °C. At the final stage the ingots were annealed at 700 °C and hot pressed into pellets.

In all processes based on reactant melting, the final product contains some amounts of $CoSb_2$, which originates from the peritectic decomposition at 876 °C. It is indispensable to perform every stage of the high-temperature synthesis in an oxygen-free environment.

Mechanical Alloying

Mechanical alloying (MA) consists in high-energy ball milling of powder mixtures, which gradually transform into strongly supersaturated solid solutions. Upon subsequent annealing, precipitation of new phases takes place. First developments date from 1966 and are connected with the manufacturing of nickel-based superalloys [13]. A characteristic feature of MA is that it produces highly defected materials through severe plastic deformation and the appearance of metastable phases is energetically favored.

Because of the mentioned peritectic decomposition, MA seems an attractive alternative for the cobalt triantimonide synthesis. Based on DTA studies, it has been stated [14] that MA in a planetary mill for less than 40h, favors the formation of $CoSb_3$, however pure $CoSb_3$ product cannot be obtained, if the starting mixture is strictly stoichiometric. Even after several dozen hours, the XRD analysis of the reaction product reveals significant amounts of $CoSb_2$, Co and Sb. Essentially pure $CoSb_3$ was received after an additional heat treatment at 700 °C in vacuum. A remarkable concentration of $CoSb_2$ in the synthesis of $Co_{(4-x)}Fe_xSb_{12}$ was attributed to very similar formation enthalpies of $CoSb_3$ and $CoSb_2$ [15]:

$$Co + 2Sb \rightarrow CoSb_2,\ \Delta H^0_f = -54\ kJ/mol$$

$$Co + 3Sb \rightarrow CoSb_3,\ \Delta H^0_f = -67\ kJ/mol$$

MA is not accompanied by excessive grain growth and is considered suitable for the production of nanopowders. First attempts described in the literature [16], have shown that in optimized conditions (5h, ball-to-powder ratio 20:1 and working voltage 110V in a high-energy mill), it is possible to obtain $CoSb_3$ crystals with nanometric sizes (20-35nm).

Spark Plasma Sintering

Spark plasma sintering (SPS) is a powder consolidation method resembling hot pressing (HP): in that it is conducted in graphite dies at elevated pressures and temperatures. The principal difference is that dc pulses are passed through the die and conducting powder compacts. The high intensity current pulses, play a role in purification and activation of powder surfaces, which leads to instantaneous heat and mass transfer, and consequently densification of the material [17]. Compared with HP, the sintering time is much shorter, temperatures lower and so is the energy consumption.

It has been reported that sintering parameters affect thermoelectric properties of $CoSb_3$ [18]. A stoichiometric mixture of elemental powders was first milled and next sintered in an evacuated graphite die under 30 MPa at temperatures 400 °C - 700 °C for 5–30 min. Systematic XRD and TEM analyses indicated that sintering temperature was decisive for the phase composition and density of the final product. but also for the type of electrical conductivity. The higher the temperature and the longer the milling stage, the better was the n-type conductivity. It turned out that $CoSb_3$ synthesis was not possible below 400 °C. An almost single-phase product was received at temperatures 500 °C -700 °C; its density increasing with temperature. $CoSb_3$ with relative density of 97.4% and grain size of about 100 nm sintered at 600 °C for 5 min had the best thermoelectric properties.

In general, combination of SPS with other advanced processing methods brings about the most satisfactory results. Extensive studies on the effect of process parameters on composition and properties of final products were conducted in order to assess the usefulness of MA and SPS in the manufacturing of nanocrystalline $CoSb_3$ bulk materials [19-21]. A Co+3Sb mixture was homogenized in a planetary ball mill (Ar, 450 rpm, 15 h) [19], and sintered under 50 MPa at 300 °C – 600 °C for 5 min. When the powders were subjected to mechanical alloying only, the product always contained some Sb and $CoSb_2$ [13-16], whereas after an additional SPS, all the reactants were consumed and the product was single-phase $CoSb_3$. Apparently, the positive effects of MA in combination with SPS are connected with better densification upon sintering, due to the high concentration of defects in the MA product and very efficient heat transfer in SPS due to electrical discharges within the powder bed. The relative densities of the obtained materials were in the range 94-99%, and grain sizes of 300 nm and 50 nm at temperatures of 600 °C and 300 °C, respectively. In spite of the same structure, there were some differences in lattice parameters and exact atomic ratio, indicating Sb deficit, i.e. composition of the final product could be described by a formula $CoSb_{3-x}$, where x increased with SPS temperature. This behavior was attributed to partial evaporation of antimony. To compensate for the antimony loss, the initial powder mixture was purposely enriched in antimony, the optimum being 6 at% [20]. The unwanted effects of that procedure were: grain size increase to 500 nm, density decrease from 7.33 to 6.95 g/cm^3 and electrical conductivity change from n-type to p-type.

Solid-Vapor Reaction

According to recent studies [22], the synthesis of $CoSb_3$ from elemental powders is feasible at temperatures lower than the melting point of antimony or the $CoSb_3$/Sb eutectic.

(a) (b) (c)

Figure 2. Microphotographs of $CoSb_3$ after the solid-state reaction at temperatures lower than 630 °C [22], showing powder shapes and sizes (a), grain microstructure (b), and cross-section with a small area of unreacted Co.

The reaction is probably initiated on the surface of cobalt grains due to relatively high vapor pressure of antimony. The XRD and SEM/EDS analyses (Fig.2) indicate that, cobalt powder grains transform into $CoSb_3$ owing to the inward diffusion of antimony. The product is nanostructured and can be sintered by any available method.

Solution Chemistry

Otherwise, $CoSb_3$ can be prepared by a solution chemistry, which is based on co-precipitation of Sb_2O_3 and CoC_2O_4, from the aqueous solutions containing the following ions: Co^{2+}-Sb^{3+}-NH_4^+-H^+-$C_2O_4^{2-}$-Cl^- [23,24]. A mixture of $CoCl_2$, $SbCl_3$ and $H_2C_2O_4$ solutions was used to receive the desired precursors at a controlled pH. The precursor powders, Sb_2O_3 and CoC_2O_4, were next calcined at 350 °C in flowing nitrogen for 2h, and then reduced at 450 °C - 600 °C in flowing hydrogen for 4h. The product was composed of 95% $CoSb_3$, with some Sb and $CoSb_2$ impurities. The method is suitable for the synthesis of doped materials [25-27].

Another route is solvothermal synthesis, where the precursors are $CoCl_2 \cdot 5H_2O$ and $SbCl_3$; ethanol being the solvent [28-30]. The process is held in a Teflon-lined autoclave, in the temperature range 190 °C - 270 °C, with $NaBH_4$ as reducer. In this procedure, some amounts of $CoSb_2$ are always identified in the reaction products, which results from the proposed two-stage mechanism. Pure Co and Sb are received by reduction of Co^{2+} and Sb^{3+}:

$$CoCl_2+2NaBH_4 \rightarrow Co+2BH_3+2NaCl+H_2$$

$$3SbCl_3+9NaBH_4 \rightarrow 3Sb+9BH_3+9NaCl+9/2H_2$$

At the first stage, Co reacts with Sb to form $CoSb_2$. At the second stage $CoSb_2$ reacts with Sb to form $CoSb_3$. Reaction yield increases with temperature [28] and with $NaBH_4$ concentration, until some limiting value is reached [30]. A similar procedure is used with tetra-ethylene glycol (TEG) as a solvent [31]. When the temperature at the initial stage is too low (165 °C), the reaction product is CoSb. This compound is thermodynamically the most stable in the Co-Sb system, so it cannot transform into $CoSb_3$. When the temperature is raised to 240 °C, the final product is $CoSb_3$. The $CoSb_3$ powders produced by this method are often characterized by irregular particle shapes and sizes in the range 20 nm - 80 nm. The most recent studies [32] indicate that by using specific chemicals, e.g. Poly(vinyl pyrrolidone) (PVP) as both structure directing and mild reducing agent, it is possible to go down with the particle size to 10 nm.

Single Crystals

Single crystals are generally grown in Sb-rich systems. For illustration, $CoSb_3$ was received from the melt with Co:Sb atomic ratio of 1:6.7 [33]. The process was conducted at temperatures from 1150 °C to 650 °C, at two cooling rates, initially 30 °C/h, and next 16.5 °C/h. $CoSb_3$ single crystals were grown by a Bridgman directional crystallization from the melt containing 93 at.% Sb in a vertical furnace with a temperature gradient of approximately 50 °C/cm at the growth interface [34,35]. Cooling rate was about 0.7 °C/h. Relative densities of the obtained single crystals were 99.5% with a slight antimony excess. In order to remove this antimony excess, the $CoSb_3$ specimens grown by the vertical Bridgman method were annealed at 800 °C for 20h, but this brought about reconstruction of the $CoSb_3$ grain externals and formation of voids in the crystal [36]. The $CoSb_3$ single crystals were also obtained by chemical vapor transport in closed silica ampoules, using iodine as the transporting agent, and from Sb-rich melts by a flux method [37].

Thin Films and Layered Structures

The choice of deposition methods is limited because of the antimony low melting point and $CoSb_3$ decomposition (Sb sublimation) at temperatures in excess of 500 °C. One of the feasible routes is pulse magnetron sputtering. A multilayer $CoSb_3$/Cu coating was prepared in a dual magnetron system operated with a mirror B field [38]. One target was made of $CoSb_3$ dense sinter and the other was metallic copper. At a relatively low cathodic current of 100 mA, the deposition rate of $CoSb_3$ was about 3.3 nm/s, while copper deposition rate was only 0.6 nm/s at a current of 850 mA. The multilayer coating was deposited onto glass, AlN and Cu substrates (Fig.3). Maximum thickness of adherent $CoSb_3$ coatings was 10μm. Substrates were maintained at a temperature of about 50 °C, which ensured formation of polycrystalline multilayer coatings with a very smooth surface.

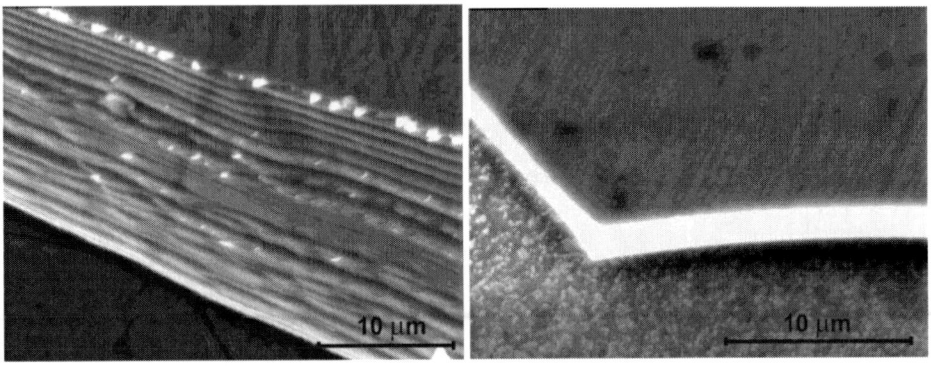

(a) (b)

Figure 3. $CoSb_3$/Cu multilayers deposited by pulse magnetron sputtering: 32 pairs (a), 1200 pairs (b).

In another embodiment, a dc magnetron sputtering technique was used to deposit binary skutterudite thin films of $CoSb_3$ [39] onto Si(100) and Al_2O_3 substrates. The target was hot-pressed $CoSb_3$, and substrate temperatures ranged from 20 °C to 200 °C. Deposition rate was 6.3 nm/min, and final layer thickness was about 190 nm. The polycrystalline layers had preferential (310) orientation, with densely packed crystallites, 50-150 nm in size.

Nanometric $CoSb_3$ layers, can be obtained by electrodeposition from an aqueous solution. The principal difficulty in this process is related to disparate reduction potentials of Co(II) and Sb(I). Under acidic conditions, the standard reduction potentials of the redox couples vs. SHE: Co^{2+}/Co^0 and Sb^{+}/Sb^0 are −0.28 and 0.212V, respectively, which is 0.492 V apart [40].

Electrodeposition of $CoSb_3$ thin layers and nanowires was conducted in the citrate bath by a pulse plating technique [41]. For thin film deposition, the electrolyte consisted of 0.003

M Sb_2O_3, 0.688 M $CoSO_4{\cdot}7H_2O$, 0.125 M potassium citrate, and 0.196 M citric acid. For nanowire depositions, the $CoSO_4{\cdot}7H_2O$ concentration was reduced to 0.172 M. $CoSb_3$ was successfully plated at a potential of -1.35 V with respect to the Hg/Hg_2SO_4 electrode, but XRD of the films showed the presence of an elemental antimony. Cobalt excess was also detected by EDS. In galvanostatic experiments, cobalt and antimony were deposited in an overall ratio of 1:3, however, due to strong segregation of the elements it was impossible to obtain $CoSb_3$ by subsequent annealing. The problem was resolved by pulse plating: a layer of antimony was deposited at one potential and then both, cobalt and antimony, at a more negative potential. Composition of the deposited layer could be controlled by oscillating between the two potentials and adjusting the deposition time,. The layers were subsequently annealed for 24h at 400 °C in vacuum to yield $CoSb_3$. EDS of the annealed samples showed some antimony loss but no cobalt-rich phases were identified by XRD, indicating either their very small amounts or amorphous form.

Electrodeposition was also performed in citric-based solutions vs. an Ag/AgCl reference electrode [42]. The electrolyte solution contained a citrate complex of cobalt and antimony. Different deposition potentials and cobalt concentrations were used to assess the influence of those parameters on chemical and phase composition of the deposited layers. Co-deposition of Co and Sb in a ratio close to 1:3 was successful at a relatively low potential and low Co concentration, however, some amounts of Sb and $CoSb_2$ were present together with $CoSb_3$. Morphology of the deposited layers was dependent on their chemical composition.

$IrSb_3$ and $CoSb_3$ thin films (multilayers) were grown by pulsed-laser deposition at 270 °C on Si/SiO_2 substrates (and fused quartz substrates) from skutterudite targets with antimony excess [43]. To increase adhesion, multilayers were deposited on a thin layer (1 nm) of chromium. $CoSb_3$ was grown on <100> oriented InSb single crystals using a target with a 5 at% Sb excess. The films were polycrystalline with small columnar grains penetrating through the layers.

Thin films of $CoSb_3$ were deposited by Excimer Laser Ablated Molecular Beam Evaporation (ELA-MBE) in ultra-high vacuum [44]. The laser ablation target with an overall Co:Sb atomic ratio of 1:3, contained the following phases: $CoSb_3$, $CoSb_2$, Sb and Co. At substrate temperature of less than 380 °C and vacuum of 10^{-6} Pa, the stoichiometry of the deposits was the same as that of the targets. Their very specific structure, characteristic by square-shaped crystallites of skutterudite, 200 nm in size, embedded in a smooth matrix, was related to direct deposition in the plume of laser ablation followed by growth of small crystallites due to the migration of deposited species.

Reactivity

Co-Sb-O System

The Co-Sb-O system is quite complex. Already in the Sb-O binary system, there are several oxide phases [45]:

1) Sb_2O_3 – oxide occurring in two polymorphs: senarmontite and valentinite;

A senarmontite to valentinite transition is a multistage process, which according to different sources takes place at temperatures ranging from 556 °C to 655 °C [46];

2) Sb_2O_4 – also represented as $Sb_2O_3{\cdot}Sb_2O_5$, occurring in two polymorphs: cervantite (α) and clinocervantite (β);
3) Sb_6O_{13} - a double oxide with a formula $Sb_2O_3{\cdot}2Sb_2O_5$;
4) Sb_2O_5.

Among all known antimony oxides, Sb_2O_3 is the most volatile. Vapor pressure over this phase increases 600 times (from 0.33 to 213.30 Pa) in the temperature range 500 °C - 600°C [47].

At 600°C the pressure of Sb_4O_6 over a mixture of solid oxides Sb_2O_3+Sb_2O_4 amounts to $6.88{\cdot}10^2$ Pa, while over solid Sb_2O_4 itself, merely $2.68{\cdot}10^{-6}$ Pa [45].

Cobalt, reacting with oxygen produces two oxide phases, CoO and Co_3O_4, whereas in the Co-Sb-O system there are three double oxides [48]:

1) $CoSb_2O_4$;
2) $CoSb_2O_6$;
3) $Co_7Sb_2O_{12}$.

The only experimental Co-Sb-O ternary phase diagram was investigated at 600 °C by solid electrolyte emf method [48]. The following reactions were taken into account:

$$6CoO + CoSb_2O_4 + O_2 \rightarrow Co_7Sb_2O_{12}$$

$$CoSb_2O_4 + O_2 \rightarrow CoSb_2O_6$$

$$CoO_{(s)} + Sb_2O_{3(s)} \rightarrow CoSb_2O_{4(s)}$$

$$7CoO + Sb_2O_5 \rightarrow Co_7Sb_2O_{12}$$

$$CoO + Sb_2O_5 \rightarrow CoSb_2O_6$$

$$CoSb_2O_4 + 2Sb_2O_4 \rightarrow CoSb_2O_6 + 2Sb_2O_3$$

$$Co_7Sb_2O_{12} + 6Sb_2O_3 \rightarrow CoSb_2O_6 + 6CoSb_2O_4.$$

Based on the experimental work, the authors proposed the conditions of thermodynamic equilibrium between different oxide phases in the Co-Sb-O system. The most stable oxide in the system is CoO, and the least stable are antimony oxides, Sb_2O_4, and Sb_2O_5.

Oxidation Behavior

Degradation of $CoSb_3$ in vacuum is mainly related to the loss of antimony by sublimation. In air, the process is more complex, due to the formation of solid phases on the surface. Not many papers have been published on the oxidation kinetics ad mechanism of $CoSb_3$ [49-53].

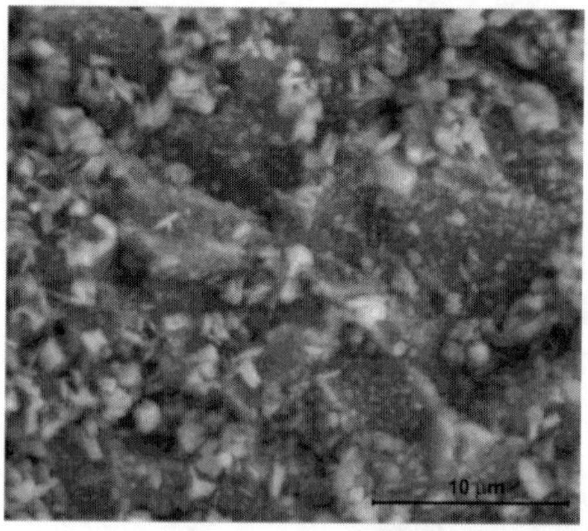

(a)

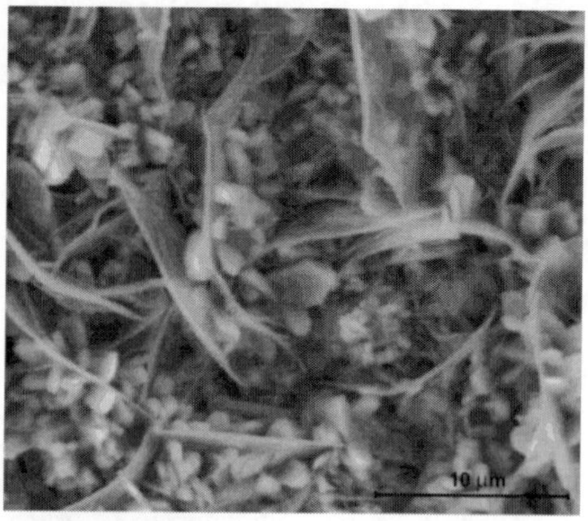

(b)

(c)

Figure 4. Micrographs (BEI) of $CoSb_3$ surface after exposure to air a) 500 °C/4x20 h b) 600 °C/4x20 h c) 700 °C/3x20h.

Platinum and palladium doped $CoSb_3$ produced by MA followed by SPS was exposed to air at 600 °C for 12, 48 and 134 h [49]. After exposure, density of the specimens decreased due to antimony loss and compositional changes in the surface layers. The outer layer was composed of β-Sb_2O_4, Sb_2O_4 and Sb_2O_3, and the inner one - of $CoSb_2O_6$; $CoSb_2O_4$, Sb_2O_3 and β-Sb_2O_4. The oxide layer growth followed a parabolic rate law.

Systematic studies on thermal stability of $CoSb_3$ [54] have shown that antimony diffusivity and volatility play major roles in its oxidation and thermal decomposition in air. DTA-TG measurements over a temperature range 20 °C - 850°C, provided valuable information on the oxidation kinetics in a powder mixture. Oxidation of $CoSb_3$ started already at 380 °C and involved several stages. Total weight change was measured at different temperatures after the same exposure times. At temperatures not exceeding 650°C, the oxidation products consisted of $CoSb_2O_4$ and Sb_2O_4, while at 700 °C and 750°C - of $CoSb_2O_4$ and $CoSb_2O_6$. The $CoSb_3$ powder after oxidation at 550 °C and 700°C was composed of Sb_2O_4, $CoSb_2O_6$ and $CoSb_2O_4$. The overall oxidation process could be divided in two stages. In the first stage, the product contained mixed oxides Sb_2O_4, $CoSb_2O_6$ and $CoSb_2O_4$. In the second stage, two layers appeared, the outer one built of Sb_2O_4 and the inner one built of $CoSb_2O_6$/$CoSb_2O_4$. Parabolic kinetics was observed in both stages.

Recent systematic studies on the oxidation of $CoSb_3$ dense sinters in isothermal and cyclic conditions over the temperature range 500 °C – 700 °C provided new data on the oxide scale structure and its evolution [52].

Oxidation products at 500°C were the following: Sb_2O_3 (senarmontite), α-Sb_2O_4 and $CoSb_2O_4$, at 600°C: Sb_2O_3 (valentinite), β-Sb_2O_4, α-Sb2O4, $CoSb_2O_4$ and $CoSb_2O_6$ and at 700°C: β-Sb_2O_4, α-Sb_2O_4, $CoSb_2O_4$ and $CoSb_2O_6$. Typical morphologies of the oxide scales are presented in Fig.4.

At 500°C the major oxides identified on the scale surface were: Sb_2O_3 (senarmontite) and α-Sb_2O_4, at 600°C: Sb_2O_3 (valentinite), α-Sb_2O_4 and β-Sb_2O_4, while at 700°C: α-Sb_2O_4 and β-Sb_2O_4.

The scales had a layered structure, as shown in Fig.5. The outermost layer was composed of antimony oxides exclusively, while two layers beneath contained cobalt and antimony in a ratio of 1:2, but the darker one was richer in oxygen. In XRD patterns, two double oxides were identified: $CoSb_2O_6$ and $CoSb_2O_4$. Lower cobalt antimonides, $CoSb_2$ and CoSb, were not detected.

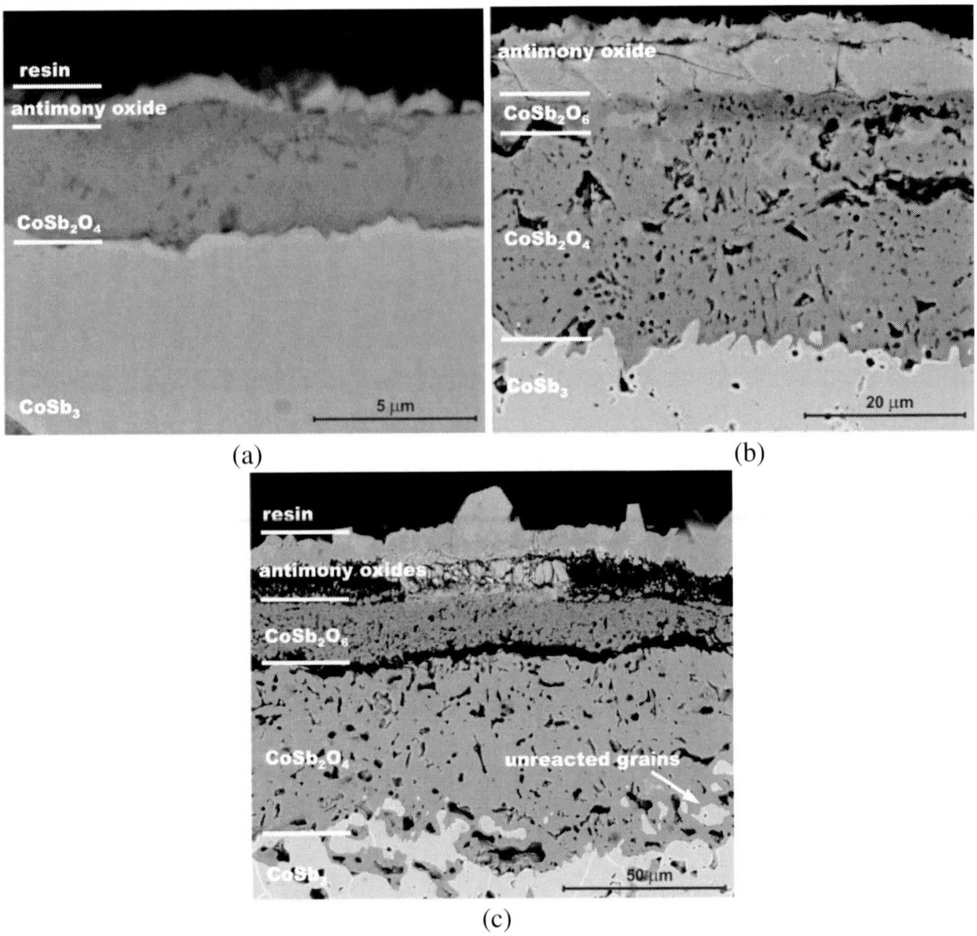

(a) (b) (c)

Figure 5. Cross-sections (BEI) of $CoSb_3$ specimens after exposure to air: 4x20 h at 500 °C (a), 4x20 h at 600 °C (b) and 2x20 h at 700°C (c).

Based on the analyses of chemical and phase composition, and the oxide scale evolution, the overall reaction on the specimen surface at 500 °C and 600 °C can be written as follows:

$$4CoSb_3\,(s) + 11O_2\,(g) = 4CoSb_2O_4\,(s) + 2Sb_2O_3(s).$$

However, two stages, i.e. decomposition of $CoSb_3$ into $CoSb_2$ and Sb and subsequent oxidation of the decomposition products are probably involved:

$$CoSb_2(s) + 2O_2(g) = CoSb_2O_4(s)$$

$$4Sb(s) + 3O_2(g) = 2Sb_2O_3(s)$$

The vapor pressure of antimony at 527°C is quite low, 0.933Pa [47], so its loss by evaporation can be neglected. Formation of solid oxide phases is, however, accompanied by evaporation of volatile species [45] and vapors are mainly composed of Sb_4O_6:

$$2Sb_2O_3(s) = Sb_4O_6(g)$$

$$4Sb(s) + 3O_2 = Sb_4O_6(g)$$

$$Sb_2O_4(s) = 1/2Sb_4O_6(g) + 1/2O_2$$

$$4Sb_2O3(s) = 2/n\ Sb_n(g) + 3\ Sb_2O_4(s).$$

The transparent flakes and spongy forms on specimen surface (Fig.4) after exposure are probably condensed antimony oxides. The $CoSb_2O_6$ layer was found in the scale after a longer oxidation time. According to the thermodynamics of the Co-Sb-O system [48], $CoSb_2O_4$ reacts with Sb_2O_4 to yield Sb_2O_3 and $CoSb_2O_6$. The fine-grained structure of the $CoSb_2O_6$ layer and porosity at the Sb_xO_y/ $CoSb_2O_6$ interface (Fig.6) might be taken as experimental proofs for this reaction mechanism. A direct oxidation: $CoSb_2O_4(s) + O_2$ -> $CoSb_2O_6(s)$, is also possible, provided oxygen can penetrate toward the inner scale layer.

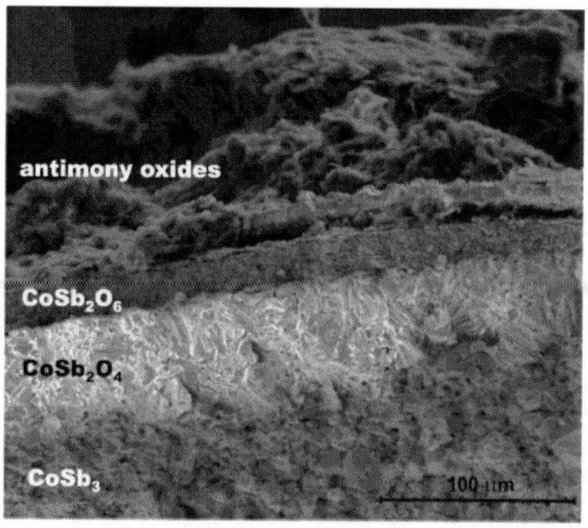

Figure 6. Fractured specimen (2 x 20 h at 700 °C) with visible fine-grained structure of $CoSb_2O_6$, acicular $CoSb_2O_4$ and porosity at the Sb-O/double oxides interface.

The sequence of the oxide layers in the scale is consistent with thermodynamics of the system, i.e. oxides with the highest dissociation pressure are present next to the solid/gas interface: Sb_2O_4, Sb_2O_3, $CoSb_2O_6$, $CoSb_2O_4$. The most stable CoO phase has not been detected at the interface with $CoSb_3$ at 500 °C or 600 °C, probably for the kinetic reasons. However, CoO is not present in the scale even after a prolonged oxidation (300 h) at 500 °C, which may be related to the low activity of cobalt in $CoSb_3$ at this temperature. In general, the mechanism of scale formation is complex and involves an outward diffusion of antimony and an inward diffusion of oxygen. Moreover, the diffusion/oxidation processes at the grain boundaries of $CoSb_3$, both seem to be of significance (Fig.5).

PROTECTIVE COATINGS

As shown in the previous section, poor oxidation resistance and tendency to antimony loss can remarkably influence the properties of $CoSb_3$. Many attempts have been made to overcome this deficiency by a range of coatings. The principle requirements were that the coating should be impermeable for oxygen and chemically inert, resistant to thermal shocks and should not deteriorate electrical properties of $CoSb_3$. First experiments were directed at metallic coatings, e.g. niobium or titanium foil [53, 55]. Specimens with the metallic coating were heat treated in a dynamic vacuum at 700 °C. That protective system turned out quite efficient since no antimony loss was found on the cross-section. Several other versions of such coatings were proposed and all were effective in preventing antimony sublimation up to 700 °C. In the case of titanium, however, the foil interacted with the material and produced a layer of $TiSb_2$, which brought about undesirable changes in ductility [55,56]. In general, metallic coatings have one serious drawback, regarding thermoelectric applications of $CoSb_3$-based materials, i.e. they provide easy paths for charge transport (electrical shorting). An alternative approach was to develop ceramic coatings with insulating properties in relation to charge and heat transport. Among those, promising results were reported for BN, $MgAl_2O_4$ and aerogels [53, 57-63].

Ceramic coatings usually have coefficients of thermal expansion closer to $CoSb_3$, which positively affects their adherence in thermal cycling conditions. Particularly good results were obtained with thick aerogel coatings, which are known for very low thermal and electrical conductivity. At 99% porosity, their density ranges from 30 to 300 mg/cm^3. Pore sizes on the order of 10^{-10} – 10^{-8} m dramatically reduce the extent of antimony diffusion within the coating and thereby the degree of surface degradation of $CoSb_3$. The whole thermoelectric element can be immersed in a vessel containing the sol solution. The synthesis begins from the polymerization of precursors, such as TEOS (tetraethyl orthosilicate) or tetramethyl orthosilicate (TMOS) in a solution containing volatile solvent (acetonitrile, ethanol or methanol). Ammonium hydroxide is used as a gelation catalyst. Subsequent aging takes typically 48 hours. Capillary forces associated with the evaporation of the solvent bring about shrinkage and formation of a xerogel (porosity of 50%). When the solvent is removed in supercritical conditions in an autoclave (295 °C, 5.5 MPa) the final product is aerogel with uniformly distributed micropores. Already first tests in a dynamic vacuum indicated antimony loss over 10 times lower compared with the unprotected material. Further improvements involved modification of the sol solution with some fillers, such as alumina or titania powder in order to raise the aerogel density. Those modifications lowered antimony sublimation by several orders of magnitude but at the same time increased shrinkage of the gel and its tendency to crack. When TMOS was used instead of TEOS and SiO_2 powder together with colloidal silica were added to enhance crosslinking, the aerogel appeared free of cracks and the rate of antimony sublimation after a 3-day annealing in vacuum was on the level of 10^{-5} g/cm^2h. By further enrichment of the sol with titania (opacifier and thickening agent), the aerogel density attained $0.1 g/cm^3$ at a 2.3% shrinkage and the coating was still more effective, i.e. the rate of antimony sublimation at temperatures up to 700 °C decreased as much as 500 times.

All the protection systems mentioned above were tested in a dynamic vacuum since space applications of $CoSb_3$-based materials were the main objective. As for the terrestrial applications, very few studies were reported. Protective coatings, which initially seemed promising were cost-effective hybrid polymer-metal coatings, which successfully prevented high-temperature oxidation of steel tubing in pcf boilers [64] The coatings were deposited by simple slurry spraying [65]. Methylphenyl siloxane resin constituted the matrix, Al and Cr powders were the fillers. Sand blasting was used to increase surface roughness of specimens and improve coating adherence. Polymerization was conducted at 250 °C for two hours. Fig.7 presents the surface and cross-section of a Cr-containing coating. Bright spots are particles of chromium. During the exposure to air, the metallic components reacted with oxygen to form the corresponding oxides, Al_2O_3 and Cr_2O_3, so the coating was actually a polymer-ceramic composite.

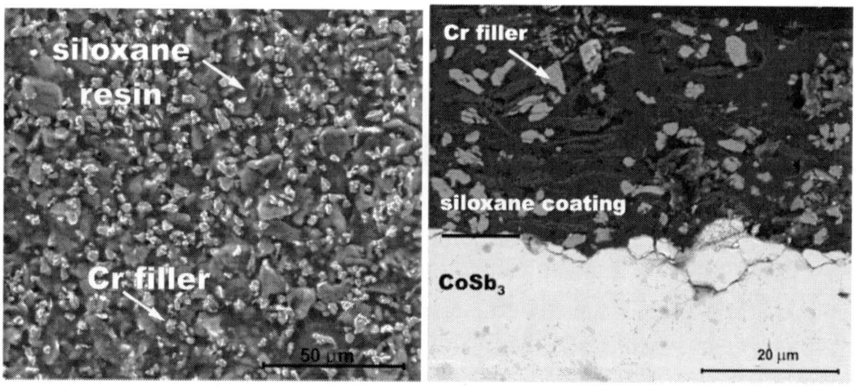

(a) (b)

Figure 7. Surface (a) and cross-section (b) of an as-received polymer-metal coating (methylphenyl siloxane resin with a chromium filler) on a $CoSb_3$ substrate.

The polymer-ceramic coatings with the chromium filler effectively reduced antimony loss up to 600 °C, however, they were not gas-tight and oxide phases formed under the coating, as shown in Fig. 8.

Another concept of surface protection was based on amorphous thick coatings which would be impermeable for all potentially diffusing species. Two types of such coatings were applied and tested [66]: one from commercial licensed frit with recrystallized TiO_2 and one experimental prepared from chemically pure reagents: SiO_2, $Na_2B_4O_7 \cdot 10H_2O$, TiO_2. $Al(OH)_3$, Li_2CO_3, Na_2CO_3, CaF_2 and $BaCO_3$. Chemical composition of the latter was tailored to match thermal expansion coefficient of $CoSb_3$ and to receive a low-melting thermally stable glass. The coating prepared from commercial frit was applied by double dipping in a slurry composed of frit, water and kaolin. Firing was conducted at different temperatures in the range 650 °C – 780 °C for 1-3 min. The coating, about 200 μm thick, had excellent protective properties, i.e. during a 20-h exposure to air at 600 °C, it totally stopped inward penetration of oxygen and outward diffusion of antimony. At the firing temperature of 780 °C, only a small amount of antimony escaped from the material and induced porosity at the $CoSb_3$/coating interface. Antimony vapors were, however, trapped in the inner part of the coating inside the round-shaped pores (bubbles), and the increased vapor pressure of antimony near the substrate/coating interface prevented further decomposition of the substrate. The coating was amorphous with some amount of TiO_2 crystallites dispersed in its outer part (Fig.9). Exclusively closed pores were present in the coating and pore sizes decreased toward the external surface. Coating adhesion to the substrate was very good and no symptoms of oxidation were identified. The experimental frit was prepared to improve mechanical properties of the coating and to reduce antimony depletion upon firing. Its softening temperature was lower and porosity at the substrate/coating interface markedly decreased. Antimony sublimation upon firing at different temperatures in the range 650 °C - 740 °C enhanced crystallization of the coating. Application of the second layer followed by firing led to satisfactory results (Fig.10a). Oxidation of the substrate under the coating was not observed after the exposure to air (Fig.10b).

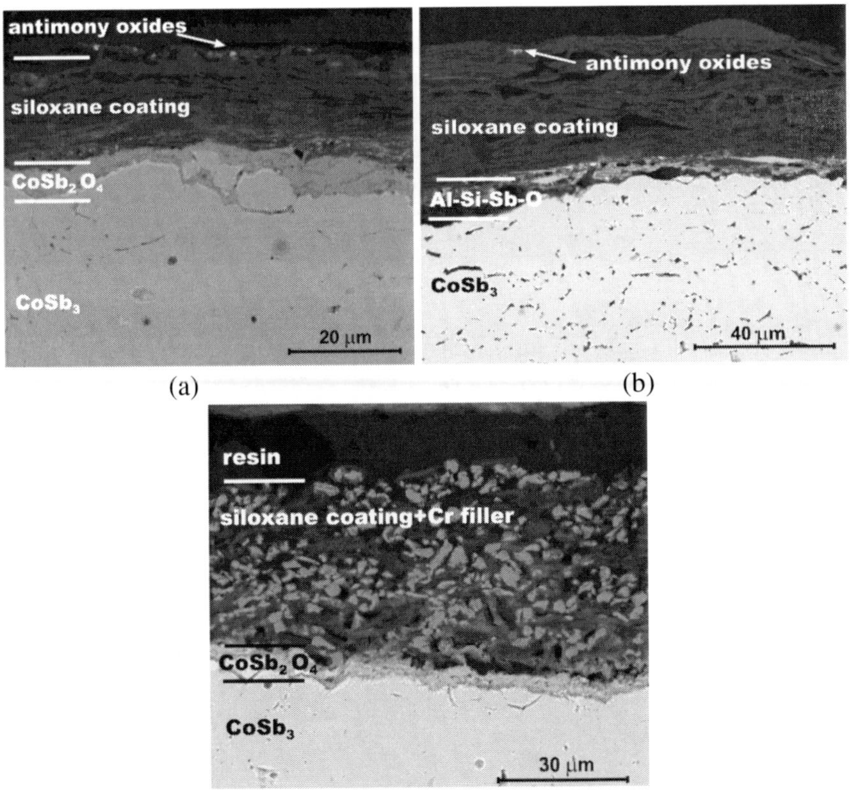

(a) (b)

(c)

Figure 8. Cross-sections (BEI) of $CoSb_3$ specimens with polymer-metal coatings after exposure to air showing Co-Sb double oxides under the coating with an Al filler (a),(b), and with a Cr filler (c); antimony oxides are present on the surface of coating (a), (b).

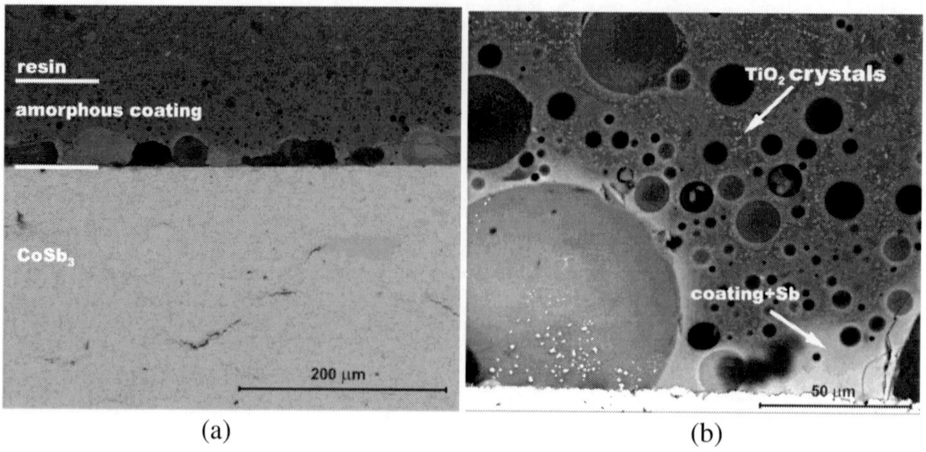

Figure 9. Cross-sections of $CoSb_3$ samples with as-received amorphous coatings (commercial frit).

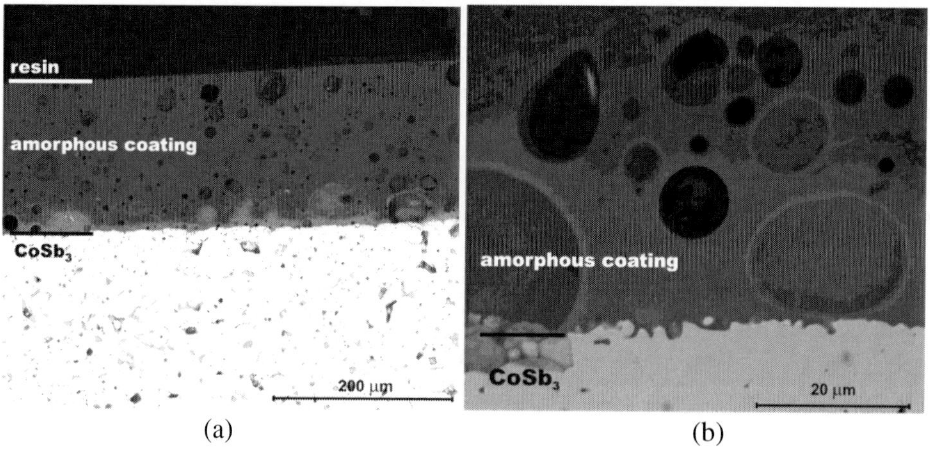

Figure 10. Cross-sections of $CoSb_3$ specimens with an amorphous coating from the experimental frit as-received (a) and after a 20-h exposure to air at 600 °C (b).

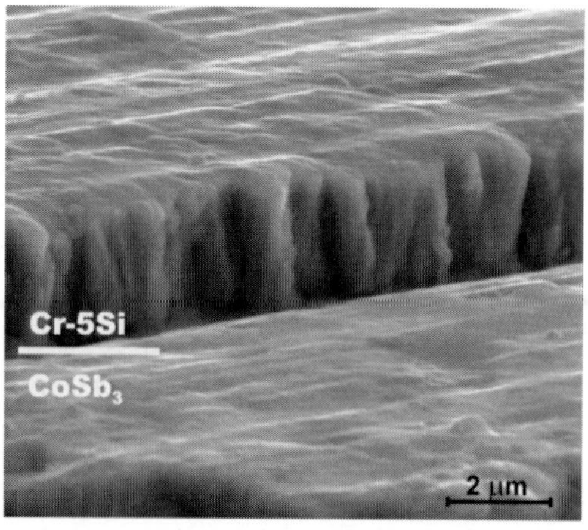

Figure 11. Fractured $CoSb_3$ specimen with as-deposited Cr-5Si coating (SEI).

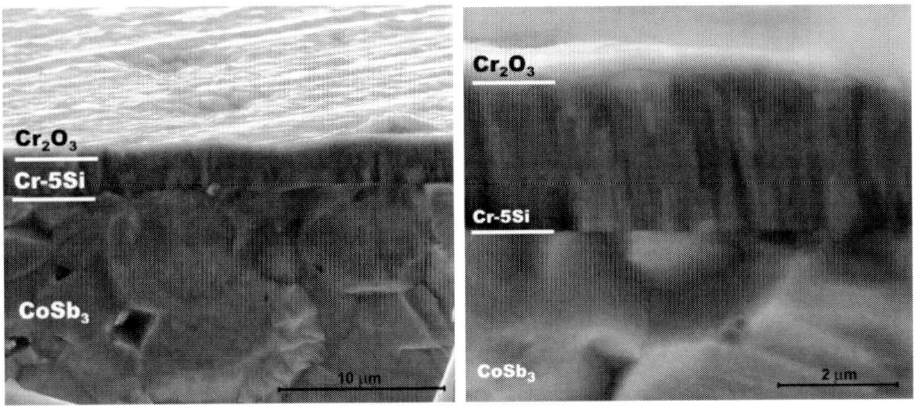

Figure 12. Fractured $CoSb_3$ specimen with a Cr-5Si coating after exposure to air at 600 °C at different magnifications.

In another approach, a relatively thin Cr-5Si layer was deposited on the surface of $CoSb_3$ by magnetron sputtering (Fig.11). The Cr-5Si coatings on γ-TiAl-based alloys showed very good mechanical properties, resistance to thermal shocks and negligible interdiffusion with the substrate up to 700 °C [67]. It was expected that the Cr-5Si layer would react with oxygen and produce an adherent oxide scale with very good protective properties. On the other side the coating would chemically bind antimony. The as-received Cr-5Si layer, approximately 2.5 μm thick (Fig.11) was built of tightly packed fibrous α-Cr grains with a relatively smooth surface. At 500 °C, oxidation of chromium thin films follows approximately a parabolic rate law and chromium is the predominant diffusing species [68]. Thickness of the Cr_2O_3 scale on the Cr-5Si coating after an 80-h oxidation (Fig.12) was in fairly good agreement with the kinetic parameters of the Cr thin film oxidation.

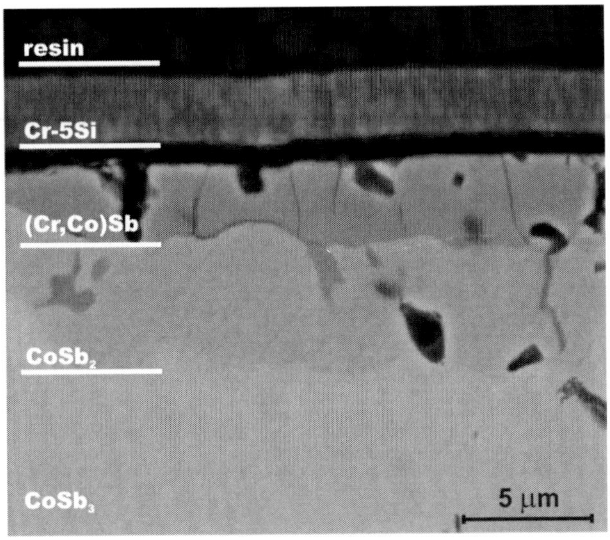

Figure 13. Cross-section of a CoSb3 specimen with a Cr–5Si layer after annealing at 400 °C for 6 h and a 20-h exposure to air at 600 °C.

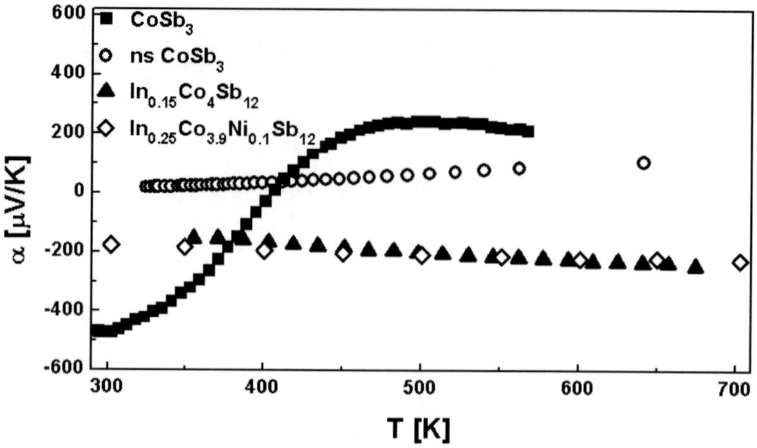

(a)

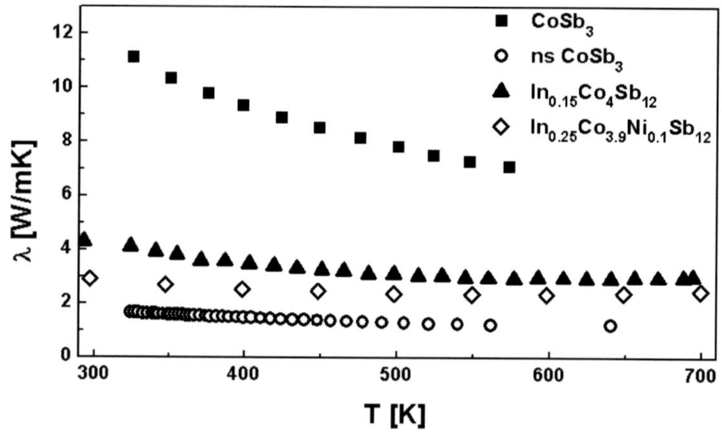

(b)

Figure 14. Continued on next page.

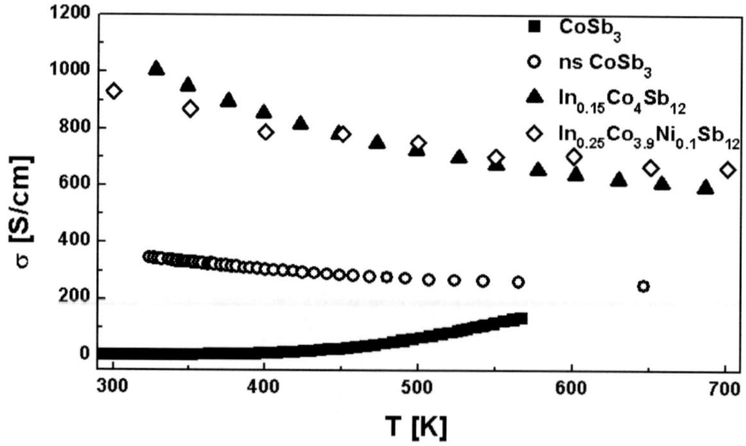

(c)

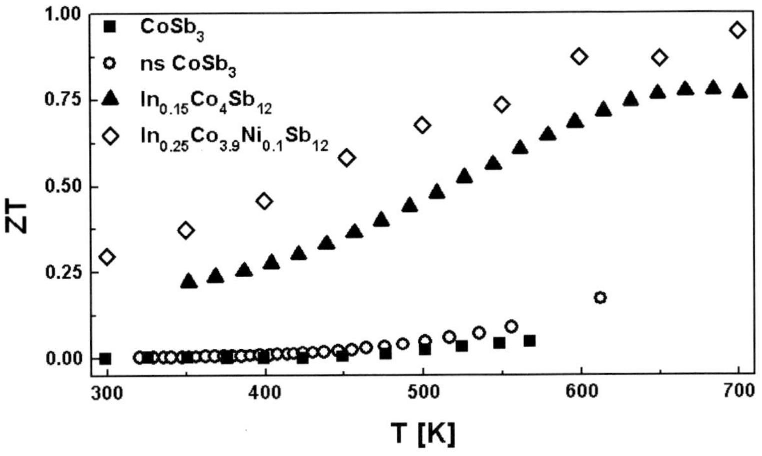

(d)

Figure 14.Temperature-dependence of thermoelectric characteristics: thermopower (a), thermal conductivity (b), electrical conductivity (c), and thermoelectric figure of merit (d) of different $CoSb_3$-based materials: $CoSb_3$-dense sinter as reference (authors' results), ns – nanostructured $CoSb_3$ [90], $In_{0.15}Co_4Sb_{12}$ [91], and $In_{0.25}Co_{3.9}Ni_{0.1}Sb_{12}$ [92].

Good oxidation resistance of specimens with the magnetron-sputtered layers is related to slow diffusion of reactants across the growing chromia scale. SiO_2 and Cr_2O_3 are thermodynamically more stable than cobalt and antimony oxides [47], so reduction of these oxides by the elements from the substrate is not possible. The in situ formed Cr–5Si/Cr_2O_3 bilayer system appeared an effective barrier to the inward diffusion of oxygen at 500 °C. Further improvement of the protective properties of the magnetron-sputtered Cr–5Si coating was achieved after an additional annealing of specimens under an inert atmosphere. According to the Cr–Sb phase diagram, the solubility of antimony in α-Cr is about 5 at.% at 600 °C [69]. The solubility limit of chromium in the crystal lattice of $Co_{1-x}Cr_xSb_3$ is about x = 0.01 at 700 °C [70].

The extent of chromium diffusion into the substrate after annealing was very small. A considerable difference was visible, however, after a 20-h oxidation at 600 °C (Fig.13). The diffusion-bonded layer had better adherence to the substrate and was practically impermeable for oxygen. An interdiffusion zone composed of (Cr,Co)Sb and $CoSb_2$, developed around the interface. The interdiffusion zone, although depleted in antimony, was much narrower than thickness loss due to oxidation of the unmodified specimens. Antimony, which otherwise would be lost by sublimation and oxidation, reacted with chromium under the magnetron-sputtered layer to yield a non-volatile compound. The in situ formed graded multilayer system Cr_2O_3/SiO_2–Cr/Si–(Cr,Co)Sb–$CoSb_2$ has potential to provide good protection for $CoSb_3$ in an oxidizing environment at elevated temperatures.

APPLICATIONS

Antimonides of group 13 elements, such as indium antimonide, aluminum antimonide and gallium antimonide, are intermetallic compounds having semiconductor properties and the highest room-temperature electron mobility. They are used in thermography, magnetoresistive devices, infrared detectors and transistors.

Alkali antimonides with a general formula M_3Sb, where M is either a single alkali metal or an alkali alloy are known for great photoemissivity and find application in the construction of photocathodes and photomultipliers. They are small direct and indirect gap semiconductors for device applications including detectors, infrared lasers and possibly also transport devices such as transistors [71].

Antimonide based compound semiconductors can be used in advanced microelectronics, optoelectronic devices, e.g. active-array space-based radars, satellite communications, ultra-high-speed and ultra-low power integrated circuits, portable mobile devices, gas monitoring, chemical detection, bio-medical diagnosis, drug analysis and other fields [72-79].

The narrow bandgap antimonide based compound semiconductors are considered as the first candidate materials for the construction of infrared photon detectors and integrated circuits with ultra-high speed and ultra-low power consumption and also as the important materials for middle and far infrared quantum cascade lasers and thermophotovoltaic cells suitable for medium and low temperature heat sources [80].

Skutterudites have been widely studied for their potential applications in thermoelectric power generation at intermediate temperatures from 600 K to 900 K (327-627 °C). The investigations are mainly focused on filling skutterudites in order to reduce their lattice thermal conductivity (λ_L) [81,82]. A great variety of metal atoms, e.g. rare earths, alkaline earths, and alkalis are introduced into the interstitial voids of the skutterudite structure. These filler atoms behave as independent oscillators, referred to as "rattlers", since their phonon modes interfere with normal phonon modes of the lattice and thus contribute to scattering of the low frequency (acoustic) phonons that are mainly responsible for heat conduction in solids. Filling of cobalt triantimonide is extensively studied because this compound exhibits high carrier mobility and large electron effective mass, the lowest thermal conductivity of binary skutterudites and is relatively inexpensive [83]. It has been found that double-filled skutterudites scatter broader spectra of acoustic phonons than the single-filled ones, and therefore have lower lattice thermal conductivity. Theoretical studies of the potential fillers for $CoSb_3$ have indicated that Ba and Yb appear to differ most in the resonant frequency [84]. Filling large interstitial voids of $CoSb_3$ structure with both Ba and Yb enhances phonon scattering more than any other combination of two fillers. Excellent thermoelectric properties have been achieved in a series of n-type $Ba_xYb_yCo_4Sb_{12}$ compounds (where $0.03 \leq x \leq 0.15$ and $0 \leq y \leq 0.12$). Highest ZT obtained in this system is 1.36 at about 800 K for $x = 0.08$ and $y = 0.09$.

Theoretically, significant improvements in thermoelectric properties can be attained by nanostructuring [85,86]. In particular, thermal conductivity can be decreased in partly- or all-nanostructured materials [87,88]. In spite of many attempts, spectacular results of practical importance have not been reported, so far. For illustration, the effects attained by compositional modifications of bulk $CoSb_3$ and by nanostructuring are shown in Figs.14a-d. The $CoSb_3$ dense sinter (reference) is characterized by a sharp and reversible n-p transition at about 460 K (187

°C), with large absolute values of thermopower of about 500 μV/K at room temperature and about 200 μV/K at 600 K (327 °C) [89].This peculiar behavior is related to the unusual band structure of $CoSb_3$ with heavy electrons in the conduction band and lightweight holes in the valence band [90]. The result is a p-type intrinsic semiconductor with a very low level of n-doping. The change of the major charge carriers from electrons to holes at about 460 K (187 °C) is consistent with earlier reports [90]. The In-filled and Ni-doped $CoSb_3$ materials exhibit small linear changes in thermopower, from -200 μV/K to -300 μV/K, in the temperature range 300 K – 700 K (27 – 427 °C) [91,92]. The nanosized $CoSb_3$ dense sinter has small positive thermopower, slightly increasing with temperature up to about 100 μV/K at 650 K (377 °C). Electrical conductivity of the chemically modified and nanostructured materials is distinctly higher and slightly decreases with temperature, while thermal conductivity is much lower, between 1 W/mK and 4 W/mK, compared with those of $CoSb_3$ reference, i.e. 11 W/mK – 7 W/mK over the temperature range 300 K – 600 K (27-327°C). Better picture of thermoelectric properties is given by the temperature-dependence of thermoeletric figure of merit (ZT) which combines thermopower (α), electrical conductivity (σ) and thermal conductivity (λ): $ZT = \alpha^2 \cdot \sigma \cdot \lambda^{-1} \cdot T$ (Fig 14d). The single-filled or double filled and doped $CoSb_3$ materials have much better characteristics compared with the dense sinter reference and the nanostructured $CoSb_3$. In the case of nanostructuring, additional grain-boundary effects are probably responsible for the unsatisfactory experimental results.

REFERENCES

[1] Ishida, K.; Nishizawa, T. *Bull. Alloy Phase Diagrams* 1990, 11, 243-248.

[2] Nolas, G. S.; Cohn, J. L.; Slack, G. A. *Phys. Rev. B* 1998, 58, 164-170.

[3] Anno, H.; Tashiro, H.; Notohara, Y.; Sakakibara, T.; Hatada, K.; Motoya, K.; Shimizu, H.; Matsubara, K. *Proc. 16th Int. Conf. Thermoelectrics* 1997, 343-346.

[4] Kawaharada, Y.; Kurosaki, K.; Uno, M.; Yamanaka, S. *J. Alloys Compd.* 2001, 315, 193-197.

[5] Kuznetsov, V. L.; Kuznetsova, L. A.; Rowe, D. M. *J. Phys.: Condens. Matter* 2003, 15, 5035-5048.

[6] Nakagawa, H.; Tanaka, H.; Kasama, A.; Anno, H.; Matsubara, K. *Proc. 16th Int. Conf. Thermoelectrics* 1997, 351-355.

[7] Zhou, Z.; Uher, C.; Jewell, A.; Caillat, T.; *Phys. Rev. B* 2005, 71, 235209- 235214.

[8] Sales, B. C.; Williams, R. K.; Martin, P. M. *Phys. Rev. B* 1997, 56, 15081-15089.

[9] Sales, B. C.; Mandrus, D.; Williams, R. K. *Science* 1996, 272, 1325-1328.

[10] Chen, B. X.; Xu, J. H.; Uher, C.; Morelli, D. T,; Meisner, G. P.; Fleurial, J. P.; Caillat, T., Borshchevsky, A. *Phys. Rev. B* 1997, 55, 1476–1480.

[11] Uher, C.; Hu, S. Q.; Yang, J. H. *Proc. 17th Int. Conf. Thermoelectrics* 1998, 306-309.

[12] Morelli, D. T.; Meisner, G. P. *J. Appl. Phys.* 1995, 77, 3777-3781.

[13] Benjamin J. S, *Met. Powder Rep.* 1990, 45, 122-127.

[14] Yang, J.; Chen, Y.; Peng, J.; Song, X.; Zhu, W.; Su, J.; Chen R. *J. Alloys Compd.* 2001, 375, 229–232.

[15] Peng, J.; Yang, J.; Song, X.; Chen, Y.; Bao, S.; Zhang, T. *Front. Mater. Sci. China* 2007, 1, 177-180.

[16] Liu, K.; Zhang, J.; Xiang, D. *J. Mater. Process. Technol.* 2007, 184, 257–260.
[17] Omori, M. *Mater. Sci. Eng.* 2000, A287, 183-188.
[18] Zhang, J. X.; Lu, Q. M.; Liu, K.G.; Zhang, L.; Zhou, M. L. *Mater. Lett.* 2004, 58, 1981– 1984.
[19] Liu, W. S.; Zhang, B. P.; Li, J. F.; Zhao, L. D. *J. Phys. D: Appl. Phys.* 2007, 40, 566–572.
[20] Liu, W. S.; Zhang, B. P.; Li, J. F.; Zhao, L. D. *J. Phys. D: Appl. Phys.* 2007, 40, 6784–6790.
[21] Liu, W. S.; Zhang, B. P.; Li, J. F.; Zhang, H. L.; Zhao, L. D. *J. Appl. Phys.* 2007, 102, 103717-1-103717-7.
[22] Godlewska, E.; Zawadzka, K.; Mars, K. unpublished results.
[23] Wang, M.; Zhang, Y.; Muhammed, M. *Nanost. Mater.* 1999, 12, 237-240.
[24] Bertini, L.; Billquist, K.; Christensen, M.; Gatti, C.; Holmgren, L.; Iversen, B.; Mueller, E.; Muhammed, M.; Noriega, G.; Palmqvist, A.; Platzek, D.; Rowe, D.M.; Saramat, A.; Stiewe, C.; Toprak, M.; Williams, S. G.; Zhang, Y. *Proc. 22nd Int. Conf. Thermoelectrics* 2003, 93-96.
[25] Toprak, M.; Zhang, Y.; Muhammed, M.; Zakhidov, A. A.; Baughman, R. H.; Khayrullin, I. *Proc. 18th Int. Conf. Thermoelectrics* 1999, 382-385.
[26] Bertini, L.; Stiewe, C.; Toprak, M.; Williams, S. G.; Platzek, D.; Mrotzek, A.; Zhang, Y.; Gatti, C.; Muller, E.; Muhammed, M.; Rowe, M.; *J. Appl. Phys.* 2003, 93, 438-447.
[27] Bertini, L.; Billquist, K.; Christensen, M.; Gatti, C.; Holmgren, L.; Iversen, B.; Mueller, E.; Muhammed, M.; Noriega, G.; Palmqvist, A.; Platzek, D.; Rowe, D. M.; Saramat, A.; Stiewe, C.; Toprak, M.; Williams, S. G.; Zhang, Y. *Proc. 22nd Int. Conf. Thermoelectrics* 2003, 48-51.
[28] Xie, J.; Zhao, X.; Mi, J. *J. Zhejiang Univ. SCI* 2004, 5, 1504-1508.
[29] Mi, J. L.; Zhao, X. B.;. Zhu, T. J.; Tu, J. P.; Cao, G. S. *J. Alloys Compd.* 2006, 417, 269–272.
[30] Ji, X.; He, J.; Alboni, P. N.; Tritt, T. M.; Kolis, J.W. *Mater. Res. Soc. Symp. Proc.* 2008, 1044, 1044-U01-02.
[31] Yang, L.; Hng, H. H.; Sun, T.; Cheng, H.; Ma, J. *Proc. 26th Int. Conf. Thermoelectrics* 2007, 187-188.
[32] Kumari, L.; Li, W.; Huang, J. Y.; Provencio, P. P. *Nanoscale Res. Lett.* 2010, 5, 1698–1705.
[33] Mandrus, D.; Migliori, T.; Darling, T. W.; Hundley, M. F.; Peterson, E. J.; Thompson, J. D. *Phys. Rev. B* 1995, 52, 4926-4931.
[34] Morelli, D. T.; Caillat, T.; Fleurial, J. P.; Borshchevsky, A.; Vandersande, J.; Chen, B.; Uher, C. *Phys. Rev. B* 1995, 51, 9622-9628.
[35] Caillat, T.; Borshevsky, A.; Fleurial, J. P. *J. Appl. Phys.* 1996, 80, 4442-4449.
[36] Akasaka, M.; Iida,T.; Sakuragi, G.; Furuyama, S.; Noda, M.; Matsui, S.; Ota, M.; Suzuki, H.; Sato, H.; Takanashi, Y.; Sakuragi, S. *J. Alloys Compd. 2005*, 386, 228-233.
[37] Arushanov, E.; Fess, K.; Kaefer, W.; Kloc, Ch.; Bucher, E. *Phys. Rev.B* 1997, 56, 1911-1917.
[38] Mania, R.; Mars, K.; Wojciechowski, K. *Bull. Pol. Acad. Sci.: Ceram.* 2001, 66, 975-980.
[39] Savchuk, V.; Boulouz, A.; Chakraborty, S.; Schumann, J.; Vinzelberg, H. *J. Appl. Phys.* 2002, 92, 5319-5326.

[40] Xiao, F.; Hangarter, C.; Yoo, B.; Rheem, Y.; Lee, K. H.; Myung, N.V. *Electrochim. Acta* 2008, 53, 8103–8117.

[41] Behnke, J. F.; Prieto, A. L.; Stacy, A.M.; Sands, T. *Proc. 18th Int. Conf. Thermoelectrics* 1999, 451-453.

[42] Cheng, H.; Hng, H. H.; Ma, J. *Solid State Phenom.* 2008, 136, 75-82.

[43] Caylor, J. C.; Stacey, A. M.; Bloom, B.; Gronsky, R.; Sands, T.; Fuller-Mora, W. W.; Ehrlich, A.; Song, D.; Chen, G. *Proc. 18th Int. Conf. Thermoelectrics* 1999, 656-661.

[44] Durand, H.-A.; Nishimoto, K.; Ito, K.; Kataoka I. *Appl. Surf. Sci.* 2000, 154–155, 387–392.

[45] Asryan, N. A.; Alikhanyan, A. S.; Nipan, G. D. *Inorg. Mater.* 2004, 40, 720-725.

[46] Orman, R. G.; Holland, D. *J. Solid State Chem.* 2007, 180, 2587-2596.

[47] Kubashevski, O.; Alcock, C. B.; Spencer, P. J. *Materials Thermochemistry,* 6th ed., Pergamon Press, Oxford: New York, NY, 1993.

[48] Swaminathan, K.; Sreedharan, O. *J. Mater. Sci.* 2001, 36, 1031-1037.

[49] Hara, R.; Inoue, S.; Kaibe, H.; Sano, S. *J. Alloys Compd.* 2003, 349, 297-301.

[50] Wojciechowski, K.; Leszczynski, J.; Gajerski, R. *Proc. 7th Eur. Workshop Thermoelectrics* 2002.

[51] Leszczynski, J. Malecki, A. Wojciechowski, K.T. *Proc. 5th Eur. Conf. Thermoelectrics* 2007, 50.

[52] Godlewska, E.; Zawadzka, K.; Adamczyk, A.; Mitoraj, M.; Mars, K. *Oxid. Met.* 2010, 74, 113-124.

[53] Sakamoto, J.; Gabriel, S.; Calliat, T.; Fleurial, J. Method of suppressing sublimation in advanced thermoelectric devices Patent No. US7480984 B1, 2009.

[54] Leszczynski, J., PhD Thesis, AGH University of Science and Technology: Krakow 2006, in Polish.

[55] Sakamoto, J.; Calliat, T.; Fleurial, J.; Snyder, G. Coating Thermoelectric Devices To Suppress Sublimation, *NASA Tech Briefs NPO-40040,* NASA's Jet Propulsion Laboratory: Pasadena, CA, 2007.

[56] Caillat, T.; Sakamoto, J.; Jewell, A.; Cheng, J.; Paik, J.; Gascoin, F.; Snyder, J.; Blair, R.; Huang, C. K.; Fleurial, J. P. *Advanced Radioisotope Power Systems; Technology development at JPL,* presented on 22nd Int. Conf. Thermoelectrics 2005.

[57] Hara, R.; Tomita, K.; Inoue, S.; Kaibe, H.; Miaukami, H. *Proc. 20th Int. Conf. of Thermoelectrics* 2001, 84-88.

[58] Dynys, F.; Nathal, M.; Nesbitt, J.; Opila, E.; Sayir, A. Sublimation Suppression Coatings Evaluated for Advanced Thermoelectric Materials, *Reaserch and technology,* Glenn Research Center: Cleveland, OH 44135–3191 NASA/TM—2007-214479, 2006.

[59] Sakamoto, J.; Fleurial, J. P.; Snyder, J.; Jones, S.; Caillat, T. *Aerogels for Thermal Insulation of Thermoelectric Devices,* NASA Tech Briefs NPO-40630, NASA's Jet Propulsion Laboratory: Pasadena, CA, 2006.

[60] Sakamoto, J.; Palk, J.; Jones, S.; Dong, W.; Caillat, T.; Fleurial, J. P.; *Improving thermoelectric technology performance and durability with aerogel,* presented on Sol-Gel Conf. 2005.

[61] Sakamoto, J.; Snyder, G.; Gabriel, S.; Calliat, T.; Fleurial, J. P.; Jones, S.; Palk J. *System and method for fabrication of high-efficiency durable thermoelectric devices* Patent Appl. No. US 2006/0157101 A1.

[62] Sakamoto, J.; Snyder, G.; Calliat, T.; Fleurial, J. P.; Jones, S.; Palk J. System and method for suppressing sublimation using opacified aerogel, Patent No. US 7461512 B2.

[63] Paik, J.; Sakamoto, J.; Jones, S. *Aerogel/Particle Composites for Thermoelectric Devices*, NASA Tech Briefs NPO-42031, NASA's Jet Propulsion Laboratory: Pasadena, CA, 2006.

[64] Tyliszczak – Halibozek, E., PhD Thesis, AGH University of Science and Technology: Krakow 2007, in Polish.

[65] Godlewska, E.; Zawadzka, K.; Gajerski, R.; Mitoraj, M.; Mars, K. *Proc. 11th Int. Conf. Exh. Eur. Cer. Soc.* 2009, 370-375.

[66] Godlewska, E.; Zawadzka, K.; Mars, K.; Nocun, M.; Opoka, A.; Wojciechowski, K.; *Ext. Abst. 1st Int. Conf. Mat. Energy* 2010, 554-556.

[67] Godlewska, E.; Mitoraj, M.; Mania, R.; Zimowski, S.; Kot, M. *Ext. Abst. 1st Int. Conf. Mat. Energy* 2010, 236-238.

[68] Salomonsen, G.; Norman, N.; Lonsjo, O.; Finstad, T. G.; *J. Phys.: Condens. Matter* 1989, 1, 7843-7850.

[69] Venkatraman, M.; Neumann, J. P. *Bull. Alloy Phase Diagrams* 1990, 11, 435-440.

[70] Yang, J.; Endres, M. G.; Meisner, G. P. *Phys. Rev. B* 2002, 66, 014436-1-014436-1.

[71] Glosser, R.; Estrera, J.P.; Bourree, L.E. Patent No.: US 6,992,441, B2 Jan. 31 2006.

[72] Bennett, B.R., Magno R., Boos, J.B. *Solid-State Electronics* 2005, 49, 12, 1875-1895.

[73] Razeghi, M. *The European Physical Journal Applied Physics* 2003, 23, 3, 149-205.

[74] Kroemer, A. H. *Physica E* 2004, 20, 3-4, 196-203.

[75] Biefeld, R.M. *Materials Science and Engineering R* 2002, 36, 4, 105-142.

[76] Wang, C.A. *Journal of Crystal Growth* 2004, 272, 1-4, 664-681.

[77] Rogalshi, A. *Infrared Physics & Technology* 2007, 50, 2-3, 240-252.

[78] Nesher, O.; Klipstein, P.C. *Opto-Electronics Review* 2006, 14, 1, 61-70.

[79] Mauk, M.G.; Andreev, V.M. *Semiconductor Science and Technology* 2003, 18, 5, S191-S201.

[80] Liu C.; Li, Y.; Zeng, Y. *Engineering* 2010, 2, 617-624.

[81] Uher, C. In *Recent Trends in Thermoelectric Materials Research I, Semiconductors and Semimetals*; Tritt, T.M.; Ed.; Academic Press, San Diego, 2001, Vol. 69, pp139-253.

[82] Slack, G.A.; Tsoukala, V.G. *J. Appl. Phys.* 1994, 76, 1665.

[83] Long, G.J.; Hermann, R.P.; Grandjean, F.; Alp, E.E.; Sturhahn, W.; Johnson, C.E.; Brown, D.E.; Leupold, O.; Rüffer, R. *Phys. Rev. B* 2005, 71, 140302 (R).

[84] Yang, J.; Zhang, W.; Bai, S.Q.; Mei, L; Chen, L.D. *Appl. Phys. Lett.* 2007, 90, 192111.

[85] Minnich, A.J.; Dresselhaus, M.S.; Ren, Z.F.; Chen, G. *Energy Environ. Sci.* 2009, 2, 466–479.

[86] Snyder, G.J.; Toberer, E.S. *Nature Materials* 2008, 7, 2, 106-114.

[87] Alboni, P.N.; Ji, X.; He, J.; Gothard, N.; Hubbard,J.; Tritt, T.M. *Journal of Electronic Materials* 2007, 36, 7, 711-715.

[88] Zhang, L.; Grytsiv, A.; Kerber, M.; Rogl, P.; Bauer, E.; Zehetbauer, M.J.; Wosik, J.; Nauer, G.E. *Journal of Alloys and Compounds* 2009, 481, 106–115.

[89] Godlewska, E.; Zawadzka, K; Mars K. *unpublished results.*

[90] Toprak, M.S.; Stiewe, C.; Platzek D.; Williams, S.; Bertini, R.; Muller, E.; Gatti, C.; Zhang, Y.; Rowe, M.; Muhammed, M. *Advanced Functinal Materials* 2004, 14, 12, 1189-1196.

[91] Sesselmanna, A.; Hassdorf, R.; Zastrow, S.; Müller, E. *Advances in Science and Technology* 2010, 74, 54-59.

[92] Jung J-Y; Park K-H; Kim I-H; Choi S-M; Seo, W-S *Journal of the Korean Physical Society* 2010, 57, 4, 773-777.

In: Antimony: Characteristics, Compounds and Applications ISBN 978-1-62100-598-8
Editor: Manijeh Razeghi

Chapter 12

ANTIMONY COMPLEXES WITH O-QUINONATO LIGANDS AND RELATED O,N-, O,N,O-, O,N,O,O-LIGANDS

Andrey I. Poddel'sky
G.A. Razuvaev Institute of Organometallic Chemistry of Russian Academy of Sciences, Nizhniy Novgorod, Russia

ABSTRACT

Antimony compounds attract the attention of scientific community due to their high activity in different reactions and the possibilities of wide use in a variety of chemical and biochemical processes, as well as in medical purposes. Coordination chemistry of antimony reveals a variety of bonding and structural types. Redox-active o-quinones and related ligands (o-iminoquinones, alpha-diimines etc.) are promising ligands in coordination chemistry due to the features of their molecular and electronic structure such as the ability to change oxidation state (from neutral to radical-anion and dianion forms) being bound to metal, the formation of strong chelate cycle that allows to realize different reactions with retention of such ligand in coordination sphere; radical-anionic forms of ligands are model objects in investigation of intramolecular magnetic interactions etc. In spite of the fact that coordination chemistry of transition and non-transition metals with o-quinonato type ligands is extensively studied and well reviewed, less attention was paid to the antimony compounds with o-quinonato type ligands. The present review deals with coordination chemistry of antimony compounds bearing o-quinonato and related ligands. The synthetic aspects, properties, structural features and some applications of quinonato antimony(III),(V) compounds and antimony complexes with N,O-, N,O,O'- and N,O,O',O''-donor ligands are described.

INTRODUCTION

Antimony compounds, particularly coordination and organometallic derivatives, are widely used in the various purposes. Antimony organometallics is a prospective and interesting topic due to the possibilities of its useful utilization for a wide range of chemical (components of polymerization catalysts; alkylation and arylation agents; intermediates and starting agents in different synthetic processes, etc.), biochemical and medical purposes (antioxidants, biocides, fungicides; antiparasitic, anticancer agents; etc.), the manufacture of semiconductors [1-11]. From the viewpoint of structure, coordination and organometallic compounds of antimony reveal a diversity of bonding and structural types [1, 2, 12, 13].

The coordination chemistry of metals with redox-active ligands on the basis of o-quinones and their analogs has developed in an independent direction of scientific research for the last 30-40 years. A number of interesting and unusual properties was discovered in this field. Among them are the metallotropy and elementotropy [14], the unique sensitivity of EPR spectral parameters of o-semiquinonato metal complexes to the nature and solvating abilities of solvent [15]; the intramolecular redox-isomerism and the phenomenon of photo-/thermomechanical effect (effect of reversible bending of crystals under the influence of visible radiation or heat) [16-18]; the intramolecular electron transfer induced by a replacement of neutral ligands [19]; the unusual magnetic properties caused by different types of magnetic interactions "metal-ligand" and "ligand-ligand" [20-22]. o-Quinonato, o-iminoquinonato and alpha-diimine complexes can act as perspective catalysts in various processes of the oxidations modeling a number of natural oxidizers [23].

Coordination compounds of o-quinones and their analogues demonstrate a diverse structural and electronic composition that is connected with features of such ligands. (i) o-Quinones, o-iminoquinones etc., are redox-active ligands and their oxidation (reduction) level can be changed in a metal coordination sphere from neutral o-quinones (Q) to radical-anions o-semiquinones (SQ) and to dianion catecholate (Cat). This fact permits the possibility of different combinations of metal's and ligand's electronic states. (ii) The application of sterical hindrances in o-quinones, o-iminoquinones allows to modify structural types from polymeric, oligomeric to monomeric up to unsaturated structures. (iii) Ligands of such type are capable to form firm metallocycles remaining at variety of chemical transformations in coordination sphere of metal. (iv) The application of polydentate o-quinonato ligands expands essentially a range of redox states, and also change effectively a sterical hindrances and structural rigidity of complexes.

Compounds of antimony bearing redox-active o-quinonato type ligands are prospective from the viewpoint of fundamental and applied chemistry. In the present chapter the synthesis, structure and some properties of antimony(V,III) compounds based on o-quinones and related di- and polydentate ligands are reviewed.

1. Catecholates of Antimony(V)

These compounds are the most widely presented class in a coordination chemistry of antimony compounds with o-quinonato type ligands. The main synthetic methods for the preparation of such complexes are the following: the interaction of Ar_3SbX_2 compounds (where X = halogen) with catechol derivatives in the presence of base (or without the latter); the oxidation of metallic antimony with different o-quinones; one-stage method - the reaction of triaryl/trialkylstibines with catechols and hydroperoxides; the oxidative addition of o-quinones to low-valent antimony compounds. Each of the presented methods has the advantages, restrictions and lacks. Below we will consider the synthesis and properties of antimony(V) catecholates in more details.

The oxidation of 2-organo- and 2-halogeno-1,3,2-dioxastibolanes with tetrachloro-o-quinone Cl_4Q in ether solution gave catecholato derivatives of antimony(V) (Scheme 1) [24].

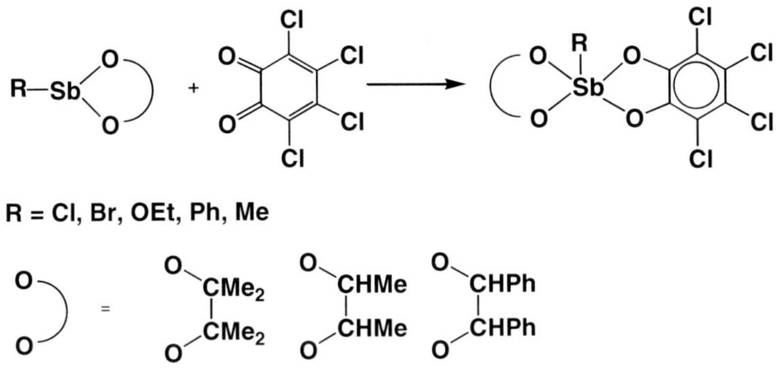

Scheme 1.

In the case of 2-tert-butyl-1,3,2-dioxastibolane, authors have reported about the formation of rather different compound - a product of a loss of tert-butyl group (see part 3 of this chapter) [24]. The reaction rate depends on the nature of substituent R and increases with a decreasing electronegativity of R (Cl,Br<OEt<Ph<Me<tBu).

Methyldialkoxystibines $MeSb(OR)_2$, where R is methyl, ethyl or isopropyl, are able to react with o-quinones (tetrachloro-o-quinone Cl_4Q and unsubstituted o-benzoquinone) producing the similar catecholate complexes (Scheme 2). Such complexes were found to be sensitive to moisture and undergo hydrolysis giving alcohols and derivatives of antimony(III) (see part 3 of this chapter).

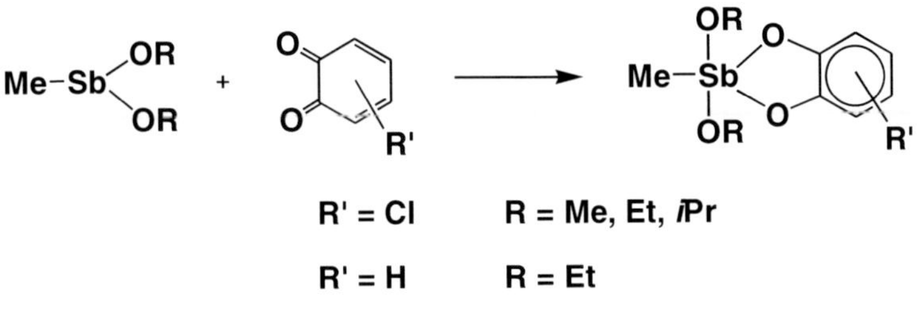

Scheme 2.

These reactions allow to prepare aimed catecholates in good yields (70-90 %). Worthy to note that this method of oxidative addition was applied to synthesize cyclic phosphoranes for the first time [25] and later it was expanded on organoantimonials [24].

For the first time, the oxidation of V groups elements by o-quinones was performed by Prokof'ev et al. [26]. The hexacoordinate mixed-ligand o-semiquinonato-catecholato complexes $(3,6\text{-DBCat})_2M(3,6\text{-DBSQ})$ (M = Sb, As) were synthesized. However, this reaction performed with an excess of amalgam of the corresponding element (As, Sb or Bi) gives complexes of metal(III) (See part 3 of this chapter).

Tuck et al. have investigated the oxidation of elemental antimony by tetrachloro/tetrabromo-o-benzoquinone (Hal_4Q) and 3,5-di-tert-butyl-o-benzoquinone (3,5-DBBQ) [27]. This reaction does not proceeds in refluxing toluene for any of three o-quinones. However reaction of Hal_4Q in diethyl ether gives products of the following formula $(Hal_4Cat)_{2.5}Sb{\cdot}nEt_2O$, (Hal = Cl, n = 1.5; Hal = Br, n = 1). The IR and NMR data of these products were consistent with data on catecholates of some other elements [28]. Quinone 3,5-DBBQ does not oxidize elemental antimony in diethyl ether. Authors were not successful to grow X-ray suitable crystals but they have supposed the ionic structure of complexes formed (Scheme 3) or the polymeric composition with bridging catecholato ligands.

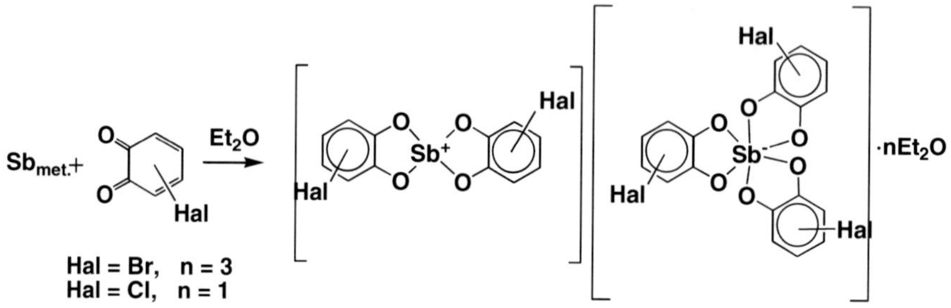

Scheme 3.

Sowerby et al. [29] have prepared catecholato complex of triphenylantimony(V) using the reaction between triphenylantimony dichloride and catechol in the presence of gaseous ammonia. Complex formed was found to be partially hydrated and contains both penta- and hexacoordinated species (Scheme 4).

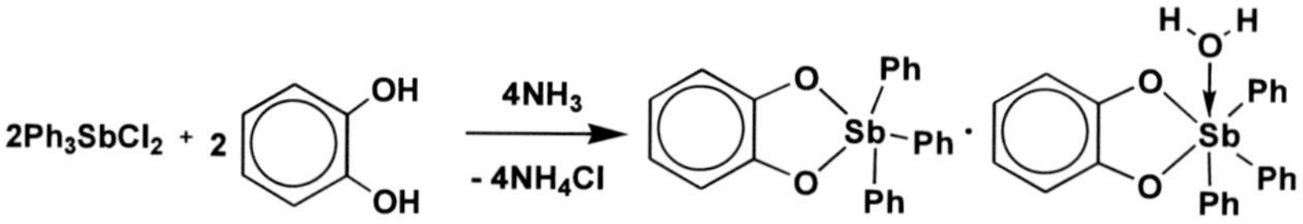

Scheme 4.

X-ray data have shown that the first molecule - pentacoordinate $(Cat)SbPh_3$ - has a distorted square pyramidal environment with catecholato ligand in basal plane while the second molecule has the distorted octahedral geometry completed by a weakly coordinated water molecule (the distance Sb-O(water) is 2.512(4) Å). This complex decomposes slowly in solution [29].

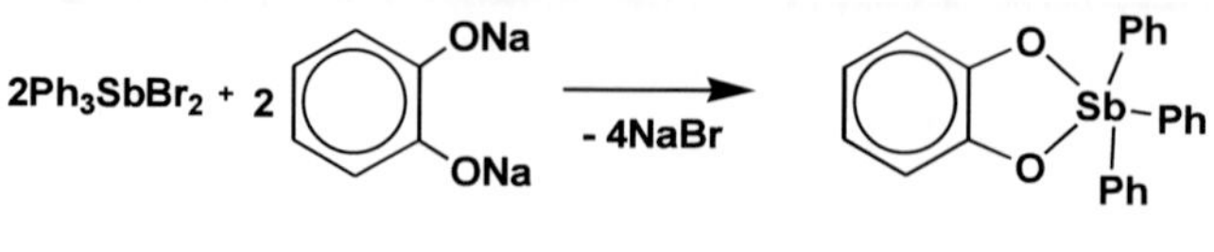

Scheme 5.

The pure pentacoordinated compound (Cat)SbPh_3 without water was prepared by the exchange reaction of triphenylantimony dibromide with disodium catecholate (Scheme 5) [30].

Holms et al. [31] have synthesized two related catecholates - (Cl_4Cat)SbPh_3 and (Cl_4Cat)SbPh(biphen) - by an oxidative addition of tetrachloro-o-benzoquinone to the corresponding stibines (Scheme 6).

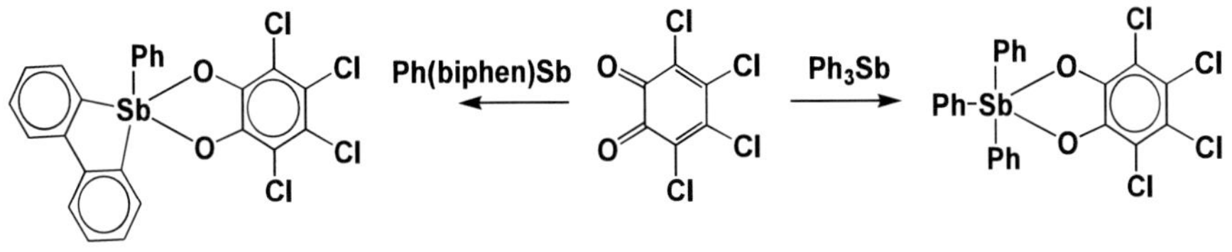

Scheme 6.

X-ray analysis of these complexes has shown different geometry of complexes: (Cl_4Cat)SbPh_3 was found to be of trigonal bipyramidal geometry with one of catecholato oxygens and two carbons of phenyl groups in the base of bipyramid, while in (Cl_4Cat)SbPh(biphen) the central antimony atom adopts a distorted square pyramidal geometry and catecholato ligand lies in the basal plane. The latter structure is less usual for the chemistry of antimony than the trigonal bipyramidal geometry.

Six years later Tuck [27] reported the synthesis of the same catecholates of formula (Hal_4Cat)SbPh_3 (Hal = Cl, Br) and catecholate (3,5-di-tert-butyl-catecholato)triphenylantimony(V), (3,5-DBCat)SbPh_3 via the identical reaction of oxidative addition of the corresponding o-quinone to triphenylstibine. Such interactions proceed easily in toluene solution.

Sowerby et al. [32] have applied the oxidative addition method for the preparation of catecholato derivatives of phenylantimony(III) dichloride. Hexacoordinate octahedral complexes (Cl_4Cat)SbPhCl_2·THF and (Cl_4Cat)SbPhCl_2·Et_2O were obtained in the corresponding donor solvents (Scheme 7) while all attempts to isolate pure products from reactions in nonpolar media failed. However, the similar pentacoordinate complex of 3,5-DBCat was successfully synthesized in a reaction of PhSbCl_2 with 3,5-DBBQ in dichloromethane.

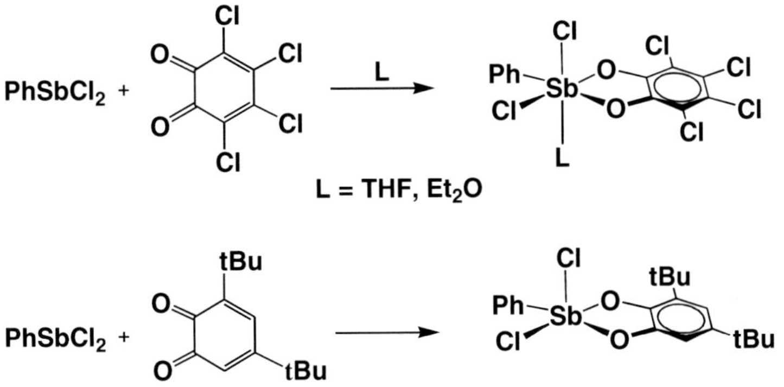

Scheme 7.

Authors [32] also made the attempts to synthesize catecholates of the type (Cat)SbPhHal$_2$ by the oxidation of (Cat)SbPh with bromine but they led to the mixtures of unidentified products.

An interesting catecholate $[Et_3NH]^+$ $[(naphtCat)SbPh_3Cl]^-$ containing complex anion with redox-active ligand was prepared by the metathesis reaction between triphenylantimony(V) dichloride and naphthalenediol-2,3 with triethylamine in the molar ratio 1:1:2 (Scheme 8) [31].

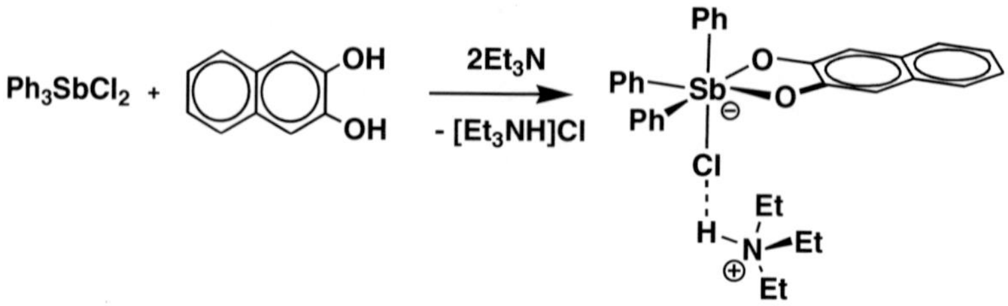

Scheme 8.

Authors proposed two ways for a formation of the ionic product $[Et_3NH]^+$ $[(naphtCat)SbPh_3Cl]^-$ (Scheme 9). The first was a condensation reaction of triphenylantimony(V) dichloride with naphthalenediol-2,3 with the subsequent removal of proton by triethylamine and final cyclization. Alternative mechanism of this reaction assumed the initial formation of a pentacoordinated complex of antimony(V) with the subsequent binding of a halogenide anion.

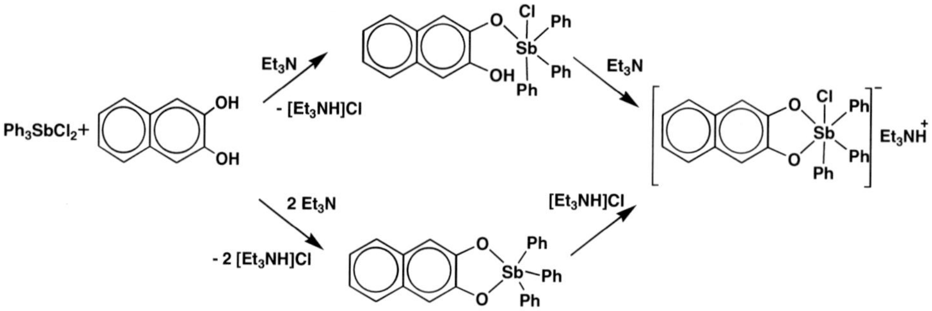

Scheme 9.

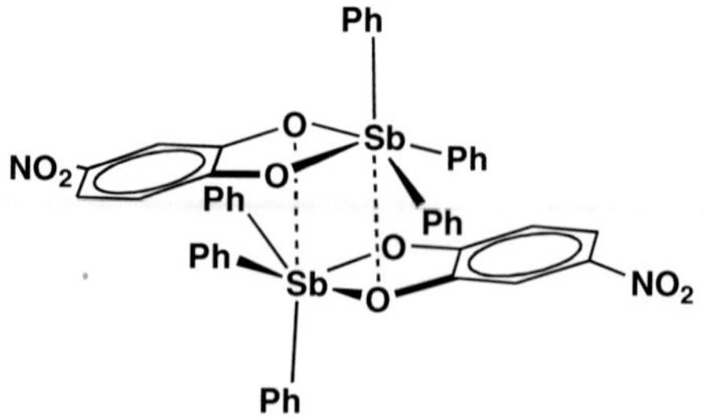

Scheme 10.

The analogous exchange reaction of triphenylantimony(V) dichloride with 4-nitrocatechol was applied by Holms et al. [33] for a preparation of complex (4-nitrocatecholato)triphenylantimony(V). The geometry of complex is close to the distorted trigonal bipyramidal but pentacoordinate environment of central antimony atom is completed by the additional weak contact between one oxygen in catecholate ligand of neighboring molecule and the central antimony (intermolecular Sb-O distance is 3.341(3) Å) (Scheme 10).

Worthy of note that the synthetic method to triarylantimony(V) catecholates starting from catechols and Ar_3SbHal_2 seems to be quite laborious and the preparative yields of aimed products rarely exceed 50 %.

Different complexes of phosphorus(V), antimony(V) and bismuth(V) can be prepared by the one-stage oxidation method - the oxidation of organoelement compounds R_3E (E = P, Sb, Bi) by the hydroperoxides (with a formation of adducts $R_3E(OH)(OR')$) followed by the reaction of nucleophilic substitution with substrates containing an active hydrogen [34-36]. This one-pot synthesis was successfully applied to synthesize antimony(V) catecholates. A number of hexacoordinate antimony(V) catecholates with additional neutral coordinated ligand was obtained (Scheme 11) [36].

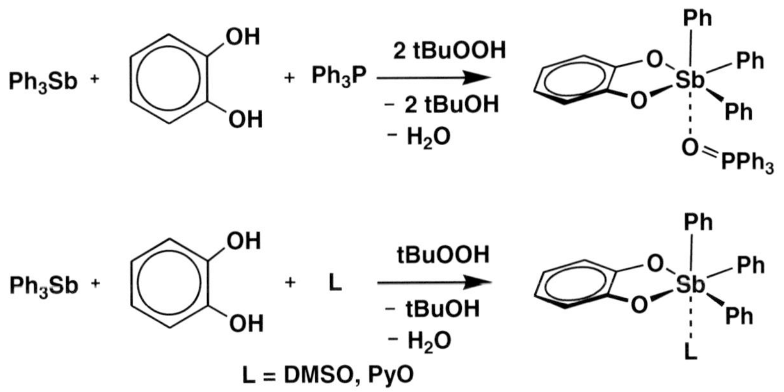

Scheme 11.

It is interesting that authors have used triphenylphosphine as a neutral ligand in the first reaction in Scheme 11, however the synthesized complex contains triphenylphosphine oxide. In the case of reaction of PPh_3 with catechol in the presence of tert-butyl hydroperoxide (i.e. without Ph_3Sb) the resulting product is a complex of triphenylphosphine oxide Ph_3PO with catechol via hydrogen bonding [36].

Analogous catecholates of trialkylantimony(V) $(Cat)SbR_3L$ (R = Me, L = DMSO, PyO, Py, Ph_3PO; R = Et, L = Ph_3PO) were also synthesized [37].

According to X-ray data, complexes have distorted octahedral geometry. The basal plane is occupied by bidentate catecholate ligand and two phenyl or methyl groups at antimony atom, additional neutral donor ligand occupies an apical site. Such complexes may be considered as products of the coordination of ligand L to square pyramidal species of $(Cat)SbR_3$ type. The distances Sb...O(N) in complexes vary in the range of 2.266(5) - 2.595(3) Å. Authors have found some consistency between the length of donor-acceptor Sb...O(N) bond and ionization potential of donor molecules; they have concluded that the donor-acceptor Sb...O(N) bond distance depends first of all on the electron factors (donor-acceptor properties of substituents at antimony atom).

A combination of hydroperoxyl group and hydroxyl (as a source of active hydrogen) in one molecule - 2-hydroxy-3,4,6-tri-isopropyl-4-hydroperoxy-cyclohexa-2,5-dienone allowed to obtain unusual complex (3,4,6-tri-isopropyl-5-hydroxycatecholato)triphenylantimony(V) ($iPr_3(OH)Cat)SbPh_3$ [38]. The reaction proceeds through the formation of unstable adduct A (Scheme 12) which is analogue of intermediates $R_3E(OH)(OR')$ [36]. Adduct A looses water to form complex derivative which transforms to derivative of 2,2,2-triphenyl-1,3,2-benzodioxastibol B during one hour. The rearrangement of B gives catecholate ($iPr_3(OH)Cat)SbPh_3$. X-ray suitable crystals were grown from methanol to show hexacoordinate complex with methanol solvate ($iPr_3(OH)Cat)SbPh_3\cdot MeOH$.

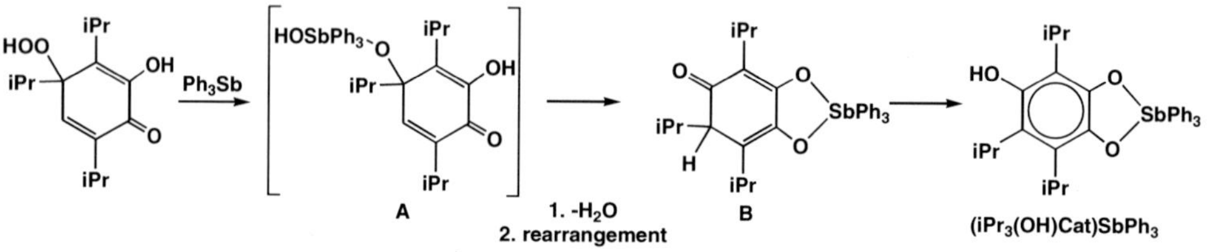

Scheme 12.

Noteworthy, catecholate ($iPr_3(OH)Cat)SbPh_3$ is also unstable and decomposes slowly (during 3-4 weeks in $CDCl_3$ or benzene-d_6) into triphenylstibine and 2-hydroxy-3,5,6-triisopropyl-1,4-benzoquinone (Scheme 13).

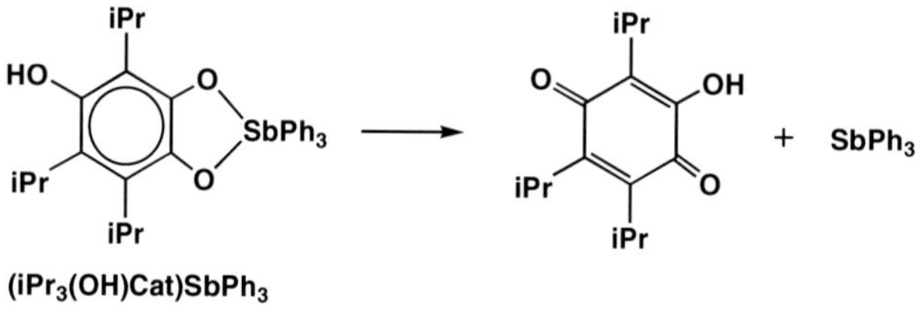

Scheme 13.

A related catecholate ($iPr_3Cat)SbPh_3$ (without hydroxyl-group in Cat ligand) was synthesized by oxidation of triphenylstibine with 3,4,6-triisopropyl-1,2-benzoquinone and was found to be stable for a long time both in solid and in solution [38]. These two complexes ($iPr_3(OH)Cat)SbPh_3$ and ($iPr_3Cat)SbPh_3$ demonstrate rather different stability in the presence of air. Catecholate ($iPr_3Cat)SbPh_3$ is air-stable while ($iPr_3(OH)Cat)SbPh_3$ can be easily oxidized with air oxygen resulting in the formation of different products depending on the oxidation conditions. The slow addition of oxygen to a solution of ($iPr_3(OH)Cat)SbPh_3$ allows to obtain a binuclear derivative containing μ-oxo group (Scheme 14, top) while the rapid addition of an excessing oxygen gives mononuclear hepta-coordinated derivative (Scheme 14, bottom). In both cases chelating ligand presents in the form of deprotonated 2-hydroxy-para-quinone.

So, the presence of hydroxyl group in catecholato ligand reduces the air-resistance of the corresponding complex. The reaction of oxidation by oxygen in this case is not reversible in contrast to the oxidative addition of molecular oxygen to some substituted catecholates and o-amidophenolates of triphenylantimony(V) (see bellow).

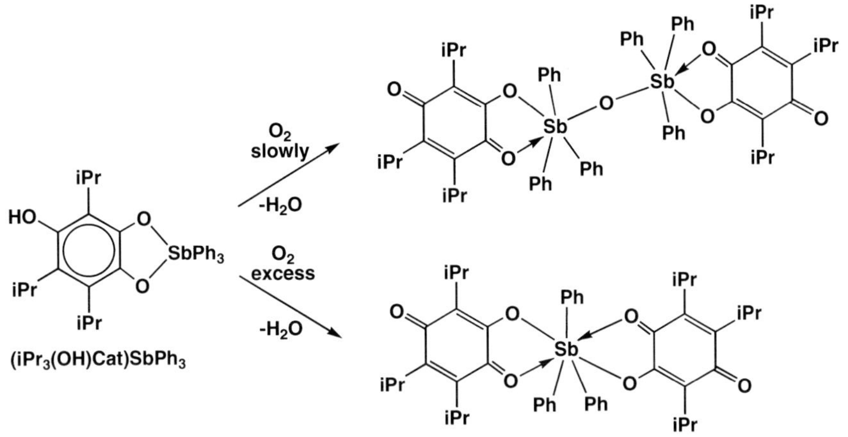

Scheme 14.

In general, the oxidative addition of substituted o-benzoquinones to organometallic derivatives of antimony(III) seems to be the most favorable for preparation of antimony(V) catecholates because in the most cases this reaction leads to the aimed product without by-products and allows to obtain considerable amounts of compound in one synthesis. Historically, the exchange reaction between transition/non-transition metal halide and catecholato or o-semiquinonato salt of alkali metal or thallium(I) is one of the most widely used synthetic method in the coordination chemistry of catecholates and o-semiquinonates. This way is also applicable for triphenylantimony(V) catecholates: f.e. dithallium(I) 3,6-di-tert-butyl-catecholate or dithallium(I) perchloroxanthrenecatecholate react with Ph_3SbCl_2 yielding catecholates (3,6-DBCat)$SbPh_3$ or ($^{ox}Cat_{Cl}$)$SbPh_3$, respectively (Scheme 15) [39]. However, in the case of the reaction between Ph_3Sb and the corresponding o-benzoquinone (Scheme 15) the synthesis was less laborious and yields were higher.

The oxidative addition of o-quinones to triphenylantimony was applied for the synthesis of the row of sterically hindered catecholato complexes (Scheme 16) containing different electron-donor and electron-acceptor substituents in 3,5-(3,6-)di-tert-butyl-catecholato ligand [40-43].

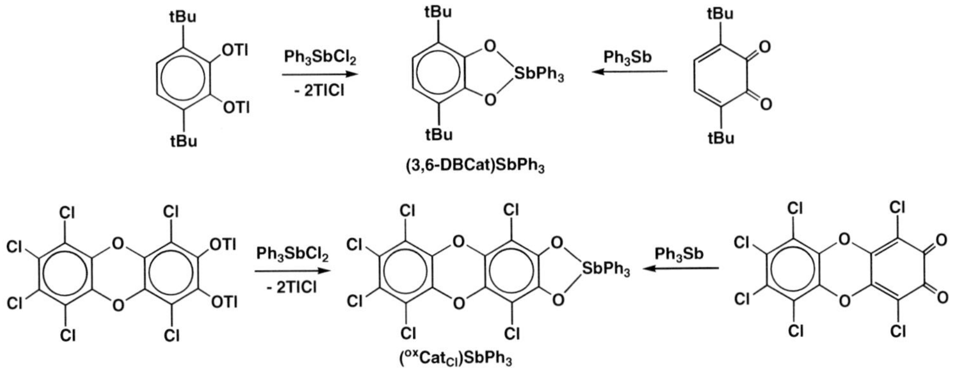

Scheme 15.

Molecular structures of pentacoordinate catecholates (3,6-DBCat)$SbPh_3$ or ($^{ox}Cat_{Cl}$)$SbPh_3$ in crystal state were confirmed by X-ray analysis [39]. Catecholate (3,6-DBCat)$SbPh_3$ has a distorted square pyramidal geometry with 3,6-DBCat ligand in the pyramid base while the geometry of ($^{ox}Cat_{Cl}$)$SbPh_3$ is distorted from square pyramidal to trigonal bipyramidal (Scheme 17, top).

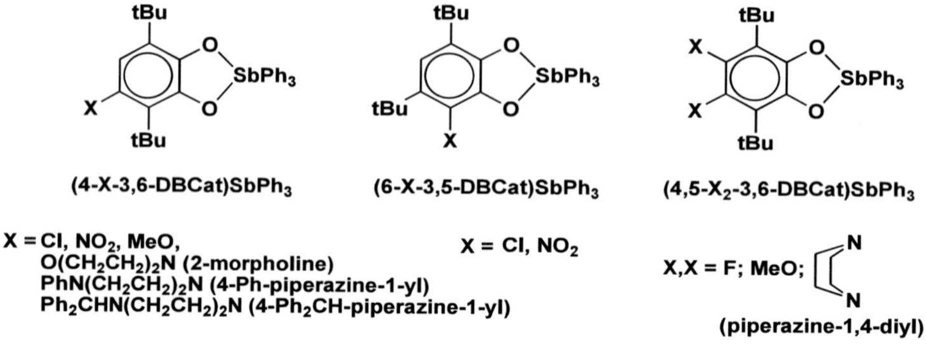

Scheme 16.

We should to note that weak intermolecular interactions take place in structures of these complexes (Scheme 17, bottom): the coordination of one chlorine atom in $^{ox}Cat_{Cl}$ ligand to antimony atom of adjacent molecule with Sb…Cl distance 3.966(5) Å in ($^{ox}Cat_{Cl}$)$SbPh_3$ [39] and weak contact between central antimony atoms and carbons of Ph rings in adjacent molecules with Sb…C distances 4.120(3) and 4.084(4) Å in (3,6-DBCat)$SbPh_3$ [44]. The sums of van der Waals radii of couples Sb, Cl and Sb, C are 4.0 and 3.9 Å, respectively. So, one can conclude that complexes tend to complete the coordination sphere of antimony atom to be hexacoordinate in solid state.

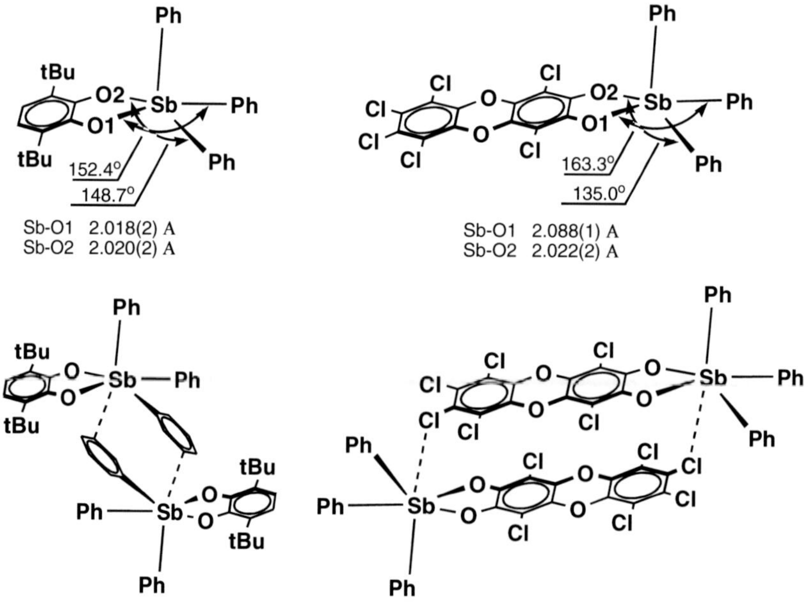

Scheme 17.

Some other catecholates were crystallized from donor coordinating solvents (methanol, acetonitrile) and they are hexacoordinate in solid state; the sixth (apical) position is occupied by a solvent molecule (f.e. (4-MeO-3,6-DBCat)$SbPh_3$·CH_3OH, (4,5-$(MeO)_2$-3,6-DBCat)$SbPh_3$·CH_3CN, (4,5-(piperazine-1,4-diyl)-3,6-DBCat)$SbPh_3$·CH_3OH) [40, 41, 43].

A complex analysis of the molecular structures of antimony(V) and antimony(III) catecholates in crystals was performed in work [44] in order to reveal the criteria determinating the tendency of complexes to the dimerization or the coordination of additional ligand (nucleophile).

One of the main features of redox-active ligands is the ability to change the oxidation state in complex derived [20-22]. Such ability of catecholato ligands allows to extend the limits of redox transitions for non-transition metal complexes and obtain new chemical properties of complexes. A few years ago the reversible binding of molecular oxygen by some triphenylantimony(V) catecholates and o-amidophenolates was found (see part 4 of this chapter) [40, 41]. Catecholates with electron donor groups (f.e. methoxy-substituted 3,6-di-tert-butyl-catecholates) are able to bind molecular oxygen with a formation of cyclic spiroendoperoxides (Scheme 18).

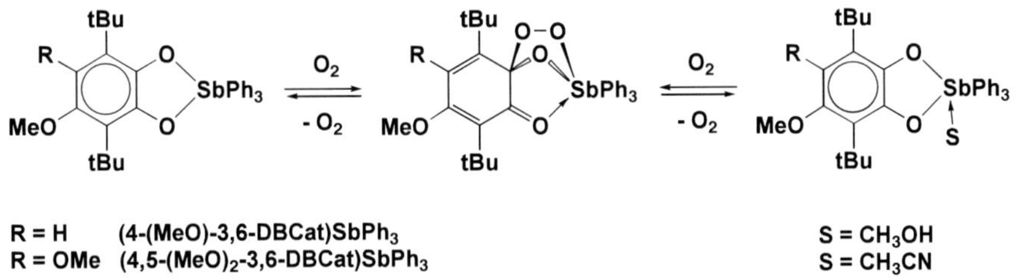

Scheme 18.

The reaction of triphenylantimony catecholates with molecular oxygen proceeds at ambient temperature with a formation of penta-membered trioxastibolane cycle. In this reaction, catecholate ligand looses the aromaticity, one of carbons of C-O group in initial catecholato ligand becomes sp^3-hybridized, another C-O bond get double character C=O but keeps the coordination to antimony atom.

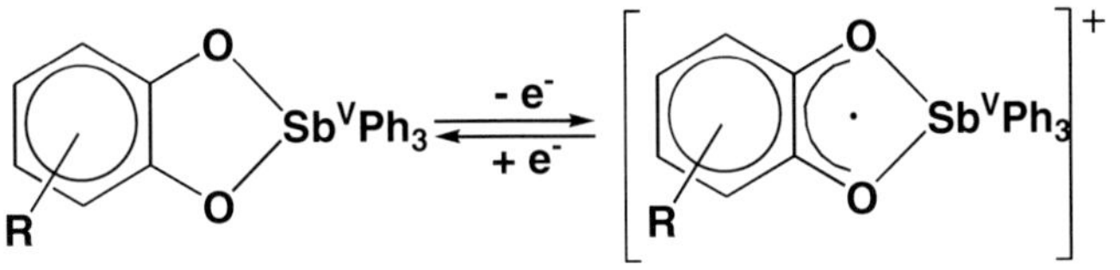

Scheme 19.

According to the proposed mechanism, one of key stages is the one-electron oxidation of dianionic catecholato ligand to anion-radical o-semiquinolate by molecular oxygen. In this case, the reactivity should depend on the first oxidation potential of catecholate (Scheme 19). The insertion of different substituents in catecholate ligand leads to the shift of redox potential of the transition "catecholate-semiquinolate" for ligand to the electropositive or electronegative region resulting in the facilitation or prevention of its oxidation. The oxidation potential of the first redox transition varies in the range from 0.65 V for catecholate (4,5-$(MeO)_2$-3,6-DBCat)$SbPh_3$ to 0.89 V for (3,6-DBCat)$SbPh_3$ and 1.16 V for (4-NO_2-3,6-DBCat)$SbPh_3$ (data are given *vs.* Ag/AgCl/$KCl_{(sat.)}$) [42, 43, 45].

It was found that triphenylantimony(V) catecholates with different donor/acceptor groups in Cat ligand have different reactivity toward O_2. Complexes with acceptor groups (f.e. haloids, NO_2) don't react with O_2 [42,45]. The insertion of weak electron-donor groups (like morpholine, piperazine, see Scheme 16) in catecholato ligand results in the appearance of reaction between the corresponding triphenylantimony(V) catecholate and O_2, the conversion degree varies in the range of 25-37 % compared with more than 90% for (4-MeO-3,6-DBCat)$SbPh_3$ or (4,5-$(MeO)_2$-3,6-DBCat)$SbPh_3$ [41].

Catecholates with strong donor groups (f.e. (4-MeO-3,6-DBCat)$SbPh_3$) forms the spiroendoperoxides where peroxo-fragment -O-O- is para-positioned towards donor group in catecholato ligand (Scheme 20), while for catecholates with weak donor groups the formation of para- and meta-positioned isomers was observed [43].

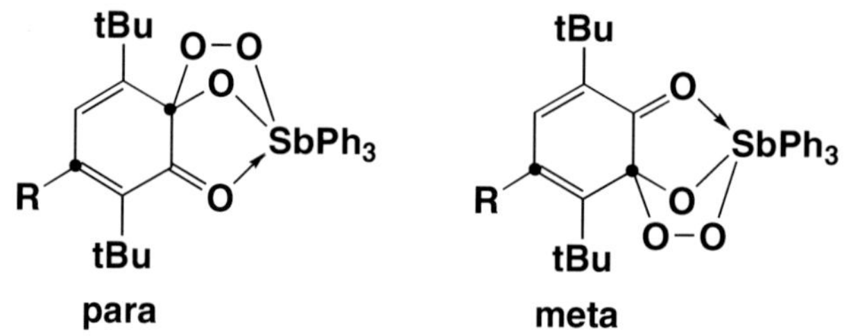

Scheme 20.

The reaction of molecular oxygen binding is reversible and the elimination of O_2 (under the heating or removing of air by the degassing of spiroendoperoxide solution) leads to the restoring of initial catecholates.

Thus, the variation of donor-acceptor properties in quinonato ligand allows to change the redox potential of derived catecholato antimony complex in a wide region (0.6-1.2 V *vs.* Ag/AgCl/$KCl_{(sat.)}$) [42, 43, 45]. At the same time, the more precision variation (tweaking) of redox potential of antimony(V) catecholato complexes is possible via the variation of substituents at antimony atom. In the following row (3,6-DBCat)Sb(*p*-F-Ph)$_3$, (3,6-DBCat)$SbPh_3$, (3,6-DBCat)Sb(pTol)$_3$ the first oxidation potentials are 0.904, 0.888, 0.838 V, respectively (in CH_2Cl_2, *vs.* Ag/AgCl/$KCl_{(sat.)}$) [46].

The reactions of stibines with o-quinones - derivatives of condensed polyaromatics - was also investigated. Tuck et al. [27] have reported that triphenylstibine reacts with anthraquinone in toluene to give a yellow colored solution typical for catecholato complexes but product undergoes some rearrangements accompanied by color change to green etc. and no individual products were collected. The reaction of phenanthrenequinone-9,10 with triphenylstibine gives very air-sensitive (phenanthrene-9,10-diolato)triphenylantimony(V) (PhenCat)$SbPh_3$ [41] which is green colored in toluene solution and in solid. This complex was found to exist in equilibrium with ionic complex $[Ph_4Sb]^+$ $[(PhenCat)_2SbPh_2]^-$ in degassed toluene solution (Scheme 21).

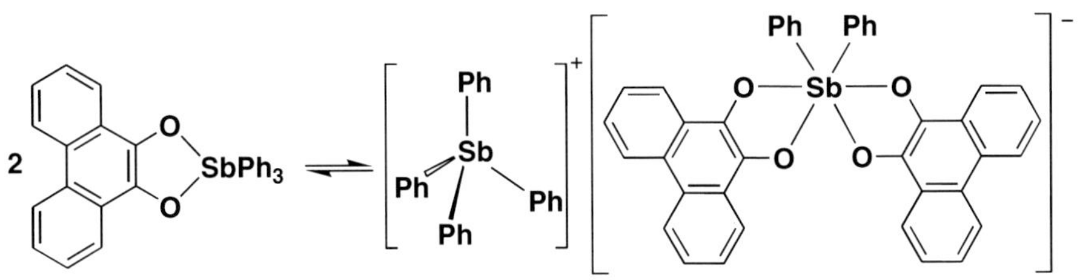

(PhenCat)SbPh$_3$ [Ph$_4$Sb]$^+$[(PhenCat)$_2$SbPh$_2$]$^-$

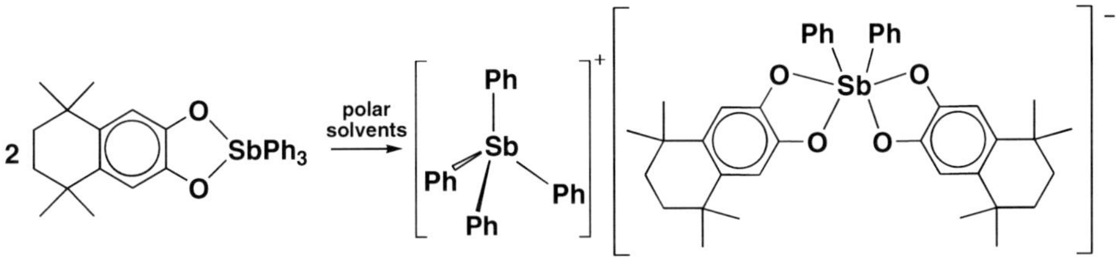

(4,5-Cat)SbPh$_3$ [Ph$_4$Sb]$^+$[(4,5-Cat)$_2$SbPh$_2$]$^-$

Scheme 21.

In complex anion [(PhenCat)$_2$SbPh$_2$]$^-$ phenyl groups occupy cis-positions and Sb-O distances with phenanthrene-9,10-diolate ligands are different (2.030(3), 2.072(3) Å for one ligand and 2.031(3), 2.069(3) Å for another ligand) reflecting trans-effect of phenyl groups.

The same rearrangement takes place in the case of (4,5-(1,1,4,4-tetramethyl-butane-1,4-diyl)-catecholato)triphenylantimony(V): complex (4,5-Cat)SbPh$_3$ transforms to ionic complex [Ph$_4$Sb]$^+$ [(4,5-Cat)$_2$SbPh$_2$]$^-$ containing diphenyl-bis- [4,5-(1,1,4,4-tetramethyl-butane-1,4-diyl)-catecholato]antimonate anion (Scheme 21) [47].

The formation of ionic complexes with [R$_4$Sb]$^+$ cations and antimonate anions is known [48-50]. F.e. triphenylantimony dichloride reacts with silver oxalate in toluene to yield anionic complex [Ph$_4$Sb$^+$] [(ox)$_2$SbPh$_2$]$^-$ (where ox is $C_2O_4^{2-}$) [48].

Unlike (PhenCat)SbPh$_3$, the transformation of (4,5-Cat)SbPh$_3$ to its ionic derivative [Ph$_4$Sb]$^+$ [(4,5-Cat)$_2$SbPh$_2$]$^-$ takes place in polar solvents (chloroform, acetone) and is not reversible [47]. The ionic complex [Ph$_4$Sb]$^+$ [(4,5-Cat)$_2$SbPh$_2$]$^-$ was electrochemically oxidized to a neutral (4,5-Cat)(4,5-SQ)SbPh$_2$ and cationic [(4,5-SQ)$_2$SbPh$_2$]$^+$ complexes.

Worthy of note, if catecholate (4,5-Cat)SbPh$_3$ is air-stable both in solid and in solution, phenanthrenecatecholate (PhenCat)SbPh$_3$ reacts with molecular oxygen in the same manner as triphenylantimony(V) catecholates with donor groups: a reversible molecular binding with a formation of spiroendoperoxide takes place (Scheme 22) [41].

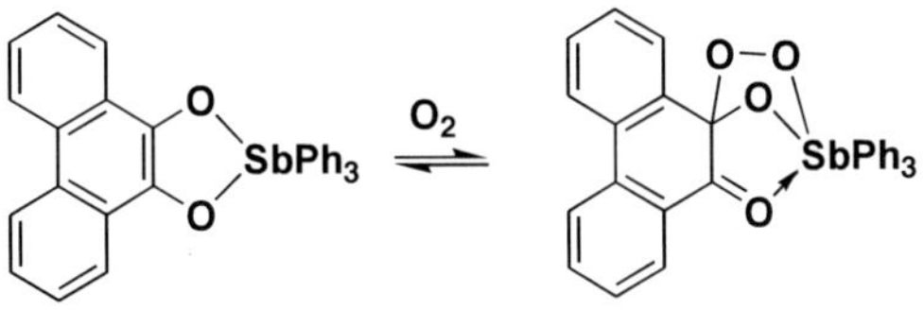

(PhenCat)SbPh$_3$

Scheme 22.

Triphenylantimony(V) dichloride reacts with quercetin Que(OH)$_3$ in the presence of Et_3N in toluene producing catecholato complexes (Que(OH)$_3$Cat)SbPh$_3$·H_2O and (Que(OH)$_2$(SbPh$_3$Cl)Cat)SbPh$_3$·H_2O (Scheme 23). In accordance with spectroscopic data, antimony atoms in mononuclear and dinuclear complexes are hexacoordinated [51]. Comparison of these compounds allowed authors to suggest that organoantimony(V) prefers the formation of five rather than six membered rings.

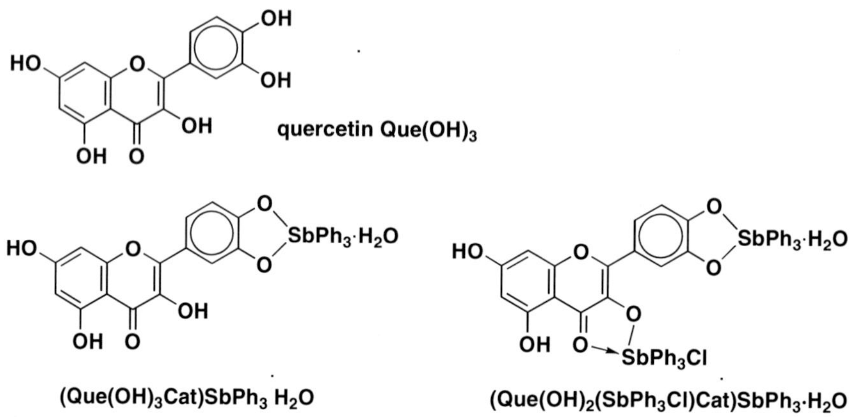

Scheme 23.

Complexes of o-quinones with alkylstibines can be prepared by the oxidative addition reaction of the corresponding o-quinone to trialkylstibine. By this way (3,6-di-tert-butyl-catecholato)triethylantimony(V), (3,6-DBCat)SbEt$_3$, [52] and (4,5-(1,1,4,4-tetramethyl-butane-1,4-diyl)-catecholato)triethylantimony(V), (4,5-Cat)SbEt$_3$, [47] have been synthesized (Scheme 24). This reaction proceeds as two-electron oxidation (as with triarylstibines) but not as 1,2- / 1,4-nucleophylic addition observed in reactions of II group metals organometallic derivatives (R_2Zn, R_2Cd, etc.) [53,54].

In contrast to triphenylantimony complex (4,5-Cat)SbPh$_3$, its triethylantimony analogue (4,5-Cat)SbEt$_3$ does not undergo rearrangement to the corresponding ionic salt $[Et_4Sb]^+$ $[(4,5\text{-Cat})_2SbEt_2]^-$ [47].

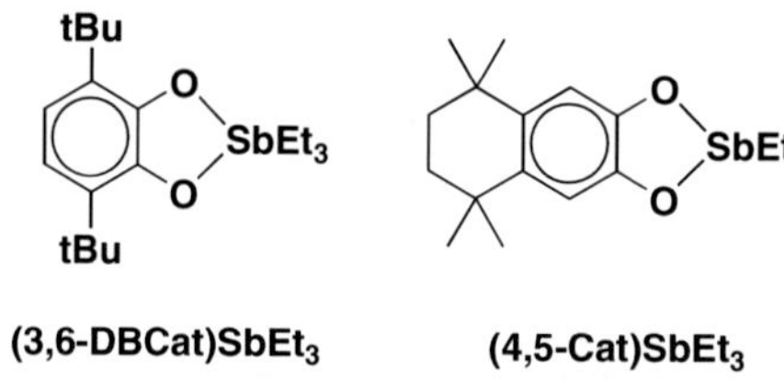

Scheme 24.

2. Antimony(V) Catecholates Based on Di-o-Quinones

Di-o-quinones are the promising objects from the viewpoint of design of di- and polynuclear compounds, molecular chains etc. [55-62]. A combination of organoantimony with di-o-quinones allows to develop dinuclear antimony catecholates and geteronuclear structures containing another O,O-chelated metal besides antimony. However, there are few literature data on antimony complexes with di-o-quinones.

The first dinuclear antimony dicatecholate was described in [39]. Diantimony complex $Ph_3Sb(Cat\text{-}Cat)SbPh_3$ was prepared by the oxidative addition of 4,4'-di-(3-methyl-6-tert-butyl-o-benzoquinone), Q-Q, to triphenylstibine in the molar ratio of 1:2 (Scheme 25). Later, the same dinuclear triethylantimony(V) and trimethylantimony(V) complexes were also synthesized by the same way [63].

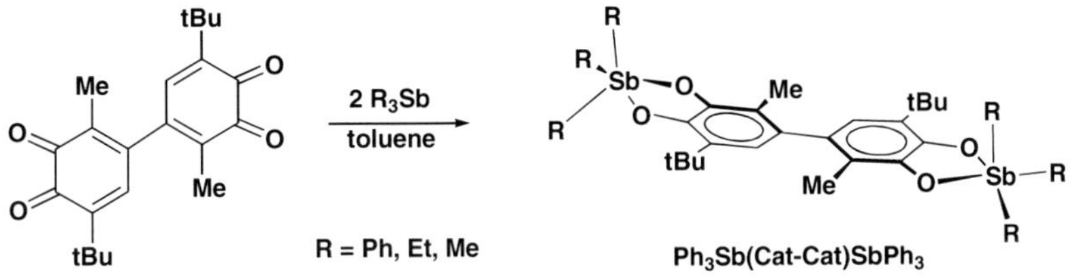

Scheme 25.

Complexes have non-planar structure: the dihedral angle between catecholate planes is 74.3° for $Et_3Sb(Cat\text{-}Cat)SbEt_3$ [63], 82.6 and 79.9° in both crystallographically independent molecules of triphenylantimony analogue $Ph_3Sb(Cat\text{-}Cat)SbPh_3$ [39]. Both antimony atoms in $Et_3Sb(Cat\text{-}Cat)SbEt_3$ as well as in $Ph_3Sb(Cat\text{-}Cat)SbPh_3$ adopt slightly distorted square pyramidal geometry.

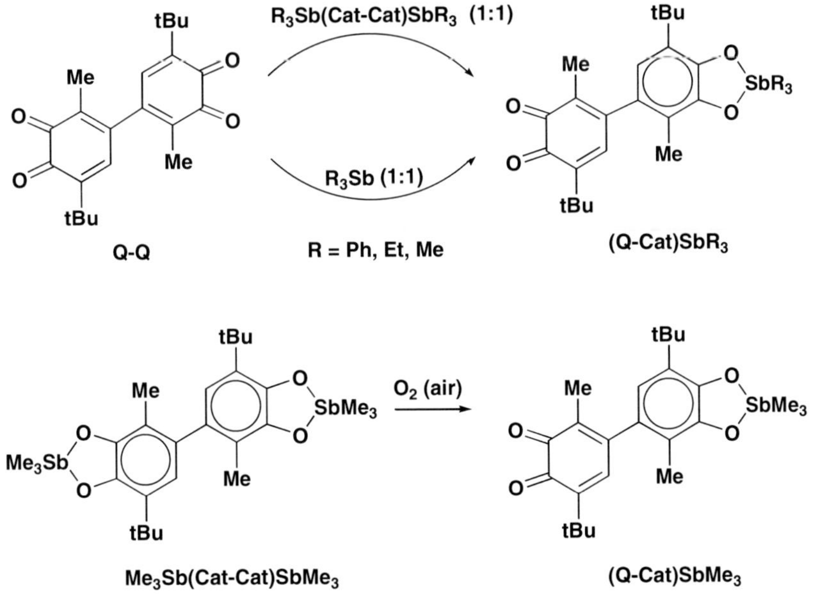

Scheme 26.

The same reaction in the equimolar ratio of Q-Q and Ph_3Sb gives mononuclear quinone-catecholato complex $(Q\text{-}Cat)SbPh_3$ which can also be synthesized by the exchange reaction between dicatecholate $Ph_3Sb(Cat\text{-}Cat)SbPh_3$ and di-o-quinone Q-Q (Scheme 26) [39].

Mononuclear trimethylantimony derivative (Q-Cat)$SbMe_3$ was also obtained via the oxidation of the corresponding dinuclear derivative with air oxygen (Scheme 26) [63]. Triphenylantimony complex Ph_3Sb(Cat-Cat)$SbPh_3$ is air-stable. As a whole, dinuclear trialkylantimony dicatecholates are less stable than triphenylantimony analogue: they loose easily one of two coordinated trialkylantimony(V) moieties under the oxidation forming mononuclear semiquinone-catecholato species $[(SQ\text{-}Cat)SbR_3]^-$ or quinone-catecholate (Q-Cat)SbR_3.

It is interesting that the irradiation of mononuclear quinone-catecholate (Q-Cat)$SbPh_3$ in toluene with the light of wavelength 546 nm (the absorption band with λ_{max} = 546 nm presents in UV-vis spectrum of (Q-Cat)$SbPh_3$) causes the counter reaction leading to the formation of di-o-quinone Q-Q and dicatecholate Ph_3Sb(Cat-Cat)$SbPh_3$ (Scheme 27) [64]. After the irradiation shut-down products formed again react to give initial quinone-catecholate.

$$2\ \text{(Q-Cat)SbPh}_3 \underset{\text{light off}}{\overset{h\nu\ (546\ \text{nm})\ \text{toluene}}{\rightleftharpoons}} \text{Q-Q} + \text{Ph}_3\text{Sb(Cat-Cat)SbPh}_3$$

Scheme 27.

An unusual mononuclear complex (Q-TTF-Cat)$SbPh_3$ containing redox-units of different nature was prepared by the reaction between 4,4',7,7'-tetra-tert-butyl-2,2'-bi(1,3-benzodithialidene)-5,5',6,6'-tetraone (di-o-quinone with tetrathiafulvalene bridge, Q-TTF-Q) with triphenylstibine as 1:1 (Scheme 28) [65].

In mononuclear catecholates - derivatives of di-o-quinones, o-quinone part keeps all features and reactivity of o-quinones. Basing on these compounds, different geterometallic derivatives are possible to be obtained. By this way, a wide row of complexes of L_nM(SQ-Cat)$SbPh_3$, where L_nM = Ph_3Sn; Tl(I); $(Ph_3P)_2Cu$(I); (dppfc)Cu(I); Li; $(CO)_4Mn$, $(CO)_3(Ph_3P)Mn$; $(CO)_2(Ph_3P)_2Mn$; $(CO)_4Re$; $(CO)_3(Ph_3P)Re$; $(CO)_3(Ph_3As)Re$; Cp_2Co, or L_nM(Cat-Cat)$SbPh_3$ types [66] and L_nM(SQ-TTF-Cat)$SbPh_3$ type, where L_nM = Ph_3Sn; Tl(I); $(Ph_3P)_2Cu$(I); Na; Li; $(CO)_4Mn$, $(CO)_3(Ph_3P)Mn$, [65] was synthesized. Some of reactions are represented in Scheme 29.

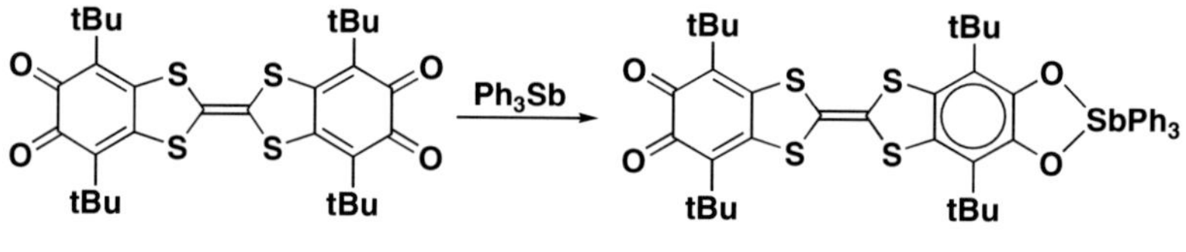

Q-TTF-Q (Q-TTF-Cat)$SbPh_3$

Scheme 28.

However, not all of di-o-quinones give mononuclear complexes with stibines. A reaction of triphenylstibine with di-o-quinone Q-CH=CH-Q containing ethylene bridge gives dinuclear derivative Ph_3Sb(Cat-CH=CH-Cat)$SbPh_3$ independently of the ratio of reagents (1:1 or 2:1) (Scheme 30) [67].

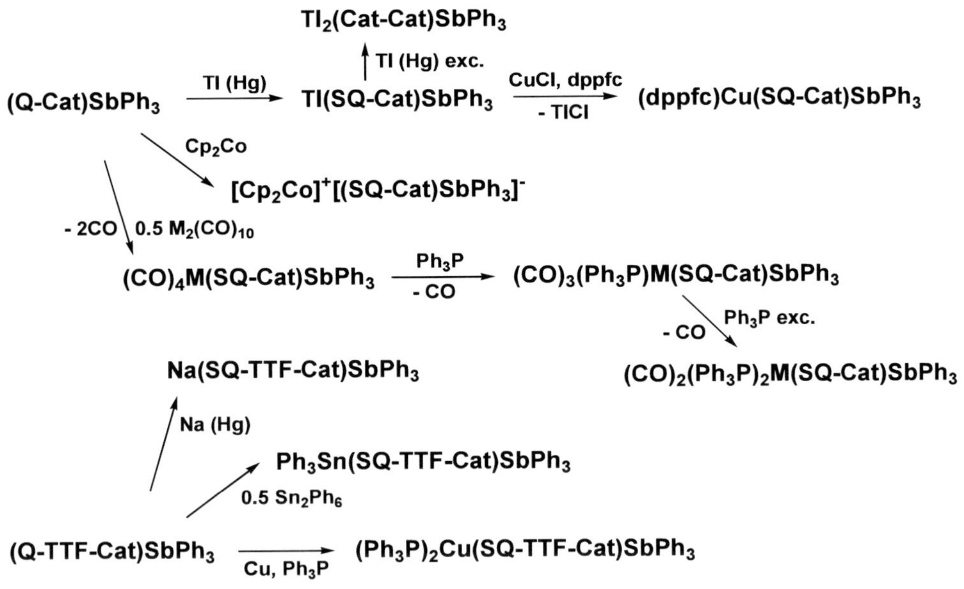

Scheme 29.

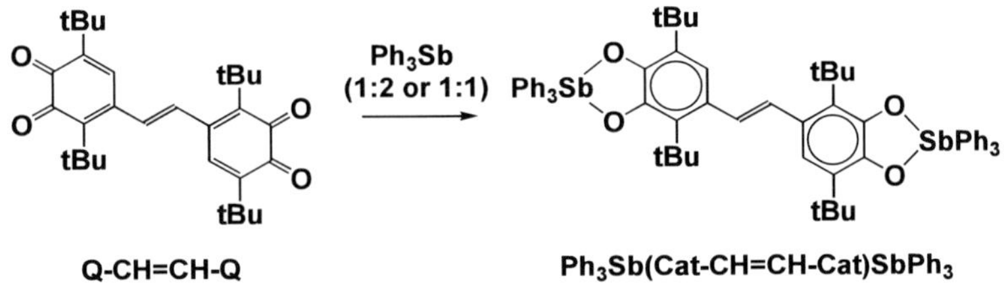

Scheme 30.

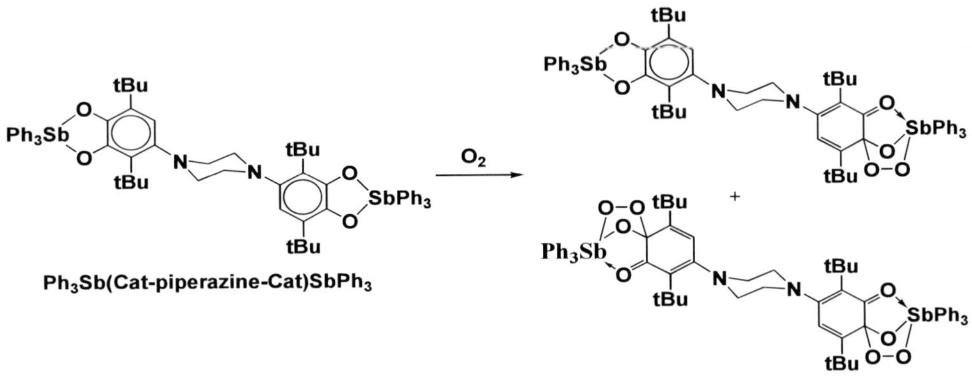

Scheme 31.

Both antimony atoms again adopt a distorted square planar geometry; two oxygen atoms of catecholate ligand and two of three phenyl carbons form the pyramid base. The planes of catecholato ligands in (Cat)$SbPh_3$-fragments were found to be nearly coplanar while the plane formed by ethylene bridge is rotated at 47.13(5)° regarding planes of catecholato ligands.

Complex $Ph_3Sb(Cat\text{-}CH{=}CH\text{-}Cat)SbPh_3$ is air stable both in solid and in solution. The insertion of piperazine bridge into di-catecholate results in new dinuclear complex Ph_3Sb(Cat-piperazine-Cat)$SbPh_3$ (Scheme 31) which being exposed to air behaves as mononuclear catecholates with weak donor groups [43]. It forms products of molecular oxygen binding - mono- and bis-spiroendoperoxides.

3. CATECHOLATES OF ANTIMONY(III)

The chemistry of antimony(III) o-quinonato complexes is quite diverse that, in many cases, is connected with the tendency of such complexes to oligo- or polymerization. However, there are many examples of monomeric structures of antimony(III) catecholates in literature.

Phenyldichlorostibine reacts with catechol in the presence of base (Et_3N) to form (catecholato)phenylantimony(III) (2-phenyl-1,3,2-dioxastibole) (Cat)SbPh (Scheme 32) [68]. Triethoxystibine reacts with catechol to give the catecholate (Cat)Sb(OEt) without the base. Both products are polymeric in contrast to the analogous derivatives of 1,2-dithiols [68]. Authors [69] report about the formation of (catecholato)methylantimony(III) in reaction of methyldialkoxystibines with catechol. The reaction of catechol with tert-butyldiethoxystibine gives monomeric (catecholato)tert-butylstibine(III) (Cat)SbtBu [70].

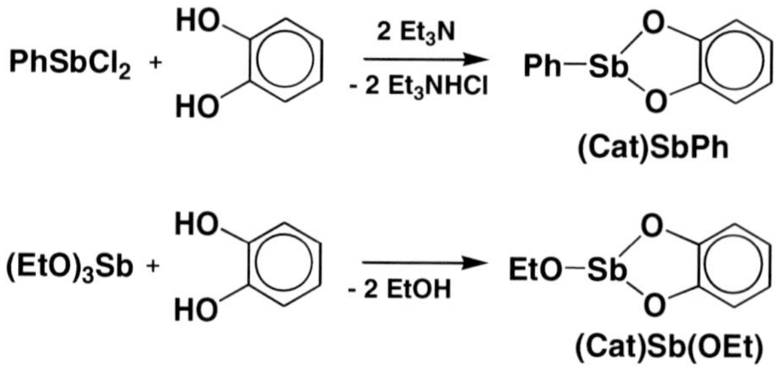

Scheme 32.

The tert-butyldialkoxystibines $tBuSb(OR)_2$ (R = Me, Et) and theirs cyclic analogue 2-tert-butyl-1,3,2-dioxastibole interact with o-quinone with a loss of tert-butyl group [24]. Reaction of $tBuSb(OR)_2$ with Cl_4Q yields antimony(III) alkoxy-catecholates, while in the case of 2-tert-butyl-1,3,2-dioxastibolane, authors reported about the formation of rather different dinuclear compound (Scheme 33). Complexes were not structurally characterized.

Mesityl-containing antimony(III) catecholate is a product of interaction of dimesitylchlorostibine with 3,5-di-tert-butyl-catechol (3,5-DBCat)H_2 in the presence of triethylamine (Scheme 34) [71].

The reaction proceeds with an elimination of one equivalent of mesitylene. However if the base is absent, both mesityl groups go away as a result of acidic cleavage of Sb-C bonds to form chloro-catecholato derivative (3,5-DBCat)SbCl.

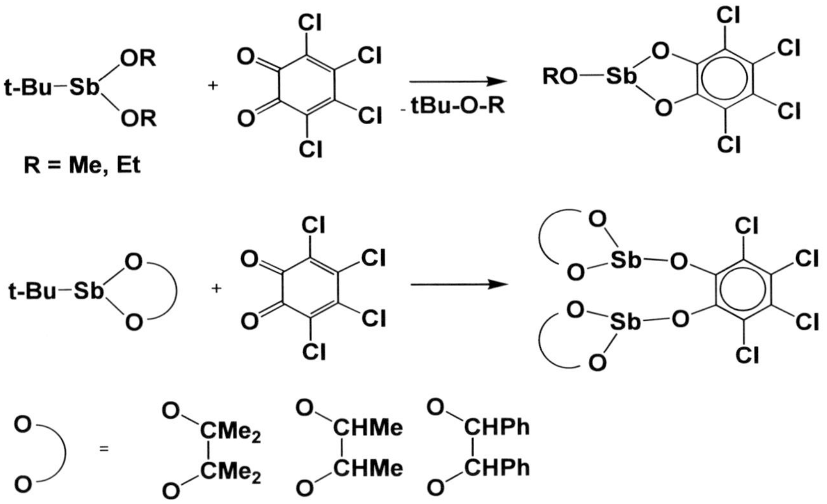

Scheme 33.

Mixed halogeno-catecholato complexes of antimony(III) (Cat)SbHal (Hal = Cl, Br, F) were investigated in a number of publications [27, 72-76]. For the first time, catecholato complexes of antimony(III) (Cat)SbX (X = OH, F) were described in 1941 [77].

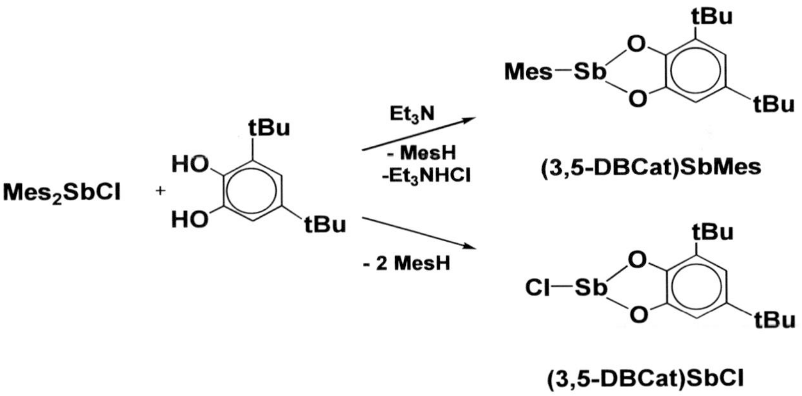

Scheme 34.

Catecholates (Cat)SbX (X = F, Cl or Br) without any additional neutral ligands were prepared by the interaction of antimony(III) halides with catechol in methanol with subsequent recrystallization of the products from acetonitrile [72]. Authors have shown (basing on X-ray diffraction data on (Cat)SbCl and the vibrational spectra of the series of halides) that the composition of these compounds is consistent with the polymer structure $[(Cat)SbX]_m$ rather than with the monomer (Scheme 35, left). Each antimony atom is tetracoordinated and has a donor-acceptor bond with halogen atom connecting (Cat)SbX monomeric units.

The addition of 1,10-phenanthroline to catecholates (Cat)SbHal (Hal = F, Cl) leads to mixed ligand compounds (Cat)SbHal(Phen) (Scheme 35, right) [73]. The bromide and iodide analogs (Hal = Br, I) were synthesized by the metathesis reaction of (Cat)SbF(Phen) with an excess of the corresponding sodium halide in refluxing methanol. Complexes are monomeric, crystal structure of fluoride derivative was reported in ref. [74]. In the accordance with data on conductivity for the solutions of complexes (Cat)SbHal(Phen) in methanol, fluoride derivative is weakly ionized whereas the other halides behave as 1:l electrolytes.

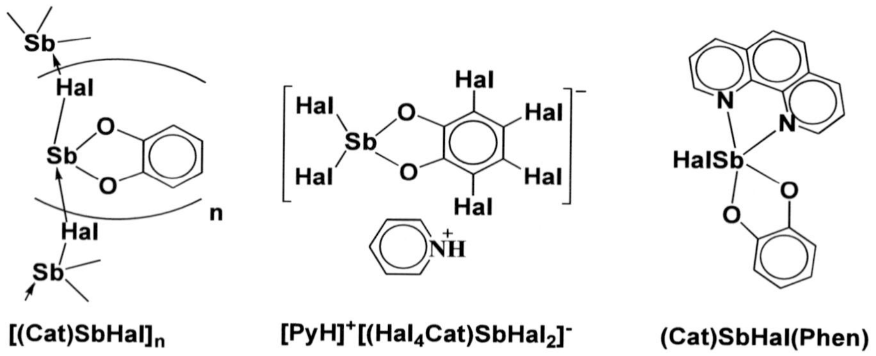

Scheme 35.

Moessbauer spectra of a number of ionic complexes of common formulation pyridinium/ quinolinium dihalogeno-(tetrahalogeno-catecholato)antimonate(III) (Scheme 35) are discussed in ref. [78]. Complexes were prepared from antimony(III) halides and halogen-substituted catechols in the presence of base (substituted pyridine, quinoline). In the crystal structures of complexes $[PyH]^+$ $[(Cl_4Cat)SbCl_2]^-$ and $[PyH]^+$ $[(Br_4Cat)SbBr_2]^-$ the central antimony is a tetracoordinated [75].

The oxidation of metallic antimony can be used not only for a preparation of antimony(V) catecholates but also for a synthesis of catecholato complexes of antimony(III). Antimony dissolves slowly (during 6 h for reaction with bromine or 10-24 h for reaction with iodine) in a reaction mixture of o-quinone (Cl_4Q, Br_4Q or 3,5-DBBQ) with bromine or iodine [27] to give the corresponding (catecholato)halogenoantimony(III) compound with the yield from 82% to quantitative. Upon addition of bidentate donor ligand bipy or Phen to a reaction mixture, the formation of pentacoordinate antimony(III) catecholate took place (Scheme 36). However, when the oxidation of excessing metallic antimony by o-quinones is performed without halogens, the reaction products are mixed ligand catecholato-semiquinonato complexes of antimony(III) [26].

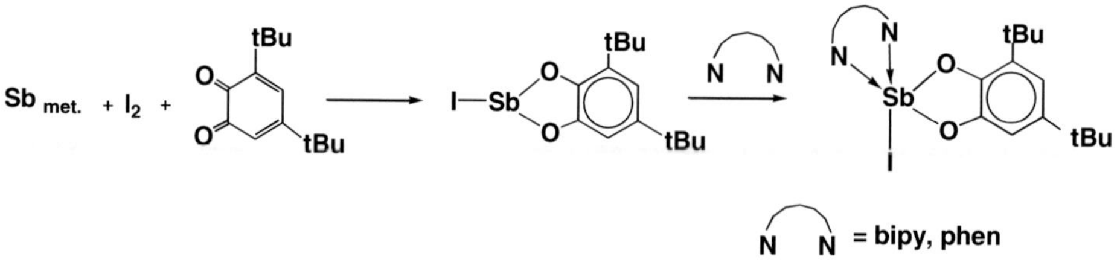

Scheme 36.

Crystal structure of (3,5-DBCat)SbI(bipy) was established by X-ray analysis. Antimony atom possesses square pyramidal geometry (which, nevertheless, can be describes as pseudo-octahedral with a lone pair). In IR spectra of these complexes the stretching vibrations of C-O bonds lie in the range of 1200-1400 cm^{-1} that is common for dianionic catecholate form of o-quinones (in different metal catecholato complexes).

An ionic antimony(III) catecholate was synthesized by treatment of (Cat)SbCl(Phen) with sodium tetrafluoroborate in the presence of Phen ligand (the molar ratio is 1:1:1) in hot methanol solution (Scheme 37) [76].

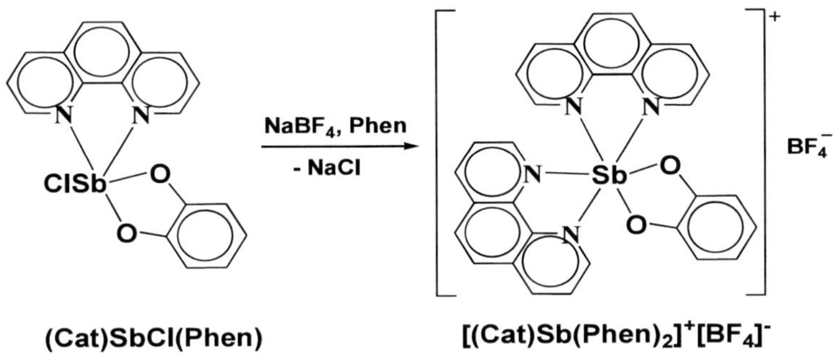

Scheme 37.

The complex cation is hexacoordinated but not octahedral: the ligands mainly located in one half of the coordinating sphere and leave free space in the other half for a lone electron pair.

A very intriguing crystal structure was determined for a dinuclear catecholate with μ-oxo-bridge $[(Cat)SbPy]_2O$ [79]. Complex was synthesized by reaction of antimony(III) acetate with catechol, pyridine and water (Scheme 38). Each antimony atom has a distorted trigonal pyramidal geometry, oxo ligand is in the equatorial plane, pyridine occupies apical position. The additional bonding of two types between molecules of $[(Cat)SbPy]_2O$ is observed: Sb1 atoms interacts with aromatic ring of catecholate of another $[(Cat)SbPy]_2O$ molecule (Sb...C distances are av. 3.7 Å) while Sb2 atoms are bonded to oxygen atoms (Sb...O distances are 3.43 Å) of adjacent catecholate ligand of another molecule (Scheme 38). Also authors have reported on the second crystalline modification of this compound with pyridine ligands in trans positions to each other.

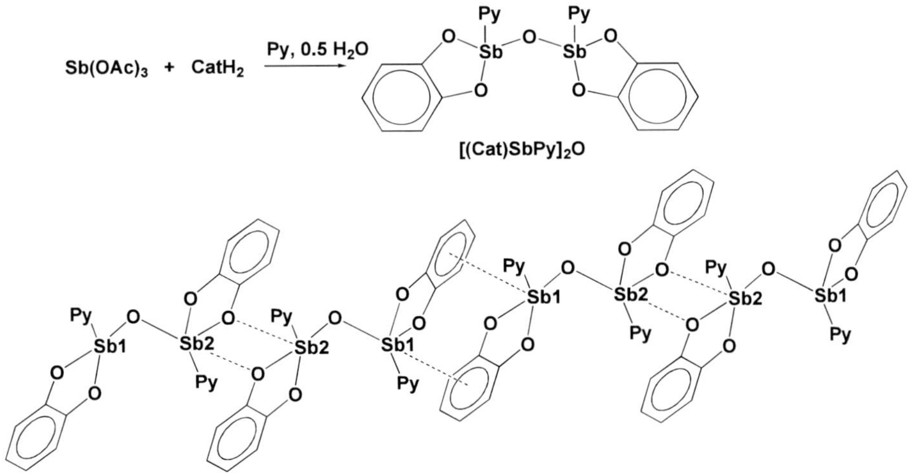

Scheme 38.

The heating of this compound at 90ºC allows to remove pyridine under vacuum. The corresponding dicatecholate $[(Cat)Sb]_2O$ was proposed to be used as a layered material for intercalation of pyridine [80].

The same reaction but without water gives another complex $[PyH]^+$ $[(Cat)_2Sb]^-$ containing dicatecholatoantimonate(III) anion [81, 82]. The crystal structure of this compound was reported in [79]. Ionic complexes $[PyH]^+$ $[(NO_2Cat)_2Sb]^-$ - products of the antimony(III) chloride reaction with 3- and 4-nitro-substituted catechols and pyridine were reported by Barnard and Alamgir [82].

Alonzo et al. have reported on the synthesis and characterization of adducts of (catecholato)fluoroantimony(III) with tridentate nitrogen-containing ligands (Cat)SbF(L) where L = pyridine-2-carboxaldehyde-2-pyridylhydrazone (paphyH); 2,2':6',2"-terpyridine (terpy) and 2,4,6-tris(2-pyridyl)-1,3,5-triazine(tpta) (Scheme 39) [83]. These complexes contain ligand L in tridentate coordination mode providing a pseudo-coordination number of Sb atom to be 7 (the seventh place is occupied by a lone pair of electrons) basing on Mossbauer and IR data.

(Catecholato)fluoroantimony(III) also reacts with di-2-pyridyl ketone (dpk) in warm methanol (Scheme 39) with an insertion of methoxy group to dpk ligand forming new catecholato complex $(Cat)Sb(dpk.OCH_3)$ [84].

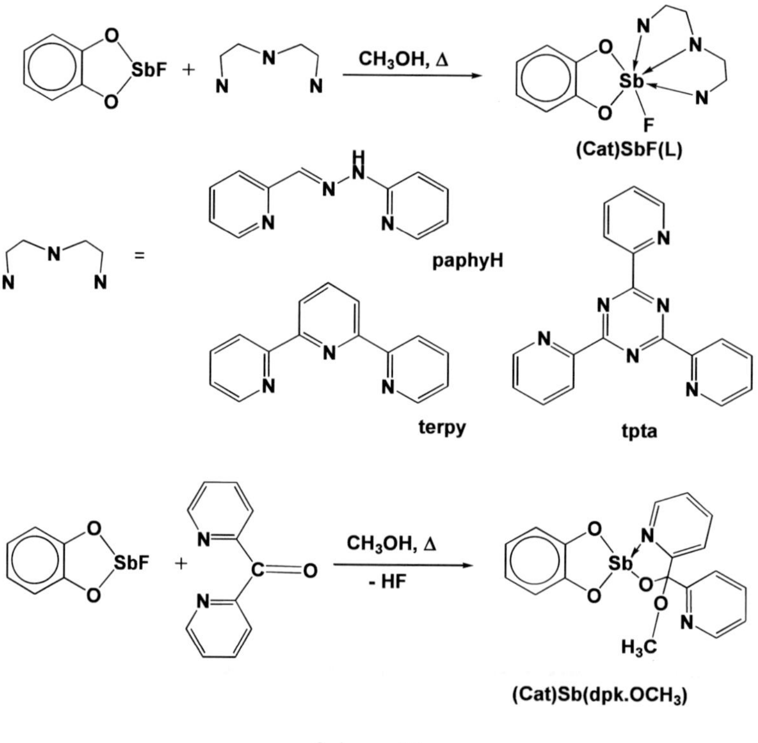

Scheme 39.

The $dpk.OCH_3$ ligand binds antimony in a bidentate fashion. Owing to the intermolecular interaction between antimony atoms and two oxygens (one is of Cat ligand and another one is of $dpk.OCH_3$ ligand) the chain structure forms.

The nucleophilic displacement reactions of (catecholato)chloroantimony(III) with relatively weak bases (tert-butylamine, phenylmagnesium bromide, diisopropylamine, diethylamine, piperidine, morpholine, dipropylamine) were investigated in [85].

Antimony(III) dicatecholate complex - a derivative of 4,5-dihydroxybenzene-3,5-disulfonate ligand - was communicated to be used as antiparasitic agent (stibophen) [4, 86].

An unusual twenty-four membered macrocycle Cyclo- $[3\text{-}(Me\text{-}Cat)_2SbNa(THF)_2]_6$ containing antimony(III) dicatecholate moieties is a product of reaction of (3-Me-CatH)Na (monosodium derivative of 3-methylcatechol $(3\text{-}Me\text{-}Cat)H_2$) with $(Me_2N)_2Sb(CH_2)_3Sb(NMe_2)_2$ in dry toluene [87]. Macrocycle contains the six dicatecholatoantimonate(III) anions connected with a sodium ditetrahydrofuranates (Scheme 40). The coordination number of each antimony atom is four. Authors have ascertained that the formation of this metallocycle proceeds through the intermediate tetraanionic complex cyclo- $[(3\text{-}Me\text{-}Cat)_2Sb(CH_2)_3Sb(3\text{-}(Me\text{-}Cat)_2)]^{4-}Na_4^{+}$.

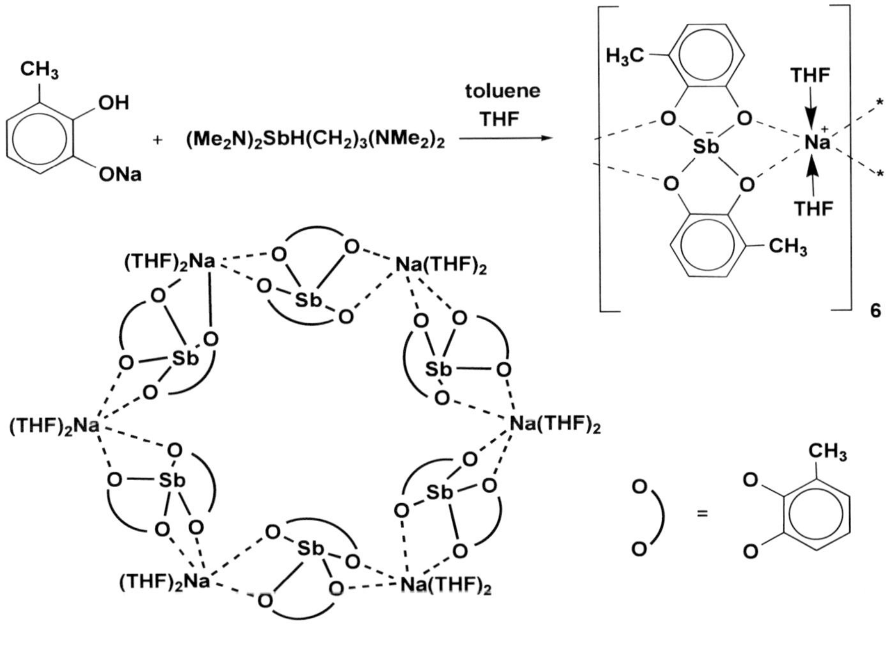

Scheme 40.

4. O-AMIDOPHENOLATES OF ANTIMONY

The coordination chemistry of antimony with redox-active bidentate O,N-ligands is much less studied than the chemistry of antimony catecholates. There are only few examples of o-amidophenolato complexes derived from bidentate o-iminoquinones.

Stegmann et al. have reported on the first examples of o-amidophenolato antimony complexes (benzo-1,3,2-oxaazastiboles) [88]. A number of complexes was synthesized like catecholates from o-aminophenols, triarylantimony dibromides and triethylamine (1st route, Scheme 41). Authors also used another synthetic way to these complexes through the corresponding 2-amino-phenoxyl radicals and triarylstibines (2nd route, Scheme 41).

The thermolysis of compound $(AP^{H})SbPh_3$ gives complex containing nine-membered metallocycle, namely 2,4,8,10-tetra-tert-butyl-6,6,6-triphenyldibenzo-1,3,6,7,2-

dioxadiazastibonine (Scheme 42) which gives hydrazobenzene after acid hydrolysis. The mechanism was not discussed but it is possible if triphenylantimony(V) o-amidophenolate reacts with o-aminophenol during the formation of nine-membered metallocycle.

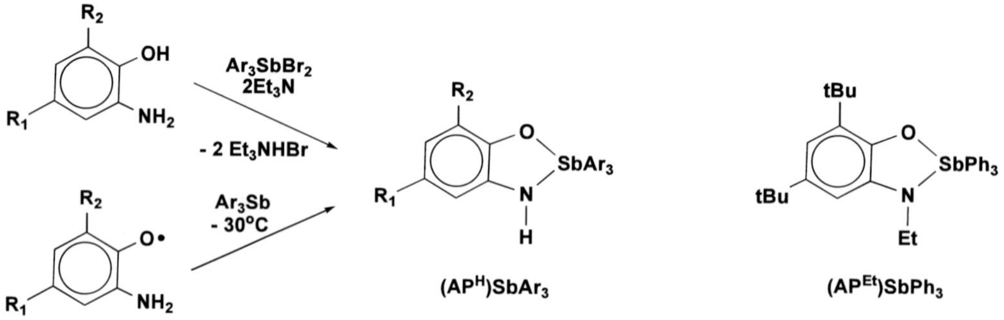

· R_1 = tBu, $C(Ph)_3$, Me; R_2 = tBu, $C(Ph)_3$, Me

Ar = Ph, 4Me - C_6H_4, 4 - DC_6H_4, 3,5-$D_2C_6H_3$, D_5C_6

Scheme 41.

The sterically hindered N-aryl-o-iminobenzoquinones are quite stable to be used as starting materials to synthesize the corresponding o-iminobenzosemiquinonato or o-amidophenolato derivatives [22]. So, the oxidative addition of o-benzoquinones to stibines can also be applied in synthesis of o-amidophenolato antimony compounds. Triphenylstibine reacts with 4,6-di-tert-butyl-N-aryl-o-iminobenzoquinones in toluene within few minutes to yield orange-yellow solutions of the corresponding o-amidophenolates of triphenylantimony(V) (AP-R)$SbPh_3$ (Scheme 43) [41, 89, 90]. Complexes geometry is intermediate between trigonal bipyramidal and tetragonal pyramidal with o-amidophenolate ligand in the basal plane of polyhedron.

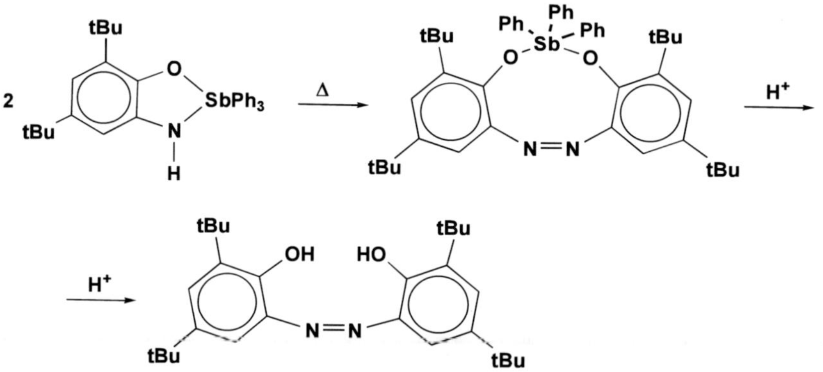

Scheme 42.

These complexes are the first examples of non-transition metal complexes with an ability of reversible dioxygen binding [89]. The reaction with O_2 (Scheme 44) proceeds under mild conditions and gives spiroendoperoxides [(L-R)O_2]$SbPh_3$ containing trioxastibolane cycle (like as triphenylantimony(V) catecholates, see part 1 of this chapter). Spiroendoperoxides were characterized structurally. The oxygen-oxygen distances O–O are 1.461(3) Å in spiroendoperoxide $[(L\text{-}iPr)O_2]SbPh_3$ - derivative of $(AP\text{-}iPr)SbPh_3$, and 1.466(2) Å in $[(L\text{-}Me)O_2]SbPh_3$ (dioxygen complex from $(AP\text{-}Me)SbPh_3$). In spiroendoperoxides derived from catecholates $(4\text{-}MeO\text{-}3,6\text{-}DBCat)SbPh_3$ and $(4,5\text{-}(MeO)_2\text{-}3,6\text{-}DBCat)SbPh_3$ the O-O bonds are close (1.475(2) Å and 1.464(2) Å, respectively). These peroxo-bonds are close to ordinary O-O bonds in known peroxo-fragments (~ 1.469 Å [91-94]). F.e. the O-O distance in endoperoxide 3,3-dihydro-5,5-dimethyl-3,3,3-triphenyl-1,2,4,3-triopxastibolane is 1.468 Å [91]; in tetrakis(triphenylstibine)di-μ^4-peroxo-tetraoxide $(Ph_2SbO)_4(O_2)_2$ is 1.47 Å [92, 93].

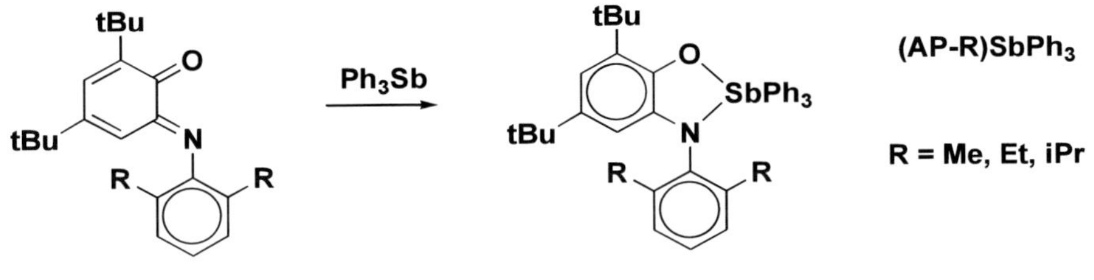

Scheme 43.

In the case of triphenylantimony(V) o-amidophenolates with unsymmetrically substituted N-aryl group (2-methyl-6-ethylphenyl, 2-methyl-6-isopropylphenyl), the corresponding spiroendoperoxides have two isomers with different relative position of peroxo fragment -O-O- and N-aryl (Scheme 45) [90]. Isomer A where substituent R_1 (methyl) and -O-O- group are located on the same side of C_6H_2ON moiety prevails over more shielded isomer B where peroxo group lies on the same side with more bulky substituent R_2 (ethyl or isopropyl).

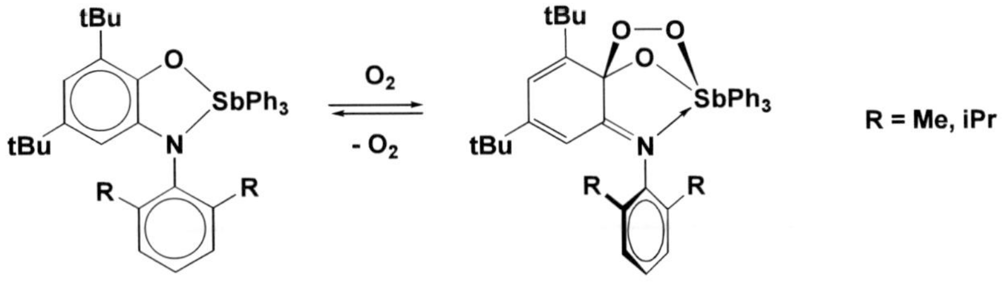

Scheme 44.

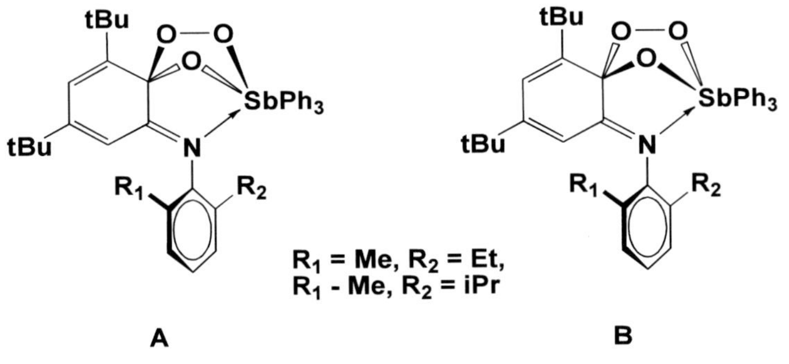

Scheme 45.

Such complexes are promising dioxygen carriers for a variety of reactions requiring molecular oxygen.

Alkyl analogues (4,6-di-tert-butyl-N-aryl-o-amidophenolato)triethylantimony(V) were also synthesized from o-iminobenzoquinones and triethylstibine [52]. Alternative method - the reaction of triethylantimony(V) dibromide with disodium salt of o-amidophenolate - is also productive (Scheme 46) [52].

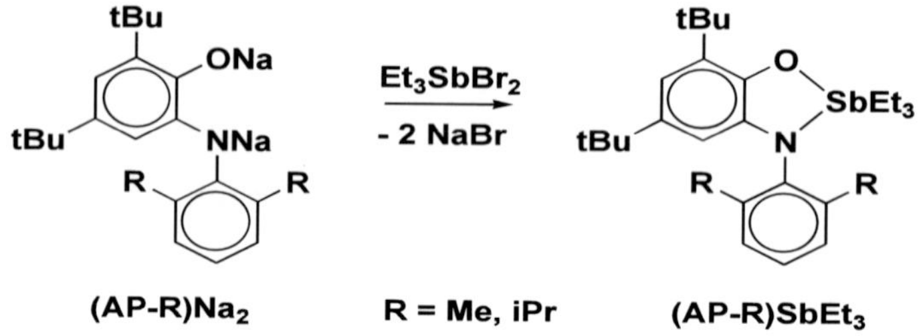

Scheme 46.

However, the change of phenyl groups with ethyls in complexes (AP-R)SbR'$_3$ results in decrease of conversion degree of o-amidophenolate to its spiroendoperoxide from 94–98% (for (AP-R)$SbPh_3$ complexes) to 15–20% (for (AP-R)$SbEt_3$ complexes) [52].

5. Antimony Complexes with Related O,N,O-, N,O,N-, O,N,N,O- and N,O,O,O- Donor Ligands

There are many different complexes of antimony with O,N- or O,N,O-ligands which do not possess redox activity (f.e. complexes of 8-quinolinol, alkoxyamines, different salicylaldehyde Schiff bases etc.). In this part we will consider complexes with redox-active or potentially redox-active ligands.

Pentacoordinated amido-bisphenolates of diaryantimony(V) were synthesized by Akiba et al. by the reactions of 2,2'-diphenolamines with arylantimony(III) dichlorides or diarylantimony(V) trichlorides in the presence of triethylamine (Scheme 47) [95]. Authors have proposed a mechanism of reaction with $ArSbCl_2$ through the formation of bicyclic anionic intermediate of $[(AP\text{-}N\text{-}AP)SbAr^{-}]$ type. Complexes are air- and moisture-stable.

Di-p-tolylantimony complexes (AP-N-AP)Sb(pTol)$_2$ react with tolyllithium to give hexacoordinated amino-bisphenolates (AP-NH-AP)Sb(pTol)$_3$ which can be deprotonated and methylated with the formation of (methylamino-bisphenolato)tri-p-tolylantimony derivatives (Scheme 48).

Nothworthy, like o-iminobenzoquinones, the sterical hindrances of this tridentate 2,2'-diphenolamine ligand permit the existence of its o-iminoquinone form N-hydroxyphenyl-o-iminoquinone (IBQ-N-AP)H [52]. The reaction of this N-hydroxyphenyl-o-iminoquinone with triphenylstibine produces the related o-amino-bisphenolate (AP-NH-AP)$SbPh_3$ as a result of proton migration from hydroxyl group to nitrogen atom (Scheme 49). In complexes (AP-NH-AP)$SbAr_3$ the coordination sphere of antimony is fulfilled by a donor-acceptor bond Sb-NH.

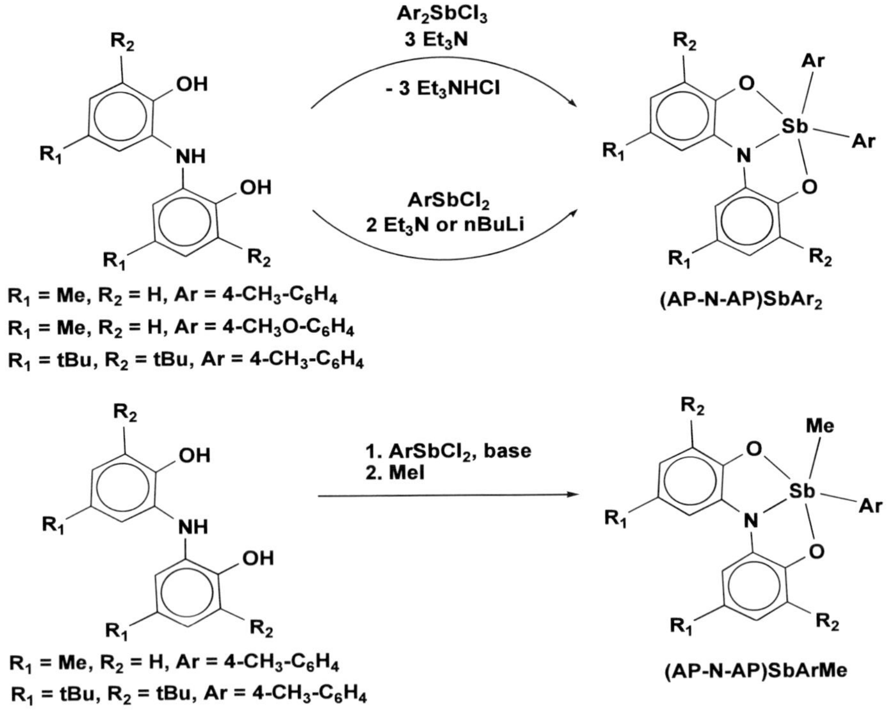

Scheme 47.

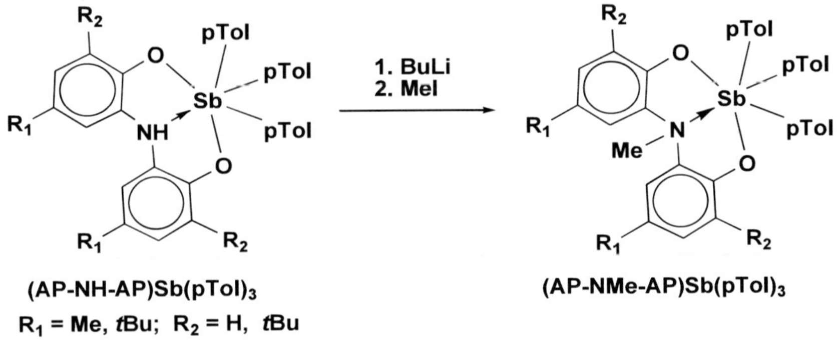

Scheme 48.

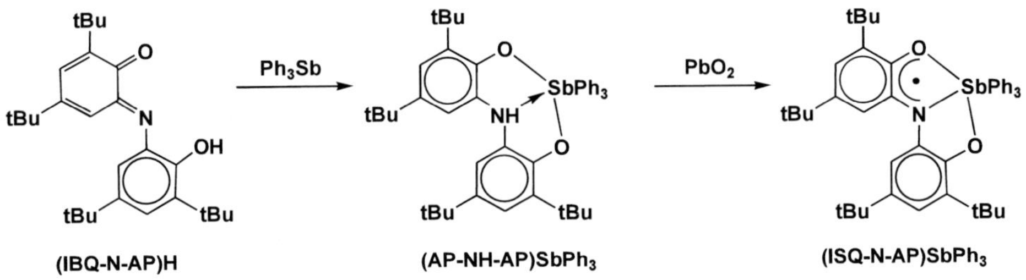

Scheme 49.

Amino-bisphenolato complexes can also be prepared from 2,2'-diphenolamine and triaryl/alkylantimony(V) dihalides. For example, (amino-bisphenolato)triethylantimony(V)

(AP-NH-AP)SbEt$_3$. Complex is unstable and looses one ethyl group and hydrogen atom to give (amido-bisphenolato) diethylantimony(V) (AP-N-AP)SbEt$_2$ (Scheme 50) [52]. The reaction accelerates in the presence of air. In contrast to (amino-bisphenolato)triethylantimony(V) (AP-NH-AP)SbEt$_3$, its triphenylantimony analogue is very stable, and its deprotonation with lead(IV) oxide proceeds slowly and requires heating. The product of deprotonation is paramagnetic triphenylantimony(V) complex (ISQ-N-AP)SbPh$_3$ containing dianion-radical ligand (Scheme 49).

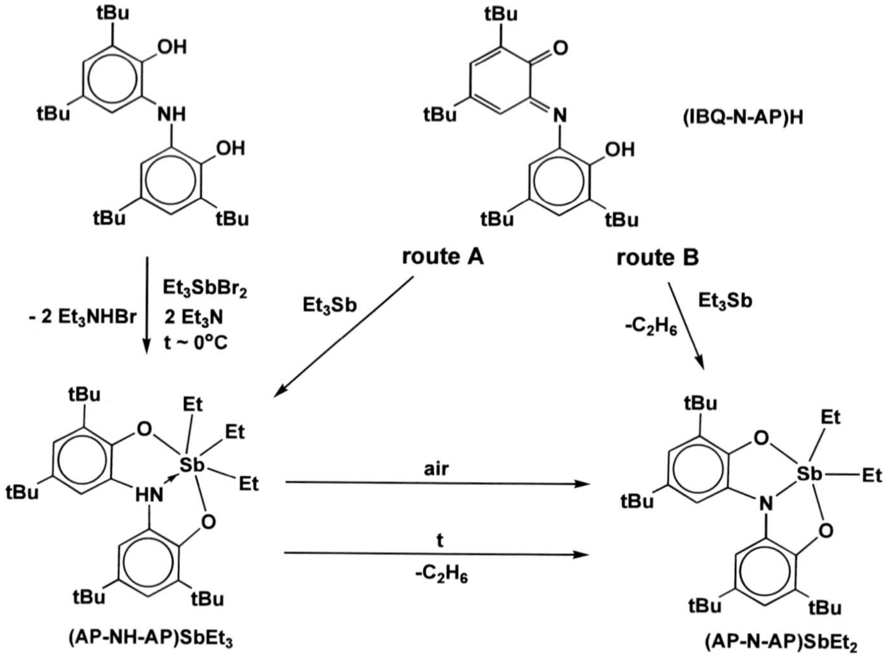

Scheme 50.

It is interesting that the order of reagent addition in reaction of 4,6-di-tert-butyl-N-(3,5-di-tert-butyl-2-hydroxyphenyl)-o-iminoquinone (IBQ-N-AP)H with triethylstibine allows to prepare different products (routes A and B in Scheme 50): a slow addition of ligand to triethylstibine leads to amino-bisphenolate (AP-NH-AP)SbEt$_3$ (route A) while a reverse order of reagents addition gives deprotonated deethylated species (AP-N-AP)SbEt$_2$ (route B) [52].

According to X-ray studies, the (AP-N-AP)$^{3-}$ ligand in (AP-N-AP)SbEt$_2$ is not planar, the angle between the planes of phenyl rings is 25.8(1)° that is significantly more than the same angles in transition metal complexes with this O,N,O-ligand (f.e. this angle is 13.4° for Mn^{IV}(ISQ-N-AP)$_2$ complex [96] and this value is the highest among transition metal derivatives) and close to that for tin complex Sn(ISQ-N-AP)$_2$ (the corresponding angles are 30.4(7)° and 28.7(7)° for two ligands, [97]).

In contrast to the reaction of (Cat)SbF with di-2-pyridyl ketone (dpk) in methanol which gives (Cat)Sb(dpk.OCH$_3$) where dpk.OCH$_3$ is bidentate ligand (see part 3 of this chapter) [84], the same reaction with SbF$_3$ proceeds also as nucleophilic addition of methanol to dpk but gives a pentacoordinate complex of antimony(III) where dpk.OCH$_3$ behaves as tridentate N,O,N-ligand (dpk.OCH$_3$)SbF$_2$ (Scheme 51) [98]. In (dpk.OCH$_3$)SbF$_2$ the central antimony(III) atom adopts square pyramidal geometry; the Sb-N bonds are of donor-acceptor nature (2.580(7) Å).

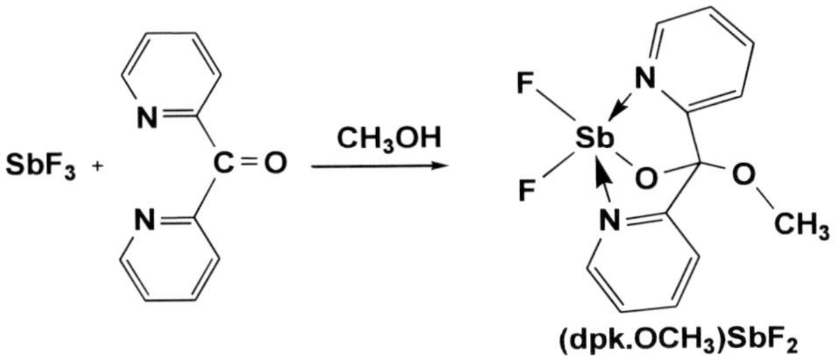

Scheme 51.

The unusual monohalidodiorganoantimony(V) complexes (AP-N=CH-CH=N-AP)SbR_2Br with tetradentate Schiff base O,N,N,O-ligand (where R is phenyl, p-tolyl or methyl) are described to be monomeric and hepta-coordinated (Scheme 52) [99].

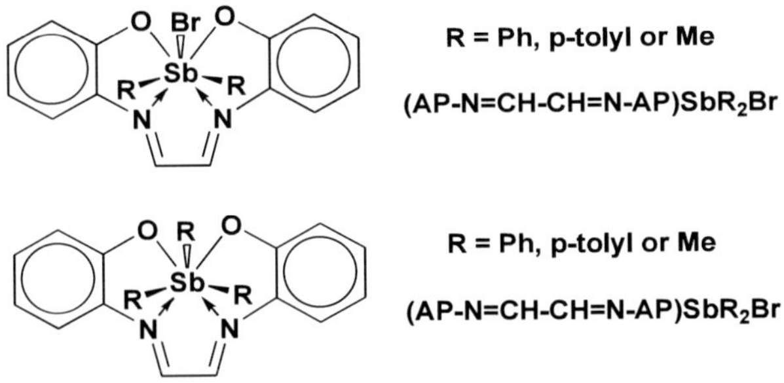

Scheme 52.

Previously, the similar triorganoantimony(V) complexes (AP-N=CH-CH=N-AP)SbR_3 (Scheme 52) were also synthesized [100].

A number of antimony(III) complexes with N,O,O,O- and N,O,O-ligands was prepared by Parkin et al. from alkoxystibine and ligands N(o-$C_6H_4OH)_3$, PhN(o-$C_6H_4OH)_2$ [101].

The ligand N(o-$C_6H_4OH)_3$ is an insufficiently explored ligand in organoelemental and coordination chemistry. There are data on aluminum, germanium, tin and phosphorus with this ligand in literature. In aluminum, germanium and tin complexes this ligand is tetradentate (N,O,O,O-), while it is tridentate (O,O,O-) in phosphorus compound. Complex { [η^4-N(o-$C_6H_4O)_3$]Sb}$_n$ was synthesized from triethoxystibine and ligand N(o-$C_6H_4OH)_3$ and crystallized from DMSO as the corresponding DMSO-solvate. Antimony(III) compound [η^4-N(o-$C_6H_4O)_3$]Sb($OSMe_2$) also contains this ligand as tetradentate (atrane structure) [101]. Authors point out more metallic character of antimony in comparison with phosphorus. But it also may be caused by the bigger covalent radius of antimony(III) in comparison with phosphorus. A coordination number of the central antimony atom in [η^4-N(o-$C_6H_4O)_3$]Sb(SMSO) is five with a distorted square pyramidal environment and a lone electron pair is trans-located toward the donor-acceptor Sb-N bond (2.433(2) Å).

The related complex [η^4-N(CH_2-o-C_6H_2(tBu)$_2$O)$_3$]Sb and its DMSO derivative [η^4-N(CH_2-o-C_6H_2(tBu)$_2$O)$_3$]Sb(DMSO) (Scheme 53) are described in [102]. The interesting reactivity of this complex was observed: a reaction of [η^4-N(CH_2-*o*-C_6H_2(tBu)$_2$O)$_3$]Sb with acetic acid gives acetate [η^3-N(CH_2-*o*-C_6H_2(tBu)$_2$OH)(CH_2-*o*-C_6H_2(tBu)$_2$O)$_2$]Sb(OAc) where one of phenolato group is decoordinated and protonated, a reaction with bromine gives complex of antimony(V) [η^4-N(CH_2-*o*-C_6H_2(tBu)$_2$O)$_3$]SbBr_2, the oxidation of this complex with triethylamineoxide proceeds with formation of a number of μ-oxo bridged and hydroxyl-complexes.

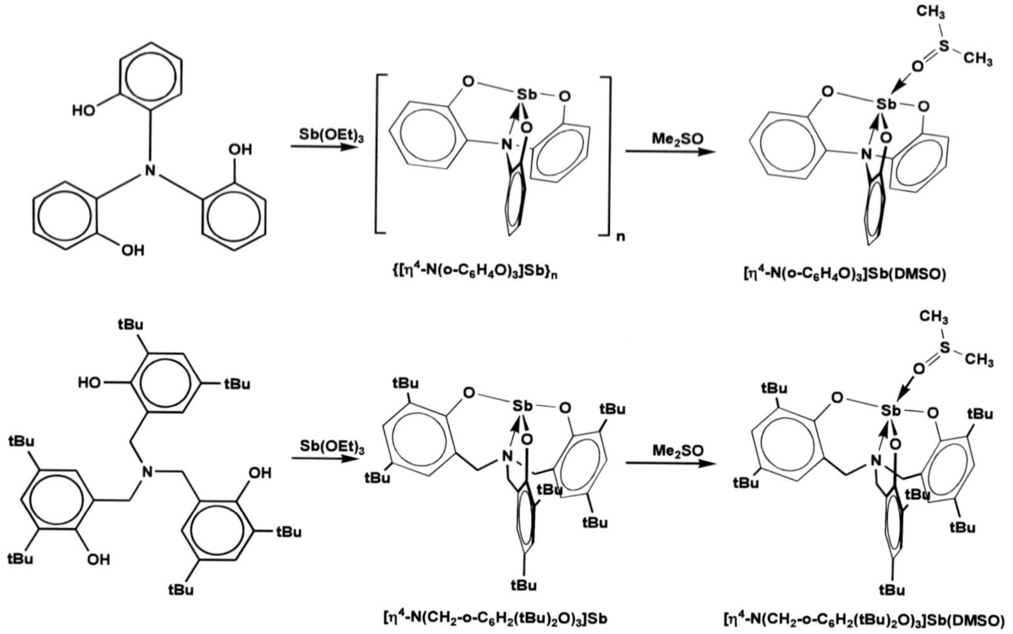

Scheme 53.

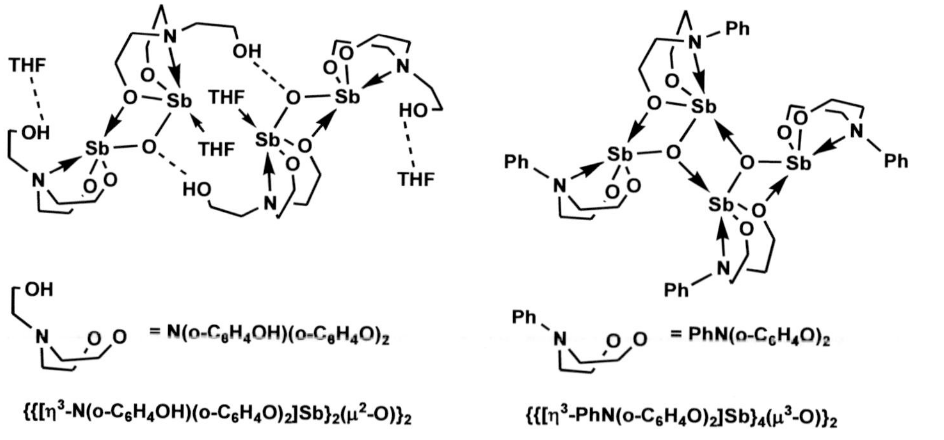

Scheme 54.

The addition of controlled amount of water to a { [η^4-N(*o*-C_6H_4O)$_3$]Sb}$_n$ and crystallization from THF gives crystals of complex {{ [η^3-N(*o*-C_6H_4OH)(*o*-C_6H_4O)$_2$]Sb}$_2$(μ_2-O)}$_2$ with μ_2-oxo bridge [101]. In this compound ligand N(*o*-C_6H_4OH)(*o*-C_6H_4O) contains one phenol group and demonstrates a N,O,O-tridentate fashion of coordination to antimony. Antimony atoms are pentacoordinated but reveal different environment: one of them is additionally coordinated by THF molecule while the second one - by aryloxy-group of the adjacent [N(*o*-C_6H_4OH)(*o*-C_6H_4O)$_2$]Sb fragment (Scheme 54). In crystals dimeric moieties { [η^3 - N(*o*-C_6H_4OH)(*o*-C_6H_4O)$_2$]Sb}$_2$(μ_2-O) are connected by intermolecular hydrogen bonding to an extended tetraantimony structure.

The replacement of ligand N(*o*-C_6H_4OH)$_3$ with ligand PhN(*o*-C_6H_4OH)$_2$ in reaction with triethoxystibine followed by water addition also yields tetranuclear derivative - complex {{ [η^3 - PhN(*o*-C_6H_4O)$_2$]Sb}$_4$(μ_3-O)}$_2$ (Scheme 54) [101]. However, there is no hydrogen bonding and μ_3-oxo-atoms play a role of bridging groups between two { [η^3 - PhN(*o*-C_6H_4O)$_2$]Sb}$_2$ moieties through the donor-acceptor Sb-O interaction (Sb-O is 2.446(6) Å).

CONCLUSIONS

The present chapter collects the information on the antimony(V,III) compounds bearing the redox-active bidentate o-quinonato and related nitrogen-containing o-iminoquinonato ligands as well as the resembling polydentate amino-bisphenolato, amino-trisphenolato ligands. Literature data shows that catecholato and o-amidophenolato complexes of antimony demonstrate a variety of structural bonding and reactivity depending on the nature of redox-active ligand, additional ligand, solvent etc. Antimony(V) complexes reveal different types of antimony coordination sphere (square-pyramidal structures are found along with trigonal bipyramidal ones) and, moreover, tends to form intermolecular contacts in the case of the absence of sterical shielding. Antimony(III) complexes form both covalent and ionic structures and have even more extended composition with different bonding pattern (di-, tri-, tetra- and polymeric, linear and cyclic etc.). The variation of redox-properties of chelating ligand (catecholate, o-amidophenolate) leads to the widening of the ranges of redox-transformations with participation of the derived complexes. In some cases, these features result in versatility of complexes and the unusual structures or reactivity.

From the viewpoint of chemical properties, antimony(V, III) complexes with redox-active o-quinonato type ligands represent the extremely perspective objects for modeling of various chemical and biochemical systems, search of new catalysts for some important chemical processes, carriers of small molecules etc. Non-transition metal complexes with redox-active ligands reveal the substantial importance in a creation of new conceptions concerning the chemical behavior of coordination and organometallic compounds.

ABBREVIATIONS

Q, o-quinone; SQ, its one-electron reduced form - radical-anion o-semiquinone; Cat, two-electron reduced form - dianion catecholate; radical-anionic and dianionic forms of other quinonato ligands are also abbreviated by change of letter "Q" with "SQ" and "Cat" in the corresponding quinone abbreviations; 3,5-DBBQ, 3,5-di-tert-butyl-o-benzoquinone, 3,6-DBBQ, 3,6-di-tert-butyl-o-benzoquinone; (3,5-DBCat)H_2, 3,5-di-tert-butyl-catechol; 4-X-3,6-DBBQ, 3,6-di-tert-butyl-o-benzoquinone with substituent X in 4th position; 6-X-3,5-DBBQ,

3,5-di-tert-butyl-o-benzoquinone with substituent X in 6th position; 4,5-X_2-3,6-DBBQ, 3,6-di-tert-butyl-o-benzoquinone with substituents X in 4th and 5th positions; Cl_4Q - tetrachloro-o-benzoquinone; Br_4Q - tetrabromo-o-benzoquinone; $^{ox}Cat_{Cl}$, perchloroxanthrenecatecholate; PhenCat, phenanthrene-9,10-diolate; naphtCat, naphthalene-2,3-diolate; 3-Me-Cat, 3-methyl-catecholate; iPr_3Cat, 3,4,6-tri-isopropyl-catecholate; $iPr_3(OH)Cat$, 3,4,6-tri-isopropyl-5-hydroxycatecholate; 4,5-Cat, 4,5-(1,1,4,4-tetramethyl-butane-1,4-diyl)-catecholate; Q-Q, 4,4'-di-(3-methyl-6-tert-butyl-o-benzoquinone); Q-TTF-Q, 4,4',7,7'-tetra-tert-butyl-2,2'-bi (1,3-benzodithialidene)-5,5',6,6'-tetraone, Q-CH=CH-Q, 4,4'-(ethene-1,2-diyl)–bis (3,6-di-tert-butyl-o-benzoquinone); Q-piperazine-Q, 4,4'-(piperazine-1,4-diyl)-bis (3,6-di-tert-butyl-o-benzoquinone); $Que(OH)_3$, quercetin; Phen, 1,10-phenanthroline; bipy, 2,2'-dipyridyl; Py, pyridine; PyO, pyridine oxide; $[PyH]^+$, pyridinium cation; biphen, 2,2'-biphenyl; paphyH, pyridine-2-carboxaldehyde-2-pyridylhydrazone; terpy, 2,2':6',2"-terpyridine; tpta, 2,4,6-tris(2-pyridyl)-1,3,5-triazine; ox, oxalate; dpk, di-2-pyridyl ketone; AP^H, 4,6-di-substituted-o-amidophenolate (substituents tert-butyl, triphenylmethyl, methyl); AP^{Et}, 4,6-di-tert-butyl-N-ethyl-o-amidophenolate; AP-R, 4,6-di-tert-butyl-N-(2,6-dialkyl(R)-phenyl)-o-amidophenolate (R = Me, Et, iPr); other abbreviations are given in text.

References

[1] McAuliffe, C. A. In Comprehensive Coordination Chemistry; Wilkinson, G.; Ed.-in-Chief; Volume 3: *Main Group and Early Transition Elements*; Gillard, R.D., McCleverty, J. A.; Eds.; Chapter 28. Arsenic, Antimony and Bismuth; Pergamon Press; 1987, pp. 237-298.

[2] Maeda, S. In Chemistry of Organic Arsenic, Antimony and Bismuth Compounds; Patai, S. Ed; Chapter 20. *Synthesis of organoantimony and organobismuth compounds*; Wiley: New York, 1994, pp 725–759.

[3] Matano Y.; In Comprehensive Organometallic Chemistry III; Crabtree, R. H., Mingos, D. M. P.; Eds.-in-Chief; Volume 9: Applications I: Main Group Compounds in Organic Synthesis; Knochel, P.; Ed.; Chapter 9.10. *Antimony and Bismuth*; Elsevier, 2007; pp. 425-456.

[4] Farrell, N. In Comprehensive Coordination Chemistry II; McCleverty, J. A.; Meyer, T. J.; Eds.-in-Chief; Volume 9: Applications of Coordination Chemistry; Ward, M. D.; Ed.; Chapter 9.18. *Metal Complexes as Drugs and Chemotherapeutic Agents*; Elsevier: Amsterdam, 2005; pp. 809-841.

[5] Jenkins, R. O.; Craig, P. J.; Francesconi, K. A.; Harrington, C. F. In Comprehensive Organometallic Chemistry III; Crabtree, R. H., Mingos, D. M. P.; Eds.-in-Chief; Volume 12: Applications III: Functional Materials, Environmental and Biological Applications; O'Hare, D.; Ed.; Chapter 12.13. *Environmental and Biological Aspects of Organometallic Compounds*; Elsevier, 2007; pp. 603-661.

[6] Reglinski, J. In The Chemistry of Arsenic, Antimony and Bismuth; Norman, N. C., Ed.; Chapter 8. *Environmental and medicinal chemistry of arsenic, antimony and bismuth*; Blackie: London, 1998.

[7] Wormser, U.; Nir, I. In Chemistry of Organic Arsenic, Antimony and Bismuth Compounds; Patai, S.; Ed.; Chapter 18. *Pharmacology and toxicology of organic bismuth, arsenic and antimony compounds*; Wiley: New York, 1994, pp 715–723.

[8] Ge, R.; Sun, H., *Acc. Chem. Res.* 2007, 40, 267-274.

[9] Tiekink, E. R. T., *Crit. Rev. Oncol./Hematol.* 2002, 42, 217-224.

[10] O'Brien, P.; Pickett, N. L. In Comprehensive Coordination Chemistry II; McCleverty, J. A.; Meyer, T. J.; Eds.-in-Chief; Volume 9: Applications of Coordination Chemistry; Ward, M. D.; Ed.; Chapter 9.23. *Coordination Complexes as Precursors for Semiconductor Films and Nanoparticles*; Elsevier: Amsterdam, 2005; pp.1006-1063.

[11] Carmalt, C. J.; Basharat, S. In Comprehensive Organometallic Chemistry III; Crabtree, R. H., Mingos, D. M. P.; Eds.-in-Chief; Volume 12: Applications III: Functional Materials, Environmental and Biological Applications; O'Hare, D.; Ed.; Chapter 12.01. *Precursors to Semiconducting Materials*; Elsevier, 2007; pp. 1-34.

[12] Levason, W.; Reid, G. In Comprehensive Coordination Chemistry II; McCleverty, J. A.; Meyer, T. J.; Eds.-in-Chief; Volume 3, Coordination Chemistry of the s, p, and f Metals; Parkin, G. F.; Ed.; Chapter 3.6. *Arsenic, Antimony, and Bismuth*; Elsevier: Amsterdam, 2005; pp. 465-544.

[13] Breunig, H. J.; Wagner, R. In Comprehensive Organometallic Chemistry III; Crabtree, R. H., Mingos, D. M. P.; Eds.-in-Chief; Volume 3: Compounds of Groups 13 to 15; Housecroft, C. E.; Ed.; Chapter 3.16. *Arsenic, Antimony, and Bismuth Organometallics* Elsevier, 2007; pp. 905-929.

[14] Kabachnik, M. I.; Bubnov, N. N.; Prokof'ev, A. I.; Solodovnikov, S. P. *Science Rev.* 1981, XXIV, 197-230.

[15] Muraev, V. A.; Abakumov, G. A.; Razuvaev, G. A. *Dokl. Akad. Nauk* USSR 1974, 217, 1083 - 1086.

[16] Abakumov, G. A.; Nevodchikov, V. I. *Dokl. Akad. Nauk* USSR 1982, 266, 1407-1410.

[17] Lange, C. W.; Foldeaki, M.; Nevodchikov, V. I.; Cherkasov, V. K.; Abakumov, G. A.; Pierpont, C. G. *J. Am. Chem.* Soc. 1992, 114, 4220-4222.

[18] Jung, O.-S.; Pierpont, C. G. *J. Am. Chem. Soc.* 1994, 116, 2229-2230.

[19] Nevodchikov, V. I.; Abakumov, G. A.; Cherkasov, V. K.; Razuvaev, G. A. *J. Organomet. Chem.* 1981, 214, 119-124.

[20] Pierpont, C. G. *Coord. Chem. Rev.* 2001, 219-221, 415-433.

[21] Pierpont, C. G.; Buchanan, T. M. *Coord. Chem. Rev.* 1981, 38, 45-87.

[22] Poddel'sky, A. I.; Cherkasov, V. K.; Abakumov, G. A. *Coord. Chem. Rev.* 2009, 253, 291-324.

[23] Chaudhuri, P.; Hess, M.; Muller, J.; Hildenbrand, K.; Bill, E.; Weyhermuller, T.; Wieghardt, K. *J. Am. Chem. Soc.* 1999, 121, 9599-9610.

[24] Wieber, Von M.; Baumann, N. *Z. Anorg. Allg. Chem.* 1975, 418, 279-286.

[25] Denney, D. B.; Jones, D. H. *J. Am. Chem. Soc.* 1969, 91, 5821-5825.

[26] Prokof"ev, A. I.; Malysheva, N. A.; Tumanskii, B. L.; Bubnov, N. N.; Solodovnikov, S. P.; Kabachnik, M. I. *Izv. Akad. Nauk* USSR. Ser. Chim. 1978, 27, 1738-1743.

[27] Tian, Z.; Tuck, D. G. *J. Chem. Soc. Dalton Trans.* 1993, 1381-1385.

[28] see for example, Annan, T. A.; Tian, Z. *J. Chem. Soc. Dalton Trans.* 1991, 1, 19-22.

[29] Hall, M.; Sowerby, D. B. *J. Am. Chem. Soc.* 1980, 102, 628-632.

[30] Shindo, M.; Okawara, R. *Inorg. Nucl. Chem. Lett.* 1969, 5, 77.

[31] Holmes, R. R.; Day, R. O.; Chandrasekhar, V.; Holmes, J. M. *Inorg. Chem.* 1987, 26, 157-163.

[32] Gibbons, M. N.; Begley, M. J.; Blake, A. J.; Sowerby, D. B. *J. Chem. Soc. Dalton Trans.* 1997, 2419-2425.

[33] Holmes, R. R.; Day, R. O.; Chandrasekhar, V.; Holmes, J. M. *Inorg.Chem.* 1987, 26, 163-168.

[34] Dodonov, V. A.; Zinov'eva, T. I. *Organomet. Chem.* USSR 1992, 5, 616.

[35] Razuvaev, G. A.; Brilkina, T. G.; Filimonov, A. I.; Krasilnikova, V.; Zinov'eva, T. I. *J. Organomet. Chem.* 1972, 152.

[36] Dodonov, V. A.; Fedorov, A. Yu.; Fukin, G. K.; Zaburdyaeva, S. N.; Zakharov, L. N.; Ignatenko, A. V. *Main Group Chem.* 1999, 3, 15-22.

[37] Fukin, G. K.; Zakharov, L. N.; Domrachev, G. A.; Fedorov, A. Yu.; Zaburdyaeva, S. N.; Dodonov, V.A. *Rus. Chem. Bull.* 1999, 48, 1722-1731.

[38] Abakumov, G. A.; Vavilina, N. N.; Kurskii, Yu. A.; Abakumova, L. G.; Fukin, G. K.; Cherkasov, V. K.; Shavyrin, A. S.; Baranov, E. V. *Russ. Chem. Bull.* 2007, 56, 1813-1820.

[39] Cherkasov, V. K.; Grunova, E. V.; Poddel'sky, A. I.; Fukin, G. K.; Kurskii, Yu. A.; Abakumova, L. G.; Abakumov, G. A. *J. Organomet. Chem.*, 2005, 690, 1273-1281.

[40] Abakumov, G. A.; Cherkasov, V. K.; Grunova, E. V.; Poddel'sky, A. I.; Abakumova, L. G.; Kurskii, Yu. A.; Fukin, G. K.; Baranov. E. V. *Dokl. Chem.* 2005, 405, 222-225.

[41] Cherkasov, V. K.; Abakumov, G. A.; Grunova, E. V.; Poddel'sky, A. I.; Fukin, G. K.; Baranov, E. V.; Kurskii, Yu. A.; Abakumova, L. G. *Chem. Eur. J.*, 2006, 12, 3916-3927.

[42] Poddel'sky, A. I.; Smolyaninov, I. V.; Kurskii, Yu. A.; Berberova, N. T.; Cherkasov, V. K.; Abakumov, G.A. *Rus. Chem. Bull.*, 2009, 58, 532-537.

[43] Poddel'sky, A. I.; Smolyaninov, I. V.; Kurskii, Yu. A.; Fukin, G. K.; Berberova, N. T.; Cherkasov, V. K.; Abakumov, G. A. *J. Organomet. Chem.*, 2010, 695, 1215-1224.

[44] Fukin, G. K.; Guzei, I. A.; Baranov, E. V.; Domrachev, G. A. *Struct. Chem.* 2009, 20, 643–654.

[45] Smolyaninov, I. V.; Poddel'sky, A. I.; Berberova, N. T.; Cherkasov, V. K.; Abakumov, G. A. *Russ. J. Coord. Chem.* 2010, 36, 644-650.

[46] Poddel'sky, A. I.; Smolyaninov, I. V.; *Russ. J. Gen. Chem.* 2010, 80, 538–540.

[47] Poddel'sky, A. I.; Smolyaninov, I. V.; Somov, N. V.; Berberova, N. T.; Cherkasov, V. K.; Abakumov, G. A. *J. Organomet. Chem.* 2010, 695, 530-536.

[48] Millington, P. L.; Sowerby, D. B. *J. Chem. Soc. Dalton Trans.* 1992, 1199-1204.

[49] Michael, H.; Sowerby, D. B. *J. Chem. Soc. Dalton Trans.* 1983, 1095-1099.

[50] Breunig, H. J.; Ebert, K. H.; Kaller, R.; Mourad, Y.; Atmani, A.; Mugnier, Y. *J. Organomet. Chem.* 1994, 483, 167-172.

[51] Demicheli, C.; Frezard, F.; Malheiros, C.; Nagem, T. J. *Main Group Met. Chem.* 2001, 24, 85-88.

[52] Poddel'sky, A. I.; Vavilina, N. N.; Somov, N. V.; Cherkasov, V. K.; Abakumov, G. A. *J. Organomet. Chem.* 2009, 694, 3462-3469.

[53] Abakumov, G. A.; Cherkasov, V. K.; Abakumova, L. G.; Druzhkov, N. O.; Nevodchikov, V. I.; Kurskii, Yu. A.; Makarenko, N. P. *Organomet. Chem.* USSR 1991, 4, 454-459.

[54] Abakumov, G. A.; Cherkasov, V. K.; Abakumova, L. G.; Nevodchikov, V. I.; Druzhkov, N. O.; Makarenko, N. P. Kurskii, Yu. A. *J. Organomet. Chem.* 1995, 491, 127-133.
[55] Shultz, D. A. *Comments Inorg. Chem.* 2002, 23, 1–21.
[56] Shultz, D. A. *Polyhedron* 2001, 20, 1627–1631.
[57] Abakumov, G. A.; Nevodchikov, V. I.; Kuropatov, V. A.; Noll, B. C.; Pierpont, C. G. *Inorg. Chem.* 1998, 37, 6117–6119.
[58] Abakumov, G. A.; Cherkasov, V. K.; Nevodchikov, V. I.; Kuropatov, V. A.; Yee, G. T.; Pierpont, C. G. *Inorg. Chem.* 2001, 40, 2434–2436.
[59] Horner, L.; Weber, K.- H. *Chem. Ber.* 1967, 100, 2842–2853.
[60] Shultz, D. A.; Boal, A. K.; Farmer, G. T. *J. Org. Chem.* 1998, 63, 9462–9469.
[61] Abakumov, G. A.; Nevodchikov, V. I.; Druzhkov, N. O.; Zakharov, L. N.; Abakumova, L. G.; Kurskii, Yu. A.; Cherkasov, V. K. *Russ. Chem. Bull.* 1997, 46, 771–775.
[62] Abakumov, G. A.; Cherkasov, V. K.; Nevodchikov, V. I.; Druzhkov, N. O.; Kuropatov, V. A. *Russ. Chem. Bull.* 1998, 47, 875–878.
[63] Poddel'sky, A. I.; Somov, N. V.; Druzhkov, N. O.; Cherkasov, V. K.; Abakumov, G. A. *J. Organomet. Chem.*, 2011, 696, 517-522.
[64] Ilyakina, E. V.; Mishchenko, O. G.; Poddel'skii, A. I.; Maslennikov, S.V.; Spirina, I. V. *Russ. J. Gen. Chem.* 2009, 79, 1483–1486.
[65] Kuropatov, V. A.; Klementieva, S. V.; Poddel'sky, A. I.; Cherkasov, V. K.; Abakumov, G. A. *Russ. Chem. Bull.* 2010, 59, 1698-1706.
[66] Cherkasov, V. K.; Grunova, E. V.; Abakumov, G. A. *Russ. Chem. Bull.*, 2005, 54, 2067-2072.
[67] Poddel'sky, A. I.; Piskunov, A. V.; Druzhkov, N. O.; Fukin, G. K.; Cherkasov, V. K.; Abakumov, G. A. *Z. Anorg. Allg. Chem.* 2009, 635, 2563-2571.
[68] Wieber, M.; Baumann, N. *Z. Anorg. Allg. Chem.* 1973, 402, 43-46.
[69] Baumann, N.; Wieber, M.; *Z. Anorg. Allg. Chem.* 1974, 408, 261-265.
[70] Wieber, M.; Baumann, N. *Z. Anorg. Allg. Chem.* 1975, 418, 167-170.
[71] Benmmaarouf, Z.; Riviere, P.; Riviere-Baudet, M.; Castel, A.; Khallaayoun, A.; Ahbala, M. *Phosphorus. sulph. and silicon*, 1997, 128, 19-29.
[72] Barnard, P. W. C.; Donaldson, J. D.; Tricker, M. J.; *Inorg. Chim. Acta.* 1979, 32, L77-78.
[73] Alonzo, G.; Bertazzi, N.; Maccotta, A. *Inorg. Chim. Acta*, 1982, 62, 167-169.
[74] Preut, H.; Huber, F.; Alonzo, G.; Bertazzi, N. *Acta Crystallogr.*, Sect. B 1982, 38, 935.
[75] Alamgir, M.; Allen, N.; Barnard, P. W. C.; Donaldson, J. D.; Silver, J. *Acta Crystallogr., Sect. B*1981, 37, 1284.
[76] Huber, P.; Preut, H.; Alonzo, G.; Bertazzi, N. *Inorg. Chim. Acta*, 1985, 102, 181-186.
[77] Brown, H. P.; Austin, J. A. *J. Am. Chem. Soc.* 1941, 63, 2054-2055.
[78] Alamgir, M.; Barnard, P. W. C.; Donaldson, J. D. *J. Chem. Soc. Dalton* 1980, 1542-1544.
[79] Biros, Sh. M.; Bridgewater, B. M.; Villeges-Estrada, A.; Tanski, J. M.; Parkin, G. *Inorg. Chem.* 2002, 41, 4051-4057.
[80] Alonzo, G.; Bertazzi, N.; Galli, P.; Marci, G.; Massacci, M. A.; Palmisano, L.; Patrono, P.; Saiano, F. *Mater. Res. Bull.* 1998, 33, 1233-1240.

[81] Sen Gupta, A. K.; Bohra, R.; Mehrotra, R. C.; Das, K. *Main Group Met. Chem.* 1990, 13, 321-339.
[82] Barnard, P. W. C.; Alamgir, Nuclear Science and Applications, Series B: *Physical Sciences* 1981, 12-13, 41-46.
[83] Alonzo, G.; Consiglio, M.; Maggio, F.; Bertazzi, N. *Inorg. Chim. Acta* 1988, 147, 217-220.
[84] Alonzo, G.; Bertazzi, N.; Maggio, F.; Benetollo, F.; Bombier, G. *Polyhedron*, 1996, 15, 4269-4273.
[85] McEwen, W. E.; Tien, J. H.; Rembetsy, L. M.; Kalbacher, B. J.; Lubinkowski, J.J.; Mari, F.; DeVincenzo, S.; Salgueiro, R. M.; Hou, J.; Alvarez, J. A. *Phosph. Sulf. and the Related Elem.*, 1988, 36, 153-164.
[86] Sharma, P.; Perez, D.; Cabrera, A.; Rosas, N.; Arias, J. L. *Acta Pharmacologica Sinica* 2008, 29, 881-890.
[87] Paver, M. A.; Joy, J. S.; Hursthouse, M. B.; *Chem. Comm.* 2002, 2150-2151.
[88] Bauer, G.; Scheffler, K.; Stegmann, H. B.; *Chem. Ber.* 1976, 109, 2231-2241.
[89] Abakumov, G. A.; Poddel'sky, A. I.; Grunova, E. V.; Cherkasov, V. K.; Fukin, G. K.; Kurskii, Yu. A.; Abakumova, L. G. *Angew. Chem. Int. Ed.* 2005, 44, 2767 –2771.
[90] Poddel'sky, A. I.; Kurskii, Yu. A.; Piskunov, A. V.; Somov, N. V.; Cherkasov, V. K.; Abakumov, G. A. *Appl. Organomet. Chem.* 2011, 25, 163–172.
[91] Bordner, J.; Doak, G. O.; Everett, T. S.; *J. Am. Chem. Soc.* 1986, 108, 4206-4213.
[92] Breunig, H. J.; Kruger, T.; Lork, E. *J. Organomet. Chem.* 2002, 648, 209-213.
[93] Sharutin, V. V.; Pakusina, A. P.; Smirnova, S. A.; Sharutina, O. K.; Platonova, T. A.; Pushilin, M. A.; Gerasimenko, A.V. *Rus. J. Coord. Chem.* 2004, 30, 314-321.
[94] Allen, F. H.; Kennard, O.; Watson, D. G.; Brammer, L.; Orpen, A. G.; Taylor, R. *J. Chem. Soc. Perkin Trans.* II. 1987, 12, 1-19.
[95] Ohkata, K.; Yako, T.; Kuwaki, T.; Akiba, K. *Chem. Lett.* 1990, 1721-1724.
[96] Larsen, S. K.; Pierpont, C. G. *J. Am. Chem. Soc.* 1988, 110, 1827-1832.
[97] Bencini, A.; Ciofini, I.; Giannasi, E.; Daul, C. A.; Doclo, K. *Inorg. Chem.* 1998, 37, 3719-3725.
[98] Alonzo, G.; Bertazzi, N.; Bombieri, G.; Bruno, G. *J. Chem. Soc. Dalton Trans.* 1987, 2337-2340.
[99] Sharma, P.; Rosas, N.; Ramashankar; Cabrera, A. *Main Group Met. Chem.* 1995, 18, 621-627.
[100] Sharma, P.; Ugal, J.R.; Jha, N.K., *Main Group Met. Chem.* 1994, 17, 631-639.
[101] Tanski, J. M.; Kelly, V.; Parkin, G. *Dalton Trans.* 2005, 2442-2447.
[102] Kelly, B. V.; Weintrob, E. C.; Buccella, D.; Tanski, J. M.; Parkin, G., *Inorg. Chem. Commun.* 2007, 10, 699-704.

About the Contributor

Manijeh Razeghi received her Docteur d'État ès Sciences Physiques, Université de Paris in September 1980. After receiving her doctoral degree, she joined Thomson-CSF (Orsay, France) as a Senior Research Scientist and then the Head of the Exploratory Materials Laboratory at from 1980 to 1991. While at Thomson, Dr. Razeghi developed and implemented modern metalorganic chemical vapor deposition (MOCVD) epitaxial growth for entire compositional ranges of III-V compound semiconductors and heterostructures. She realized the first InP Quantum wells and Superlattices and demonstrated the marvels of quantum mechanics in the low dimensional world. Her pioneering work on InP based compound semiconductors culminated in her 1989 publication of *The MOCVD Challenge Volume 1: A Survey of GaInAsP-InP for Photonic and Electronic Applications.* She joined Northwestern University, Evanston, IL, as a Walter P. Murphy Professor and Director of the Center for Quantum Devices in Fall 1991, where she created the undergraduate and graduate program in solid-state engineering and the Center for Quantum devices. She is one of the leading scientists in the field of semiconductor science and technology, pioneering in the development and implementation of major modern epitaxial techniques. Her current research interest is in nanoscale optoelectronic quantum devices. She has authored or coauthored more than 1000 papers, 15 books, and more than 30 book chapters. She holds 30 U.S. patents and has given more than 1000 invited and plenary talks. She received the IBM Europe Science and Technology Prize in 1987, the Achievement Award from the SWE in 1995, the R.F. Bunshah Award in 2004 and many best paper awards. Dr. Razeghi is an elected Fellow of SWE (1995), SPIE (2000), IEC (2003), OSA (2004), APS (2004) IOP (2005), IEEE (2005) and MRS (2008).

CONTRIBUTOR BIOGRAPHICAL SKETCH

DR. MANIJEH RAZEGHI

Name: Razeghi, Manijeh Ph.D.
Titles: Walter P. Murphy Professor of Electrical Engineering and Computer Science, and Director, Center for Quantum Devices
Date of Birth: April 28, 1943

Education:

1980	Docteur d'État ès Sciences Physiques, Université de Paris, France
1977	Docteur 3ème Cycle, Solid State Physics, Université de Paris, France
1976	DEA, Science des Matériaux, Université de Paris, France

Contact Points:

Northwestern University
Center for Quantum Devices (CQD)
Electrical Engineering and Computer Science (EECS) Department
2220 Campus Drive, Cook Hall Room 4051
Evanston, IL 60208
Tel: 847-491-7251
FAX: 847-467-1817
Email: razeghi@eecs.northwestern.edu

Research and Professional Experience:

1993-present	Adjunct Professor, College of Optical Sciences, University of Arizona

1991-present	Walter P. Murphy Professor of Electrical Engineering and Computer Science, and Director, Center for Quantum Devices, Northwestern University, Evanston, IL
2008-present	Global Ambassador for France, McCormick School of Engineering and Applied Science, Northwestern University, Evanston, IL
	Head, Exploratory Materials Lab, Thomson-CSF, Orsay, France
1989-1990	Invited Professor, Ecole Polytechnique Federale de Lausanne, France
1986	Invited Professor, University of Michigan-Ann Arbor, MI
1981-1985	Senior Research Scientist, Thomson-CSF, Orsay, France

Professional Appointments:

1993-present	Adjunct Professor, College of Optical Sciences, University of Arizona
1991-present	Walter P. Murphy Professor of Electrical Engineering and Computer Science, and Director, Center for Quantum Devices, Northwestern University, Evanston, IL
2008-present	Global Ambassador for France, McCormick School of Engineering and Applied Science, Northwestern University, Evanston, IL
	Head, Exploratory Materials Lab, Thomson-CSF, Orsay, France
1989-1990	Invited Professor, Ecole Polytechnique Federale de Lausanne, France
1986	Invited Professor, University of Michigan-Ann Arbor, MI
1981-1985	Senior Research Scientist, Thomson-CSF, Orsay, France

Honors:

- Elected Fellow of Materials Research Society (MRS) – 2008
- Featured in SPIE 2008 Women in Optics Planner Calendar and on cover
- Supervised Best Student Paper Award – Yanbo Bai, "Electrically pumped photonic crystal distributed feedback quantum cascade lasers," SPIE Photonics West Symposium, San Jose, CA, January 2008
- Best Paper Award at the 25th Army Science Conference, Orlando, FL, "Infrared Imaging with Self-Assembled InGaSb Quantum Dot Infrared Detectors" – November 2006
- Supervised research training period award, "Prix de la Fondation from Ecole Polytechnique, France," presented to Pierre-Yves Delaunay for research project entitled, "Infrared Focal Plane Arrays" – November 2005
- Elected Fellow of Institute of Electrical and Electronics Engineers (IEEE) - 2005
- Elected Fellow of Institute of Physics (IOP) - 2005
- Elected Fellow of American Physical Society (APS) - 2004
- Elected Fellow of Optical Society of America (OSA) - 2004

- R.F. Bunshah Award from Int[al] Conference on Metallurgical Coatings and Thin Films - 2004
- Photonics West Optoelectronics '04 Symposium Best Paper Award - 2004
- Nominated for Northwestern University McCormick Teacher of the Year Award - 2003
- Elected Fellow of International Engineering Consortium (IEC) - 2003
- Elected Fellow of Society of Photo-Optical Instrumentation Engineers (SPIE) - 2000
- Photonics West Optoelectronics '98 Symposium Best Paper Award - 1998
- Elected Fellow of Society of Women Engineers - 1995
- Society of Women Engineers (SWE) Achievement Award - 1995
- IBM Europe Science and Technology Prize - 1987

Publications Last 3 Years:

E. K. Huang, P. Y. Delaunay, B. M. Nguyen, S. Abdollahi Pour, and M. Razeghi, "Photovoltaic MWIR Type-II Superlattice Focal Plane Array on GaAs Substrate," *IEEE Journal of Quantum Electronics (JQE),* December 2010, Vol. 46, No. 12, p. 1704-1708.

C. Bayram, Z. Vashaei, and M. Razeghi, "Reliability in room temperature negative differential resistance characteristics of low-aluminium-content AlGaN/GaN double-barrier resonant tunneling diodes," to appear in *Applied Physics Letters.*

P. Manurkar, S. R. Darvish, B. M. Nguyen, M. Razeghi, and J. Hubbs, "High performance long wavelength Infrared mega-pixel focal plane array based on type-II superlattices," to appear in *Applied Physics Letters.*

J. S. Yu, S. Slivken, and M. Razeghi, "Injector doping level dependent continuous-wave operation of InP-based QCLs at $\lambda \sim 7.3$ μm above room temperature," to appear in *Semiconductor Science and Technology.*

M. Razeghi, Y. Bai, S. Slivken and S. R. Darvish, "High performance mid-infrared quantum cascade lasers at Northwestern University," to appear in *SPIE Special Issue of Optical Engineering on Quantum and Interband Cascade Lasers,* November 2010, Vol. 49, No. 11.

B. Gökden, Y. Bai, N. Bandyopadhyay, S. Slivken, and M. Razeghi, "Broad area photonic crystal distributed feedback quantum cascade lasers emitting 34 W at $\lambda \sim 4.36$ μm," *Applied Physics Letters,* 27 September 2010, Vol. 97, No. 13, p. 131112-1.

N. Bandyopadhyay, Y. Bai, B. Gokden, A. Myzaferi, S. Tsao, S. Slivken and M. Razeghi, "Watt level performance of quantum cascade lasers in room temperature continuous wave operation at $\lambda \sim 3.76$ mm," *Applied Physics Letters,* 27 September 2010, Vol. 97, No. 13, p. 131117-1.

Z. Vashaei, C. Bayram, P. Lavenus, and M. Razeghi, "Photoluminescence characteristics of polar and nonpolar AlGaN/GaN superlattices," *Applied Physics Letters,* 20 September 2010, Vol. 97, No. 12, p. 121918-1.

C. Bayram, Z. Vashaei, and M. Razeghi, "Room temperature negative differential resistance characteristics of polar III-nitride resonant tunneling diodes," *Applied Physics Letters,* 30 August 2010, Vol. 97, No. 9, p. 092104-1.

R. McClintock, E. Cicek, Z. Vashaei, C. Bayram, M. Razeghi and M. Ulmer, "III-nitride based avalanche photo detectors," SPIE Conference, August 1-5, 2010, San Diego, CA *Proceedings* – Detectors and Imaging Devices: Infrared, Focal plane and Single Photon, Vol. 7780, p. 77801B.

E. Cicek, Z. Vashaei, C. Bayram, R. McClintock, M. Razeghi and M. Ulmer, "Comparison of ultraviolet APDs grown on free-standing GaN and sapphire substrates," SPIE Conference, August 1-5, 2010, San Diego, CA *Proceedings* – Detectors and Imaging Devices: Infrared, Focal plane and Single Photon, Vol. 7780, p. 77801P.

E. Cicek, Z. Vashaei, R. McClintock, C. Bayram and M. Razeghi, "Geiger-mode operation of ultraviolet avalanche photodiodes grown on sapphire and free-standing GaN substrates," *Applied Physics Letters,* 28 June 2010, Vol. 96, No. 26, p. 261107-1.

Z. Vashaei, E. Cicek, C. Bayram, R. McClintock and M. Razeghi, "GaN avalanche photodiodes grown on m-plane freestanding GaN substrate," *Applied Physics Letters*, 17 May 2010, Vol. 96, No. 20, p. 201908-1.

M. Razeghi, E. K. Huang, B. M. Nguyen, S. Abdollahi Pour and P. Y. Delaunay, "Type-II antimonide-based superlattices for the third generation infrared focal plane arrays," SPIE Conference, April 5-9, 2010, Orlando, FL *Proceedings* – Infrared Technology and Applications XXXVI, Vol. 7660, p. 76601F.

M. Razeghi, C. Bayram, R. McClintock, F. Hosseini Teherani, D. J. Rogers and V. E. Sandana, "Novel Green Light Emitting Diodes: Exploring Droop-Free Lighting Solutions for a Sustainable Earth," *Journal of Light Emitting Diodes*, Vol. 2, No. 1, April 2010, p. 1-33.

Z. Vashaei, C. Bayram, and M. Razeghi, "Demonstration of negative differential resistance in GaN/AlN resonant tunneling diodes at room temperature," *Journal of Applied Physics*, 15 April 2010, Vol. 107, no. 8, p. 083505.

P. Y. Delaunay and M. Razeghi, "Spatial noise and correctability of Type-II InAs/GaSb focal plane arrays," *IEEE Journal of Quantum Electronics*, April 2010, Vol. 46, No. 4, p. 584-588.

Y. Bai, S. Slivken, S. Kuboya, S. R. Darvish, and M. Razeghi, "Quantum cascade lasers that emit more light than heat," *Nature Photonics*, February 2010, Vol. 4, p. 99-102.

J. L. Pau, J. Piqueras, D. J. Rogers, F. Hosseini Teherani, K. Minder, R. McClintock, and M. Razeghi, "On the interface properties of ZnO/Si electroluminescent diodes," *Journal of Applied Physics,* 1 February 2010, Vol. 107, No. 3, p. 033719-1.

M. Razeghi and B.M. Nguyen, "Band gap tenability of Type II Antimonide-based superlattices," *Proceedings* of 14th International Conference on Narrow Gap Semiconductors and Systems\(NGSS-14), Sendai, Japan, July 13-17, 2009, *Physics Procedia,* Vol. 3, Issue 2, 31 January 2010, p. 1207-1212.

C. Bayram, Z. Vashaei, and M. Razeghi, "AlN/GaN double-barrier resonant tunneling diodes grown by metal-organic chemical vapor deposition," *Applied Physics Letters,* 25 January 2010, Vol. 96, No. 4, p. 042103-1.

S. Tsao, A. Myzaferi, and M. Razeghi, "High performance quantum dot-quantum well infrared focal plane arrays," SPIE Photonics West, January 22-28, 2010, San Francisco, CA *Proceedings* – Optoelectronic Integrated Circuits XII, Vol. 7605, p. 76050J-1.

D. J. Rogers, F. Hosseini Teherani, V. E. Sandana, and M. Razeghi, "ZnO thin films and nanostructures for emerging optotlectronic applications," SPIE Photonics West, January

22-28, 2010, San Francisco, CA *Proceedings* – Optoelectronic Integrated Circuits XII, Vol. 7605, p. 76050K-1.

B. Gokden, S. Slivken, and M. Razeghi, "High power photonic crystal distributed feedback quantum cascade lasers emitting at 4.5 mm," SPIE Photonics West, January 22-28, 2010, San Francisco, CA *Proceedings* – Quantum Sensing and Nanophotonic Devices VII, Vol. 7608, p. 760806-1.

S. Slivken, Y. Bai, B. Gokden, S. R. Darvish and M. Razeghi, "Current status and potential of high power mid-infrared intersubband lasers," SPIE Photonics West, January 22-28, 2010, San Francisco, CA *Proceedings* – Quantum Sensing and Nanophotonic Devices VII, Vol. 7608, p. 76080B-1.

Y. Bai, S. Slivken, S. R. Darvish and M. Razeghi, "Very high wall plug efficiency of quantum cascade lasers," SPIE Photonics West, January 22-28, 2010, San Francisco, CA *Proceedings* – Quantum Sensing and Nanophotonic Devices VII, Vol. 7608, p. 76080F-1.

M. Razeghi, B. M. Nguyen, P. Y. Delaunay, S. Abdollahi Pour, E. K. W. Huang, P. Manukar, S. Bogdanov, and G. Chen, "High operating temperature MWIR photo detectors based on Type II InAs/GaSb superlattice," SPIE Photonics West, January 22-28, 2010, San Francisco, CA *Proceedings* – Quantum Sensing and Nanophotonic Devices VII, Vol. 7608, p. 76081Q-1.

B. M. Nguyen, S. Abdollahi Pour, S. Bogdanov, and M. Razeghi, "Minority electron unipolar photodetectors based on Type II InAs/GaSb/AlSb superlattices for very long wavelength infrared detection," SPIE Photonics West, January 22-28, 2010, San Francisco, CA *Proceedings* – Quantum Sensing and Nanophotonic Devices VII, Vol. 7608, p. 760825-1.

M. Razeghi, S. Slivken, Y. Bai, B. Gokden, S. R. Darvish, "High Power Quantum Cascade Lasers," special issue "Advanced Semiconductor Heterostructures for Optoelectronics," *New Journal of Physics* (NJP), Volume 11, December 2009, p. 125017.

Y. Bai, S. Slivken, S. R. Darvish, A. Haddadi, B. Gokden, and M. Razeghi, "High power broad area quantum cascade lasers," *Applied Physics Letters*, 30 November 2009, Vol. 95, No. 22, p. 221104-1.

C. Bayram, N. Pere-Laperne, and M. Razeghi, "Effects of growth temperature and well width on optical and structural characteristics of AlN/GaN superlattices grown by metal-organic chemical vapor deposition," *Applied Physics Letters*, 16 November 2009, Vol. 95, No. 20, p. 201906.

B. M. Nguyen, S. Bogdanov, S. Abdollahi Pour, and M. Razeghi, "Minority electron unipolar photodetectors based on Type II superlattice for very long wavelength infrared detection," *Applied Physics Letters*, 2 November 2009, Vol. 95, No. 18, p. 183502.

S. Abdollahi Pour, B-M. Nguyen, S. Bogdanov, E. K. Huang and M. Razeghi, "Demonstration of high performance long wavelength infrared Type II InAs/GaSb superlattice photodiode grown on GaAs substrate," *Applied Physics Letters*, 26 October 2009, Vol. 95, No. 17, p. 173505.

P. Y. Delaunay and M. Razeghi, "Noise Analysis in Type-II InAs/GaSb Focal Plane Arrays," *Virtual Journal of Nanoscale Science and Technology*, 5 October 2009, Vol. 20, No. 14.

N. Pere-Laperne, C. Bayram, L. Nguyen-The, R. McClintock and M. Razeghi, "Tunability of intersubband absorption from 4.5 to 5.3 μm in a $GaN/Al_{0.2}Ga_{0.8}N$ superlattices grown by metalorganic chemical vapor deposition, " *Applied Physics Letters*, 28 September 2009, Vol. 95, No. 13, p. 131109.

P. Y. Delaunay and M. Razeghi, "Noise Analysis in Type-II InAs/GaSb Focal Plane Arrays," *Journal of Applied Physics*, 15 September 2009, Vol. 106, Issue 6, p. 063110.

T. Yamanaka, B. Movaghar, S. Tsao, S. Kuboya, A. Myzaferi and M. Razeghi, "Gain-length scaling in quantum dot/quantum well infrared photodetectors," *Virtual Journal of Nanoscale Science and Technology*, 14 September 2009, Vol. 20, No. 11.

C. Bayram and M. Razeghi, "Ultraviolet Detectors: Nitrides push performance of UV Photodidoes," *Laser Focus World*, September 2009, Vol. 45, Issue 9, p. 47-51.

J. S. Yu, H. K. Lee, S. Slivken, and M. Razeghi, "Thermal characteristics and analysis of quantum cascade lasers for biochemical sensing applications," SPIE Conference, August 2-6, 2009, San Diego, CA *Proceedings* – Biosensing II, Vol. 7397, p. 739705F-1.

T. Yamanaka, B. Movaghar, S. Tsao, S. Kuboya, A. Myzaferi and M. Razeghi, "Gain-length scaling in quantum dot/quantum well infrared photodetectors," *Applied Physics Letters*, 31 August 2009, Vol. 95, No. 9, p. 093502-1.

M. Razeghi, "Superlattices see in the dark," SPIE Newsroom http://spie.org/x36271.xml?highlight=x2410&ArticleID=x36271, 28 July 2009.

Y. Bai, B. Gokden, S. R. Darvish, S. Slivken, and M. Razeghi, "Photonic crystal distributed feedback quantum cascade lasers with 12 W output power," *Applied Physics Letters*, 20 July 2009, Vol. 95, No. 3, p. 031105-1.

C. Bayram and M. Razeghi, "Stranski-Krastanov growth of InGaN quantum dots emitting in green spectra," *Applied Physics A: Materials Science and Processing,* Vol. 96, No. 2, August 2009, p. 403-408.

B. M. Nguyen, D. Hoffman, E. K. Huang, S. Bogdanov, P. Y. Delaunay, M. Razeghi and M. Tidrow, "Demonstration of mid-infrared type II InAs/GaSb superlattice photodiodes grown on GaAs substrate," *Applied Physics Letters*, 8 June 2009, Vol. 94, No. 22, p. 223506-1.

M. Razeghi, D. Hoffman, B. M. Nguyen, P. Y. Delaunay, E. K. Huang, M. Tidrow, and V. Nathan, "Recent Advances in LWIR Type-II InAs/GaSb Superlattice Photodetectors and Focal Plane Arrays at the Center for Quantum Devices, " *Proceedings of the IEEE Journal on Optics and Photonics for Defense and Security*, Vol. 97, No. 6, June 2009, p. 1056-1066.

M. Razeghi and C. Bayram, "Material and design engineering of (Al)GaN for high-performance avalanche photodiodes and intersubband applications," SPIE Conference, May 4-6, 2009, Dresden, Germany *Proceedings* – Photonic Materials, Devices and Applications III, Vol. 7366, p. 73661F-1.

M. Razeghi and R. McClintock, "A Review of III-Nitride Research at the Center for Quantum Devices," *Proceedings* of 2nd International Symposium on Growth of III-Nitrides (ISGN-2), Laforet Shuzenji Izu, Japan, July 6-9, 2008, *Journal of Crystal Growth,* 1 May 2009, Vol. 311, No. 10, p. 3067-3074.

M. Razeghi, "High Performance InP-Based Mid-Infrared Quantum Cascade Lasers," *IEEE Journal of Selected Topics in Quantum Electronics: Semiconductor Lasers*, Vol. 15, Issue 3, May/June 2009, p. 941-951.

D. J. Rogers, F. Hosseini Teherani, T. Moudakir, S. Gautier, F. Jomard, M. Molinari, M. Troyon, D. McGrouther, J. N. Chapman, M. Razeghi and A. Ougazzaden, "Microstructural compositional, and optical characterization of GaN grown by metal organic vapor phase epitaxy on ZnO epilayers," *Journal of Vacuum Science and Technology B*, Vol. 27, No. 3, May/June 2009, p. 1655-1657.

C. Bayram, D. J. Rogers, F. Hosseini Teherani, and M. Razeghi, "Fabrication and Characterization of Novel Hybrid Green LEDs based on Substituting n-type ZnO for n-type GaN in an Inverted p-n Junction," *Journal of Vacuum Science and Technology B*, Vol. 27, No. 3, May/June 2009, p. 1784-1788.

V. E. Sandana, D. J. Rogers, F. H. Teherani, R. McClintock, C. Bayram, M. Razeghi, H.-J. Drouhin, M. C. Clochard, V. Sallet, G. Garry, and F. Falyouni, "Comparison of ZnO Nanostructures Grown Using pulsed layer deposition, metalorganic chemical vapor deposition, and physical vapor transport", *Journal of Vacuum Science and Technology B*, Vol. 27, No. 3, May/June 2009, p. 1678-1683.

B. M. Nguyen, D. Hoffman, E. K. Huang, P. Y. Delaunay, and M. Razeghi, "High performance antimony based type-II superlattice photodiodes on GaAs substrate," SPIE Conference, April 13-17, 2009, Orlando, FL *Proceedings* – Infrared Technology and Applications XXXV, Vol. 7298, p. 72981T.

P. Y. Delaunay, B. M. Nguyen and M. Razeghi, "Background limited performance of long wavelength infrared focal plane arrays fabricated from type-II InAs/GaSb M-structure superlattice," SPIE Conference, April 13-17, 2009, Orlando, FL *Proceedings* – Infrared Technology and Applications XXXV, Vol. 7298, p. 72981Q.

C. Bayram, N. Pere-Laperne, R. McClintock, B. Fain and M. Razeghi, "Pulsed metalorganic chemical vapor deposition of high quality AlN/GaN superlattices for near-infrared intersubband transitions, " *Applied Physics Letters*, 23 March 2009, Vol. 94, No. 12, p. 121902-1.

P. Y. Delaunay, B. M. Nguyen, D. Hoffman, E. K. Huang and M. Razeghi, "Background limited performance of long wavelength infrared focal plane arrays fabricated from M-structure InAs-GaSb superlattices," *IEEE Journal of Quantum Electronics*, Vol. 45, No. 2, February 2009, p. 157-162.

E. K. Huang, D. Hoffman, B. M. Nguyen, P. Y. Delaunay and M. Razeghi, "Surface leakage reduction in narrow band gap type-II antimonide-base superlattice photodiodes," *Applied Physics Letters*, 94 (5), 2 February 2009, p. 053506-1.

H. K. Lee, K. S. Chung, J. S. Yu and M. Razeghi, "Thermal analysis of buried heterostructure quantum cascade lasers for long wavelength infrared emission using 2D anisotropic heat dissipation model," *Physica Status Solidi (a)*, Vol. 206, February 2009, p. 356-362.

R. McClintock, J. L. Pau, C. Bayram, B. Fain, P. Giedratis, M. Razeghi and M. Ulmer, "III-Nitride avalance photodiodes," SPIE Conference, January 25-29, 2009, San Jose, CA *Proceedings* – Quantum Sensing and Nanophotonic Devices VI, Vol. 7222, p. 7222-0U.

Y. Bai, B. Gokden, S. Slivken, S. R. Darvish, S. A. Pour and M. Razeghi, "Mid-infrared quantum cascade lasers with high wall plug efficiency," SPIE Conference, January 25-29, 2009, San Jose, CA *Proceedings* – Quantum Sensing and Nanophotonic Devices VI, Vol. 7222, p. 7222-00.

C. Bayram, B. Fain, N. Pere-Laperne, R. McClintock and M. Razeghi, "Pulsed metalorganic chemical vapor deposition of high quality AlN/GaN superlattices for intersubband transitions," SPIE Conference, January 25-29, 2009, San Jose, CA *Proceedings* – Quantum Sensing and Nanophotonic Devices VI, Vol. 7222, p. 7222-12.

J. L. Pau, C. Bayram, P. Giedraitis, R. McClintock, and M. Razeghi, "GaN-based nanostructured photodetectors," SPIE Conference, January 25-29, 2009, San Jose, CA *Proceedings* – Quantum Sensing and Nanophotonic Devices VI, Vol. 7222, p. 7222-14.

D. Hoffman, B. M. Nguyen, E. K. Huang, P. Y. Delaunay, S. Bogdanov, P. Manurkar, M. Razeghi, and V. Nathan, "The importance of band alignment in VLWIR type-II InAs/GaSb heterodiodes containing the M-structure barrier," SPIE Conference, January 25-29, 2009, San Jose, CA *Proceedings* – Quantum Sensing and Nanophotonic Devices VI, Vol. 7222, p. 7222-15.

P. Y. Delaunay, B. M. Nguyen, D. Hoffman, E. K. Huang, P. Manurkar, S. Bogdanov and M. Razeghi, "Background limited performance of long wavelength infrared focal plane arrays fabricated from M-structure InAs/GaSb superlattices," SPIE Conference, January 25-29, 2009, San Jose, CA *Proceedings* – Quantum Sensing and Nanophotonic Devices VI, Vol. 7222, p. 7222-0W.

E. K. Huang, B. M. Nguyen, D. Hoffman, P. Y. Delaunay and M. Razeghi, "Inductively coupled plasma etching and processing techniques for type-II InAs/GaSb superlattices infrared detectors toward high fill factor focal plane arrays," SPIE Conference, January 25-29, 2009, San Jose, CA *Proceedings* – Quantum Sensing and Nanophotonic Devices VI, Vol. 7222, p. 7222-0Z.

C. Bayram, F. H. Teherani, D. J. Rogers and M. Razeghi, "Hybrid green LEDs based on *n*-ZnO/(InGaN/GaN) multi-quantum-wells/*p*-GaN," SPIE Conference, January 25-29, 2009, San Jose, CA *Proceedings* – Zinc Oxide Materials and Devices IV, Vol. 7217, p. 7217-0P.

S. Tsao, T. Yamanaka, S. A. Pour, I-K. Park, B. Movaghar and M. Razeghi, "Quantum dot in a well infrared photodetectors for high operating temperature focal plane arrays," SPIE Conference, January 25-29, 2009, San Jose, CA *Proceedings* – Quantum Dots, Particles and Nanoclusters VI, Vol. 7224, p. 7224-0V.

M. Razeghi, "High-power high-wall plug efficiency mid-infrared quantum cascade lasers based on InP/GaInAs/InAlAs material system," SPIE Conference, January 25-29, 2009, San Jose, CA *Proceedings* – Novel In-Plane Semiconductor Lasers VIII, Vol. 7230, p. 7230-11.

J. L. Pau, C. Bayram, R. McClintock and M. Razeghi, "GaN nanostructred p-i-n photodiodes," *Applied Physics Letters*, 93 (22), 1 December 2008, p. 221104-1.

C. Bayram, J. L. Pau, R. McClintock, and M. Razeghi, "Comprehensive study of blue and green multi-quantum-well light emitting diodes grown on conventional and lateral epitaxial overgrowth GaN," *Applied Physics B-Lasers and Optics*, Vol. 95, 29 November 2008, p. 307-314.

C. Bayram, J. L. Pau, R. McClintock, M. Razeghi, M. P. Ulmer, and D. Silversmith, "High quantum efficiency back-illuminated GaN avalanche photodiodes," *Applied Physics Letters,* 93 (21), 24 November 2008, p. 211107-1.

J. S. Yu, S. Slivken, A. Evans, and M. Razeghi, "High-Performance, Continuous-Wave Quantum-Cascade Lasers Operating up to 85^0C at 1 ~ 8.8 mm," *Applied Physics A: Materials Science and Processing*, Vol 93, No. 2, November 2008, p. 405-408.

B. M. Nguyen, D. Hoffman, P. Y. Delaunay, E. K. Huang, M. Razeghi and J. Pellegrino, "Band edge tunability of M-structure for heterojunction design in Sb based Type II superlattice photodiodes," *Applied Physics Letters,* 93 (16), 20 October 2008, p. 163502-1.

C. Bayram, J. L. Pau, R. McClintock, and M. Razeghi, "Delta-doping optimization for high quality *p*-type GaN," *Journal of Applied Physics,* 104 (8), 15 October 2008, p. 083512-1.

S. Slivken, Y. Bai, S.R. Darvish and M. Razeghi, "Powerful QCLs eye remote sensing," *Compound Semiconductor Magazine*, Vol. 14 No. 9, October 2008, p. 21-23.

B. Movaghar, S. Tsao, S. Abdollahi Pour, T. Yamanaka, and M. Razeghi, "Gain and recombination dynamics in photodetectors made with quantum nanostructures: the quantum dot in a well and the quantum well," *Virtual Journal of Nanoscale Science and Technology*, Vol. 18, No. 14, 6 October 2008.

M. Razeghi, "Quantum Devices based on Modern Band Structure Engineering and Epitaxial Technology," *Modern Physics Letters B*, Vol. 22 No. 24, 20 September 2008, p. 2343-2371.

B. M. Nguyen, D. Hoffman, E. K. Huang, P. Y. Delaunay, and M. Razeghi, "Background limited long wavelength infrared Type-II InAs/GaSb superlattice photodiodes operating at 110K," *Applied Physics Letters,* 93 (12), 22 September 2008, p. 123502-1.

B. Movaghar, S. Tsao, S. Abdollahi Pour, T. Yamanaka, and M. Razeghi, "Gain and recombination dynamics in photodetectors made with quantum nanostructures: the quantum dot in a well and the quantum well," *Physical Review B*, Vol. 78, No. 11, 15 September 2008, p. 115320-1.

C. Bayram, F. H. Teherani, D. J. Rogers and M. Razeghi, "A hybrid green light-emitting diode comprised of *n*-ZnO/(InGaN/GaN) multi-quantum wells/*p*-GaN," *Applied Physics Letters,* 93 (8), 25 August 2008, p. 081111-1.

J. S. Yu, S. Slivken, A. Evans, and M. Razeghi, "High-Performance Continuous-Wave Operation of 1 ~ 4.6 mm Quantum-Cascade Lasers Above Room Temperature," *IEEE Journal of Quantum Electronics,* Vol. 44, No. 8, August 2008, p. 747-754.

M. Razeghi, B. M. Nguyen, D. Hoffman, P. Y. Delaunay, E. K. Huang, M. Z. Tidrow, and V. Nathan, "Development of material quality and structural design for high performance type-II InAs/GaSb superlattice photodiodes and focal plane arrays," SPIE Conference, August 11-14, 2008, San Diego, CA *Proceedings* – Infrared Spaceborne remote Sensing and Instrumentation XVI, Vol. 7082, p. 708204.

B. M. Nguyen, D. Hoffman, P. Y. Delaunay, E. K. Huang, and M. Razeghi, "Very high performance LWIR and VLWIR type-II InAs/GaSb superlattice photodiodes with M-structure barrier," SPIE Conference, August 11-14, 2008, San Diego, CA *Proceedings* – Infrared Spaceborne remote Sensing and Instrumentation XVI, Vol. 7082, p. 708205.

D. Hoffman, B. M. Nguyen, E. K. Huang, P. Y. Delaunay, M. Razeghi, M. Tidrow, and J. Pellegrino, "The Effect of Doping the M-Barrier in VLWIR Type-II InAs/GaSb Heterodiodes," *Applied Physics Letters*, 93 (3), 21 July 2008, p. 031107-1.

Y. Bai, S. Slivken, S. R. Darvish and M. Razeghi, "Room temperature continuous wave operation of quantum cascade lasers with 12.5% wall plug efficiency," *Applied Physics Letters*, 93 (2), 14 July 2008, p. 021103-1.

M. Razeghi, "Type II superlattices could be the next solution for fast and uniform infrared imaging," *SPIE Professional Magazine*, July 2008.

D. J. Rogers, D. C. Look, F. H. Teherani, K. Minder, M. Razeghi, A. Largeteau, G. Demazeau, J. Morrod, K. A. Prior, A. Lusson, and S. Hassani, "Investigations of ZnO thin films grown on c-Al_2O_3 by pulsed laser deposition in N_2 + O_2 ambient," *Physica Status Solidi (c)*, Vol. 5, No. 9, July 2008, p. 3084-3087.

M. Razeghi, S. Slivken, Y. Bai, and S. R. Darvish, "The Quantum Cascade Laser: A Versatile and Powerful Tool," *Optics and Photonics News (OPN) Magazine*, July/August 2008, Vol 19, No. 7/8, p. 42-47.

C. Bayram, J. L. Pau, R. McClintock, and M. Razeghi, "Performance enhancement of GaN ultraviolet avalanche photodiodes with p-type d–doping," *Applied Physics Letters*, 92 (24), 16 June 2008, p. 241103-1.

S. Tsao, H. Lim, H. Seo, W. Zhang and M. Razeghi, "InP-based quantum-dot infrared photodetectors with high quantum efficiency and high temperature imaging," *IEEE Sensors Journal, Special Issue on Nanosensors for Defense and Security*, Vol. 8, No. 6, June 2008, p. 936-941.

P. Y. Delaunay, B. M. Nguyen, D. Hoffman, A. Hood, E. K. Huang, M. Razeghi and M. Z. Tidrow, "High quantum efficiency two color type-II InAs/GaSb n-i-p-p-i-n photodiodes, "*Applied Physics Letters*, 92 (11), 17 March 2008, p. 111112-1.

M. Razeghi, D. Hoffman, B. M. Nguyen, P. Y. Delaunay, E. K. Huang, and M. Z. Tidrow, "Recent advances in LWIR type-II InAs/GaSb superlattice photodetectors and focal plane arrays at the Center for Quantum Devices," SPIE Conference, March 17-19, 2008, Orlando, FL *Proceedings* – Infrared Technology and Applications XXXIV, Vol. 6940, p. 694009.

P. Y. Delaunay, B. M. Nguyen, D. Hoffman and M. Razeghi, "High performance focal plane array based on InAs/GaSb superlattices with a 10 mm cutoff wavelength," *IEEE Journal of Quantum Electronics,* Vol. 44, No. 5, May 2008, p. 462-467.

J. L. Pau, R. McClintock, C. Bayram, K. Minder, D. Silversmith and M. Razeghi, "High Optical Response in Forward Biased (In, Ga)N-GaN Multiquantum Well Diodes under Barrier Illumination," *IEEE Journal of Quantum Electronics*, Vol. 44, No. 4, April 2008, p. 346-353.

J. L. Pau, C. Bayram, R. McClintock, M. Razeghi, and D. Silversmith "Back-illuminated separate absorption and multiplication GaN avalanche photodiodes," *Applied Physics Letters*, 92 (10), 10 March 2008, p. 101120-1.

Y. Bai, S. Slivken, S. R. Darvish, W. Zhang, A. Evans, J. Nguyen and M. Razeghi, "Room temperature continuous wave operation of quantum cascade lasers with watt-level optical power," *Applied Physics Letters*, 92 (10), 10 March 2008, p. 101105-1.

R. McClintock, J. L. Pau, K. Minder, C. Bayram and M. Razeghi, "III-Nitride photon counting avalanche photodiodes," SPIE Conference, January 20-25, 2008, San Jose, CA *Proceedings* – Quantum Sensing and Nanophotonic Devices V, Vol. 6900, p. 69000N-1-11.

P. Y. Delaunay and M. Razeghi, "High performance focal plane array based on type-II InAs/GaSb superlattice heterostructures," SPIE Conference, January 20-25, 2008, San Jose, CA *Proceedings* – Quantum Sensing and Nanophotonic Devices V, Vol. 6900, p. 69000M-1-10.

S. Slivken, A. Evans, J. Nguyen, Y. Bai, P. Sung, S. R. Darvish, W. Zhang and M. Razeghi, "Overview of Quantum Cascade Laser Research at the Center for Quantum Devices," SPIE Conference, January 20-25, 2008, San Jose, CA *Proceedings* – Quantum Sensing and Nanophotonic Devices V, Vol. 6900, p. 69000B-1-8.

Y. Bai, P. Sung, S. R. Darvish, W. Zhang, A. Evans, S. Slivken, and M. Razeghi, "Electrically pumped photonic crystal distributed feedback quantum cascade lasers," SPIE Conference, January 20-25, 2008, San Jose, CA *Proceedings* – Quantum Sensing and Nanophotonic Devices V, Vol. 6900, p. 69000A-1-8.

A. Hood, A. Evans and M. Razeghi, "Type-II Superlattices and Quantum Cascade Lasers for MWIR and LWIR Free-Space Communications," SPIE Conference, January 20-25, 2008,

San Jose, CA *Proceedings* – Quantum Sensing and Nanophotonic Devices V, Vol. 6900, p. 690005-1-9.

V. E. Sandana, D. J. Rogers, F. H. Teherani, R. McClintock, M. Razeghi, H. J. Drouhin, M. C. Clochard, V. Sallett, G. Garry and F. Fayoud, "MOCVD Growth of ZnO Nanostructures Using Au Droplets as Catalysts," SPIE Conference, January 20-25, 2008, San Jose, CA *Proceedings* – Zinc Oxide Materials and Devices III, Vol. 6895, p. 68950Z-1-6.

INDEX

#

20th century, 6

A

absorption spectra, 130, 135, 139, 140, 154, 163, 165
absorption spectroscopy, 24, 26, 220
access, 22
accounting, 78, 167
acetic acid, 300
acetone, 126, 145, 283
acetonitrile, 257, 281, 289
acid, 4, 7, 22, 24, 26, 28, 48, 56, 59, 146, 151, 201, 202, 209, 211, 216, 217, 218, 219, 220, 221, 252, 294, 300
acidic, 4, 22, 126, 127, 208, 251, 288
acidity, 208
ACL, 44, 46
activation energy, 78, 80, 81, 84, 85, 216, 219, 221
active compound, 47
additives, 182, 200
adhesion, 230, 237, 252, 259
adlayers, 202, 206, 216, 217
adsorption, 20, 22, 23, 26, 191, 192, 193, 194, 195, 201, 202, 205, 206, 207, 208, 209, 210, 212, 214, 216, 217, 218, 219, 220, 221, 222
aerogels, 257
aerosols, 1, 2, 3, 4, 5, 9, 10, 11, 12, 13, 14, 15
Africa, 6
age, 4, 177
agencies, 3, 117
alcohols, 273
algae, 28, 37
Algeria, 46, 56
alkylation, 272
alters, 48
aluminium, 171, 311
amalgam, 273
ambient air, 17
amine, 26
amino, 293, 296, 297, 298, 301
ammonia, 274
ammonium, 104
ancient world, 2
aneuploidy, 61
animal husbandry, 55
annealing, 126, 128, 129, 130, 131, 134, 136, 139, 166, 167, 173, 247, 252, 258, 261, 263
anthroponotic visceral leishmaniasis (AVL), 44
antibiotic, 43, 44, 57
antibiotic resistance, 57
aquatic systems, 19, 20, 21, 22, 23, 25, 37, 39
aqueous solutions, 160, 250
argon, 27, 247
arsenic, 1, 2, 3, 4, 5, 7, 9, 13, 15, 16, 17, 49, 50, 171, 175, 177, 192, 198, 199, 302, 303
arsenic poisoning, 4
asbestos, 8, 15
assessment, 8, 14, 16, 19, 101, 185, 198
asymmetry, 158
atmosphere, 4, 10, 11, 132, 135, 149, 159, 167, 263
atmospheric pressure, 3
atoms, 63, 64, 66, 67, 68, 70, 73, 74, 76, 78, 80, 83, 86, 88, 89, 94, 104, 147, 152, 158, 167, 183, 195, 202, 203, 205, 207, 210, 212, 214, 217, 218, 219, 246, 264, 280, 284, 285, 287, 291, 292, 301
ATP, 48, 58
attachment, 20, 145
Au nanoparticles, 159, 160, 161, 167, 168, 169, 171, 174
Austria, 244
Automobile, 8
awareness, vii

B

bacteria, 28, 37, 55
band gap, 95, 96, 98, 99, 121, 130, 135, 136, 139, 140, 150, 153, 315
bandgap, 96, 97, 103, 104, 264
Bangladesh, 4, 17, 46, 56
barriers, 95
base, 5, 70, 75, 76, 89, 121, 144, 151, 158, 162, 167, 176, 225, 247, 272, 275, 280, 287, 288, 290, 299, 315
basic research, 144
basicity, 151, 152, 158, 169, 172
batteries, 7, 227, 239, 241, 243
behaviors, 20, 22, 23, 214, 237
bending, 104, 147, 272
benefits, 101
benzene, 278
bias, 108, 109, 113, 115, 195, 206, 210, 213
binding energy, 172
bioaccumulation, 28, 32, 36, 37
bioavailability, 19, 20, 22, 29, 35, 36, 37
biochemical processes, 271
bioconversion, 47
biological processes, 20, 21
biomarkers, 54
biomass, 175, 176, 183, 196
biomedical applications, 227
biosphere, 3
biosynthesis, 53
biotic, 21
birefringence, 166
bismuth, 143, 167, 243, 277, 302, 303
boilers, 258
Bolivia, 6
Boltzmann constant, 106
bonding, 151, 206, 246, 271, 272, 277, 291, 301
bonds, 64, 68, 70, 74, 78, 79, 89, 104, 147, 151, 152, 202, 210, 288, 291, 295, 299
bounds, 66, 85
brake-linings, 1, 2, 8, 11, 13, 14, 15
Brazil, 46, 49, 60
breathing, 5
Britain, 177
bromine, 13, 276, 290, 300
bulk interactions, 175
bulk materials, 98, 203, 229, 236, 248
by-products, 279

C

cadmium, 13, 177
calcium, 7
calibration, 115
cancer, 4
candidates, 149, 161, 166, 169, 239
carbon, 28, 60, 125, 130, 135, 144, 145, 175, 176, 201, 220, 221, 235, 244
carbon dioxide, 175, 176
carbon monoxide, 175, 176, 201
carcinogenicity, 5
catabolism, 56
catalysis, 163, 199, 201, 202, 216
catalyst, 7, 183, 199, 202, 218, 221, 257
catalytic activity, 211, 218, 221
catalytic effect, 218
catalytic properties, 181, 183
cation, 143, 150, 152, 202, 208, 241, 291, 302
CBD, 126
cell death, 62
cell line, 57
ceramic, 245, 257, 258
Ceramics, 245
chalcogenides, 125
challenges, 27, 115
chemical, 1, 2, 11, 13, 14, 19, 20, 21, 23, 24, 25, 26, 30, 35, 38, 63, 65, 66, 72, 73, 74, 76, 93, 103, 104, 126, 127, 128, 131, 135, 136, 139, 155, 175, 201, 202, 208, 212, 222, 229, 241, 245, 251, 252, 256, 264, 271, 272, 281, 301, 307, 312, 313, 315
chemical bonds, 74
chemical interaction, 202
chemical properties, 281, 301
chemical stability, 245
chemical vapor deposition, 307, 312, 313, 315
chemical vapour deposition, 229
chemicals, 250
chemisorption, 183, 195, 199, 217
chemotherapy, 49
children, 7, 60
China, 6, 7, 10, 176, 198, 201, 222, 265
chlorine, 280
chloroform, 283
chromatography, 26, 28
chromium, 252, 258, 261, 263
chromosome, 59, 61
classification, 2, 5
cleanup, 176, 193, 196, 197
cleavage, 288
climate, 56
closure, 7
clothing, 7
clustering, 54, 208
clusters, 63, 64, 66, 67, 68, 70, 72, 73, 74, 75, 76, 78, 83, 84, 86, 88, 89, 139, 203, 208, 209, 213
CO2, 177, 187, 217, 218, 221

coal, 9, 23, 175, 176, 177, 178, 181, 183, 185, 186, 187, 190, 191, 192, 193, 194, 196, 197, 198, 199
coatings, 7, 245, 251, 257, 258, 259, 260, 261
cobalt, 245, 246, 247, 248, 250, 252, 255, 257, 263, 264
coding, 53
coke, 195
coke formation, 195
collaboration, 4
Colombia, 61
color, 4, 108, 109, 116, 120, 131, 145, 153, 158, 159, 164, 165, 166, 173, 189, 282, 318
combustion, 7, 9, 10, 13, 176
commercial, 103, 182, 221, 243, 258, 260
communication, 16, 93
community, 125, 271
compensation, 121
competition, 10
complex interactions, 80
complications, 46
composites, 165
composition, 4, 8, 9, 10, 11, 14, 17, 22, 25, 35, 51, 64, 72, 74, 80, 81, 83, 84, 88, 89, 126, 144, 146, 148, 149, 150, 152, 153, 158, 160, 161, 162, 163, 164, 165, 167, 176, 177, 199, 208, 225, 234, 235, 236, 237, 246, 248, 252, 256, 258, 272, 274, 289, 301
compound semiconductors, 126, 264, 307
compounds, vii, 2, 4, 5, 6, 9, 22, 25, 43, 44, 47, 48, 50, 63, 65, 66, 72, 73, 74, 76, 80, 82, 85, 88, 95, 96, 103, 126, 136, 151, 183, 190, 191, 243, 246, 264, 271, 272, 277, 284, 286, 289, 294, 301, 302, 303
condensation, 1, 3, 9, 13, 276
conductance, 99
conduction, 95, 98, 99, 101, 104, 105, 125, 130, 153, 160, 164, 165, 192, 233, 264, 265
conductivity, 137, 139, 192, 228, 230, 236, 237, 241, 243, 245, 248, 249, 257, 263, 264, 290
conductor, 8
conference, 244
configuration, 27, 100, 108, 114, 155, 185, 221
conflict, 44
Congress, 243
congruence, 88
conjugation, 53
connectivity, 191, 193, 194
consensus, 30, 51
consolidation, 248
constant rate, 187
constituents, 25, 162, 177, 209
construction, 264
consumers, 7
consumption, 7, 33, 55, 93, 248, 264
contact time, 48
contaminant, 20, 61
contaminated sites, 33, 34
contaminated soil, 25, 26
contaminated soils, 25, 26
contamination, vii, 7, 11, 19, 20, 27, 30, 31, 50, 58, 61, 104, 147, 235
Continental, 17, 197
controversial, vii
controversies, 167
cooling, 9, 74, 83, 144, 225, 227, 239, 247, 250
coordination, 77, 78, 147, 271, 272, 277, 279, 280, 281, 292, 293, 296, 299, 301
copper, vii, 3, 10, 13, 14, 15, 74, 75, 126, 139, 145, 251
correlation, 16, 29, 33, 36, 55, 58, 61, 63, 65, 75, 76, 77, 78, 79, 85, 86
correlation function, 75, 76, 77, 86
corrosion, 7
cosmetic, 8
cosmetics, 6, 7
cost, 51, 54, 57, 61, 103, 115, 126, 156, 159, 258
covalency, 158
covalent bond, 64, 70, 78, 79, 89, 151, 152, 246
covalent bonding, 151, 246
covering, 220
crabs, 37
cracks, 191, 193, 258
crop, 29
crust, 1, 3, 4, 8, 9, 13, 17, 176
crystal growth, 191
crystal structure, 2, 91, 245, 290, 291, 292
crystalline, 25, 35, 63, 73, 76, 78, 128, 168, 185, 234, 243, 291
crystallinity, 125, 128
crystallites, 129, 183, 188, 251, 252, 259
crystallization, 74, 76, 83, 131, 135, 145, 147, 162, 250, 259, 300
crystals, 94, 95, 100, 148, 162, 180, 181, 185, 189, 190, 193, 231, 244, 248, 250, 252, 272, 274, 278, 281, 300
CSF, 307, 310
Cuba, 60
cultivation, 58
culture, 53
cure, 46, 56
CV, 203, 209, 214, 219
cycles, 4, 16, 48, 211, 214, 247
cycling, 19, 21, 28, 30, 38, 214, 241, 257

D

damages, 4
damping, 164
database, 178
DEA, 309
decay, 153, 167
decomposition, 10, 127, 247, 248, 251, 254, 256, 259
decoration, 217
defects, 54, 95, 148, 249
defence, 47, 48
deficiency, 257
deficit, 249
deformation, 144, 159, 161, 193, 247
degenerate, 166
degradation, 103, 116, 184, 187, 189, 190, 191, 192, 193, 194, 196, 245, 257
degradation process, 194
Delta, 316
dendritic cell, 58
density functional theory, 220
Department of Energy, 197
depolarization, 166
deposition, 15, 16, 34, 126, 131, 159, 195, 201, 202, 205, 209, 219, 220, 221, 225, 229, 230, 231, 234, 241, 251, 252, 307, 315, 317
deposition rate, 231, 251
deposits, 3, 4, 6, 176, 252
depth, 33, 38, 52, 185, 195
derivatives, 56, 245, 271, 272, 273, 275, 279, 282, 284, 286, 288, 294, 296, 298
desorption, 19, 20, 21, 22, 23, 187, 208, 211, 214, 217
detectable, 117
detection, 28, 93, 96, 98, 99, 100, 109, 111, 116, 120, 122, 166, 228, 231, 233, 264, 313
detector performance, 94, 100
deviation, 67, 75, 76, 78, 80, 81, 82, 88, 106, 220
DFT, 200, 220
dielectric constant, 143, 150, 152, 163, 164, 169
dielectrics, 160
diffraction, 63, 64, 66, 68, 70, 73, 75, 76, 78, 79, 102, 103, 119, 128, 136, 137, 145, 175, 176, 225, 234, 289
diffusion, 86, 106, 107, 108, 139, 140, 187, 192, 194, 203, 250, 257, 259, 263
diffusivity, 106, 254
digestion, 36
dimensionality, 94, 98, 243
dimerization, 281
diodes, 95, 104, 105, 106, 166, 170, 311, 312
disaster, 16
discharges, 249
discs, 185
diseases, 4, 5, 7, 48
dislocation, 100, 206
disorder, 72
dispersion, 20, 21, 22, 74, 145, 150, 173
displacement, 293
dissociation, 126, 127, 219, 257
distribution, 16, 21, 22, 31, 33, 34, 35, 37, 63, 64, 65, 66, 68, 69, 70, 73, 75, 76, 78, 88, 107, 155, 156, 159, 160, 190, 198, 220
distribution function, 65
divergence, 88
diversity, vii, 61, 272
DNA, 58, 61, 62
dogs, 56
dominance, 1
dopants, 131
doping, 107, 121, 135, 169, 174, 265, 311, 316, 318
drainage, 30
drinking water, 4
drug resistance, 44, 46, 53, 54, 55, 56, 57, 60, 62
drugs, 7, 44, 47, 57, 60, 62
drying, 145
ductility, 257
durability, 267
dusts, 15
dynamic viscosity, 85

E

earthworms, 29, 37
effluent, 30
EIS, 209
election, 49, 52, 53
electric charge, 214
electric current, 226
electric field, 107, 152, 155, 161, 169
electrical conductivity, 192, 243, 245, 248, 249, 257, 263, 265
electrical properties, 132, 139, 141, 257
electricity, 176
electrocatalysis, 201, 216, 217, 221
electrocatalyst, 216
electrochemical behavior, 214
electrochemical deposition, 229
electrochemical impedance, 209
electrochemistry, 222
electrode surface, 201, 202, 214, 221
electrodeposition, 202, 208, 251
electrodes, 145, 195, 201, 202, 203, 205, 210, 214, 216, 217, 218, 219, 220, 221
electrolyte, 182, 184, 185, 189, 190, 191, 192, 193, 194, 221, 240, 241, 244, 251, 252, 253

electromagnetic, 93, 160, 161, 163, 166, 167, 168
electromagnetic waves, 168
electron, 2, 72, 98, 103, 105, 106, 122, 127, 128, 129, 133, 138, 145, 151, 152, 160, 161, 164, 167, 173, 176, 182, 206, 213, 215, 236, 264, 272, 278, 279, 281, 282, 284, 291, 299, 301, 313
electron diffraction, 128, 176, 182
electron microscopy, 145, 176
electronic circuits, 227, 239
electronic structure, 183, 201, 271
electrons, 93, 95, 101, 147, 150, 152, 160, 164, 165, 210, 214, 265, 292
elliptical nanoparticles, 159, 161, 163
emergency, 17
emission, 1, 4, 5, 6, 7, 8, 9, 10, 11, 13, 14, 15, 28, 36, 145, 154, 155, 156, 157, 158, 167, 168, 169, 172, 174, 315
encoding, 47, 161
endothermic, 183
energy consumption, 248
energy transfer, 156, 157, 167, 168, 169, 174
engineering, 95, 96, 97, 100, 119, 167, 307, 314
England, 90
entropy, 77
environment, vii, 4, 19, 20, 21, 28, 35, 48, 49, 105, 155, 158, 166, 167, 168, 173, 247, 263, 274, 277, 299, 301
environmental characteristics, 23
environmental conditions, 24
environmental control, 7, 15
environmental impact, 20, 28
environmental issues, vii
Environmental Protection Agency, 40
environmental standards, 10
environmental sustainability, 239
environments, 27, 30, 37, 50, 55, 155, 168, 184
enzymatic activity, 48
enzyme, 60
enzymes, 47, 53
EPA, 36
epidemiology, 56, 61
epitaxial growth, 307
epithelium, 49, 50
EPR, 272
equilibrium, 26, 86, 149, 160, 177, 183, 199, 202, 253, 282
equipment, 28
erbium, 144, 171, 174
erosion, 8, 11, 13
ERS, 166
etching, 103, 121, 145, 235, 316
ethanol, 216, 250, 257
ethylene, 250, 286, 287
ethylene glycol, 250
Europe, 3, 4, 7, 10, 15, 307, 311
European Union, 8, 16
europium, 144, 174
evaporation, 136, 139, 225, 229, 230, 231, 233, 235, 236, 237, 244, 249, 256, 257
evidence, 1, 26, 37, 50, 58, 78, 152, 162, 194
evolution, 16, 60, 167, 210, 217, 255, 256
excitation, 145, 153, 154, 155, 156, 157, 158, 160, 166, 167, 168, 169, 174
exclusion, 26
experimental condition, 207
experimental design, 25
expertise, viii
exploitation, 6
exposure, 37, 48, 50, 106, 131, 159, 180, 181, 182, 186, 187, 189, 190, 191, 192, 196, 254, 255, 256, 258, 259, 260, 261
extinction, 163, 165
extraction, 6, 7, 24, 25, 29, 33, 241

F

fabrication, 100, 103, 106, 111, 113, 115, 159, 225, 227, 267
Fabrication, 103, 116, 125, 171, 315
families, 104
fauna, 29, 36
FEM, 239
Fermi level, 104
fiber, 134
fillers, 257, 258, 264
film thickness, 229
films, 125, 126, 127, 128, 129, 130, 131, 132, 133, 134, 135, 136, 137, 138, 139, 140, 141, 173, 174, 225, 227, 229, 230, 233, 234, 235, 236, 237, 240, 241, 243, 244, 251, 252, 261
filters, 9, 173
filtration, 9, 20, 21, 34
financial, 222
financial support, 222
fingerprints, 1, 13, 14
Finland, 198
fires, 4
fish, 29, 33, 37
fitness, 54
flame, 7, 10
flame retardants, 7, 10
flexibility, 95, 97, 175
flooding, 36
flora, 29, 36
flora and fauna, 29, 36
flow field, 20

fluctuations, 20, 57, 78, 86, 87, 88
fluid, 20, 199
fluidized bed, 177
fluorescence, 145, 154, 155, 156, 157, 158, 166, 168, 171, 172, 174
food, 55
force, 149, 165, 183
forest fire, 4
formula, 122, 127, 241, 246, 249, 253, 264, 274, 275
fragments, 287, 295
France, 25, 49, 307, 309, 310
free energy, 86, 160, 177, 182, 183
freedom, 94
freshwater, 19, 30, 38
friction, 8, 11, 13, 14
fructose, 47
FTIR, 145, 202, 217, 218, 219, 222
FTIR spectroscopy, 218, 219, 222
fuel cell, 175, 176, 177, 181, 184, 196, 197, 198, 199, 200, 202, 216, 217, 221
fungi, 28

G

gallium, 171, 264
gasification, 175, 176, 177, 197, 198
gel, 159, 165, 167, 174, 244, 258
gelation, 257
gene amplification, 53, 59, 61
gene expression, 55, 58, 61
genes, 47, 52, 53, 55
genetic background, 54
genetic mutations, 53
geometry, 274, 275, 277, 280, 285, 287, 290, 291, 294, 299
germanium, 198, 299
Germany, 1, 5, 7, 8, 10, 17, 100, 144, 145, 314
glass transition, 144, 150, 151
glass transition temperature, 144, 150, 151
glasses, 7, 143, 144, 145, 146, 147, 148, 149, 150, 151, 152, 153, 154, 155, 156, 157, 158, 159, 160, 164, 165, 166, 167, 169, 170, 171, 172, 173, 174
global warming, 57
glucose, 48, 56
glutathione, 48, 53, 60
glycol, 250
gold nanoparticles, 173, 174
graduate program, 307
grain boundaries, 194, 257
grain size, 8, 9, 10, 128, 248, 249
grants, 222
graphite, 130, 248
grass, 15, 29
grazing, 132, 138
Great Britain, 177
Greece, 49
groundwater, 27, 30, 31
growth, 55, 93, 95, 96, 98, 99, 100, 101, 103, 113, 116, 128, 134, 139, 191, 207, 230, 248, 250, 252, 254, 307, 313, 314
growth temperature, 313
GSA, 154, 157, 158
guidance, 101

H

halogen, 135, 137, 272, 289, 290
halogens, 290
hardness, 7
harvesting, 227, 244
healing, 43, 44
health, vii, 4, 5, 15, 16, 17, 33, 44
health risks, 15
heat transfer, 249
heating rate, 144
heavy metals, 5, 32, 49, 177
height, 68, 70, 205, 209
hemisphere, 10, 58
heterogeneous catalysis, 202
hexagonal lattice, 208
HFP, 244
highlands, 4
histone, 61
history, 159
HIV, 44, 58
homeland security, 93
homogeneity, 76
host, 44, 47, 48, 49, 50, 52, 56, 58, 60, 145, 155, 158, 166, 167, 246
hot pressing, 247, 248
hot spots, 168
human, 7, 33, 49, 57, 60, 62, 93, 237, 240, 243
human activity, 240, 243
human body, 7
human health, 33
husbandry, 55
hybrid, 159, 165, 166, 169, 174, 245, 258, 317
hydrogen, 23, 26, 147, 159, 167, 175, 176, 178, 180, 181, 182, 183, 186, 187, 193, 195, 210, 215, 216, 217, 218, 220, 250, 277, 278, 298, 301
hydrogen sulfide, 178
hydrolysis, 273, 294
hydroperoxides, 272, 277
hydroxide, 146, 257
hydroxyl, 26, 147, 278, 296, 300
hydroxyl groups, 147

I

ID, 61
ideal, 50, 86, 88, 151, 159, 239
identification, 26, 38, 53, 116
identity, 160, 212
illumination, 111, 114, 115, 135, 136
image, 104, 111, 116, 129, 160, 161, 184, 186, 187, 188, 205, 206, 208, 212
images, 104, 133, 145, 146, 147, 162, 163, 186, 189, 203, 206, 207, 208, 209, 212, 234
immersion, 144
immune response, 60
immune system, 48
immunosuppression, 44
improvements, 15, 257, 264
impurities, 175, 176, 177, 178, 181, 182, 183, 184, 186, 187, 197, 250
in transition, 298
in vitro, 48, 50, 52, 53, 54, 55, 56, 57, 60, 61, 62
in vivo, 43, 44, 47, 48, 49, 50, 56, 57, 62
incidence, 44, 55, 59, 132, 138
incompatibility, 3
Independence, 192
India, 4, 43, 46, 48, 49, 54, 56, 59, 60, 61, 143, 170
indirect effect, 19
indium, 264
individuals, 54
induction, 247
inductor, 241
industrial emissions, 5
industry, vii, 1, 8, 93, 159, 169, 227
infection, 48, 55, 60
inflation, 134
ingredients, 5
inhibition, 47, 52, 56, 57, 210, 218, 220, 221
inhibitor, 221
inhomogeneity, 115
insertion, 241, 281, 282, 288, 292
insulators, 104
integrated circuits, 264
integration, 111, 112, 113, 114, 115, 227, 241
interface, 95, 100, 102, 105, 113, 119, 168, 182, 189, 191, 192, 194, 196, 202, 250, 256, 257, 259, 263, 312
interference, 215
intermetallic compounds, 72, 264
intermetallics, 217
intermolecular interactions, 280
inversion, 120, 158
invertebrates, 30
iodine, 208, 251, 290
ion implantation, 159, 173
ion-exchange, 159, 167, 173
ionization, 277
ions, 23, 126, 127, 143, 144, 145, 148, 149, 150, 152, 153, 154, 155, 156, 158, 159, 166, 167, 168, 169, 171, 174, 205, 211, 213, 241, 250
IR spectra, 147, 290
IR transmission, 147, 152
Iran, 46
iron, 6, 7, 11, 14, 25, 26, 56, 111, 198
irradiation, 125, 131, 132, 133, 134, 135, 136, 159, 173, 286
irregular shaped clusters, 139
irrigation, 21
island formation, 210
islands, 206
isolation, 103
isomers, 282, 295
isotherm of viscosity, 83
isotherms, 80, 82
Israel, 46, 49
issues, vii, 7, 19, 176, 202
Italy, 49, 57
I-V curves, 106, 108

J

Japan, 16, 144, 170, 172, 312, 314

K

K^+, 145, 151
KBr, 147
kinetic model, 202
kinetic parameters, 261
kinetics, 35, 218, 219, 245, 253, 254

L

laboratory tests, 184, 194
lakes, 5
lanthanide, 174
laser ablation, 252
lasers, 153, 169, 264, 310, 311, 312, 313, 314, 315, 316, 317, 318
lattice parameters, 212, 249
leaching, 17
lead, vii, 5, 7, 8, 10, 11, 13, 15, 16, 52, 53, 64, 72, 107, 143, 150, 171, 172, 177, 181, 182, 184, 187, 193, 219, 298
leakage, 98, 100, 104, 105, 121, 315
LEED, 176, 182, 192

leishmaniasis, 43, 44, 45, 46, 48, 49, 55, 56, 57, 58, 59, 60, 61, 62
lens, 114
lesions, 46, 56
leucine, 57
life cycle, 49, 60, 240, 243
lifetime, 50, 97, 103, 108, 194
ligand, 26, 158, 167, 271, 272, 273, 274, 275, 276, 277, 278, 279, 280, 281, 282, 283, 287, 289, 290, 291, 292, 293, 294, 296, 298, 299, 301
light, 11, 24, 86, 153, 159, 160, 161, 163, 166, 169, 172, 174, 186, 189, 217, 219, 246, 286, 312, 316, 317
light emitting diode, 166, 170, 316
light scattering, 166
linear dependence, 67, 75, 220
linear function, 66, 73
liquid chromatography, 28
liquid phase, 65, 78
liquids, 72, 78, 81
lithium, 240, 241, 243, 244
liver, 4
localization, 59
Louisiana, 243
low risk, 7
low temperatures, 8, 126
low-energy electron diffraction, 182
luciferase, 57
lumen, 49
luminescence, 153, 158, 166, 167, 168, 169, 171, 172, 174

M

macrophages, 50, 51, 55, 57, 59
magnetic field, 80
magnetic properties, 272
magnitude, 4, 103, 104, 106, 107, 108, 113, 153, 193, 258
majority, 43, 44, 160
malnutrition, 44
mammals, 30, 49
man, 16
manganese, 25, 183
mantle, 2, 3, 16
manufacturing, 247, 248
mapping, 54, 58
marine environment, 4
market share, 1
Mars, vi, 245, 266, 267, 268
mass, 2, 9, 26, 28, 53, 97, 98, 99, 101, 115, 149, 160, 164, 214, 215, 230, 231, 248, 264
mass spectrometry, 26, 28
materials, vii, 7, 8, 14, 64, 76, 89, 94, 95, 97, 98, 99, 100, 103, 116, 125, 126, 131, 136, 144, 147, 148, 149, 150, 152, 153, 158, 166, 170, 174, 183, 203, 216, 225, 226, 227, 228, 229, 230, 231, 236, 237, 243, 247, 249, 250, 257, 258, 263, 264, 265, 294
materials science, 166
matrix, 24, 74, 159, 161, 162, 163, 165, 167, 172, 252, 258
matter, 22, 23, 24, 25, 26, 28, 35, 38, 91, 177
maximum sorption, 22
measurement, 1, 11, 14, 64, 115, 220
measurements, 13, 14, 63, 74, 137, 139, 145, 225, 235, 254
mechanical properties, 259, 261
media, 135, 143, 164, 275
median, 113
medical, vii, 5, 6, 7, 8, 264, 271, 272
medicine, vii, 93
Mediterranean, 44, 46, 49, 58
Mediterranean countries, 58
melt, 66, 67, 69, 70, 73, 83, 85, 160, 161, 250
melting, 2, 3, 64, 66, 68, 70, 73, 74, 77, 78, 83, 89, 91, 134, 144, 147, 148, 159, 166, 169, 171, 247, 249, 251, 258
melting temperature, 73
melts, 3, 63, 64, 65, 67, 68, 69, 70, 71, 73, 74, 75, 76, 79, 80, 81, 82, 83, 84, 85, 86, 87, 88, 90, 143, 160, 251
memory, 159
mercury, 177, 198
Mercury, 94, 96, 98, 178
mesitylene, 288
meta-analysis, 61
metabolism, 48, 50, 62
metal complexes, 272, 281, 294, 298, 301
metal ion, 167
metal ions, 167
metal nanoparticles, 143, 145, 158, 159, 160, 167, 168, 170, 173
metal oxides, 21, 22, 23, 24, 171
metals, vii, 4, 5, 9, 10, 11, 16, 20, 24, 28, 29, 32, 37, 49, 65, 72, 74, 78, 90, 127, 177, 246, 271, 272, 284
meteorites, 3
meter, 144
methanol, 216, 257, 278, 281, 289, 290, 291, 292, 298
methodology, 51
methyl group, 277
methyl groups, 277
methylation, 28
Mexico, 38, 39
Miami, 16

microelectronics, 241, 264
microheterogeneity, 83
microorganisms, 28, 48
microscopy, 102, 176, 202
microstructure, 147, 162, 163, 249
microstructures, 146, 147
migration, 252, 296
milligrams, 143
Miocene, 176
missions, 157
mitogen, 59
mixing, 63, 66, 72, 76, 85, 86, 87, 88, 89, 231
mobile device, 264
MOCVD, 307, 319
model system, 222
models, 48, 50, 56, 63, 85, 202, 205, 212
moderate activity, 192
modifications, 135, 202, 258, 264
modules, 120, 227, 243
modulus, 231
moisture, 146, 273, 296
molecular beam, 101
molecular beam epitaxy, 101
molecular mass, 53
molecular oxygen, 279, 281, 282, 283, 288, 296
molecular structure, 281
molecules, 26, 43, 44, 47, 50, 52, 53, 145, 147, 201, 202, 216, 217, 219, 277, 280, 285, 291, 301
mollusks, 30, 38
molybdenum, 8, 13, 14, 230
Mongolia, 176
monks, vii
monolayer, 192, 194, 201, 206, 212, 216
monomers, 26
morbidity, 43, 44
Morocco, 43
morphology, 125, 129, 132, 133, 134, 139, 190, 206, 234
mortality, 43, 44
Moscow, 89, 90
motivation, 216
MR, 60
multiplication, 318
mussels, 36
mutation, 54
mutations, 53, 54

N

Na^+, 167
NaCl, 23, 35, 136
nanocomposites, 145, 158, 159, 160, 161, 162, 163, 164, 165, 166, 167, 169, 171
nanocrystals, 162, 173
nanodots, 173
nanomaterials, 64, 89, 159, 174
nanoparticles, 143, 144, 145, 158, 159, 160, 161, 162, 163, 165, 166, 167, 168, 169, 170, 171, 173, 174
nanorods, 173, 174
nanostripes, 209
nanostructured materials, 264
nanostructures, 165, 166, 169, 174, 227, 312, 317
Nanostructures, 315, 319
nanotechnology, 144, 159
nanotube, 60, 244
nanowires, 251
naphthalene, 302
natural compound, 4
Netherlands, 145
neutral, 23, 26, 203, 271, 272, 277, 283, 289
nickel, 175, 176, 177, 180, 181, 182, 183, 184, 185, 186, 194, 195, 196, 197, 198, 199, 200, 247
niobium, 257
NIR, 149, 153, 154, 163, 169, 172
nitric oxide, 56, 58, 59, 60, 61
nitrogen, 27, 127, 129, 130, 131, 134, 250, 292, 296, 301
NMR, 171, 274
nodes, 197, 199, 200, 241
non-linear optics, 166
non-polar, 144
North America, 16
Norway, 246
Nuclear Magnetic Resonance, 171
nucleation, 163, 173, 207
nuclei, 163, 209
Nuevo León, 125

O

obstacles, 187
OH, 22, 23, 26, 27, 147, 152, 157, 172, 202, 214, 258, 267, 277, 278, 284, 289, 302
oil, 9, 257
one dimension, 95
operations, 30, 31, 240, 243
optical communications, 153
optical properties, 98, 119, 135, 143, 158, 159, 165, 170, 173
optical systems, 158
optimization, 100, 106, 108, 316
optoelectronic properties, 125
orbit, 97
ores, 4, 6, 7, 8, 9, 10, 13, 15
organ, 7, 271, 279, 284, 301

organic matter, 22, 23, 24, 25, 26, 28, 35, 38, 177
ornithine, 58, 59
oscillation, 160, 231
oscillators, 264
overlap, 98, 152
overlay, 169
ox, 203, 206, 211, 283, 302
oxalate, 283, 302
oxidation products, 245, 254
oxidative damage, 48
oxygen, 48, 59, 147, 148, 150, 151, 152, 202, 205, 211, 216, 245, 247, 253, 255, 256, 257, 258, 259, 261, 263, 277, 278, 281, 282, 283, 286, 287, 288, 291, 295, 296

P

PACs, 29
Pakistan, 46
palladium, 254
parallel, 103, 108, 161, 236
parasite, 43, 44, 46, 47, 48, 49, 50, 52, 53, 55, 56, 57, 59, 62
parasites, 43, 44, 46, 47, 48, 49, 50, 52, 54, 56, 58, 61
parasitic infection, 48
parity, 158
partition, 95, 177
passivation, 103, 104, 105, 121, 183, 199
pasture, 29
patents, 307
pathways, 48, 153
PCR, 62
peat, 4, 5, 16
percentile, 33
percolation, 192
Periodic Table, 66, 70
periodicity, 209
permeability, 164
permission, 180, 182, 184, 186, 188, 189, 190, 191, 192, 193, 195, 196
permit, 296
Peru, 46, 49, 59
PET, 7, 10, 17
pH, 4, 22, 23, 24, 25, 27, 28, 30, 31, 32, 35, 126, 127, 250
pharmaceutical, 60
phase diagram, 65, 66, 70, 71, 74, 76, 78, 79, 83, 85, 88, 89, 180, 245, 247, 253, 263
phase transitions, 171
phenol, 301
phenotype, 54, 58, 60
phenoxyl radicals, 28, 293
phonons, 158, 264
phosphate, 143, 165, 172, 241
phosphate glasses, 143
phosphorescence, 166
phosphorus, 177, 192, 199, 241, 277, 299
photoconductivity, 139
photodetector, 94, 106
photodetectors, 93, 94, 96, 100, 104, 106, 120, 121, 122, 313, 314, 315, 316, 317, 318
photoelectron spectroscopy, 176, 222, 225
photoemission, 196
photographs, 241
photoluminescence, 153, 166, 167, 170, 172, 174
photonics, 144
photons, 117, 153, 158
photooxidation, 28
photoresponse, 139
photosensitivity, 126
photovoltaic devices, 126
photovoltaic structures, 125, 126
physical properties, 64, 90, 125, 126, 135, 145
physicochemical properties, 81
physics, 94, 100
phytoremediation, 29
plants, 7, 29, 50
plasmodium, 53
plastic deformation, 247
plastics, 7, 10
platform, 113
platinum, 199, 218, 219, 220
Platinum, 254
playing, 78
PM, 56, 57
point of zero charge (PZC), 23
poison, 201, 218, 219
polar, 144, 283, 311
polarity, 103
polarizability, 150, 152, 155, 165, 167, 172
polarization, 150, 151, 152, 161, 166, 173, 182, 187, 192
policy, 61
polluters, 13
pollution, 16, 37
polyamines, 58
polycondensation, 201
polyethyleneterephthalate, 7
polyimide, 105, 106, 121, 230
polymer, 7, 104, 258, 259, 289
polymer structure, 289
polymerase, 57
polymerase chain reaction, 57
polymerization, 257, 272, 288
polymers, 7, 165, 173

polymorphism, 145
polymorphisms, 53
polypeptide, 53
population, 5, 7, 48, 49, 53, 54, 153, 156, 158
population group, 5
population structure, 54
porosity, 194, 256, 257, 259
Portugal, 49, 225, 243
positive correlation, 33
potassium, 8, 22, 144, 146, 252
power generation, 198, 264
precipitation, 20, 126, 247, 250
predators, 37
preparation, viii, 38, 125, 126, 139, 141, 143, 159, 272, 275, 277, 279, 290
preservation, 21
prevention, 281
principles, 221
probability, 48, 54, 107
probe, 144, 222
producers, 8
project, viii, 310
promoter, 220
propagation, 54
protection, 3, 245, 258, 263
protective coating, 240
proteins, 47, 48, 53, 58, 59
proteome, 54
proteomics, 62
prototype, 170, 241, 242
public health, 17
purification, 175, 248
purity, 144
PVP, 250

Q

quantification, 50, 54
quantization, 95
quantum devices, 307
quantum dot, 94, 109, 312, 314, 317
quantum dots, 109, 314
quantum mechanics, 94, 97, 307
quantum state, 99
quantum well, 94, 95, 97, 101, 109, 113, 312, 314, 317
quartz, 209, 231, 247, 252
quercetin, 284, 302
quinone, 273, 275, 278, 284, 285, 286, 288, 290, 301
quinones, 271, 272, 273, 274, 279, 282, 284, 286, 290, 291

R

radial distribution, 65
radiation, 93, 116, 145, 154, 155, 157, 159, 172, 228, 272
Radiation, 17, 239
radicals, 28, 293
radius, 3, 163, 166, 203, 299
rainfall, 19, 30, 33, 34, 38
Raman spectroscopy, 225
ratio analysis, 129
raw materials, 144, 147, 148
RE, 144, 153, 159, 166, 167, 169
reactant, 247
reactants, 247, 249, 263
reaction mechanism, 78, 217, 256
reaction rate, 273
reaction time, 202
reactions, 20, 182, 187, 196, 201, 202, 216, 253, 271, 273, 275, 282, 284, 286, 293, 296
reactive oxygen, 48, 52, 59
reactivity, 27, 281, 282, 286, 300, 301
reagents, 258, 286, 298
real time, 231
reciprocal temperature, 79
recombination, 53, 96, 97, 108, 120, 317
reconstruction, 205, 206, 251
recovery, 11, 15
recrystallization, 289
rectification, 241
recycling, 159
red shift, 130, 135, 156, 165
reflexes, 73
refractive index, 125, 126, 143, 150, 155, 163, 164, 165, 166, 169, 171, 174
refractive indices, 144, 150, 160
regeneration, 200
regions of the world, 44
relaxation, 153, 155, 156, 157, 158, 164, 173, 205
relaxation process, 158
relaxation rate, 153
relevance, 5, 57, 62
remediation, 30
remote sensing, 317
repulsion, 208
requirements, 7, 194, 257
researchers, vii, 27, 28, 29, 76, 167
resistance, 43, 44, 46, 48, 49, 50, 51, 52, 53, 54, 55, 56, 57, 59, 60, 62, 100, 104, 106, 115, 121, 159, 175, 187, 189, 191, 192, 194, 257, 261, 263, 278, 311, 312
resolution, 34, 103, 145, 196, 205, 207, 208

response, 48, 49, 50, 55, 60, 98, 105, 140, 161, 186, 189, 191, 194
restrictions, 273
Richland, 175
rings, 246, 280, 284, 298
risk, 4, 5, 7, 8, 14, 33, 43, 44, 54, 59
risk assessment, 8, 14
risk factors, 59
risks, 4, 5, 7, 8, 10, 14, 15
river systems, 27, 30
room temperature, 63, 119, 126, 136, 144, 145, 160, 225, 229, 237, 239, 265, 311, 312
root, 98
roots, 29
roughness, 103, 258
routes, 251, 298
rubber, 7
rules, 54, 158
runoff, 29, 34
Russia, 6, 89, 90, 176, 271

S

safety, 7, 8, 61
salts, 139
samarium, 171
sapphire, 312
saturation, 112, 214, 216, 218
saturation coverage, 214, 218
scaling, 314
scandium, 9
scanning electron microscopy, 145
scatter, 175, 264
scattering, 150, 163, 166, 264
science, vii, 93, 159, 166, 307
scope, 117, 239
security, 93, 161
sediment, 19, 20, 21, 22, 24, 25, 29, 30, 31, 33, 34, 35, 36, 38
sediments, 3, 4, 5, 19, 21, 22, 24, 25, 29, 30, 31, 33, 34, 35, 37, 38
segregation, 64, 75, 183, 192, 252
selectivity, 24
selenium, 5, 13, 136, 177
semiconductor, 71, 78, 90, 94, 96, 98, 104, 105, 119, 130, 136, 143, 243, 264, 265, 307
semiconductors, 64, 76, 94, 95, 96, 97, 126, 139, 264, 272, 307
semimetals, 64, 96
senses, 93
sensing, 93, 172, 228, 244, 314
sensitivity, 52, 62, 70, 272
sensors, 153, 226, 230, 231, 241
shape, 47, 64, 88, 159, 160, 166, 169, 173
shear, 231
showing, 14, 55, 69, 70, 103, 139, 147, 153, 157, 161, 162, 206, 233, 237, 238, 240, 249, 259
Si3N4, 240
signals, 231
silica, 144, 147, 152, 155, 167, 183, 251, 258
silicon, 230, 239, 240, 305
silver, vii, 126, 136, 145, 166, 167, 173, 174, 283
simulation, 239
simulations, 239
sine wave, 231
Singapore, 174
single crystals, 244, 250, 252
sintering, 247, 248, 249
SiO2, 104, 105, 174, 252, 258, 263
SiO2 films, 174
skin, 43, 44
slag, 25, 30
smoothing, 209
sodium, 56, 57, 58, 61, 62, 104, 135, 290, 291, 293
software, 144
solar cells, 139, 166, 170
sol-gel, 159, 165, 167, 174
Solid oxide fuel cell, 176, 197, 200
solid oxide fuel cells, 175, 176, 197, 199, 200
solid phase, 181, 192, 193, 194, 253
solid solutions, 63, 181, 194, 245, 247
solid state, 64, 65, 66, 70, 72, 74, 76, 79, 82, 132, 149, 166, 280, 281
solidification, 63, 64, 66, 72, 73, 89
solubility, 2, 4, 10, 27, 29, 65, 66, 74, 181, 185, 192, 263
solution, 22, 23, 24, 28, 59, 68, 80, 81, 104, 126, 136, 160, 180, 183, 188, 190, 191, 192, 202, 203, 206, 209, 212, 214, 215, 216, 221, 241, 250, 251, 252, 257, 273, 274, 275, 278, 282, 283, 288, 291, 317
solvents, 275, 281, 283
solvothermal synthesis, 250
sorption, 19, 20, 21, 22, 23, 26, 196
South Africa, 6
South America, 44, 46
SP, 144
Spain, 49, 56, 244
speciation, 19, 20, 22, 27, 28, 37, 61, 176, 177, 184
98
specific adsorption, 22, 23
spectroscopy, 24, 26, 166, 171, 174, 176, 192, 202, 209, 218, 219, 220, 222, 225
spin, 97
SS, 60, 183

stability, 22, 27, 66, 68, 70, 115, 159, 180, 181, 182, 191, 195, 241, 245, 254, 278
stabilization, 23
standard deviation, 106
standardization, 51, 52
state, 27, 28, 33, 60, 63, 64, 65, 66, 67, 70, 71, 72, 74, 75, 76, 77, 78, 79, 82, 83, 84, 86, 87, 91, 94, 96, 98, 100, 101, 103, 109, 117, 132, 149, 152, 153, 154, 156, 157, 158, 166, 187, 190, 191, 209, 225, 239, 240, 243, 244, 249, 271, 280, 281, 307
states, 2, 21, 27, 28, 99, 158, 167, 243, 272
steel, 7, 36, 184, 258
sterile, 36
STM, 203, 207, 208, 209, 212
stoichiometry, 82, 151, 212, 236, 245, 252
storage, 17, 153, 154, 241
stress, 48, 193, 230
stretching, 147, 148, 158, 159, 290
strong force, 183
strong interaction, 185, 187
structural changes, 150, 151
structural characteristics, 313
structural transformations, 78
structure formation, 66
substitutes, 169
substitution, 68, 277
substrate, 36, 96, 100, 103, 115, 120, 121, 126, 130, 136, 202, 203, 208, 217, 225, 229, 231, 233, 234, 235, 236, 237, 240, 251, 252, 258, 259, 261, 263, 312, 313, 314, 315
substrates, 23, 94, 100, 103, 115, 116, 119, 127, 129, 136, 139, 209, 214, 220, 222, 230, 231, 237, 244, 251, 252, 277, 312
succession, 4
Sudan, 46, 55
sulfate, 205
sulfide-rich materials, 14
sulfides, 2, 3, 8, 9, 25, 198
sulfur, vii, 3, 9, 104, 177, 194, 196, 197, 200
sulphur, 127
Sun, 80, 90, 198, 214, 219, 222, 223, 224, 266, 303
superlattice, 93, 95, 97, 98, 99, 100, 101, 102, 103, 104, 105, 106, 107, 108, 110, 111, 113, 114, 115, 116, 117, 119, 120, 121, 122, 123, 313, 314, 315, 316, 317, 318
suppliers, 8
suppression, 96, 100, 108, 120, 121, 122
surface area, 194
surface layer, 203, 254
surface modification, 202, 217, 222
surface structure, 202, 203, 207, 219, 221, 222
surveillance, 53
survival, 36, 62
susceptibility, 46, 47, 48, 49, 50, 52, 53, 54, 55, 56, 58, 61, 62, 153
sustainability, 239
sustainable development, vii
Sweden, 16
Switzerland, 16
symmetry, 147, 148, 152, 155, 158, 167, 168, 207, 214
symptoms, 4, 259
synergistic effect, 24
synthesis, 53, 144, 159, 167, 169, 175, 176, 177, 178, 179, 181, 182, 183, 184, 185, 186, 187, 191, 197, 245, 246, 247, 248, 249, 250, 257, 272, 273, 275, 277, 279, 290, 292, 294
synthetic methods, 272

T

Tajikistan, 6
target, 24, 48, 53, 93, 116, 227, 240, 251, 252
technical support, 170
techniques, 8, 21, 24, 28, 29, 113, 121, 126, 159, 202, 225, 227, 307, 316
technologies, 50, 93, 94, 113, 143, 239
technology, 93, 94, 100, 111, 117, 136, 176, 177, 197, 227, 267, 307
TEG, 250
tellurium, 229, 234, 236, 237, 243
TEM, 145, 160, 161, 170, 248
temperature dependence, 65, 70, 78, 85, 111
temporal variation, 32, 33
TEOS, 257
terminals, 229
terraces, 206, 209, 219, 220
testing, 115, 186, 187, 199
textiles, 7
thallium, 198, 279
therapy, 43, 46, 48, 54, 56, 61
thermal decomposition, 254
thermal energy, 226
thermal evaporation, 136, 139
thermal expansion, 144, 150, 169, 230, 257, 258
thermal properties, 170
thermal resistance, 159
thermal stability, 70, 254
thermal treatment, 134
thermodynamic equilibrium, 253
thermodynamic properties, 86, 91
thermodynamics, 245, 256, 257
thermolysis, 293
thin films, 125, 126, 127, 128, 129, 130, 131, 132, 133, 135, 136, 137, 138, 139, 140, 141, 173, 229, 235, 236, 237, 251, 252, 261, 312, 317

thin-film deposition, 225
thiosulphate, 126, 127
tin, 1, 7, 8, 11, 70, 76, 174, 197, 199, 241, 298, 299
tissue, 29, 36, 37, 57
titania, 257
titanium, 257
TNF, 61
TNF-alpha, 61
toluene, 144, 274, 275, 282, 283, 284, 286, 293, 294
toxic effect, 4
toxic substances, 31, 32
toxicity, vii, 27, 28
toxicology, 303
toys, 7
trace elements, 37, 197, 198
training, 310
transcription, 57
transfection, 51
transformation, 20, 21, 26, 28, 37, 61, 66, 73, 85, 132, 188, 189, 190, 246, 283
transformations, 28, 29, 78, 272, 301
transition metal, 246, 271, 279, 281, 294, 298, 301
transition rate, 156
transition temperature, 144, 150, 151, 152
transmission, 43, 44, 143, 145, 147, 148, 149, 150, 152
Transmission Electron Microscopy, 102
transport, 19, 20, 21, 22, 24, 25, 26, 27, 28, 34, 191, 198, 243, 244, 251, 257, 264, 315
transport processes, 22
transportation, 4
treatment, 43, 44, 46, 48, 54, 55, 56, 57, 58, 59, 60, 61, 62, 134, 159, 167, 173, 207, 248, 291
trial, 60
triphenylphosphine, 277
TSH, 47
tuberculosis, 54, 56, 57
tungsten, 135, 137
tunneling, 95, 96, 98, 104, 107, 108, 202, 212, 311, 312
Turkey, 46, 61, 176, 198
turnover, 20
Type-II superlattices, 94, 95, 96, 98, 100, 103, 104, 117

U

UK, 244
Ukraine, 63
underlying mechanisms, 24
uniform, 129, 134, 139, 185, 190, 203, 205, 206, 317
unique features, 95
United, 7, 15, 25, 40, 145
United States, 7, 15, 25, 40, 145
urban, 1, 17, 30, 50
urbanization, 44
USA, 89, 91, 93, 145, 197, 243
USSR, 199, 303, 304
UV, 145, 149, 150, 152, 154, 159, 160, 163, 165, 169, 172, 173, 174, 286, 314

V

vacancies, 205
vacuole, 53
vacuum, 128, 130, 136, 138, 139, 140, 164, 202, 230, 245, 248, 252, 253, 257, 258, 292
valence, 95, 96, 97, 98, 99, 101, 104, 153, 265
Valencia, 60, 244
vanadium, 177
vapor, 247, 250, 251, 256, 259, 307, 314, 315
variations, 4, 15, 31, 32, 49, 51, 53, 215
vector, 48, 49, 50, 55, 70, 73
vegetation, 30
vehicles, 8, 11
vein, 198
velocity, 20, 65, 164
ventilation, 21
versatility, 51, 301
vibration, 147, 148, 152, 158, 216
viscose, 29
viscosity, 63, 78, 79, 80, 81, 82, 83, 84, 85, 86, 88, 161
vision, 93
volatility, 3, 6, 28, 254
volatilization, 9, 10, 20
vulcanization, 7, 10

W

waste, 10
waste disposal, 10
waste incineration, 10
wastewater, 30
water, 2, 3, 4, 7, 10, 17, 19, 20, 21, 22, 23, 24, 25, 27, 29, 30, 31, 33, 34, 35, 36, 37, 38, 49, 50, 144, 147, 209, 215, 247, 259, 274, 275, 278, 291, 292, 300, 301
water chemistry, 19, 22, 23, 24, 30, 35, 38
water quality, 33
watershed, 30
wave vector, 70, 73
wavelengths, 104, 109, 116, 159, 169
weakness, 210
wells, 95, 98, 109, 307, 316
windows, 95, 158

wires, 175, 183, 187
workers, 203, 205
worldwide, 4, 167

X

XPS, 176, 185, 191, 194, 225, 235
X-ray analysis, 275, 280, 290
X-ray diffraction, 63, 66, 68, 70, 73, 75, 76, 102, 103, 128, 136, 137, 145, 225, 234, 289
X-ray diffraction (XRD), 225, 234
X-ray diffraction data, 68, 289
X-ray photoelectron spectroscopy (XPS), 225
XRD, 127, 128, 132, 138, 139, 145, 146, 148, 161, 162, 225, 234, 248, 250, 252, 255

Y

yield, 149, 182, 250, 252, 256, 263, 283, 290, 294

Z

zinc, 5, 10, 14, 15, 48, 57, 171, 172
zirconia, 176, 197, 199
ZnO, 171, 312, 314, 315, 316, 317, 319